Polynomial Invariants of
Finite Groups

Research Notes in Mathematics

Volume 6

Polynomial Invariants of Finite Groups

Larry Smith
Mathematisches Institut
Georg-August-Universität
Göttingen, Germany

CRC Press
Taylor & Francis Group
Boca Raton London New York

CRC Press is an imprint of the
Taylor & Francis Group, an **informa** business

AN A K PETERS BOOK

First published 1995 by A K Peters, Ltd.

Published 2018 by CRC Press
Taylor & Francis Group
6000 Broken Sound Parkway NW, Suite 300
Boca Raton, FL 33487-2742

© 1995 by Taylor & Francis Group, LLC
CRC Press is an imprint of Taylor & Francis Group, an Informa business

First issued in paperback 2019

No claim to original U.S. Government works

ISBN 13: 978-0-367-44913-1 (pbk)
ISBN 13: 978-1-56881-053-9 (hbk)

Visit the Taylor & Francis Web site at
http://www.taylorandfrancis.com

and the CRC Press Web site at
http://www.crcpress.com

Typeset by $\mathcal{LS}$TEX

Library of Congress Cataloging-in-Publication Data
Smith, L. (Larry) , 1942–
    Polynomial invariants of finite groups / Larry Smith.
      p. cm.
    Includes bibliographic references and index.
    ISBN 1-56881-053-9
    1. Invariants. 2. Finite groups. 3. Algebraic topology.
  I. Title.
QA201.S65 1995   94-46590
512'.5—dc20   CIP

To the better craftsmen,
Nelson Killius, Peter Landweber, Bob Stong and Bryan Wills,
and the better craftswoman,
Mara D. Neusel.

Preface

This is a book about the invariant theory of finite groups, written by an algebraic topologist. Algebraic topology is a creation of the 20th century, invariant theory was supposed to have died in the 19th century, and we are almost in the 21st century.

I imagine that these remarks require some clarification, so I suppose I should say something about the subject of the book and how an algebraic topologist comes to be doing invariant theory. Of course I could simply say it is part of an apprenticeship one serves here in Göttingen in the shadows of Klein, Hilbert and Noether. But that isn't all of the story.

I first encountered invariant theory as an undergraduate when I tried to read Weyl's book THE CLASSICAL GROUPS. After a ten year recuperation period I was again confronted by invariant theory. This time in 1973/74 at the hands of John Ewing in the form of his paper with Alan Clark [56] and many personal tutorials from him. For several years (my interests were the homotopy groups of spheres at that time) I did not pay much attention to the subject, but the paper of Adams-Wilkerson [2] changed all that. Since 1981, when Bob Switzer and I took that paper apart bit by bit, I have found myself more and more involved in the invariant theory of finite groups. I have been engrossed in an attempt to understand Weyl's book THE CLASSICAL GROUPS (which is **not** just about finite groups) and do all the exercises in Bourbaki GROUPES ET ALGÈBRES DE LIE chapitres 4, 5, 6 (which contains **more** than just invariant theory).

During the past 10 years or so I have written well over a dozen papers whose contents have in part been invariant theory, and several, which have been exclusively invariant theory, such as for example [214], [221], [223], [215], [164] and [168] among others. In short I became an invariant theorist.

I have taught courses on invariant theory on several occasions: in 1986 in Göttingen, in 1987 in Charlottesville, in 1990 at Yale, in 1992 in Göttingen, and again in the winter semester 1993/94 in Göttingen. The algebraic topology seminar in Göttingen has more than once devoted a whole semester to invariant theory: on one occasion even trying to work through Weyl's book [259] THE CLASSICAL GROUPS.

I have hundreds, perhaps thousands, of pages of notes and musings on invariant theory, and have experimented with the organization of the material in different ways: what depends on what is often a question of taste. I was reluctant to get involved in a book project until the year I spent at Yale where I received encouragement to at least make my notes available in some form. One of my students (!) at Yale, Tseueno Tamagawa was very involved in my course, and one of the outcomes was a completely new way of looking at the Dickson invariants, which appears here in chapter 8, and is due to Tamagawa.

After an abortive attempt to write a jointly authored book with David Benson I finally decided to take the plunge and write about the polynomial invariants of finite groups on my own. In a subject with as long and varied a history as invariant theory it is impossible in a single volume to encompass all aspects of the theory. By confining myself to finite groups and polynomial invariants I hope to have picked a portion of the whole that is interesting and accessible to a wide audience.

Invariant theory served as one of the major motivations for the development of commutative algebra: from Hilbert's basis theorem to Noetherian rings and modules. Algebraic geometry was of course another major force providing impetus to commutative algebra; often the two threads came together, as illustrated in [126]. I have tried to use invariant theory as motivation for introducing my students to some of this algebra. This is done in the chapters on Noether's Normalization Theorem (which is needed for the proof of the Eagon-Hochster result that nonmodular rings of invariants are Cohen-Macaulay) and Hilbert's Syzygy Theorem which is perhaps the origin of homological algebra. In other words, this is more than just a book on invariant theory: it introduces some of the basic ideas of ideal theory and homological algebra in a hopefully motivated context. In addition, in chapters 10 and 11 I try to show some of the influence of invariant theory on algebraic topology: and the influence of algebraic topology on invariant theory!

Most mathematicians and mathematics students like problems, so let me try to introduce invariant theory with a simple problem. Consider the task of finding all the polynomials $f(x, y)$ that satisfy the identities

$$f(\tfrac{x+(q-1)y}{\sqrt{q}}, \tfrac{x-y}{\sqrt{q}}) = f(x, y)$$
$$f(-x, -y) = f(x, y)$$

where q is a fixed positive real number. This is a problem of invariant theory, taken from [209], that arises in connection with binary codes (see also [60]). Note that we are demanding that the polynomial be invariant under the linear change of variables given by the matrices

$$A = \frac{1}{\sqrt{q}} \begin{bmatrix} 1 & q-1 \\ 1 & -1 \end{bmatrix}, \quad B = \begin{bmatrix} -1 & 0 \\ 0 & -1 \end{bmatrix}.$$

These matrices are each of order two and commute with each other, so generate a subgroup of $GL(2, \mathbb{R})$ isomorphic to the Klein 4-group K (i.e. $\mathbb{Z}/2 \oplus \mathbb{Z}/2$). The group K acts via these matrices on the polynomial functions $\mathbb{R}[x, y]$ in two variables by linear change of variables. The problem we posed asks for a description of all polynomials whose values remain unaltered after a change of coordinates from the group K.

Let me describe some of the material in the book in more detail and how it may be applied to the preceding problem.

The first chapter introduces the themes and problems that recur through the book. The material is pretty standard; the definitions of invariants, coinvariants and relative invariants are embedded in a discussion of some basic examples. Using this language, a solution to the preceding problem is a description of the ring of invariants, denoted by $\mathbb{R}[x, y]^K$, of the Klein 4-group acting on the polynomial functions in two variables through the linear change of coordinates given by the matrices A and B.

In chapter two, we prove the Hilbert-Noether finiteness theorems. This assures us that the algebra $\mathbb{R}[x, y]^K$ is finitely generated, but gives us little help in actually finding generators. Hilbert's proof of the finiteness theorem, as modified by Noether, is best understood in the context of Noetherian rings and modules, and we see in a concrete way how a problem with its origins in invariant theory served as a model for a more general result in commutative algebra. The original problem that led Hilbert to his basis theorem was to prove the existence of a finite fundamental system of basic invariants, i.e. to show that a minimal set of algebra generators for the invariants is finite. His first paper on invariant theory [101] left the actual construction of a complete system of basic invariants open. In practice it is often very difficult to know where to look for invariants, and/or when we have found all of them. The commutative algebra introduced in succeeding chapters deals largely with the second of these problems. The first of these problems was studied by Hilbert in his second paper on invariant theory [102]. He found an upper bound for the degrees of a minimal set of algebra generators for the invariants of a reductive group. For finite groups a fundamental general bound is due to Noether [170] and appears in chapter 2 along with another finiteness proof, also due to Noether [170], which is constructive, and leads to an algorithm easily implemented on a computer. (Unfortunately, in practice, the algorithm does not seem to be very efficient.) It implies that $\mathbb{R}[x, y]^K$ can be generated as an algebra by polynomials of degree at most four.

The third chapter is based on my paper with Bob Stong [221] and portions of our correspondence with Peter Landweber. It provides a general method for constructing invariants, and results that assure that sometimes one can find all the needed invariants in this way. It is motivated by an attempt to understand Weyl's account ([259], pp. 275 – 276) of Noether's proof [170] of the Hilbert finiteness theorem. (We were not the only ones to be struck by this account. See also [23].) When these methods are applied to the problem of computing $\mathbb{R}[x, y]^K$ they quickly yield two algebraically independent invariant polynomials

$$f = x^2 + (q - 1)y^2, \quad h = y(x - y).$$

There is quite a bit of new material in this chapter (nothing earthshaking: but I hope useful), and some old forgotten results out of Weber's algebra book [257] (rearranged in contemporary terminology) which I couldn't resist including.

The fourth chapter on Poincaré series contains classical material originating with Molien [148]. The central results are the rationality of the Poincaré series of rings of invariants (the theorem of Hilbert-Serre), and the formula of Molien. The theorem of Hilbert-Serre is best understood in the context of finitely generated algebras over a field. So we again see how a problem in invariant theory serves as a model for a more general result of commutative algebra. Applying Molien's formula to compute the Poincaré series of $\mathbb{R}[x, y]^K$ *suggests* we have found enough invariants in f and h to generate the ring of invariants. An example of Richard Stanley in section 4.1 warns of a possible pitfall however.

The formula of Molien for the Poincaré series of a ring of invariants in coprime characteristic is presented essentially unchanged from [148]. This formula has a number of remarkable combinatorial consequences first noticed by Richard Stanley [235]. I have tried to indicate some of them here (and in chapter 9), but the connection with defect sums related to the Atiyah-Bott fixed point theorem would take us too far afield to describe here. (See for example [217].)

Chapter 5 contains some standard commutative algebra in graded form, and introduces localization as an alternative to grading to study rings of invariants. Localization and the use of gradings are fundamental concepts I want to introduce my students to as early as possible. The proof of graded Noether normalization is one I learned from John C. Moore or Paul F. Baum more than 25 years ago. At this point we have enough tools to examine some examples of rings of invariants in detail. In particular one can show using 5.5.5

that $\mathbb{R}[x,\ y]^K$ is generated by the two polynomials f and h, and that there are no algebraic relations between them. The problem posed at the start of this discussion may be answered by saying: any polynomial invariant under the action of the Klein group through linear change of coordinates via the matrices A and B is uniquely expressible as a polynomial in f and h. In the last section of this chapter we compute the rings of invariants of $\mathrm{GL}(2,\ \mathbb{F}_p)$ and many of its subgroups.

Chapter 6 has its origins in the Hilbert Syzygy Theorem, the Urtheorem of homological algebra. I originally wanted to present Tate's proof [247] of the converse of the syzygy theorem, but found I could not distill it down to less than nine handwritten pages, and while very interesting, it is really homological algebra per se and perhaps overkill in a book on invariant theory. So I gave the proof of Eilenberg for the case of Krull dimension 0, and my own for the case of positive dimension. The paper [247] is highly recomended reading in any case. This chapter has a fair amount of material appearing for the first time in book form. For example that the ring of coinvariants is a Poincaré duality algebra in the polynomial case, a Moore influenced proof of Macaulay's theorems, and some goodies of my own on Koszul complexes, regular sequences and other things. A bonus from the work on the Koszul complex is an upper bound, due originally to Garsia and Stanton, on the degrees of a complete system of basic invariants for permutation representations. In this chapter we also prove the theorem of Hochster and Eagon that rings of invariants in the coprime characteristic case are Cohen-Macaulay. This result has a number of applications to Poincaré series which can be useful in practical computations.

The chapter on groups generated by reflections centers around the theorem of Shephard and Todd which characterizes these groups in the complex case as exactly the groups with polynomial rings of invariants. We present a proof in the nonmodular case independent of the classification of the finite complex pseudoreflection groups used by Shephard and Todd. We follow Tits' account in Bourbaki for one direction and my own paper [214] for the other. This theorem provides us with a *theoretical* justification that (in our example taken from coding theory) $\mathbb{R}[x,\ y]^K$ is a polynomial ring on two generators of degree 2. I have included enough of a discussion of the classification of Coxeter groups (these are the real reflection groups) to prove that they are always generated by n elements, where n is the dimension of their real reflection representation. This has consequences for their invariant theory, not all of which have been generalized to the complex pseudoreflection groups. New in this chapter is a generalization of the theorem of Papadima [179] and Sperlich [230]. It deals with the group of automorphisms of the ring of coinvariants $\mathbb{F}[V]_G$ of a pseudoreflection group. There is also an account of the relative invariants of pseudoreflection groups in the nonmodular case following the characteristic zero account of Stanley [233] and Springer [231]. There are discussions of a few of the groups in the Shephard-Todd list, but one could write a whole book about them alone, and I forced myself to stop at some point. Examples of the invariant theory of some of these groups is presented in detail at several points in the chapter.

Invariant theory in the modular case is far less developed and organized than in the coprime characteristic case. Many very basic problems have not yet been solved, such as, for example, when is the ring of invariants Cohen-Macaulay? There is also no analog of Molien's theorem in the modular case, nor any estimates (such as those of Noether) for the degrees or number of a complete fundamental system of invariants. The work on modular invariant theory is divided into two chapters. The first chapter on modular invariants, chapter 8, begins with several old results of L. E. Dickson [68] with completely new proofs due to T. Tamagawa. The Dickson algebra provides a source of universal modular invariants (i.e. the Dickson polynomials are present in any ring of invariants in characteristic p) and one way to try to study a ring of invariants in characteristic p is as an integral extension of the Dickson algebra. We present some computations and examples in the hopes they will stimulate further investigations.

In general, the ring of invariants of a finite group in finite characteristic dividing its order need not be Cohen-Macaulay. However, even in the simplest of such examples (the invariants of the regular representation of $\mathbb{Z}/5$ over $\mathbb{F}_5$) there is still much left undiscovered. Landweber and Stong investigated the codimension (also called the depth) of rings of invariants in characteristic p. The difference between the dimension and the codimension is a measure of the deviation from being a Cohen-Macaulay algebra. The work of Landweber and Stong includes the study of groups G whose fixed point set is codimension 1. In a series of papers [155] [156] Nakajima investigated the dual situation for abelian groups. We present a uniform account of all these results based on the notion of transvections.

In chapter 9, we deviate from considering only polynomial invariants, and look at a few other types of invariants. By no means have we exhausted the possibilities in this chapter. See for example: [238] for groups acting on differential operators, [67] for groups acting on tensor algebras and [178] for groups acting on differential forms of various sorts. We examine the invariants of pseudoreflection groups acting on polynomial tensor exterior algebras based on Louis Solomon's paper [226] and the mod p analog from my paper with Aguadé [6]. This chapter also includes Solomon's application of the invariant theory of $\mathbb{F}[V] \otimes E[V]$ to study the representation $\mathbb{F}[V]_G$ of the algebra of coinvariants of a reflection group.

Chapter 10 is concerned with the interplay between invariant theory and algebraic topology. Much of this material appears for the first time in book form (and in some cases for the first time in publication). It begins with a way for topologists to explain the Steenrod algebra (à la Milnor-Kuhn) to their friends in commutative algebra and uses this to present a proof of a portion of the theorem of Adam-Wilkerson [2] due to R. M. Switzer and myself [223]. The portion isolated is concerned directly with invariant theory and represents an application of algebraic topology, or ideas coming from algebraic topology and algebraic topologists, back into invariant theory. The result presented here is a mod p converse to the theorem of Shephard-Todd-Chevalley: namely it characterizes which polynomial subalgebras $\mathbb{F}_p[f_1, \ldots, f_n] \subset \mathbb{F}_p[t_1, \ldots, t_n]$ can occur as rings of invariants. The important role in the study of modular rings of invariants played by the Steenrod algebra is illustrated by several results in chapters 10 and 11. We also show how Dywer, Miller and Wilkerson [75] extended the classification of Shephard-Todd to classify groups generated by pseudoreflections in characteristic p when the group order is prime to p. We give another proof of Dickson's theorem using Steenrod operations due to Switzer and myself [224].

In chapter 11 we return to the study of modular invariant theory. We examine the interplay between the Steenrod operations, the Dickson invariants and the ideal structure of rings of invariants, and apply the results of this study to the transfer homomorphism.

Throughout the book I have tried very hard to include as many examples as would fit. I made an extra effort to illustrate the theorems by applying them to concrete cases. I hope to have found the right balance. A recurrent theme throughout the book is the extension of characteristic zero results to arbitrary characteristic, or to at least the nonmodular case. This often required completely different proofs than the traditional ones already in the literature. Often the extensions are false: that too is interesting.

"Bien que nous puissions guère indiquer de référence precise, ce lemme est connu et résulte trivialement de properiétés classiques ... "

This quote, taken from [34], explains one of the problems with a mathematical theory that is so vast, and has gone through so many phases, as invariant theory. One of the reasons I wrote this book was to simplify life for *myself*. In my thousands of pages of notes and the many reprints and preprints that sit on my desk are the answers to all sorts of questions about invariants of finite groups: but finding those answers sometimes requires a search through many papers, preprints, reprints and notes.

This book represents the distillation of what I have found to be (and I hope you find it so also) the most essential and interesting results in the theory of polynomial invariants of finite groups. I hope that you have as much fun reading the book as I did writing it and that it simplifies your literature searches also.

I am very grateful to Bryan Wills, Robert E. Stong, Mara D. Neusel, Peter S. Landweber and Nelson Killius who read various versions of this manuscript and tried very hard to purge my spelling and grammatical errors, not to mention the mathematical ones!, from the final manuscript. I hope they succeeded, as the mathematics to be presented is too lovely to be marred by my slovenly Brooklynese, or with proofs obtained by jumping to conclusions (my favorite sport).

I am also grateful to Christine Bocksch and Karen Parshall for information concerning the history of invariant theory. Thanks also go to Iris Böcker who supplied me with extensive lists of misprints to working versions of the manuscript.

The appearance of this manuscript makes its origins as a collection of TeX files unmistakable. The macro package $\mathcal{LS}$TeX was written for the purpose of processing these files. Some of the commutative diagrams were coded with the Xy-pic macro package of Kristoffer H. Rose. The final filming was done at the Gesellschaft für Wissenschaftliche Datenverarbeitung in Göttingen on a Linotronic 330, and I want to particularly thank G. Koch for his assistance in the final stages.

$\mathcal{LS}$

Göttingen 12/27/94

For the second printing I have augmented the reference lists and the citations in the text as much as possible to bring them up to date without any serious changes. Much has happened since work on this book was completed in 1994, and only a major revision could take account of the most recent develpoments centered around the solution of the **depth conjecture** of Landweber and Stong by Dorra Bourguiba and Said Zarati [37].

I would like to thank the many people who supplied me with lists of misprints and lapses of mathematical rigor (sic!) for this corrected printing: in addition to those who helped me with the original edition I wish to thank H. E. A. Campbell, G. Kemper, and D. Wehlau.

$\mathcal{LS}$

Göttingen 1/31/97

Contents

Chapter 1
Invariants and Relative Invariants

In this chapter, we examine several of the basic ideas and problems of invariant theory in the context of some elementary examples. The origins of invariant theory can perhaps be found in the study of symmetric polynomials, and so we begin with this classical topic.

1.1 Symmetric Polynomials

Let $\mathbb{F}$ be a field and $\mathbb{F}[x_1, \ldots, x_n]$ the polynomial algebra in n variables over $\mathbb{F}$. A polynomial

$$f(x_1, \ldots, x_n) \in \mathbb{F}[x_1, \ldots, x_n]$$

is called **symmetric** if

$$f(x_{\sigma(1)}, \ldots, x_{\sigma(n)}) = f(x_1, \ldots, x_n)$$

for all $\sigma \in \Sigma_n$, where Σ_n denotes the symmetric group on n elements. Examples of symmetric polynomials are the power sums $p_m = x_1^m + \cdots + x_n^m$ and the product $x_1 \cdots x_n$. It is apparent that if f and h are symmetric polynomials then so are their sum and their product. Thus the subset of $\mathbb{F}[x_1, \ldots, x_n]$ consisting of the symmetric polynomials forms a subring: a subring with 1 since the constant polynomials are clearly symmetric. How can this subring be described?

Let us introduce a new variable X, and consider the identity

$$\Psi(X) = \prod_{i=1}^{n}(1 - x_i X) = \sum_{i=0}^{n}(-1)^i e_i(x_1, \ldots, x_n) X^i$$

defining $e_i(x_1, \ldots, x_n) \in \mathbb{F}[x_1, \ldots, x_n]$. Equivalently we may define e_i by the identity

$$\prod_{i=1}^{n}(X + x_i) = \sum_{i=0}^{n} e_i X^{n-i}.$$

1

Clearly the polynomials e_i are symmetric. They are called the **elementary symmetric polynomials** in $x_1, \ldots, x_n$.

The power sums $p_m = x_1^m + \cdots + x_n^m$ are also symmetric. A formula of Isaac Newton shows how to express the power sums in terms of the elementary symmetric polynomials. To obtain this formula, take the logarithmic derivative of Ψ :

$$-\frac{\Psi'(X)}{\Psi(X)} = \sum_{i=1}^{n} \frac{x_i}{1 - x_i X} = p_1 + p_2 X + p_3 X^2 + \cdots.$$

By regarding the inverse of a polynomial in X with constant term 1 as a formal power series in X, the left hand side can be expanded out to a formal power series which has as its coefficients polynomials in the elementary symmetric polynomials. Equating coefficients of powers of X, we obtain formulae for the p_m as polynomials in $e_1, \ldots, e_n$. Equivalently, we could multiply out to get

$$\Psi'(X) = -\Psi(X)(p_1 + p_2 X + p_3 X^2 + \cdots),$$

equate coefficients to obtain,

$$e_1 = p_1$$

$$2e_2 = p_1 e_1 - p_2$$

$$3e_3 = p_1 e_2 - p_2 e_1 + p_3$$

$$\vdots$$

$$ke_k = \sum_{i=1}^{k} (-1)^{i-1} p_i e_{k-i}$$

$$\vdots$$

(where $e_0 = 1$) and solve for p_i inductively. The general formula given above is often referred to as **Newton's formula**.

This is a particular case of the following theorem, which was known to Newton, Waring and probably others.

THEOREM 1.1.1 : *Every symmetric polynomial*

$$f(x_1, \ldots, x_n) \in \mathbb{F}[x_1, \ldots, x_n]$$

can be written as a polynomial in the elementary symmetric polynomials.

PROOF : We make use of the **lexicographic order** on monomials. Namely, we define a partial order $\prec$ on monomials by stipulating that

$$x_1^{a_1} \cdots x_n^{a_n} \prec x_1^{b_1} \cdots x_n^{b_n}$$

if and only if the first nonzero difference $b_i - a_i$ is positive. For example $x_1 x_2^2 \prec x_1^2 x_2$.

We proceed by induction over the lexicographic order. Let $f(x_1, \ldots, x_n)$ be a symmetric polynomial. The action of the symmetric group on $\mathbb{F}[x_1, \ldots, x_n]$ maps homogeneous polynomials[1] to homogeneous polynomials of the same degree. Hence a polynomial f is symmetric if and only if each of its homogeneous components is symmetric. Therefore there is no loss in generality if we assume that f is homogeneous. Let $x_1^{a_1} \cdots x_n^{a_n}$ be the largest monomial appearing with nonzero coefficient, say a, in $f(x_1, \ldots, x_n)$. It is not hard to see that $a_{i+1} \le a_i$ for all i, for supposing that this was not the case, there would be a smallest i such that $a_{i+1} > a_i$. The transposition $\tau(i, i+1)$ that interchanges i and $i+1$ belongs to Σ_n and since f is invariant, $x_1^{a_1} \cdots x_i^{a_{i+1}} \cdot x_{i+1}^{a_i} \cdots x_n^{a_n}$ also appears in f with coefficient a, but is larger in the lexicographic order, which is a contradiction. Next note that

$$e_1^{a_1 - a_2} e_2^{a_2 - a_3} \cdots e_n^{a_n}$$

(this product only makes sense because $a_{i+1} \le a_i$ for all i) also contains the monomial $x_1^{a_1} \cdots x_n^{a_n}$ as highest monomial. So

$$f(x_1, \ldots, x_n) - a e_1^{a_1 - a_2} e_2^{a_2 - a_3} \cdots e_n^{a_n}$$

is a symmetric polynomial whose highest monomial is lower in the lexicographic order than the highest monomial of f. The polynomial is also symmetric. By repeating this process, we must eventually reach the zero polynomial. Rearranging the resulting equality expresses f as a polynomial in the elementary symmetric polynomials. $\square$

Thus the subalgebra of symmetric polynomials is generated by $e_1, \ldots, e_n$. How unique is the expression of a symmetric polynomial in terms of the elementary symmetric polynomials? This can be answered by studying the subalgebra of $\mathbb{F}[x_1, \ldots, x_n]$ generated by the elementary symmetric polynomials more closely.

THEOREM 1.1.2 : *The polynomials $e_1, \ldots, e_n$ are algebraically independent.*

PROOF : Suppose that $f(x_1, \ldots, x_n) \in \mathbb{F}[x_1, \ldots, x_n]$ is a polynomial such that

$$f(e_1, \ldots, e_n) = 0$$

as a polynomial in $x_1, \ldots, x_n$. We claim that f is the zero polynomial. The polynomial $f(x_1, \ldots, x_n)$ is a sum of monomials, which we choose to write in the form

$$x_1^{a_1 - a_2} x_2^{a_2 - a_3} \cdots x_n^{a_n}$$

---

[1] In the classical literature homogeneous polynomials are called *forms*.

for integers $a_1 \geq a_2 \geq \cdots \geq a_n$. Provided $f \neq 0$, let

$$ax_1^{a_1-a_2} \cdots x_n^{a_n} \qquad (a \neq 0)$$

be the largest monomial in the lexicographic order among all possibilities for the n-tuple $(a_1, \ldots, a_n)$. Then $f(e_1, \ldots, e_n)$, regarded as a polynomial in $x_1, \ldots, x_n$, would have $ax_1^{a_1} \cdots x_n^{a_n}$ as its largest monomial in the lexicographic order and therefore $a = 0$. This is a contradiction and therefore $f = 0$.
□

This means that the expression of a symmetric polynomial as a polynomial in the elementary symmetric polynomials is unique.

Let us place this result in a more general context by introducing some notations and definitions.

1.2   The Basic Objects of Study

Let G be a finite group, $\mathbb{F}$ a field of coefficients, V a finite dimensional $\mathbb{F}$-vector space and $\varrho : G \longrightarrow \mathrm{Aut}(V) = \mathrm{GL}(V)$ a homomorphism of groups, i.e. a representation of G on V. Via ϱ, G acts on V through linear substitutions. Equivalently, V is a finitely generated $\mathbb{F}(G)$-module, where $\mathbb{F}(G)$ denotes the group ring of G over $\mathbb{F}$.

For a vector space V over $\mathbb{F}$ we denote by $\mathbb{F}[V]$ the **algebra of polynomial functions on** V, which we define to be the symmetric algebra on V^*, the dual of V. In other words, if V has dimension n as a vector space, and $x_1, \ldots, x_n$ is a basis for the dual space $V^* = \mathrm{Hom}_{\mathbb{F}}(V, \mathbb{F})$, then

$$\mathbb{F}[V] = \mathbb{F}[x_1, \ldots, x_n] = \mathbb{F} \oplus V^* \oplus S^2(V^*) \oplus S^3(V^*) \oplus \cdots .$$

Here, $S^m(V^*)$ denotes the m-th symmetric power of V^*, which consists of the homogeneous polynomials of degree m in $x_1, \ldots, x_n$. Thus for example $S^2(V^*)$ has a basis consisting of the $\binom{n+1}{2}$ monomials $x_i x_j$, for the $\binom{n}{2}$ choices with $i < j$, and the $\binom{n}{1}$ choices with $i = j$. In general, the dimension of $S^m(V^*)$ as a vector space is $\binom{n+m-1}{m} = \binom{n+m-1}{n-1}$. We regard $\mathbb{F}[V]$ as a graded ring by putting each x_i in degree one. (If you are a topologist, then you may occasionally wish to double all the degrees.) The homogeneous component of $\mathbb{F}[V]$ of degree n will be denoted by $\mathbb{F}[V]_n$. (See the introduction to chapter 4 for a discussion of gradings.)

It is often very convenient to think of the elements of the polynomial algebra $\mathbb{F}[V]$ as being functions. There is a problem in characteristic p in that the field $\mathbb{F}$ may not contain enough elements to separate say the function x from the function x^p. The remedy is to allow the functions to take values in a larger

field. Specifically, a homogeneous polynomial $f = \sum \alpha_{i_1,\ldots,i_n} x_1^{i_1} \cdots x_n^{i_n} \in$
$\mathbb{F}[x_1, \ldots, x_n]$ of degree d defines a function

$$e(f) : V \longrightarrow \mathbb{F} \qquad (V = \mathbb{F}^n)$$

by the formula

$$f(v) = \sum_{i_1 + \cdots + i_n = d} \alpha_{i_1,\ldots,i_n} a_1^{i_1} \cdots a_n^{i_n}$$

where $\alpha_{i_1,\ldots,i_n} \in \mathbb{F}$ and $v = (a_1, \ldots, a_n) \in V$. If $\bar{\mathbb{F}}$ is the algebraic closure of $\mathbb{F}$ then $e(f)$ extends to a function

$$\bar{e}(f) : V \otimes_{\mathbb{F}} \bar{\mathbb{F}} = \bar{V} \longrightarrow \bar{\mathbb{F}},$$

and

$$\bar{e} : \mathbb{F}[x_1, \ldots, x_n] \longrightarrow \mathrm{Fun}(\bar{V}, \bar{\mathbb{F}})$$

is a ring homomorphism. For $f \in \mathbb{F}[x_1, \ldots, x_n]$ the functions $e(f) : V \longrightarrow \mathbb{F}$ and $\bar{e}(f) : \bar{V} \longrightarrow \bar{\mathbb{F}}$ are polynomial functions.

LEMMA 1.2.1 : *Let $\mathbb{F}$ be a field and $n \in \mathbb{N}$, then*

$$\bar{e} : \mathbb{F}[x_1, \ldots, x_n] \longrightarrow \mathrm{Fun}(\bar{V}, \bar{\mathbb{F}})$$

is a monomorphism.

PROOF : The homomorphism $\bar{e}$ may be regarded as the composite

$$\bar{e} : \mathbb{F}[x_1, \ldots, x_n] \overset{\iota}{\longrightarrow} \bar{\mathbb{F}}[x_1, \ldots, x_n] \overset{e}{\longrightarrow} \mathrm{Fun}(\bar{V}, \bar{\mathbb{F}}).$$

The map ι is induced by the inclusion $\mathbb{F} \subset \bar{\mathbb{F}}$ and is a monomorphism. So it suffices to prove that e is a monomorphism. This we do by induction on $n \in \mathbb{N}$.

For $n = 1$ and $0 \neq f \in \bar{\mathbb{F}}[x]$ we have

$$f = \prod_{i=1}^{d} (x - \lambda_i) \qquad d = \deg(f)$$

where $\lambda_1, \ldots, \lambda_d \in \bar{\mathbb{F}}$ are the zeros of f. Since $\bar{\mathbb{F}}$ is algebraically closed it is infinite, so there exists $0 \neq \lambda \in \bar{\mathbb{F}}$, $\lambda \notin \{\lambda_1, \ldots, \lambda_d\}$. Then

$$e(f)(\lambda) = \prod_{i=1}^{d} (\lambda - \lambda_i) \neq 0.$$

Hence $e(f) : \bar{V} \longrightarrow \bar{\mathbb{F}}$ is not the zero map.

Assume the result established for $n - 1 \in \mathbb{N}$. Let $0 \neq f \in \bar{\mathbb{F}}[x_1, \ldots, x_n]$ and write f in the form

$$f = \sum_{i=0}^{d} h_i(x_1, \ldots, x_{n-1}) x_n^i$$

where $h_i(x_1, \ldots, x_{n-1}) \in \bar{\mathbb{F}}[x_1, \ldots, x_{n-1}]$ for $i = 1, \ldots, d$.

Since $f \neq 0$ there is a coefficient $h_j(x_1, \ldots, x_{n-1})$ that is also nonzero. By the induction hypothesis there are elements $\lambda_1, \ldots, \lambda_{n-1} \in \bar{\mathbb{F}}$ such that $h_j(\lambda_1, \ldots, \lambda_{n-1}) \neq 0 \in \bar{\mathbb{F}}$. Consider

$$\sum_{i=0}^{d} h_i(\lambda_1, \ldots, \lambda_{n-1}) x_n^i \in \bar{\mathbb{F}}[x_n]$$

and let $\mu_1, \ldots, \mu_d \in \bar{\mathbb{F}}$ be the zeros of this polynomial. Choose $0 \neq \lambda_n \in \bar{\mathbb{F}}$ such that $\lambda_n \notin \{\lambda_1, \ldots, \lambda_{n-1}, \mu_1, \ldots, \mu_d\}$. This is possible since $\bar{\mathbb{F}}$ is infinite. Then

$$e(f)(\lambda_1, \ldots, \lambda_n) = \sum_{i=0}^{d} h_i(\lambda_1, \ldots, \lambda_{n-1}) \lambda_n^i \neq 0 \in \bar{\mathbb{F}}$$

from which the lemma follows. $\square$

PROPOSITION 1.2.2 : *Let* $\mathbb{F}$ *be a field and* $n \in \mathbb{N}$, *then* $\mathrm{Im}(e) \subset$ $\mathrm{Fun}(\bar{V}, \bar{\mathbb{F}})$ *is the ring of polynomial functions on* $V = \mathbb{F}^n$ *i.e.* $\mathrm{Im}(e) = \mathbb{F}[V]$. *(N.b. If* $z_1, \ldots, z_n \in V^*$ *is a basis then* $\mathbb{F}[V] \cong \mathbb{F}[z_1, \ldots, z_n]$.) $\square$

Given a representation $\varrho : G \hookrightarrow \mathrm{GL}(V)$ the group G acts on $\mathbb{F}[V]$ via $(gf)(v) := f(\varrho(g^{-1})v)$, and the basic objects of study of invariant theory is the fixed point set of the G-action, the **ring of invariants**

$$\mathbb{F}[V]^G := \{f \in \mathbb{F}[V] \mid gf = f \ \forall g \in G\}.$$

If $\varrho : G \longrightarrow \mathrm{GL}(n, \mathbb{F})$ is not faithful (i.e. if ϱ has a nonzero kernel) then ϱ induces a faithful representation $\varrho : H := G/\ker(\varrho) \hookrightarrow \mathrm{GL}(n, \mathbb{F})$. Clearly $\mathbb{F}[V]^G = \mathbb{F}[V]^H$ and so it is no loss of generality (and sometimes a necessity) to assume that ϱ is faithful. In this language, the example of the last section may be rewritten as follows.

THEOREM 1.2.3 : *Let* V *be the defining* n-*dimensional permutation module for* Σ_n, *and let* $x_1, \ldots, x_n \in V^*$ *be the dual of the permutation basis. Let* $e_1, \ldots, e_n$ *be the elementary symmetric polynomials in* $x_1, \ldots, x_n$. *Then* $\mathbb{F}[V]^{\Sigma_n} = \mathbb{F}[e_1, \ldots, e_n]$. $\square$

We define the ring of **coinvariants** to be the quotient of $\mathbb{F}[V]$ by the ideal generated by the invariant polynomials of positive degree. In other words, if

$$\overline{\mathbb{F}[V]^G} = \{f \in \mathbb{F}[V]^G \mid \deg(f) > 0\}$$

then the coinvariants are defined by

$$\mathbb{F}[V]_G = \mathbb{F}[V]/(\overline{\mathbb{F}[V]^G}),$$

where $(-)$ denotes the ideal in $\mathbb{F}[V]$ generated by $-$. Using tensor products, this may be written as

$$\mathbb{F}[V]_G = \mathbb{F} \otimes_{\mathbb{F}[V]^G} \mathbb{F}[V]$$

where $\mathbb{F}[V]^G$ acts on $\mathbb{F}[V]$ by multiplication of polynomials, and on $\mathbb{F}$ via the **augmentation** homomorphism

$$\varepsilon : \mathbb{F}[V]^G \longrightarrow \mathbb{F} \qquad \varepsilon(f) = \begin{cases} f & \text{if } f \text{ has degree } 0 \\ 0 & \text{if } f \text{ has positive degree} \end{cases}$$

for a homogeneous polynomial f (polynomials of degree 0 may be identified with elements of $\mathbb{F}$). Already in the example above, the structure of the ring $\mathbb{F}[V]_G$ is less apparent. The classical literature talks of **covariants**: this is a different concept. See section 8.2.

Denote by $\mathbb{F}^\times$ the nonzero elements of the field $\mathbb{F}$, which may be identified with the group $\mathrm{GL}(1, \mathbb{F})$. If $\chi : G \to \mathbb{F}^\times$ is a 1–dimensional representation of G, the module of χ–**relative invariants**, denoted by $\mathbb{F}[V]^G_\chi$ is defined by

$$\mathbb{F}[V]^G_\chi = \{f \in \mathbb{F}[V] \mid g(f) = \chi(g)f \ \forall \, g \in G\}.$$

More generally, if S is a simple $\mathbb{F}(G)$-module, we set

$$\mathbb{F}[V]^G \Big|_S = S \otimes_{\mathrm{End}_{\mathbb{F}(G)}(S)} \mathrm{Hom}_{\mathbb{F}(G)}(S, \mathbb{F}[V]).$$

If $\mathbb{F}$ is algebraically closed then Schur's lemma [202] says that $\mathrm{End}_{\mathbb{F}(G)}(S) = \mathbb{F}$. The evaluation map

$$S \otimes_{\mathrm{End}_{\mathbb{F}(G)}(S)} \mathrm{Hom}_{\mathbb{F}(G)}(S, \mathbb{F}[V]) \longrightarrow \mathbb{F}[V]$$

is injective and identifies $\mathbb{F}[V]^G \Big|_S$ with the largest subspace of $\mathbb{F}[V]$ consisting of a direct sum of copies of S. $\mathbb{F}[V]^G \Big|_S$ is referred to as the S-**isotypical summand** of $\mathbb{F}[V]^G$. Multiplication in $\mathbb{F}[V]$ gives a map

$$\mathbb{F}[V]^G \times \mathbb{F}[V]^G \Big|_S \longrightarrow \mathbb{F}[V]^G \Big|_S$$

making $\mathbb{F}[V]^G \Big|_S$ into a $\mathbb{F}[V]^G$-module.

The invariants, coinvariants and relative invariants are all graded and their homogeneous components of degree n are denoted by $\mathbb{F}[V]^G_n$, $(\mathbb{F}[V]_G)_n$ and $(\mathbb{F}[V]^G \Big|_S)_n$ respectively.

EXAMPLE 1 : Consider the representation of $\mathbb{Z}/2$ given by the matrix

$$\begin{bmatrix} -1 & 0 \\ 0 & -1 \end{bmatrix} \in \mathrm{GL}(2, \ \mathbb{F})$$

where $\mathbb{F}$ has characteristic different from 2. Then the invariants of $\mathbb{Z}/2$ acting on $\mathbb{F}[x, \ y]$ are generated by

$$x^2, \ xy, \ y^2.$$

Notice that

$$x^2 \cdot y^2 = (xy)(xy)$$

and that x^2, y^2 and xy are indecomposable elements of $\mathbb{F}[x, \ y]^{\mathbb{Z}/2}$ which are not associates. Therefore $\mathbb{F}[x, \ y]^{\mathbb{Z}/2}$ is not a unique factorization domain. If we denote by $\mathbb{F}[X, \ Y, \ Z]$ a polynomial algebra on three generators all of degree 2 then the homomorphism of algebras

$$\alpha : \mathbb{F}[X, \ Y, \ Z] \longrightarrow \mathbb{F}[x, \ y]$$

defined by requiring

$$\alpha(X) = x^2, \ \alpha(Y) = y^2, \ \alpha(Z) = xy$$

is a surjection onto the invariants with kernel the ideal generated by $Z^2 - XY$. Therefore

$$\mathbb{F}[x, \ y]^{\mathbb{Z}/2} \cong \mathbb{F}[X, \ Y, \ Z]/(Z^2 - XY)$$

as algebras over $\mathbb{F}$. This ring has three generators (one more than the number of variables in the ambient polynomial algebra $\mathbb{F}[x, \ y]$) and one relation.

The expression for an invariant polynomial in terms of the complete fundamental system of invariants x^2, xy, y^2 is not unique. However note that $\mathbb{F}[x, \ y]^{\mathbb{Z}/2}$ is a free module over the subalgebra $\mathbb{F}[x^2, \ y^2]$ with generators 1, xy and so there is a unique expression for every invariant polynomial in the form $f = h' + h'' \cdot xy$, where h', $h'' \in \mathbb{F}[x^2, \ y^2]$.

EXAMPLE 2 : Consider the tautological representation of Σ_3 on $V = \mathbb{F}^3$. If $\{x, \ y, \ z\}$ is the standard basis for V^*, which is permuted by Σ_3, then

$$\mathbb{F}[x, \ y, \ z]^{\Sigma_3} = \mathbb{F}[x + y + z, \ xy + yz + zx, \ xyz]$$

and the ring of coinvariants is

$$\mathbb{F}[x, \ y, \ z]_{\Sigma_3} = \mathbb{F}[x, \ y, \ z]/(x + y + z, \ xy + yz + zx, \ xyz).$$

The relation $x + y + z = 0 \in \mathbb{F}[x, \ y, \ z]_{\Sigma_3}$ allows us to express z in terms of x and y and so we obtain

$$\mathbb{F}[x, \ y, \ z]_{\Sigma_3} \cong \mathbb{F}[x, y]/(x^2 + xy + y^2, \ x^2 y + xy^2).$$

One way to visualize the ring of coinvariants is with the aid of the following diagram

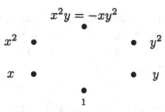

Diagram 1.2.1: $\mathbb{F}[x,\ y,\ z]_{\Sigma_3}$

where the nodes on a horizontal level indicate basis vectors for the elements of degree equal to the height of the node above the node labeled 1, which has degree 0. We will see in the next chapter that $\mathbb{F}[V]_G$ is always a finite dimensional representation of G. In the case at hand it is the regular representation of Σ_3: in degree 0 we have the trivial 1-dimensional representation, the irreducible 2-dimensional representation occurs twice, in degrees 1 and 2, and the determinant representation occurs in degree 3. This is a special case of a theorem of Chevalley concerning reflection groups (see section 7.5). Note also the symmetry of the diagram from top to bottom and how the products work out. This too is a special case of a more general phenomenon (see section 6.5).

At a coarser level, we could look at the action of G on the field of fractions $\mathbb{F}(V)$ of $\mathbb{F}[V]$. This may be regarded as the field of rational functions on V. The action of G is given by $g(f_1/f_2) := g(f_1)/g(f_2)$ and we define $\mathbb{F}(V)^G$ (as we did for $\mathbb{F}[V]^G$) as the set of elements left fixed by the G action.

DEFINITION : *If $A \supseteq B$ is an extension of commutative rings, an element $a \in A$ is called* **integral** *over B if it is a root of a monic polynomial with coefficients in B. B is said to be* **integrally closed** *in A if every element in A integral over B belongs to B.*

To illustrate these ideas let $B \subset A = \mathbb{F}[z]$ be the subalgebra of polynomials with linear term 0. Thus $\{1,\ z^2,\ z^3,\ldots,\ \}$ is a basis for B over $\mathbb{F}$. The polynomial $z \in A$ is integral over B as it is the unique zero of the monic polynomial $1 \cdot x^3 - z^2 \cdot x \in B[x]$. Note also that $z = z^3/z^2$ belongs to the field of fractions $FF(B)$ of B but not to B itself, so B is not integrally closed in its field of fractions. A is integrally closed, and the inclusion $B \subset A$ is the integral closure of B.

PROPOSITION 1.2.4 : *Suppose that V is a finite dimensional faithful representation of a finite group G over a field $\mathbb{F}$. Then $\mathbb{F}(V)$ is a Galois (i.e., normal and separable) extension of $\mathbb{F}(V)^G$ with Galois group G. The field $\mathbb{F}(V)^G$ is the field of fractions of $\mathbb{F}[V]^G$, and $\mathbb{F}[V]^G$ is integrally closed in $\mathbb{F}(V)^G$.*

PROOF : Since G acts as field automorphisms on $\mathbb{F}(V)$, it is clear that $\mathbb{F}(V)$ is a Galois extension of $\mathbb{F}(V)^G$ with Galois group G.

Every element of $\mathbb{F}(V)$ may be written in the form f_1/f_2 with f_2 G-invariant, just by multiplying top and bottom by the distinct images of f_2 under the action of G. If f_1/f_2 and f_2 are G-invariant then so is f_1. It follows that $\mathbb{F}(V)^G$ is the field of fractions of $\mathbb{F}[V]^G$.

Any element $f \in \mathbb{F}(V)^G$ which is integral over $\mathbb{F}[V]^G$ is also integral over $\mathbb{F}[V]$. Since $\mathbb{F}[V]$ is integrally closed in $\mathbb{F}(V)$, this means that $f \in \mathbb{F}[V]$. Since f is G-invariant, this means that $f \in \mathbb{F}[V]^G$. $\square$

It follows from 1.2.4 that when $H < G$ is a proper subgroup then $\mathbb{F}[V]^G \subset \mathbb{F}[V]^H$ is a proper inclusion.

The question of whether $\mathbb{F}(V)^G$ is purely transcendental over $\mathbb{F}$ is a hard one in general. For a nilpotent group in coprime characteristic, the answer is affirmative [149], as it is for finite p-groups in characteristic p [90]. The first examples with $\mathbb{F} = \mathbf{Q}$ where $\mathbb{F}(V)^G$ is not purely transcendental were produced by Swan [245], and with $\mathbb{F} = \mathbf{C}$ by Saltman [191]. See also the survey article of Kervaire and Vust [122].

When $G = \Sigma_n$, the symmetric group on n elements, and ϱ is the tautological representation, then the preceding proposition says that a rational function that is invariant with respect to permutation of the variables may be expressed as a rational function of the elementary symmetric polynomials. It is also possible, and interesting, to examine the ring of invariants of G acting on other functions on V. For example if $\mathbb{F} = \mathbb{R}$, the real field, then we could consider C^k-functions, C^∞-functions, or analytic functions. Even in the case of the symmetric group and its tautological representation, the results are neither apparent nor easy. See for example Barbançon and Raïs [23] and the references there.

1.3   The Alternating Group

As a further example let us consider the alternating group $A_n < \Sigma_n$, consisting of even permutations, and the restriction to A_n of the tautological representation of Σ_n. Suppose that the characteristic of $\mathbb{F}$ is not equal to 2, and consider the determinant representation (also called the *signum* representation)

$$\text{sgn} : \Sigma_n \to \{\pm 1\} \subset \mathbb{F}^\times$$

that associates to a permutation its sign. It is clear that A_n is the kernel of sgn and that $f \in \mathbb{F}[V]^{\Sigma_n}_{\text{sgn}}$ if and only if f is an alternating polynomial, so we have the following.

PROPOSITION 1.3.1 : *Suppose that $\mathbb{F}$ is a field which is not of characteristic two. Let V be the n-dimensional permutation module for Σ_n, and* sgn : $\Sigma_n \to \mathbb{F}^\times$ *the determinant representation. Then*

$$\mathbb{F}[V]^{\Sigma_n}_{\mathrm{sgn}} = \mathbb{F}[V]^{\Sigma_n} \cdot \Delta$$

where

$$\Delta = \det \begin{bmatrix} 1 & \cdots & 1 \\ x_1 & \cdots & x_n \\ \vdots & \cdots & \vdots \\ x_1^{n-1} & \cdots & x_n^{n-1} \end{bmatrix} = \prod_{j<i}(x_i - x_j)$$

is the Vandermonde determinant. In other words, $\mathbb{F}[V]^{\Sigma_n}_{\mathrm{sgn}}$ is a free $\mathbb{F}[V]^{\Sigma_n}$-module on the single generator Δ.

PROOF : Since Δ is clearly in $\mathbb{F}[V]^{\Sigma_n}_{\mathrm{sgn}}$, it is enough to show that if $f \in \mathbb{F}[V]^{\Sigma_n}_{\mathrm{sgn}}$ then Δ divides f. Since $\mathbb{F}[V]$ is a unique factorization domain, it suffices to show that each $x_i - x_j$ divides f, or equivalently that f vanishes on restriction to the hyperplane $x_i = x_j$ in V. Since the permutation (i, j) negates f, it follows that $f = -f$ on this hyperplane. Since $\mathbb{F}$ does not have characteristic two, it follows that f vanishes as required. □

To compute $\mathbb{F}[V]^{A_n}$ we first note that there are inclusions of rings

$$\mathbb{F}[V]^{\Sigma_n} \leq \mathbb{F}[V]^{A_n} \leq \mathbb{F}[V].$$

Moreover, since $A_n \triangleleft \Sigma_n$ is a normal subgroup, the quotient group $\mathbb{Z}/2 = \Sigma_n/A_n$ acts on $\mathbb{F}[V]^{A_n}$. Since $\mathbb{F}$ has characteristic different from 2, $\mathbb{F}[V]^{A_n}$ splits into the ± 1 eigenspaces of the nontrivial element of $\mathbb{Z}/2$. Clearly $\mathbb{F}[V]^{\Sigma_n}$ is the $+1$ eigenspace and $\mathbb{F}[V]^{\Sigma_n}_{\mathrm{sgn}}$ is the -1 eigenspace. Thus:

COROLLARY 1.3.2 : *Suppose that $\mathbb{F}$ is a field which is not of characteristic two. Let V be the n-dimensional permutation module for Σ_n, and* sgn : $\Sigma_n \to \mathbb{F}^\times$ *the determinant representation with kernel the alternating group A_n. Then the ring of invariants $\mathbb{F}[V]^{A_n}$ is a free module of rank two over the subring $\mathbb{F}[V]^{\Sigma_n}$:*

$$\mathbb{F}[V]^{A_n} = \mathbb{F}[V]^{\Sigma_n} \oplus \mathbb{F}[V]^{\Sigma_n}_{\mathrm{sgn}} = \mathbb{F}[e_1, \ldots, e_n] \oplus \mathbb{F}[e_1, \ldots, e_n] \cdot \Delta$$

where Δ is the Vandermonde determinant as above. □

REMARK : Note that $e_1, \ldots, e_n, \Delta$ form a minimal generating set for $\mathbb{F}[V]^{A_n}$, and are not algebraically independent, since $\Delta^2 \in \mathbb{F}[e_1, \ldots, e_n]$. So $\mathbb{F}[V]^{A_n}$ is not a polynomial algebra, but it is a finitely generated free module over the polynomial subalgebra $\mathbb{F}[e_1, \ldots, e_n]$. This is an example of a theorem of Eagon and Hochster [107], which we shall describe in section 6.7.

To deal with the case where $\mathbb{F}$ has characteristic two, we introduce a general technique for dealing with permutation modules.

Suppose that a finite group G acts as permutations on a finite set X. A subset $B \subset X$ is called an **orbit** (see also chapter 3) if G permutes the elements of B among themselves and the induced permutation action of G on B is transitive.

NOTATION : *If X is a G-set and R a commutative ring we denote by RX the free R-module on the set X and regard it as an $R(G)$-module via the permutation action.*

LEMMA 1.3.3 : *Let X be a finite G-set and R a commutative ring. Then the orbit sums*

$$\mathfrak{S}_B = \sum_{b \in B} b \qquad B \subset X \text{ an orbit}$$

are a basis for the fixed point set $(RX)^G$.

PROOF : Since $\mathfrak{S}_B$ is a sum over an entire orbit it belongs to the fixed point set $(RX)^G$. The orbits of G on X are disjoint. Let

$$X = X_1 \sqcup \cdots \sqcup X_s$$

be the decomposition of X into orbits and write $\mathfrak{S}_i$ for $\mathfrak{S}_{X_i}$. Denote the elements of X_i by $y_{i,1}, \ldots, y_{i,m_i}$ for $i = 1, \ldots, s$. In this notation $\{y_{i,j} \mid 1 \leq j \leq m_i, \ i = 1, \ldots, s\}$ is a basis for RX. The action of G on X, and hence on RX, maps each X_i bijectively to itself for $i = 1, \ldots, s$. Suppose that

$$y = \sum a_{i,j} y_{i,j} \in (RX)^G.$$

Rewrite this in the form

$$y = \sum_{j=1}^{m_1} a_{1,j} y_{1,j} + \cdots + \sum_{j=1}^{m_s} a_{s,j} y_{s,j}.$$

If $y_{k,j'}$, $y_{k,j''} \in X_k$, then there exists an element $g \in G$ such that $g y_{k,j'} = y_{k,j''}$ so applying g to y and equating coefficients we see that $a_{k,j'} = a_{k,j''}$. Since this holds for all $1 \leq j'$, $j'' \leq m_k$, $k = 1, \ldots, s$ we may write

$$y = a_1 \sum_{j=1}^{m_1} y_{1,j} + \cdots + a_s \sum_{j=1}^{m_s} y_{s,j}$$

$$= a_1 \mathfrak{S}_1 + \cdots + a_s \mathfrak{S}_s$$

so the orbit sums $\mathfrak{S}_1, \ldots, \mathfrak{S}_s$ generate $(RX)^G$.

To see that the orbit sums are linearly independent, suppose

$$a_1 \mathfrak{S}_1 + \cdots + a_s \mathfrak{S}_s = 0.$$

Rewrite this as

$$0 = a_1 \sum_{j=1}^{m_1} y_{1,j} + \cdots + a_s \sum_{j=1}^{m_s} y_{s,j}$$

$$= a_1 y_{1,1} + \cdots + a_1 y_{1,m_1} + \cdots + a_s y_{s,1} + \cdots + a_s y_{s,m_s}.$$

Since $\{y_{i,j} \mid 1 \leq j \leq m_i, \ i = 1, \ldots, s\}$ is a basis for RX it follows that

$$a_1 = \cdots = a_s = 0$$

as required. $\square$

NOTATION : *If X is a finite set and R is a commutative ring we write $R[X]$ for the polynomial functions on RX, i.e. the polynomial ring generated by X.*

The R-submodule of $R[X]$ of homogeneous polynomials of degree n is denoted by $R[X]_n$ and called the **homogeneous component** of $R[X]$ **of degree** n. The ring $R[X]$ is isomorphic to the direct sum of its homogeneous components. If a finite group G acts as permutations on a finite set X then the action extends to $R[X]$ as well as to RX. The action of G on $R[X]$ sends monomials to monomials. The monomials of degree n are a basis for $R[X]_n$ and hence $R[X]_n$ is the linear representation associated to the permutation representation of G on the monomial basis.

LEMMA 1.3.4 : *Let G be a finite group acting on the finite set X. If R is a commutative ring then $R[X]_n^G = R \otimes_{\mathbb{Z}} \mathbb{Z}[X]_n^G$ is a free R-module, whose rank is independent of R.*

PROOF : By 1.3.3 $R[X]_n^G$ has an R-basis consisting of the orbit sums. The number of such orbit sums does not depend on whether we count them in $R[X]_n$ or $\mathbb{Z}[X]_n$. $\square$

This means that in characteristic two, there are exactly as many homogeneous invariants of a given degree for A_n as in any other characteristic. The problem is that Δ is actually symmetric in characteristic two. However, the orbit sum of the A_n-orbit of the monomial $x_1^{n-1} x_2^{n-2} \cdots x_{n-1}$ is a nonzero A_n-invariant and so this expression makes sense in characteristic two, or indeed over any coefficient ring (see Göbel [93]). Alternatively, the coefficients of

$$\alpha = \frac{1}{2} \left(\prod_{i<j}(x_i - x_j) + \prod_{i<j}(x_i + x_j) \right)$$

are integers, and so this expression also makes sense in characteristic two and over any coefficient ring (see also Samuel [192], Bertin [29], Berlekamp [27], and Revoy [184]).

THEOREM 1.3.5 : *Suppose that R is a commutative ring and that $\{x_1, \ldots, x_n\}$ is a set of n elements acted on in the usual way by the groups A_n and Σ_n. Then*

$$R[x_1, \ldots, x_n]^{\Sigma_n} = R[e_1, \ldots, e_n]$$

$$R[x_1, \ldots, x_n]^{A_n} = R[e_1, \ldots, e_n] \oplus R[e_1, \ldots, e_n] \cdot \alpha.$$

PROOF : By 1.3.4 it suffices to prove this for $R = \mathbb{Z}$. In the case of Σ_n, the proof given in 1.1.1 works without alteration. In the case of A_n, the given invariants have the right rank, and the right reduction modulo p for p odd. So we need to check that if $f(e_1, \ldots, e_n)$ and $h(e_1, \ldots, e_n)$ are homogeneous polynomials in $e_1, \ldots, e_n$ and $u = f(e_1, \ldots, e_n) + h(e_1, \ldots, e_n) \cdot \alpha$ is divisible by two in $\mathbb{Z}[X]$ then $f(e_1, \ldots, e_n)$ and $h(e_1, \ldots, e_n)$ are divisible by two. To see this, we observe that if σ is an odd permutation then

$$\sigma(\alpha) = \alpha - \prod_{i<j}(x_i - x_j)$$

so that if u is divisible by two then so is

$$u - \sigma(u) = h(e_1, \ldots, e_n) \cdot \prod_{i<j}(x_i - x_j).$$

An examination of largest monomials in the lexicographic order (cf. the proof of Theorem 1.1.1) shows that $h(e_1, \ldots, e_n)$ is divisible by two, and hence so is $f(e_1, \ldots, e_n)$. □

Note that, as a consequence of the above theorem, every polynomial that is invariant under the (tautological) action of the alternating group may be expressed as a polynomial in the elementary symmetric polynomials $e_1, \ldots, e_n$ and the polynomial α. Since α^2 is always Σ_n-invariant, the polynomials (there are too many of them in any case!) $e_1, \ldots, e_n, \alpha$ are not algebraically independent: hence the expression of an alternating polynomial as a polynomial in $e_1, \ldots, e_n$, α is not unique.

1.4 Matrix Invariants

Although we will be studying almost exclusively the invariant theory of finite groups, it would be remiss not to explain at least one example involving infinite groups.

Let $\mathrm{Mat}_n(\mathbb{F})$ denote the ring of $n \times n$ matrices over the field $\mathbb{F}$. $\mathrm{Mat}_n(\mathbb{F})$ is an n^2-dimensional vector space over $\mathbb{F}$. The general linear group, that is the group of invertible $n \times n$ matrices $\mathrm{GL}(n,\ \mathbb{F})$, operates on $\mathrm{Mat}_n(\mathbb{F})$ via conjugation, i.e.

$$g \cdot T := gTg^{-1} \qquad g \in \mathrm{GL}(n,\ \mathbb{F}) \text{ and } T \in \mathrm{Mat}_n(\mathbb{F}).$$

The elements of $\mathrm{Mat}_n(\mathbb{F})$ fall into equivalence classes with respect to this action, two matrices T', T'' being equivalent if and only if there is an invertible matrix g such that $T' = gT''g^{-1}$, i.e. the equivalence classes are exactly the similarity classes. A fundamental goal of linear algebra is to find similarity invariants, i.e. ways to associate to each $n \times n$ matrix T a field element $f(T)$ such that $f(T') = f(T'')$ whenever T', T'' are similar. Two such simple invariants are the determinant $\det(-)$ and the trace $\mathrm{tr}(-)$. Both of these are polynomial functions of the matrix entries. A natural problem is: determine all similarity invariants which are polynomial functions of the matrix entries. This amounts to computing $\mathbb{F}[\mathrm{Mat}_n(\mathbb{F})]^{\mathrm{GL}(n,\ \mathbb{F})}$.

To this end, for $T \in \mathrm{Mat}_n(\mathbb{F})$ let

$$\Delta_T(t) = \det(T - t \cdot I) = \sum_{i=0}^{n} a_i(T)t^{n-i}$$

denote the characteristic polynomial of T, defining

$$a_i : \mathrm{Mat}_n(\mathbb{F}) \longrightarrow \mathbb{F} \qquad i = 0, \ldots, n$$

which are polynomial functions of the entries of T. Of course a_0 is the constant function at 1. There is the following neat solution to the problem.

THEOREM 1.4.1 : *For an algebraically closed field* $\mathbb{F}$

$$\mathbb{F}[\mathrm{Mat}_n(\mathbb{F})]^{\mathrm{GL}(n,\ \mathbb{F})} \cong \mathbb{F}[a_1, \ldots, a_n]$$

We will present the proof for $\mathbb{F} = \mathbb{C}$.

PROOF FOR $\mathbb{F} = \mathbb{C}$: Every matrix $T \in \mathrm{Mat}_n(\mathbb{C})$ has a Jordan normal form

$$\begin{bmatrix} \lambda_1 & \epsilon_1 & 0 & \ldots & 0 \\ 0 & \lambda_2 & \epsilon_2 & 0\ldots & 0 \\ \vdots & \vdots & \vdots & \vdots & \vdots \\ 0 & 0 & \ldots & \lambda_{n-1} & \epsilon_{n-1} \\ 0 & 0 & \ldots & \ldots & \lambda_n \end{bmatrix}$$

where $\lambda_1, \ldots, \lambda_n$ are the eigenvalues of T and $\epsilon_1, \ldots, \epsilon_{n-1} \in \{0,\ 1\}$. If a matrix has distinct eigenvalues then it is diagonalizable. A necessary condition for T **not** be diagonalizable is that it have a repeated eigenvalue. This happens if and only if the characteristic polynomial and its derivative have a common zero. Two polynomials have a common zero if and only if the resultant of the two polynomials vanishes [252]. The resultant is itself a polynomial in the coefficients of the two polynomials and hence a polynomial of the matrix entries. Thus the matrices that are not diagonalizable are contained in the zero set of this polynomial. (Alternatively, a polynomial has multiple roots if and only if the discriminant $\prod_{i<j}(\zeta_i - \zeta_j)$ vanishes, where $\zeta_1, \ldots, \zeta_n$ are the roots of the polynomial, each appearing as often as its multiplicity: the roots of a polynomial are symmetric functions in the coefficients of the polynomial and therefore the discriminant is a polynomial in the coefficients of the polynomial, and its vanishing is a necessary and sufficient condition for the polynomial to have multiple roots.) The complement of this zero set in $\mathrm{Mat}_n(\mathbb{C})$, namely the matrices with distinct eigenvalues, is dense. Therefore the diagonalizable matrices are dense in $\mathrm{Mat}_n(\mathbb{C})$.

If f is a polynomial invariant of $\mathrm{GL}(n,\ \mathbb{C})$ acting on $\mathrm{Mat}_n(\mathbb{C})$ then because of the invariance it suffices to know the value of f on the Jordan normal forms to determine f. Furthermore, since the matrices with distinct eigenvalues are dense in the set of matrices and polynomials are continuous functions, it suffices, to know the value of f on matrices of the form

$$\begin{bmatrix} \lambda_1 & 0 & 0 & \ldots & 0 \\ 0 & \lambda_2 & 0 & 0\ldots & 0 \\ \vdots & \vdots & \vdots & \vdots & \vdots \\ 0 & 0 & \ldots & \lambda_{n-1} & \\ 0 & 0 & \ldots & \ldots & \lambda_n \end{bmatrix}$$

So f must be a polynomial function of the eigenvalues. The general linear group $\mathrm{GL}(n,\ \mathbb{C})$ contains the symmetric group Σ_n. Conjugating a diagonal matrix with a permutation matrix permutes the eigenvalues. Therefore f is a symmetric polynomial function of the n eigenvalues, and hence is a polynomial in the elementary symmetric polynomials of the eigenvalues. These are exactly the $a_1, \ldots, a_n$. $\square$

In the case of a general algebraically closed field one can replace the norm topology by the Zariski topology [266] and the proof proceeds otherwise unchanged.

For example, when $n = 2$ the elementary symmetric polynomials of the eigenvalues of a matrix are:

$$\lambda_1 + \lambda_2 = \mathrm{tr}(T)$$

$$\lambda_1 \cdot \lambda_2 = \det(T)$$

and so any polynomial function of the matrix entries that is a similarity invariant must be a polynomial in tr and det.

In characteristic zero we may express the elementary symmetric polynomials $e_1, \ldots, e_n$ as polynomials in the power symmetric polynomials $p_1, \ldots, p_n$.

There is a denominator so this does not always work in nonzero characteristic. If $T \in \mathrm{Mat}_n(\mathbb{F})$ is a matrix with eigenvalues $\lambda_1, \ldots, \lambda_n$ then

$$p_i(\lambda_1, \ldots, \lambda_n) = \lambda_1^i + \cdots + \lambda_n^i = \mathrm{tr}(T^i)$$

and thus we may rephrase the preceding result as follows.

COROLLARY 1.4.2 : *Let $\mathbb{F}$ be an algebraically closed field of characteristic zero. Define polynomial functions*

$$\mathrm{tr}_i : \mathrm{Mat}_n(\mathbb{F}) \longrightarrow \mathbb{F} \quad \text{by } \mathrm{tr}_i(T) = \mathrm{tr}(T^i) \quad i = 1, \ldots, n.$$

Then

$$\mathbb{F}[\mathrm{Mat}_n(\mathbb{F})]^{\mathrm{GL}(n,\ \mathbb{F})} \cong \mathbb{F}[\mathrm{tr}_1, \ldots, \mathrm{tr}_n]. \quad \Box$$

Two matrices with the same eigenvalues need not be similar, so these results say there are no algebraic formulas involving the entries of a matrix that determine its similarity class. Notice also that $\mathbb{F}[\mathrm{Mat}_n(\mathbb{F})]$ has transcendence degree n^2 over $\mathbb{F}$ but $\mathbb{F}[\mathrm{Mat}_n(\mathbb{F})]^{\mathrm{GL}(n,\ \mathbb{F})}$ only has transcendence degree n over $\mathbb{F}$. This cannot happen for the ring of invariants of a finite group (see chapter 5). Thus whenever $n > 1$ there are nonzero matrices $T \in \mathrm{Mat}_n(\mathbb{F})$ on which all possible invariants vanish. To study the structure of the set of vectors on which all invariants vanish Hilbert proved the famous Nullstellensatz [102].

The group $\mathrm{GL}(n, \mathbb{F})$ also acts on $\mathrm{Mat}_n(\mathbb{F})$ via

$$g \cdot T := gTg^{\mathrm{tr}} \quad g \in \mathrm{GL}(n,\ \mathbb{F}) \text{ and } T \in \mathrm{Mat}_n(\mathbb{F}).$$

This is a completely different action, and corresponds to the classification of n-array forms. The ring of invariants in this case is known in only a few cases. See for example [195].

1.5   Elementary Properties of Invariants and Coinvariants

If G is a finite group acting on a vector space V over the field $\mathbb{F}$ we write V^G for the **fixed point set** of the G-action. Thus $V^G = \{v \in V \mid gv = v \ \forall \ g \in G\}$. Likewise for an element $g \in G$ we write V^g for $\{v \in V \mid gv = v\}$. Note that the homogeneous component of $\mathbb{F}[V]^G$ of degree n is nothing other than $S^n(V^*)^G$, where the action of G on the vector space $S^n(V^*)$ is induced from the action of G on V.

If $H \lhd G$ is a normal subgroup, then G/H acts on V^H, and a routine verification shows that $(V^H)^{G/H} = V^G$. In particular this is true for the induced action of G on $S^n(V^*)$ for $n = 1, 2, \ldots$, and hence we have shown:

PROPOSITION 1.5.1 : *Let $\varrho : G \hookrightarrow \mathrm{GL}(n,\ \mathbb{F})$ be a representation of the finite group G. If $H \lhd G$ is a normal subgroup then $\mathbb{F}[V]^G \cong (\mathbb{F}[V]^H)^{G/H}$.*
□

Let $\varrho' : G' \hookrightarrow \mathrm{GL}(n',\ \mathbb{F})$, $\varrho'' : G'' \hookrightarrow \mathrm{GL}(n'',\ \mathbb{F})$ be representations. Then G' acts on $V' = \mathbb{F}^{n'}$, G'' acts on $V'' = \mathbb{F}^{n''}$ and $G = G' \times G''$ acts on $V' \oplus V''$. If we identify $\mathbb{F}[V' \oplus V'']$ with $\mathbb{F}[V'] \otimes \mathbb{F}[V'']$ in the standard way we obtain an action of G on $\mathbb{F}[V'] \otimes \mathbb{F}[V'']$. The group $G' = G' \times \{1\} < G$ is a normal subgroup with quotient group $G/G' \cong G''$. Applying the preceding proposition we obtain

$$
\begin{aligned}
\mathbb{F}[V]^G = \mathbb{F}[V' \oplus V'']^{G' \times G''} &= (\mathbb{F}[V'] \otimes \mathbb{F}[V''])^{G' \times G''} \\
&= ((\mathbb{F}[V'] \otimes \mathbb{F}[V''])^{G'})^{G''} = (\mathbb{F}[V']^{G'} \otimes \mathbb{F}[V''])^{G''} \\
&= \mathbb{F}[V']^{G'} \otimes \mathbb{F}[V'']^{G''} .
\end{aligned}
$$

Hence we have proven:

PROPOSITION 1.5.2 : *Let $\varrho' : G' \hookrightarrow \mathrm{GL}(n',\ \mathbb{F})$, $\varrho'' : G'' \hookrightarrow \mathrm{GL}(n'',\ \mathbb{F})$ be representations of the finite groups G', G''. Set $V' = \mathbb{F}^{n'}$ and $V'' = \mathbb{F}^{n''}$. Then*

$$
\mathbb{F}[V' \oplus V'']^{G' \times G''} \cong \mathbb{F}[V']^{G'} \otimes \mathbb{F}[V'']^{G''} . \quad \square
$$

COROLLARY 1.5.3 : *Let $\varrho' : G' \hookrightarrow \mathrm{GL}(n',\ \mathbb{F})$, $\varrho'' : G'' \hookrightarrow \mathrm{GL}(n'',\ \mathbb{F})$ be representations of the finite groups G', G''. Set $V' = \mathbb{F}^{n'}$ and $V'' = \mathbb{F}^{n''}$. Then*

$$
\mathbb{F}[V' \oplus V'']_{G' \times G''} \cong \mathbb{F}[V']_{G'} \otimes \mathbb{F}[V'']_{G''} .
$$

PROOF : By definition of the coinvariants and 1.5.2 we have

$$
\begin{aligned}
\mathbb{F}[V' \oplus V'']_{G' \times G''} &= \mathbb{F} \otimes_{\mathbb{F}[V' \oplus V'']^{G' \times G''}} \mathbb{F}[V' \oplus V''] \\
&\cong \mathbb{F} \otimes_{\mathbb{F}[V']^{G'} \otimes \mathbb{F}[V'']^{G''}} \mathbb{F}[V'] \otimes \mathbb{F}[V''] \\
&\cong (\mathbb{F} \otimes \mathbb{F}) \otimes_{\mathbb{F}[V']^{G'} \otimes \mathbb{F}[V'']^{G''}} (\mathbb{F}[V'] \otimes \mathbb{F}[V'']) \\
&\cong (\mathbb{F} \otimes_{\mathbb{F}[V']^{G'}} \mathbb{F}[V']) \otimes (\mathbb{F} \otimes_{\mathbb{F}[V'']^{G''}} \mathbb{F}[V'']) \\
&= \mathbb{F}[V']_{G'} \otimes \mathbb{F}[V'']_{G''}
\end{aligned}
$$

(where the next to the last isomorphism follows from [54] IX.2.1) as required.
□

There are similar isomorphisms for relative invariants.

EXAMPLE 1 : Suppose that $k_1, \ldots, k_n \in \mathbb{N}$ are relatively prime positive integers. Let $\lambda_j = \exp(2\pi i/k_j)$ for $j = 1, \ldots, n$. Then the diagonal matrix

$$T = \begin{bmatrix} \lambda_1 & 0 & \ldots & 0 \\ \vdots & \vdots & \vdots & \vdots \\ 0 & \ldots & 0 & \lambda_n \end{bmatrix} \in \mathrm{GL}(n, \ \mathbb{C})$$

has order $k = k_1 \cdots k_n$, so defines a representation of the group $\mathbb{Z}/k$. From proposition 1.5.2 we see that

$$\mathbb{C}[z_1, \ldots, z_n]^{\mathbb{Z}/k} \cong \mathbb{C}[z_1]^{\mathbb{Z}/k_1} \otimes \cdots \otimes \mathbb{C}[z_n]^{\mathbb{Z}/k_n} \ .$$

Therefore to compute the invariants of the representation induced by T we need only consider the case $n = 1$. For $\mathbb{Z}/k$ acting on $\mathbb{C}[z]$ by

$$z \mapsto \lambda \cdot z \qquad \lambda = \exp(2\pi i/k)$$

we see that z^j is invariant if and only if $j \equiv 0 \bmod k$ and hence

$$\mathbb{C}[z]^{\mathbb{Z}/k} = \mathbb{C}[z^k] \ .$$

So we conclude

$$\mathbb{C}[z_1, \ldots, z_n]^{\mathbb{Z}/k} = \mathbb{C}[z_1^{k_1}, \ldots, z_n^{k_n}] \ .$$

In a similar vein, given a representation $\varrho : G \hookrightarrow \mathrm{GL}(n, \ \mathbb{F})$ of the finite group G and $m \in \mathbb{N}$ a positive integer, we may form the representation $m\varrho$ of G on the vector space $\underset{m}{\oplus} V$ given by

$$g \cdot (v_1, \ldots, v_m) = (gv_1, \ldots, gv_m) \qquad \forall \, g \in G, \quad (v_1, \ldots, v_m) \in \underset{m}{\oplus} V \ .$$

Finding the invariants $\mathbb{F}[\underset{m}{\oplus} V]^G$, where G acts via $m\varrho$, is not quite so straightforward, as the following example illustrates.

EXAMPLE 2 : Consider the permutation representation τ of $\mathbb{Z}/2$ on $V = \mathbb{F}^2$ given by the matrix

$$T = \begin{bmatrix} 0 & 1 \\ 1 & 0 \end{bmatrix} \in \mathrm{GL}(2, \ \mathbb{F}) \ .$$

For each $k \in \mathbb{N}$ there is the representation $k\tau$ of $\mathbb{Z}/2$ on $\underset{k}{\oplus} V$. For $k = 1$ we have as a special case of 1.2.3

$$\mathbb{F}[x, y]^{\mathbb{Z}/2} = \mathbb{F}[x + y, \ xy]$$

where x, y are basis vectors for V^* interchanged by the action of T. When $k = 2$ we may choose as a basis for $\underset{2}{\oplus} V^* = V^* \oplus V^*$ elements x', y', x'', y'', where x', y' is a basis for the first factor, x'', y'' is a basis for the second factor, and the action of $\mathbb{Z}/2$ is given by the matrix

$$
S = \begin{bmatrix} 0 & 1 & 0 & 0 \\ 1 & 0 & 0 & 0 \\ 0 & 0 & 0 & 1 \\ 0 & 0 & 1 & 0 \end{bmatrix} \in \mathrm{GL}(4, \ \mathbb{F}).
$$

The polynomials $x' + y'$, $x'y'$, $x'' + y''$, $x''y'' \in \mathbb{F}[V]^{\mathbb{Z}/2} \otimes \mathbb{F}[V]^{\mathbb{Z}/2} \subset \mathbb{F}[V \oplus V]^{\mathbb{Z}/2}$ are clearly invariant, but are not sufficient to generate $\mathbb{F}[V \oplus V]^{\mathbb{Z}/2}$ as there is the additional invariant $x'x'' + y'y'' \in \mathbb{F}[V \oplus V]^{\mathbb{Z}/2}$. In fact it is not hard to show (see for example section 4.3 example 4) that these five forms generate $\mathbb{F}[V \oplus V]^{\mathbb{Z}/2}$. In addition they satisfy a single quadratic relation and $\mathbb{F}[V \oplus V]^{\mathbb{Z}/2}$ is not a polynomial algebra. The group $\mathbb{Z}/2$ embeds diagonally in $\mathbb{Z}/2 \times \mathbb{Z}/2$ and is a normal subgroup with quotient $\mathbb{Z}/2$. Thus by proposition 1.5.2 we have

$$
\mathbb{F}[V \oplus V]^{\mathbb{Z}/2 \times \mathbb{Z}/2} = (\mathbb{F}[V \oplus V]^{\mathbb{Z}/2})^{\mathbb{Z}/2}
$$

which shows that at the level of quotient fields $\mathbb{F}(V \oplus V)^{\mathbb{Z}/2}$ is a quadratic field extension of $\mathbb{F}(V \oplus V)^{\mathbb{Z}/2 \times \mathbb{Z}/2}$.

This example is the simplest example of what are loosely called **vector invariants**, i.e. rings of invariants of the form $\mathbb{F}[V \oplus \cdots \oplus V]^G$ where G acts diagonally on $V \oplus \cdots \oplus V$ via an $\mathbb{F}(G)$-module structure on V. See for example section 3.4, [185], [43] and [44]

It is quite natural to try to construct new invariants from old ones. Classically such constructions[2] were referred to as **covariants**[3] (or **concomitants**). One such classical construction is the Hessian.

DEFINITION : *Let V be a finite dimensional vector space over a field and $z_1, \ldots, z_n \in V^*$ a basis. If $f \in \mathbb{F}[V]$ then the **Hessian of f**, denoted by $H(f)$, is defined by*

$$
H(f) = \det \left(\frac{\partial^2 f}{\partial z_i \partial z_j} \right).
$$

(Here $\frac{\partial}{\partial z_k}$ etc. denotes the formal partial derivative.)

PROPOSITION 1.5.4 (O. Hesse): *Let $\mathbb{F}$ be a field, V an n-dimensional vector space over $\mathbb{F}$ and $z_1, \ldots, z_n$ a basis for V^*. If $f \in \mathbb{F}[V]$ and $g \in \mathrm{GL}(n, \ \mathbb{F})$ then*

$$
H(gf) = (\det(g))^2 H(f).
$$

---

[2] This was a major occupation at the inception of invariant theory in the 19-th century.
[3] Which should not be confused with coinvariants or the covariants introduced in section 8.2.

PROOF : Recall that gf is defined as the composition

$$V \xrightarrow{g^{-1}} V \xrightarrow{f} \bar{\mathbb{F}},$$

where $\bar{\mathbb{F}}$ is the algebraic closure of $\mathbb{F}$. We may therefore apply the chain rule to compute the matrix of second order partial derivatives and we find

$$\left[\frac{\partial^2 (gf)}{\partial z_i z_j} \right] = g \left[\frac{\partial f^2}{\partial z_i z_j} \right] g^{tr}$$

from which the result follows by taking determinants. $\square$

Thus the Hessian of $f \in \mathbb{F}[V]$ is well defined up to a nonzero scalar. Moreover for representations of a special nature one has as an immediate consequence:

COROLLARY 1.5.5 : Let $\varrho : G \hookrightarrow \mathrm{GL}(n, \mathbb{F})$ be a representation of a finite group G such that $\det(\varrho(g))^2 = 1$ for all $g \in G$. If $f \in \mathbb{F}[V]^G$ then $H(f) \in \mathbb{F}[V]^G$. $\square$

If V is an n-dimensional vector space over a field $\mathbb{F}$ and $f_1, \dots, f_n \in \mathbb{F}[V]$ are n polynomials, their **Jacobian determinant** is defined by

$$J(f_1, \dots, f_n) = \det \left(\frac{\partial f_i}{\partial z_j} \right)$$

where $z_1, \dots, z_n \in V^*$ is a basis. The chain rule for formal derivatives gives

$$J(gf_1, \dots, gf_n) = \det(g) \cdot J(f_1, \dots, f_n)$$

so the Jacobian is well defined up to a nonzero scalar factor. Moreover, one has as an immediate consequence of the chain rule:

PROPOSITION 1.5.6 : Let $\varrho : G \hookrightarrow \mathrm{SL}(n, \mathbb{F})$ be a representation of a finite group over a field $\mathbb{F}$, where $\mathrm{SL}(n, \mathbb{F})$ denotes the subgroup of the general linear group $\mathrm{GL}(n, \mathbb{F})$ of elements with determinant 1. If $f_1, \dots, f_n \in \mathbb{F}[V]^G$ then $J(f_1, \dots, f_n) \in \mathbb{F}[V]^G$. $\square$

Another constant theme in invariant theory has been the study of which nice properties of $\mathbb{F}[V]$ are inherited by a ring of invariants. One such nice property is unique factorization. Already in section 1.2 example 1 we saw that $\mathbb{F}[V]^G$ can fail to be a unique factorization domain. However we do have the following positive result.

PROPOSITION 1.5.7 : Let $\varrho : G \hookrightarrow \mathrm{GL}(n, \mathbb{F})$ be a representation of a finite group. Suppose that there is no nontrivial homomorphism $G \longrightarrow \mathbb{F}^\times$. Then $\mathbb{F}[V]^G$ is a unique factorization domain.

PROOF : Let $f \in \mathbb{F}[V]^G$ and write f as a product of powers of irreducible factors $f = f_1^{k_1} \cdots f_m^{k_m}$ that are pairwise nonassociates in $\mathbb{F}[V]$. The ideals $(f_1), \ldots, (f_m)$ generated by the irreducible factors in $\mathbb{F}[V]$ are permuted by G. Let $B_1, \ldots, B_k$ be the orbits of G acting on the set of ideals $\{(f_1), \ldots, (f_m)\}$ and set

$$F_i = \prod_{f_j \in B_i} f_j^{k_j} \qquad i = 1, \ldots, k.$$

The ideals $(F_1), \ldots, (F_k)$ are stable under the action of G, i.e., if $h \in (F_i)$ and $g \in G$ then $gh \in (F_i)$. In particular,

$$g F_i = \lambda(g) F_i \qquad \forall\, g \in G \quad \lambda(g) \in \mathbb{F}^\times.$$

The function $\lambda : G \longrightarrow \mathbb{F}^\times$ is a representation, so since there are no nontrivial homomorphisms $G \longrightarrow \mathbb{F}^\times$ it follows that $\lambda(g) = 1 \,\forall\, g \in G$ and F_i is invariant for $i = 1, \ldots, k$. No factor of F_i can be invariant since F_i is a product over a G-orbit in $\mathbb{F}[V]$. Therefore $F_i \in \mathbb{F}[V]^G$ is an irreducible polynomial and

$$f = F_1 \cdots F_k,$$

is a representation in $\mathbb{F}[V]^G$ as a product of pairwise nonassociate irreducible factors. The uniqueness of this decomposition (in the usual sense) is readily established, completing the proof. $\square$

It is worthwhile noting that the conclusion of 1.5.7 holds in any of the following situations:

    (i) G is a simple group (e.g. $G = A_n$ for $n \geq 5$),

    (ii) $\mathbb{F}$ is a field of characteristic p and G is a finite p-group,

    (iii) $\mathbb{F}$ is a finite field with q elements and G a finite group whose order is prime to $q - 1$,

    (iv) $\mathbb{F} = \mathbb{Q}$ and G is a group of odd order,

    (v) $\mathbb{F} = \mathbb{F}_2$ and G is finite, or $\mathbb{F} = \mathbb{F}_3$ and G has odd order.

For further reference we specifically note:

COROLLARY 1.5.8 : If $\varrho : P \hookrightarrow \mathrm{GL}(n, \mathbb{F}_p)$ is a representation of a finite p-group P over a field of characteristic p, then $\mathbb{F}[V]^P$ is a unique factorization domain. $\square$

Chapter 2
Finite Generation of Invariants

For a finite group G, a field of coefficients $\mathbb{F}$ and a finite dimensional representation $\varrho : G \hookrightarrow \mathrm{GL}(V)$, there is the action of G on the ring of polynomial functions $\mathbb{F}[V]$ via $(gf)(v) = f(\varrho(g^{-1})v)$, and the ring of invariants $\mathbb{F}[V]^G$. A problem of basic importance is the finiteness problem: does there exist (as in the case of the tautological representation of the symmetric group and the alternating group) a finite set of polynomials $f_1, \ldots, f_s$ (called *fundamental integral invariants* or a *complete fundamental system of invariants* in the 19th century) such that every invariant polynomial $f \in \mathbb{F}[V]^G$ can be expressed as a polynomial in $f_1, \ldots, f_s$?

One of the most striking advances of the *abstract* methods of algebra developed at the close of the 19th century was Hilbert's proof [101] of the finite generation of rings of invariants in characteristic zero. Hilbert's famous basis theorem served as a stepping stone in his proof. The first proof of the finite generation in arbitrary characteristic is due many years later to E. Noether [170].

Perhaps we should begin with a simple example to show that there really is something to be proved. Consider the subalgebra of $\mathbb{F}[x, y]$ generated by the elements
$$1, xy, xy^2, \ldots, xy^n, \ldots$$
Clearly no generator xy^n can be in the subalgebra generated by the remaining generators, since the product of any two other generators, and hence any polynomial in the other generators that is divisible by y^n, must be divisible by x^2. So there are subalgebras of $\mathbb{F}[x, y]$ that are not finitely generated.

We shall present several proofs of the finiteness theorem in this chapter, (see chapter 3 for further finiteness proofs) each with its advantages and disadvantages. The first is in principle Noether's proof [172] as enhanced [13] by Artin and Tate. It is a pure existence proof, offering no clue how to actually find a finite system of generators for the invariants. It has the advantage of being characteristic free, i.e. it works for all finite groups in arbitrary characteristic. The second proof is also due to Noether [170] and is constructive in nature, offering an upper bound on the number and degrees of the generators required. It has the disadvantage that it works only if the characteristic of the ground field is zero or greater than the group order. It has the advantage that it yields an algorithm to compute a complete fundamental system of invariants when it applies. Hilbert's original proof applied not to finite groups but to *continuous* matrix groups. This proof has the disadvantage that it uses analysis, but the advantage that it applies to nonfinite groups, such as the orthogonal and unitary groups. We discuss this proof in connection with the classical form problem.

2.1 The Field of Invariants

Let $\varrho : G \hookrightarrow \mathrm{GL}(n, \mathbb{F})$ be a representation of a finite group G. It follows from 1.2.4 that $\mathbb{F}[V]^G$ can never be generated by fewer than n elements, for the field of fractions $FF(\mathbb{F}[V]^G) = \mathbb{F}(V)^G$ has transcendence degree n. Hence $\mathbb{F}[V]^G$ must contain n algebraically independent elements, and therefore cannot be generated by fewer than n elements. The minimum number of polynomials needed to generate $\mathbb{F}[V]^G$ is therefore n. If we allow ourselves to express *polynomial* invariants as *rational functions* of polynomials, we need only $n + 1$ such rational functions [102].

THEOREM 2.1.1 (D. Hilbert): *Let $\varrho : G \hookrightarrow \mathrm{GL}(n, \mathbb{F})$ be a representation of a finite group G. Then there exist n polynomials $\varphi_1, \ldots, \varphi_n \in \mathbb{F}[V]^G$ and one rational invariant $\varphi_{n+1} \in \mathbb{F}(V)^G$ such that every polynomial invariant under the action of G may be expressed as a **rational** function of $\varphi_1, \ldots, \varphi_{n+1}$. It is not possible to find strictly fewer than n polynomials with this property.*

PROOF : Denote by $FF(-)$ the field of fractions functor. Then

$$\mathbb{F}(V)^G = FF(\mathbb{F}[V]^G) < \mathbb{F}(V)$$

is a finite field extension (see 1.2.4) where $\mathbb{F}(-)$ denotes the field of rational functions generated by $-$. Therefore the transcendence degree of $\mathbb{F}(V)^G$ over $\mathbb{F}$ is n and we may choose transcendental elements $f_1, \ldots, f_n \in \mathbb{F}(V)^G$. These are rational functions on V and as in the proof of 1.2.4 we may clear denominators to obtain n algebraically independent polynomials $\varphi_1, \ldots, \varphi_n \in \mathbb{F}[V]^G$. Since $\mathbb{F}(V)$ is finite and separable over $\mathbb{F}(V)^G$ there exists a primitive element $\varphi_{n+1} \in FF(\mathbb{F}[V]^G)$ such that $FF(\mathbb{F}[V]^G) = \mathbb{F}(\varphi_1, \ldots, \varphi_n)(\varphi_{n+1})$. These elements $\varphi_1, \ldots, \varphi_n, \varphi_{n+1}$ have the desired property. By considering transcendence degrees, no fewer than n polynomials could have this property. $\square$

In fact the proof shows that for *any* n algebraically independent elements $\varphi_1, \ldots, \varphi_n \in \mathbb{F}[V]^G$ it is always possible to find a rational function $\varphi_{n+1} \in$

$FF(\mathbb{F}[V]^G)$ such that every invariant polynomial may be written as a rational function of $\varphi_1, \ldots, \varphi_n, \varphi_{n+1}$.

2.2 Noetherian Rings and Modules

The first proof of finiteness depends on a certain amount of commutative algebra, and we begin with some comments. Classically, one of the main motivations for the development of commutative algebra was to put algebraic geometry on a firm foundation. Two topics which provided impetus were intersection theory (various forms of Bézout's theorem and generalizations) and invariant theory, the topic of this book. In the course of this book, we shall need to make use of a considerable amount of commutative algebra, which we introduce as we need it. General references for commutative algebra are [14], [58], [141] and [266].

A module M over a commutative ring A is said to be **Noetherian** if every ascending chain of submodules is eventually constant. Clearly, every finite direct sum of Noetherian modules, as well as every submodule and quotient module of a Noetherian module is again Noetherian. The ring A itself is said to be Noetherian if it is so as a module over itself; in other words, if every ascending chain of ideals is eventually constant.

PROPOSITION 2.2.1 : *A module M over a Noetherian ring A is Noetherian if and only if it is finitely generated.*

PROOF : If M is finitely generated, then it is a quotient of a finite direct sum of copies of A, and hence Noetherian. Conversely, if M is Noetherian, we choose a sequence of elements $x_1, x_2, \ldots$ in M with each x_i not in the submodule generated by $x_1, \ldots, x_{i-1}$. Since M is Noetherian such a sequence must terminate and thus M is finitely generated. □

COROLLARY 2.2.2 : *A submodule of a finitely generated module over a Noetherian ring is again finitely generated.* □

LEMMA 2.2.3 : *A commutative ring A is Noetherian if and only if every ideal is finitely generated.*

PROOF : If A is Noetherian then by the corollary, every ideal is finitely generated. Conversely, suppose that every ideal of A is finitely generated. Given an ascending chain of ideals of A, the union is an ideal in A, and hence finitely generated. But any finite subset of the union lies in one of the ideals, and so the chain is constant from that ideal onwards. □

THEOREM 2.2.4 (Hilbert's Basis Theorem): *If A is a commutative Noetherian ring, then so is the polynomial ring $A[x]$.*

PROOF : By the lemma, it suffices to prove that if I is an ideal in $A[x]$ then I is finitely generated. Let I' be the ideal in A generated by the leading coefficients of the polynomials in I. Since A is Noetherian, I' is finitely generated, say by $a_1, \ldots, a_n$. Choose polynomials $f_1, \ldots, f_n \in I$ with these leading coefficients, let t be the maximum of the degrees of the f_i, and write I_0 for the ideal in $A[x]$ generated by $f_1, \ldots, f_n$. If f is any polynomial in I of degree at least t, then the leading coefficient of f is in I', and we may subtract an A-linear combination of products of f_i's with powers of x in order to get rid of the leading coefficient and hence $f = g + h$ with $g \in I_0$, $h \in I$, and $\deg(h) < \deg(f)$. By repeating the argument on h in place of f we eventually obtain $f = \bar{g} + \bar{h}$ with $\bar{g} \in I_0$, $\bar{h} \in I$ and $\deg(\bar{h}) < t$, and thus have

$$I = I_0 + (I \cap (A \oplus Ax \oplus \cdots \oplus Ax^{t-1})).$$

It follows that I is finitely generated. $\square$

COROLLARY 2.2.5 : *If A is a finitely generated commutative algebra over a field $\mathbb{F}$, then A is Noetherian.*

PROOF : A is a quotient of some polynomial ring over $\mathbb{F}$, which is Noetherian by the Hilbert basis theorem. $\square$

2.3   Finite Groups in Arbitrary Characteristic

Our first proof of the finite generation of the invariants is based on the concept of integrality. It is a special instance of a theorem of Artin and Tate [13]. Recall from section 1.2 that if $A \supseteq B$ is an extension of rings we say that an element of A is **integral** over B if it is a zero of a monic polynomial (i.e., a polynomial whose leading coefficient is one) with coefficients in B. This is the same as saying that the element lies in a subring of A which contains B and is finitely generated as a B-module. So the sum and product of integral elements is again integral. We say that A is an **integral extension** of B if every element of A is integral over B. If A is an integral extension of B and finitely generated over B as a ring, then it is finitely generated over B as a module. In this case, we say that A is a **finite extension** of B.

THEOREM 2.3.1 (E. Noether): *Suppose that $\mathbb{F}$ is a field, and G is a finite group acting as automorphisms of a finitely generated commutative $\mathbb{F}$-algebra A (for example $A = \mathbb{F}[V]$). Then A^G is also a finitely generated commutative $\mathbb{F}$-algebra and A is finitely generated as a module over A^G.*

PROOF : If $a \in A$ then a satisfies the monic polynomial

$$\prod_{g \in G} (X - g(a)) \in A^G[X]$$

and so A is an integral extension of A^G. Let B be the subalgebra of A^G generated by the coefficients of the monic polynomials satisfied by a finite set of $\mathbb{F}$-algebra generators of A. Then B is a finitely generated $\mathbb{F}$-algebra, and hence Noetherian. A is a finitely generated B-module, and hence so is A^G by Corollary 2.2.2. Thus A^G is a finitely generated $\mathbb{F}$-algebra. $\square$

Note the nonconstructive nature of the proof of 2.3.1.
  (i) We need monic polynomials satisfied by generators of A over A^G. *But* we do not know A^G yet.
  (ii) We need a set of B-module generators for A^G. Since we do not know A^G, we do not know B, so have no clue as to how to find the needed generators.

CORObook LARY 2.3.2 : *Suppose that G is a finite group, $\mathbb{F}$ is a field and V is a finite dimensional $\mathbb{F}(G)$-module. Then $\mathbb{F}[V]^G$ is a finitely generated $\mathbb{F}$-algebra and $\mathbb{F}[V]$ is a finite integral extension of $\mathbb{F}[V]^G$.* $\square$

An immediate consequence is that $\mathbb{F}[V]^G$ always contains $n = \dim_{\mathbb{F}}(V)$ algebraically independent elements. (Compare the proof of 2.1.1. It could not contain any more than n.) But as the example of the alternating group shows, these may not be sufficient to generate all of $\mathbb{F}[V]^G$.

Since $\mathbb{F}[V]$ is finite over $\mathbb{F}[V]^G$ it follows that for any element $z \in V^*$ there is an integer k, depending on z, such that z^k belongs to the ideal $(\overline{\mathbb{F}[V]^G})$ of $\mathbb{F}[V]$ generated by the invariants of strictly positive degree. In particular this holds for a basis $z_1, \ldots, z_n$ for V^*. Therefore 0 is the only element of V on which every invariant polynomial is zero. This is in stark contrast to the situation for nonfinite groups (compare section 1.4) where the null cone

$$\mathcal{N} = \{v \in V \mid f(v) = 0 \ \forall \ f \in \mathbb{F}[V]^G, \ \deg(f) > 0\}$$

plays a central role in the theory initiated by Hilbert in [102] and carried forward by Mumford (see for example [126]).

Since $\mathbb{F}[V]^G$ is Noetherian the preceding corollaries lead to the following finiteness results for relative invariants and coinvariants.

COROLLARY 2.3.3 : *Suppose that G is a finite group and $G \hookrightarrow \mathrm{GL}(n, \mathbb{F})$ is a representation of G. Then for any simple $\mathbb{F}(G)$-module S the S-isotypical summand $\mathbb{F}[V]^G\big|_S$ is finitely generated as a module over $\mathbb{F}[V]^G$.* $\square$

COROLLARY 2.3.4 : *Suppose that G is a finite group and $G \hookrightarrow \mathrm{GL}(n, \mathbb{F})$ is a representation of G. Then the ring of coinvariants $\mathbb{F}[V]_G$ is a finite dimensional G-representation.* $\square$

2.4   Noether's Bound

E. Noether[1] [170] proved that if $\mathbb{F}$ is a field of characteristic zero, then $\mathbb{F}[V]^G$ is generated by elements of degree at most $|G|$. In particular, the number of generators needed is at most $\binom{\dim_{\mathbb{F}}(V)+|G|}{|G|}$. In fact, this is a special case of a relative version, which says that if H is a subgroup of G and $\mathbb{F}[V]^H$ is generated by elements of degree at most m, then $\mathbb{F}[V]^G$ is generated by elements of degree at most $m \cdot |G : H|$. The proof we give is based on a relativization of Noether's original argument due to Barbara Schmid [194].

The basic tool we need is the **transfer**. If H is a subgroup of a finite group G and V is a finite dimensional $\mathbb{F}(G)$-module, we define

$$\mathrm{Tr}_H^G : \mathbb{F}[V]^H \longrightarrow \mathbb{F}[V]^G$$

by

$$\mathrm{Tr}_H^G(f)(x) = \sum_{gH \in G/H} g(f)(x) = \sum_{gH \in G/H} f(g^{-1}(x)).$$

The notation means that the sum runs over a set of left coset representatives of H in G. It is an easy calculation to show that for $f \in \mathbb{F}[V]^H$ and $g'H = g''H \in G/H$ then $g'f = g''f \in \mathbb{F}[V]^G$, so the right hand side makes sense independent of the choice of elements in G representing the cosets G/H. Since

$$g' \cdot \mathrm{Tr}_H^G(f) = \sum_{g''H \in G/H} g'g''(f) = \sum_{g'g''H \in G/H} g'g''(f) = \mathrm{Tr}_H^G(f),$$

(if $\{g''H\}$ runs through all cosets exactly once so does $\{g'g''H\}$ for a fixed g') we see that $\mathrm{Tr}_H^G(f) \in \mathbb{F}[V]^G$. Note that the composite

$$\mathbb{F}[V]^G \hookrightarrow \mathbb{F}[V]^H \xrightarrow{\mathrm{Tr}_H^G} \mathbb{F}[V]^G$$

is equal to multiplication by $|G : H|$. In particular, if $|G : H|$ is invertible in $\mathbb{F}$ then the transfer is surjective, and the map

$$\pi_H^G = \frac{1}{|G : H|}\mathrm{Tr}_H^G : \mathbb{F}[V]^H \longrightarrow \mathbb{F}[V]^G \hookrightarrow \mathbb{F}[V]^H$$

is an idempotent projection whose image is equal to $\mathbb{F}[V]^G$. In this case $\mathbb{F}[V]^H$ is a direct sum of $\mathbb{F}[V]^G$ and the kernel of π_H^G. If $H = 1$, we just write π^G for $\pi_1^G = \frac{1}{|G|}\sum_{g \in G} g$, so that $\mathbb{F}[V] = \mathbb{F}[V]^G \oplus \ker(\pi^G)$.

---

[1] Although this paper is only three and a half pages long, it contains two complete proofs of the finiteness theorem and a correction to a proof of Weber's. Both proofs of the finiteness theorem have been reworked many times, the account of Weyl in [259] pp. 275-276 being perhaps the most famous.

In general the transfer behaves badly with respect to products in $\mathbb{F}[V]^H$. However, note that $\mathbb{F}[V]^G$ being a subalgebra of $\mathbb{F}[V]^H$, $\mathbb{F}[V]^H$ is a $\mathbb{F}[V]^G$-module. Moreover

$$\left. \begin{array}{l} \mathrm{Tr}_H^G(f) = |G:H| \cdot f \\ \mathrm{Tr}_H^G(f \cdot h) = f \cdot \mathrm{Tr}_H^G(h) \end{array} \right\} \quad f \in \mathbb{F}[V]^G, \ h \in \mathbb{F}[V]^H, \ \deg(h) > 0.$$

Therefore, if $|G : H|$ is invertible in $\mathbb{F}$, then $\pi_H^G = \frac{1}{|G:H|}\mathrm{Tr}_H^G$ is a $\mathbb{F}[V]^G$-module homomorphism, while if $|G : H| = 0 \in \mathbb{F}$ the transfer is zero on products.

The moral is that the transfer has good properties almost exclusively when $|G : H|$ is invertible in $\mathbb{F}$. In particular, even in well studied examples where the characteristic of $\mathbb{F}$ divides the order of G rather little is known about the image of the transfer map $\mathrm{Tr} := \mathrm{Tr}_1^G : \mathbb{F}[V] \longrightarrow \mathbb{F}[V]^G$. (See for example chapter 11, [82] and the references there.)

We shall also need the following combinatorial lemma.

LEMMA 2.4.1 : *Let V be a vector space over a field $\mathbb{F}$ and $u_1, \ldots, u_j \in \mathbb{F}[V]$. If $j! \neq 0 \in \mathbb{F}$ then the monomial $u_1 \cdots u_j$ is a linear combination of j-th powers of sums of elements of $\{u_1, \ldots, u_j\}$.*

PROOF : This follows from the formula

$$(-1)^j j! \, u_1 \cdots u_j = \sum_{I \subseteq \{1,\ldots,j\}} (-1)^{|I|} \left(\sum_{i \in I} u_i \right)^j.$$

In this formula, I runs over all subsets of $\{1, \ldots, j\}$ and $|I|$ is the cardinality of I. $\square$

THEOREM 2.4.2 (E. Noether): *Let V be a finite dimensional representation of a finite group G and $H \leq G$ a subgroup of G. If $|G : H|!$ is invertible in $\mathbb{F}$ and $\mathbb{F}[V]^H$ is generated by elements of degree at most m, then $\mathbb{F}[V]^G$ is generated by elements of degree at most $m \cdot |G : H|$.*

PROOF : Let B be the subalgebra of $\mathbb{F}[V]^G$ generated by elements of degree at most $m|G : H|$. Our goal is to show that $B = \mathbb{F}[V]^G$. To this end set $d = |G : H|$ and

$$N = \mathrm{Span}_{\mathbb{F}} \left\{ f \in \mathbb{F}[V]^H \mid \deg(f) \leq m \right\}$$

$$M = \mathrm{Span}_{\mathbb{F}} \left\{ f_1^{e_1} \cdots f_k^{e_k} \, \middle| \, \begin{array}{l} k, \, e_1, \ldots, e_k \in \mathbb{N}, \, e_1 + \cdots + e_k < d \\ f_1, \ldots, f_k \in N \end{array} \right\}.$$

We are going to show that $B \cdot M = \mathbb{F}[V]^H$, i.e. that M generates $\mathbb{F}[V]^H$ as a B-module. Granted this, the proof may be completed as follows: we apply the projection π_H^G described above to $\mathbb{F}[V]^H$. Since the characteristic of $\mathbb{F}$ is relatively prime to $|G : H|$ the map π_H^G is a surjective $\mathbb{F}[V]^G$-module map, so we obtain

$$\mathbb{F}[V]^G = \pi_H^G(\mathbb{F}[V]^H) = \pi_H^G(B \cdot M) = B\pi_H^G(M) = B$$

(as $\pi_H^G(M) \subseteq B$) as was to be shown.

It remains to show $\mathbb{F}[V]^H = B \cdot M$. If $f \in \mathbb{F}[V]^H$, then f is a root of the polynomial

$$\prod_{gH \in G/H} (X - gf) = X^d + b_1 X^{d-1} + \cdots + b_d$$

(the product is over a set of coset representatives of H in G) where b_i is the value of the i-th elementary symmetric function in d variables evaluated at the d elements (counting repetitions) $\{gf \mid g \in G/H\}$. Substituting $X = f$ and solving for f^d yields

$$(*) \qquad\qquad f^d = -(b_1 f^{d-1} + \cdots + b_d).$$

If $\deg(f) \leq m$ then

$$\deg(b_1) \leq \deg(b_2) \leq \cdots \leq \deg(b_d) \leq dm$$

so $b_1, \ldots, b_d \in B$. The polynomials $f, f^2, \ldots, f^d$ belong to M so $(*)$ shows $f^d \in B \cdot M$ for any $f \in N$.

Next suppose that $f^E = f_1^{e_1} \cdots f_k^{e_k}$ with $f_1, \ldots, f_k \in N$ and $e_1 + \cdots + e_k = d$. From lemma 2.4.1 we obtain

$$(**) \qquad (-1)^d d! f^E = \sum_{I \subseteq \{1, \ldots, d\}} (-1)^{|I|} \left(\sum_{i \in I} f_i \right)^d = \sum_{I \subseteq \{1, \ldots, d\}} (-1)^{|I|} h_I^d,$$

where $h_I \in N$. Since $d! \neq 0 \in \mathbb{F}$ it therefore follows that $f^E \in B \cdot M$.

Assume inductively that all monomials $f^E = f_1^{e_1} \cdots f_k^{e_k}$ with $k, e_1, \ldots, e_k \in \mathbb{N}$, $f_1, \ldots, f_k \in N$ and $e_1 + \cdots + e_k \leq d + i$ belong to $B \cdot M$. Consider a monomial $f^E = f_1^{e_1} \cdots f_k^{e_k}$ with $k, e_1, \ldots, e_k \in \mathbb{N}$, $f_1, \ldots, f_k \in N$ and $e_1 + \cdots + e_k = d + i + 1$. Without loss of generality we may suppose $f^E = f^{E'} \cdot f_k$. By the induction hypothesis we have $f^{E'} \in B \cdot M$ and therefore we may choose $h_1, \ldots, h_l \in N$ and $d_1, \ldots, d_l \in \mathbb{N}$ with $d_1 + \cdots + d_l < d$ and $c_D \in B$ so that

$$f^{E'} = \sum c_D h^D = \sum_{|D| < d-1} c_D h^D + \sum_{|D| = d-1} c_D h^D,$$

where

$$h^D = \prod_{i=1}^{l} h_i^{d_i}.$$

If $|D| < d - 1$ then $h^D f_k \in M$ for degree reasons and hence

$$\sum_{|D|<d-1} c_D h^D f_k \in B \cdot M.$$

If $|D| = d - 1$ then by $(**)$ $h^D f_k \in B \cdot M$ and hence

$$\sum_{|D|=d-1} c_D h^D f_k \in B \cdot M.$$

Combining these inclusions gives

$$f^E = f^{E'} \cdot f_k = \sum c_D h^D f_k = \sum_{|D|<d-1} c_D h^D f_k + \sum_{|D|=d-1} c_D h^D f_k \in B \cdot M.$$

Therefore by induction any monomial $f^E = f_1^{e_1} \cdots f_k^{e_k}$ with $f_1, \ldots, f_k \in N$ belongs to $B \cdot M$. Since N generates $\mathbb{F}[V]^H$ as an algebra we have shown that $B \cdot M = \mathbb{F}[V]^H$ as required. $\square$

COROLLARY 2.4.3 (E. Noether): *Suppose that V is a finite dimensional representation of a finite group G over a field $\mathbb{F}$. If $|G|! \in \mathbb{F}^{\times}$, i.e. $\mathbb{F}$ has characteristic zero or the characteristic of $\mathbb{F}$ is greater than $|G|$, then $\mathbb{F}[V]^G$ is generated by at most $\binom{\dim_{\mathbb{F}}(V)+|G|}{|G|}$ elements of degree at most $|G|$.* $\square$

The preceding results provide us with an algorithm to compute a generating set of invariants, provided that $|G|! \in \mathbb{F}^{\times}$. Namely write down an additive basis for the polynomials of degree at most $|G|$ and apply the transfer Tr^G (or the projection π^G) to them. The resulting polynomials lie in $\mathbb{F}[V]^G$ and are a set of algebra generators. Unfortunately this algorithm is not usually very efficient, as the number of potential generators it considers, $\binom{\dim_{\mathbb{F}}(V)+|G|}{|G|}$, is rarely sharp. The algorithm is however very easy to implement on a computer. Here is a simple example to illustrate this. It also shows that the upper bound on the degrees of a complete fundamental system of invariants in Noether's theorem can be sharp.

EXAMPLE 1 : The cyclic group of order 3 acts on the vector space $\mathbb{F}^3$ by cyclic permutation of the standard basis vectors E_1, E_2, E_3. The 2-dimensional subspace V spanned by the vectors with coordinate sum 0 is $\mathbb{Z}/3$-invariant and has as a basis the vectors $E_1 - E_2$, $E_3 - E_2$. The matrix of the generator of $\mathbb{Z}/3$ with respect to this basis is

$$A = \begin{bmatrix} 0 & 1 \\ -1 & -1 \end{bmatrix} \in \mathrm{GL}(2, \mathbb{F}).$$

If $\{x,\ y\}$ is the dual basis of V^*, then

$$Ax = -y \quad Ay = x - y\ .$$

If the characteristic of $\mathbb{F}$ is different from 2 or 3 then by Noether's theorem $\mathbb{F}[x,\ y]^{\mathbb{Z}/3}$ is generated as an algebra by polynomials of degree at most 3. The polynomials

$$x,\ y \in S^1(V^*)$$
$$x^2,\ xy, y^2 \in S^2(V^*)$$
$$x^3,\ x^2y, xy^2, y^3 \in S^3(V^*)$$

are a basis for the homogeneous components of $\mathbb{F}[x,\ y]$ of degree 1, 2 and 3 respectively. The proof of Noether's theorem shows that a complete fundamental system of invariants for $\mathbb{F}[x,\ y]^{\mathbb{Z}/3}$ may be obtained by applying the projection

$$\pi^{\mathbb{Z}/3} : \mathbb{F}[x,\ y] \longrightarrow \mathbb{F}[x,\ y]^{\mathbb{Z}/3}$$

to these nine polynomials. The following table summarizes the values of $\pi^{\mathbb{Z}/3}$ on these nine polynomials.

φ	$\pi^{\mathbb{Z}/3}(\varphi)$
x	0
y	0

φ	$\pi^{\mathbb{Z}/3}(\varphi)$
x^2	$\frac{1}{3}\left(2x^2 - 2xy + 2y^2\right)$
xy	$\frac{1}{3}\left(x^2 - xy + y^2\right)$
y^2	$\frac{1}{3}\left(2x^2 - 2xy + 2y^2\right)$

φ	$\pi^{\mathbb{Z}/3}(\varphi)$
x^3	$x^2y \quad - \quad xy^2$
x^2y	$\frac{1}{3}\left(-x^3 + 3x^2y - y^3\right)$
xy^2	$\frac{1}{3}\left(-x^3 + 3xy^2 - y^3\right)$
y^3	$-x^2y \quad + \quad xy^2$

Table 2.4.1: The values of $\pi^{\mathbb{Z}/3}(\varphi)$ for $\deg(\varphi) \le 3$

From this we see that the polynomials

$$f = x^2 - xy + y^2$$
$$h = x^2y - xy^2$$
$$k = x^3 - 3xy^2 + y^3 \qquad (\text{or } x^3 - 3x^2y + y^3)$$

are a complete fundamental system of invariants for $\mathbb{F}[x, y]^{\mathbb{Z}/3}$. Since h and k have degree 3 the upper bound on the degrees given by Noether's theorem is sharp. However to obtain 3 fundamental invariants we considered 9 polynomials.

If $|G|$ is divisible by the characteristic of $\mathbb{F}$ then the transfer homomorphism $\text{Tr}^G : \mathbb{F}[V] \longrightarrow \mathbb{F}[V]^G$ is no longer ([84] theorem 2.4 and [220] theorem 4.1) an epimorphism. The bounds provided by Noether's theorem in the nonmodular case on the degrees and number of generators of a complete fundamental system of invariants also need no longer hold. The following example illustrates this.

EXAMPLE 2 : Let $\sigma : \mathbb{Z}/2 \hookrightarrow \text{GL}(2, \mathbb{F}_2)$ be the representation given by the matrix

$$S = \begin{bmatrix} 1 & 0 \\ 1 & 1 \end{bmatrix} \in \text{GL}(2, \mathbb{F}_2).$$

This is just the permutation representation of $\mathbb{Z}/2$ on a two element set in a somewhat different basis. If $\{x, y\}$ is the dual to the standard basis of $\mathbb{F}_2^2$ then

$$S(x) = x + y \qquad S(y) = y.$$

The invariants $\mathbb{F}_2[x, y]^{\mathbb{Z}/2}$ are

$$\mathbb{F}_2[x, y]^{\mathbb{Z}/2} = \mathbb{F}_2[y, x(x + y)].$$

A basis for the quadratic forms in $\mathbb{F}_2[x, y]$ is provided by x^2, xy, y^2. Since

$$\text{Tr}^{\mathbb{Z}/2}(x^2) = x^2 + (x + y)^2 = x^2 + x^2 + y^2 = y^2$$
$$\text{Tr}^{\mathbb{Z}/2}(xy) = xy + (x + y)y = xy + xy + y^2 = y^2$$
$$\text{Tr}^{\mathbb{Z}/2}(y^2) = y^2 + y^2 = 0$$

it follows that $x(x + y) \notin \text{Im}(\text{Tr}^{\mathbb{Z}/2})$. Hence

$$\text{Tr}^{\mathbb{Z}/2} : \mathbb{F}_2[x, y] \longrightarrow \mathbb{F}_2[x, y]^{\mathbb{Z}/2}$$

is not an epimorphism. (For a further discussion of the transfer see section 11.5.)

Consider next the representation $\sigma \oplus \sigma \oplus \sigma : \mathbb{Z}/2 \hookrightarrow \mathrm{GL}(6, \mathbb{F}_2)$ (compare section 1.5 example 2). We regard $\mathbb{F}_2^6$ as $\mathbb{F}_2^2 \oplus \mathbb{F}_2^2 \oplus \mathbb{F}_2^2$ and let x_i, y_i denote the dual to the standard basis for the i-th factor of $\mathbb{F}_2^2$ for $i = 1, 2, 3$.

From the preceding discussion we see that the polynomials

$$y_i \qquad i = 1, 2, 3$$
$$x_i(x_i + y_i) \qquad i = 1, 2, 3$$

are invariant. The following transfer computation

$$\mathrm{Tr}^{\mathbb{Z}/2}(x_i x_j) = x_i x_j + (x_i + y_i)(x_j + y_j) = x_i y_j + x_j y_i + y_i y_j$$

(here we have used $x_i x_j + x_i x_j = 0$) yields the invariants

$$x_i y_j + x_j y_i \qquad i, j = 1, 2, 3, \ i < j.$$

Direct computation shows that these polynomials are an $\mathbb{F}_2$-basis in degrees 1 and 2 for $\mathbb{F}_2[x_1, y_1, x_2, y_2, x_3, y_3]^{\mathbb{Z}/2}$.

Consider next the following cubic invariant

$$f = \mathrm{Tr}^{\mathbb{Z}/2}(x_1 x_2 x_3) = x_1 x_2 x_3 + (x_1 + y_1)(x_2 + y_2)(x_3 + y_3)$$
$$= x_1 x_2 x_3 + [x_1 x_2 x_3 + x_1 x_2 y_3 + x_1 y_2 x_3 + x_1 y_2 y_3 + y_1 x_2 x_3 + y_1 x_2 y_3 + y_1 y_2 x_3 + y_1 y_2 y_3]$$
$$= x_1 x_2 y_3 + x_1 y_2 x_3 + x_1 y_2 y_3 + y_1 x_2 x_3 + y_1 x_2 y_3 + y_1 y_2 x_3 + y_1 y_2 y_3.$$

We claim that in $\mathbb{F}_2[x_1, y_1, x_2, y_2, x_3, y_3]^{\mathbb{Z}/2}$ this cubic invariant is indecomposable. To verify this we need to show that f is not in the subalgebra B generated by the nine invariants

$$y_i \qquad i = 1, 2, 3$$
$$x_i(x_i + y_i) \qquad i = 1, 2, 3$$
$$x_i y_j + x_j y_i \qquad i, j = 1, 2, 3, \ i < j.$$

This may be seen as follows: if $h \in B$ has degree 3 then h must be a sum of monomials of the form

$$y_i y_j y_k \qquad x_i^2 y_j, \ x_i y_i y_j \qquad i, j, k = 1, 2, 3.$$

However f contains the monomial $x_1 x_2 y_3$ with coefficient 1 so is not of this form.

Therefore $f \in \mathbb{F}_2[x_1, y_1, x_2, y_2, x_3, y_3]^{\mathbb{Z}/2}$ is an indecomposable algebra generator with $\deg(f) = 3 > 2 = |\mathbb{Z}/2|$ and Noether's nonmodular bound does not hold in this example.

For an alternative discussion of these examples using the permutation basis due to Mara D. Neusel see [220] § 4 examples 1, 2, and 3. For further examples of this type see [185] and the references there.

If $\chi : G \longrightarrow \mathbb{F}^\times$ is a 1-dimensional representation we may also define a χ-twisted version of the transfer by

$$\mathrm{Tr}_\chi^G(f) = \sum_{g \in G} \chi(g^{-1}) \cdot gf \qquad \forall\, f \in \mathbb{F}[V].$$

If the order of G is prime to the characteristic of the field $\mathbb{F}$, then we may also introduce a projection

$$\pi_\chi^G = \frac{1}{|G|} \mathrm{Tr}_\chi^G : \mathbb{F}[V] \longrightarrow \mathbb{F}[V]_\chi^G,$$

which is an $\mathbb{F}[V]^G$-module homomorphism. The analog of the preceding discussion may be carried through for the χ-relative invariants and π_χ^G.

In addition to an upper bound for the degrees of generating invariants, we can obtain a lower bound in all characteristics.

PROPOSITION 2.4.4 : *Suppose that $\varrho : G \hookrightarrow \mathrm{GL}(n, \mathbb{F})$ is a representation of the finite group G. Let a be the order of $Z\mathrm{GL}(n, \mathbb{F}) \cap \varrho(G)$, where $Z\mathrm{GL}(n, \mathbb{F})$ denotes the center of $\mathrm{GL}(n, \mathbb{F})$. Then the degrees of the G-invariant polynomials are divisible by a and hence a is the smallest degree of an element in $\mathbb{F}[V]^G$ of positive degree.*

PROOF :   Let $A = Z\mathrm{GL}(n, \mathbb{F}) \cap \varrho(G)$. Then A is a subgroup of the center of G and hence normal in G. The center of $\mathrm{GL}(n, \mathbb{F})$ consists of scalar matrices, i.e. multiples of the identity, and therefore is cyclic. If $z_1, \ldots, z_n$ is a basis for V^* then a polynomial $f \in \mathbb{F}[V]$ is invariant under A if and only if it is a sum of monomials of the form

$$z_1^{k_1} \cdots z_n^{k_n} \qquad k_1 + \cdots + k_n \equiv 0 \ (a).$$

Any such polynomial has a degree divisible by a and since $A \triangleleft G$

$$\mathbb{F}[V]^G = (\mathbb{F}[V]^A)^{G/A}$$

and the result follows. $\square$

For a representation $\varrho : G \hookrightarrow \mathrm{GL}(n, \mathbb{F})$ the subgroup $Z\mathrm{GL}(n, \mathbb{F}) \cap \varrho(G)$ will be called the **similarity subgroup** of G and denoted by $A(G)$.

For a further discussion of bounds for the degrees of a minimal set of algebra generators for the ring of invariants see [102] and [194]. For the special case of permutation groups see [89], the theorems 6.2.9 and 6.2.10 in section 6.2, and especially [93]. Noether's bound is known to hold for solvable groups in the nonmodular case, see e.g [112] or [219]. The upper bounds provided by Noether's theorem 2.4.3 no longer hold in the modular case as shown in example 2. See also [185].

2.5   The Classical Form Problem : Linear Reductive Groups

Invariant theory is not limited to just the study of the polynomial invariants of finite groups. From the outset infinite groups enter, as in finding the polynomial invariants of matrices (see section 1.4), which involves $GL(n, \mathbb{F})$. This is further illustrated with a simplified version of the classical study of binary forms. See [195] for a detailed account of the invariant theory related to the form problem and [111] for an exposition of the classical results in the language of representation theory.

A **binary form of degree** d is a homogeneous element of $\mathbb{F}[x, y]$ of degree d. Two such forms f', f'' are called **equivalent** if there is an element $g \in SL(2, \mathbb{F})$ such that $gf' = f''$. It is a problem of long standing to classify binary forms, i.e. to determine all the equivalence classes of binary forms. We can choose to write a binary form f in the manner

$$f = \sum_{i=0}^{d} a_i x^i y^{d-i} .$$

It is then natural to look for functions of the coefficients $F(a_0, \ldots, a_d)$ that distinguish between inequivalent forms. Such a function F provides a tool for the classification of binary forms if and only if the value of F on the coefficients of a form f' are the same as the value on the coefficients of all forms f'' equivalent to f'. Finding all such polynomial functions F of the coefficients of binary forms of degree d is clearly a problem in invariant theory. Namely, the group $SL(2, \mathbb{F})$ acts on the space $S^d(x, y) = \mathbb{F}[x, y]_d$ of binary forms of degree d and we are searching for a set of generators for the ring of invariants $\mathbb{F}[S^d(x, y)]^{SL(2, \mathbb{F})}$. It would be nice to know that only finitely many generators are needed. If $\mathbb{F}$ is infinite then so is $SL(n, \mathbb{F})$ so our previous work does not apply.

More generally if $V = \mathbb{F}^n$ is an n-dimensional vector space over $\mathbb{F}$, the elements of $S^d(V^*)$ are called **homogeneous forms of degree** d **in** n **variables** or[2] n**-ary forms of degree** d. The general linear group $GL(n, \mathbb{F})$ acts on

---

[2] In the classical literature homogeneous polynomials are often called **quantics** or **forms**. A **binary quantic** or **binary form** is therefore a homogeneous polynomial in 2 variables, a **ternary form** a homogeneous polynomial in 3 variables, and so on. Thus a **binary cubic form** is a homogeneous polynomial in 2 variables of degree 3 and a **ternary quadratic form** a homogeneous polynomial in 3 variables of degree 2, etc., depending on how much Greek and Latin you know.

the vector space $S^d(V^*)$ and the classification problem for n-ary forms of degree d consists in describing the orbits[3] of SL(n, $\mathbb{F}$) (or perhaps some other subgroup of GL(n, $\mathbb{F}$)) acting on $S^d(V^*)$, or what is the same thing, the orbit space $S^d(V^*)/\text{SL}(n, \mathbb{F})$. Any polynomial $f \in \mathbb{F}[S^d(V^*)]^{\text{SL}(n, \mathbb{F})}$ is constant on each orbit of SL(n, $\mathbb{F}$) on $S^d(V^*)$, so f defines (we are ignoring subtleties of topology and continuity) a function on the orbit space

$$S^d(V^*)/\text{SL}(n, \mathbb{F}) \longrightarrow \mathbb{F}.$$

In this sense f is an *invariant* of the homogeneous forms of degree d, i.e. f takes the same values on any two forms that are equivalent. A solution to the *form problem* would be a description of $\mathbb{F}[S^d(V^*)]^{\text{SL}(n, \mathbb{F})}$. If the functions in $\mathbb{F}[S^d(V^*)]^{\text{SL}(n, \mathbb{F})}$ separate the points of $S^d(V^*)/\text{SL}(n, \mathbb{F})$, then a complete system of generators for $f_1, \ldots, f_m \in \mathbb{F}[S^d(V^*)]^{\text{SL}(n, \mathbb{F})}$ would provide a complete classification for n-ary forms of degree d in the sense that two such forms Q', $Q'' \in S^d(V^*)$ would be equivalent if and only if $f_i(Q') = f_i(Q'')$ for $i = 1, \ldots, m$. The forms on which all invariants vanish are called **null forms** and their study was inaugurated by Hilbert [102]. Unless all the null forms are equivalent to each other the invariants of homogeneous forms of degree d cannot provide a complete classification, see e.g. [102], [126] or [195].

The preceding discussion of the classical form problem shows the need for a finiteness theorem for $\mathbb{F}[V]^G$ for nonfinite groups G. The proof of finite generation using the transfer may be extended to a larger class of groups by replacing the sum used to average over the group in the finite case with an integral over the group in the compact or algebraic case. (In the compact case one can use Haar measure to construct the necessary integrals.) We say that a subgroup of GL(n, $\mathbb{F}$) is a **linear algebraic group** if it is closed in the Zariski topology, i.e. it can be written as the set of zeros of a collection of polynomials in the n^2 variables. Note that GL(n, $\mathbb{F}$) is the Zariski open subset of $\text{Mat}_n(\mathbb{F}) = \mathbb{F}^{n^2}$ given by the non-vanishing of the determinant. By adding an extra coordinate corresponding to the inverse of the determinant, one can consider GL(n, $\mathbb{F}$) as a closed subset of $\mathbb{F}^{n^2+1}$. Examples of linear algebraic groups are given by GL(n, $\mathbb{F}$), SL(n, $\mathbb{F}$), $\mathbb{Sp}(2n, \mathbb{F})$, the group of diagonal matrices T^n, the group of upper triangular matrices with ones on the diagonal, Uni(n), any finite group, and so on.

A **rational representation** of a linear algebraic group is a homomorphism $G \longrightarrow \text{GL}(m, \mathbb{F})$ with the property that the m^2 entries in the matrices are polynomials in the $n^2 + 1$ variables given by the matrix entries of G and the

---

[3] There are reasons for singling out SL(n, $\mathbb{F}$) and its invariants, but in principle any subgroup of GL(n, $\mathbb{F}$) could be considered.

inverse of the determinant. The inclusion of this last variable enables us to regard the dual of a rational representation as a rational representation.

In this generality, finite generation no longer holds for $\mathbb{F}[V]^G$, even in the case $\mathbb{F} = \mathbb{C}$. Nagata [151] (1958) showed that if G is a subgroup of $\mathrm{Uni}(2) \times \cdots \times \mathrm{Uni}(2)$ (n^2 copies, $n \geq 4$, regarded as a subgroup of $\mathrm{GL}(2n^2, \mathbb{C})$) described by 3 linear relations "in general position" then $\mathbb{C}[V]^G$ is not finitely generated. See also Dieudonné and Carrell [72], and Pommerening [180].

A linear algebraic group is said to be **linearly reductive** if every finite dimensional representation is completely reducible (i.e. a direct sum of irreducible representations).

PROPOSITION 2.5.1 : *A linear algebraic group G is linearly reductive if and only if for every rational representation V of G there is a unique G-invariant $\mathbb{F}$-linear operator $\pi_V : V \longrightarrow V$ which projects V onto the G-invariants.*

PROOF : Let $V = \overset{m}{\underset{i=1}{\oplus}} V_i$ be a decomposition of V into irreducible G-modules. Note that

$$(V_i)^G = \begin{cases} V_i & \text{if } (V_i)^G \neq \{0\} \\ \{0\} & \text{otherwise.} \end{cases}$$

Therefore we may form the submodule $V^{\text{free}} = \underset{(V_i)^G = \{0\}}{\oplus} V_i \subset V$ of the simple components of V on which G acts freely, i.e. the only fixed point in V^{free} is 0. It is easy to see that V^{free} is the unique maximal G-submodule of V on which G acts freely. The desired projection is then $V \longrightarrow V/V^{\text{free}} \cong \underset{V_i^G = V_i}{\oplus} V_i$.

For the other direction, first note that because of uniqueness of π, if $\alpha : V \longrightarrow W$ is a homomorphism of representations then $\pi_W \circ \alpha = \alpha \circ \pi_V$ (both are the projection of $\mathrm{Im}(\alpha)$ onto its G-invariants). In particular, if $\alpha : V \longrightarrow W$ is surjective, then so is the map $\alpha|_{V^G} : V^G \longrightarrow W^G$.

Suppose that $V' \leq V$ is G-invariant. We make $\mathrm{Hom}_{\mathbb{F}}(V', V)$ into a rational representation in the normal way, viz., $g(\varphi)(v') = g(\varphi(g^{-1}v'))$. The inclusion $i : V' \hookrightarrow V$ is a G-invariant element of $\mathrm{Hom}_{\mathbb{F}}(V', V)$. Consider the induced map

$$i^* : \mathrm{Hom}_{\mathbb{F}}(V, V') \longrightarrow \mathrm{Hom}_{\mathbb{F}}(V', V').$$

i^* is surjective, and so the identity element is the image of an invariant element $\varrho \in \mathrm{Hom}_{\mathbb{F}}(V, V')$. The kernel of ϱ is then an invariant complement of V' in V. $\square$

As an example, if G is the complexification of a compact real Lie group G_0 (for example, $\mathrm{GL}(n, \mathbb{C})$ is the complexification of the unitary group $\mathrm{U}(n)$) we can take

$$\pi(x) = \int_{G_0} g(x)dg$$

where the integral is with respect to normalized Haar measure on G_0, and so G is linearly reductive. This is called Weyl's **unitarian trick**.

As a particular case of this, if G is finite with order coprime to the characteristic of $\mathbb{F}$, then the above map π is the same as π^G (see section 2.4), and this shows that G is linearly reductive.

LEMMA 2.5.2 : *The map* $\pi : \mathbb{F}[V] \longrightarrow \mathbb{F}[V]$ *is an* $\mathbb{F}[V]^G$-*module homomorphism.*

PROOF : If $x \in \mathbb{F}[V]^G$ then the maps $y \mapsto x\pi(y)$ and $y \mapsto \pi(xy)$ are both projections of $\mathbb{F}[V]$ onto the invariants $\mathbb{F}[V]^G$, so by uniqueness we have $x\pi(y) = \pi(xy)$. $\square$

The proof of the following theorem is, apart from trivial adjustments in terminology, Hilbert's original proof [101].

THEOREM 2.5.3 : *If* G *is a linearly reductive algebraic group then* $\mathbb{F}[V]^G$ *is a finitely generated* $\mathbb{F}$-*algebra.*

PROOF : Let I be the ideal in $\mathbb{F}[V]$ generated by the homogeneous invariants of positive degree. By Hilbert's basis theorem (theorem 2.2.4), $\mathbb{F}[V]$ is Noetherian, and so I may be generated as an ideal by a finite number of invariant elements $f_1, \ldots, f_s$. We claim that the invariants are generated as an $\mathbb{F}$-algebra by $f_1, \ldots, f_s$. If f is any homogeneous invariant of positive degree, then $f \in I$, so we can write $f = \sum h_i f_i$ with $h_i \in \mathbb{F}[V]$. By the lemma, applying π yields

$$f = \sum \pi(h_i)f_i.$$

The $\pi(h_i)$ are invariants of lower degree than f, so by induction we may assume that they are in the subalgebra generated by $f_1, \ldots, f_s$, and hence so is f. $\square$

As a consequence we see that $\mathbb{F}[S^d(V^*)]^{\mathrm{GL}(n, \ \mathbb{F})}$ is finitely generated as an algebra, so there is a finite complete system of fundamental invariants [101] for n-array forms of degree d.

Chapter 3
Construction of Invariants

In this chapter we address the problem of constructing generators for $\mathbb{F}[V]^G$ in some concrete way when V is a finite dimensional G-representation. Some of the ideas come from algebraic topology: the theory of characteristic classes of vector bundles, in particular the splitting principle. This leads to the notion of orbit polynomials and orbit Chern classes. It was motivated by an attempt to understand Weyl's account [259] (pp. 275–276) of E. Noether's [170] proof of Hilbert's finiteness theorem. In her paper [170] E. Noether gave two proofs of the existence of a finite system of algebra generators for $\mathbb{F}[V]^G$ when G is a finite group and $\mathbb{F}$ a field of characteristic zero. We have already seen one of these, 2.4.2 in section 2.4. In this chapter we examine the second of these, which contains an algorithm for constructing a complete system of fundamental invariants. The basic references for this chapter are [170] and [221].

3.1   Orbit Polynomials and Orbit Chern Classes

Let G be a finite group acting on a set X. The set X together with the action of G is referred to as a G-**set**. If X and Y are G-sets a G-**map** $f : X \longrightarrow Y$ is a function from X to Y such that $f(g \cdot x) = g \cdot f(x)$ for all $x \in X$ and $g \in G$. The collection of all G-sets and G-maps form a category (see [136] chapter 1 for an introduction to categories and functors). We introduce some of the standard definitions and terminology associated with the category of G-sets.

Two G-sets X' and X'' are called **isomorphic** as G-sets if and only if there is a bijective map $f : X' \longrightarrow X''$ such that $f(g \cdot x') = g \cdot f(x')$ for all $x' \in X'$ and $g \in G$. A subset Y of X is called G-**invariant** if $g \cdot y \in Y$ for all $g \in G$ and $y \in Y$. If $B \subset X$ is invariant and G acts transitively on B (i.e. $\forall \ b', b'' \in B \ \exists \ g \in G \mid g \cdot b' = b''$), then we call B a G-**orbit**, or just an **orbit** if G is clear from the context. An invariant subset breaks up into a disjoint union of orbits. If X is a G-set, two points $x', \ x'' \in X$ are called G-**equivalent** if there is an element $g \in G$ such that $x' = gx''$. G-equivalence is an equivalence relation and the set of equivalence classes is denoted by X/G. The points of X/G are the orbits of the action of G on X and X/G is called the **orbit space** of X by the G action. The **isotropy group** G_x of x is the subset $G_x := \{g \in G \mid g \cdot x = x\}$, which clearly is a subgroup of G. The **orbit** of x, denoted by $G \cdot x$, $[x]_G$, or just $[x]$ when G is clear from context, is defined by $\{g \cdot x \mid g \in G\}$. The orbit of a point $x \in X$ is a G-orbit, and as a G-set is isomorphic to the set of left cosets G/G_x with G-action given by left translation.

Let V be a finite dimensional G-representation, G a finite group. For an orbit $B \subset V^*$ set

$$(*) \qquad \varphi_B(X) = \prod_{b \in B}(X + b)$$

which we regard as an element of the ring $\mathbb{F}[V][X]$, with X a new variable. The polynomial $\varphi_B(X)$ is called the **orbit polynomial** of B. Since the product is taken over an invariant subset of V^* it is clear that in fact $\varphi_B(X) \in \mathbb{F}[V]^G[X]$. More generally the formula $(*)$ makes sense for any finite subset $B \subset V$, and defines an element of $\mathbb{F}[V][X]$. If the subset B is G-invariant then $\varphi_B(X) \in \mathbb{F}[V]^G[X]$. If B', $B'' \subset V$ are disjoint then $\varphi_{B'}(X) \cdot \varphi_{B''}(X) = \varphi_{B' \cup B''}(X)$. Since any G-invariant subset is a disjoint union of orbits we may, for the most part, restrict attention to subsets B that are orbits.

If $|B|$ denotes the cardinality of the orbit B, we may expand $\varphi_B(X)$ to a polynomial of degree $|B|$ in X obtaining

$$\varphi_B(X) = \sum_{i+j=|B|} c_i(B) \cdot X^j$$

defining classes $c_i(B) \in \mathbb{F}[V]^G$ called the **orbit Chern classes** of the orbit B. Note that $\mathbb{F}[V]$ is integral over $\mathbb{F}[V]^G$ of finite type and, for $v \in V^*$ the orbit polynomial $\varphi_{G \cdot v}(X)$ is the minimal polynomial of the element $-v$ over $\mathbb{F}[V]^G$.

The first orbit Chern class $c_1(B)$ is the sum of the orbit elements and hence $c_1(B) = \mathrm{Tr}^{G/G_b}(b)$ where $b \in B$ is arbitrary and $G_b \leq G$ is the isotropy group of G. If $|B| = k$ then $c_k(B)$ is the product of all the elements in the orbit B and referred to as the **top Chern class of the orbit**. If $b \in B$ then the top Chern class of B is also referred to as the **norm** of b. The first Chern class is additive and the norm is multiplicative.

The Chern classes of the orbit are nothing but the elementary symmetric functions in the elements of the orbit. Thus if $B \subset V^*$ is an orbit and V_B^* is a vector space with basis identified with the elements of B (e.g. the space of functions from B to $\mathbb{F}$, with basis the **delta** functions

$$\delta_b : B \longrightarrow \mathbb{F} \qquad \delta_b(x) = \begin{cases} 1 & \text{for } x = b \\ 0 & \text{otherwise} \end{cases}$$

denoted by $\text{Fun}(B, \mathbb{F}))$, then there is a natural map of vector spaces $V_B^* \xrightarrow{i_B} V^*$ given by the identification. This map induces a map $\mathbb{F}[V_B]^{\Sigma_{|B|}} \longrightarrow \mathbb{F}[V]^G$ which sends the k-th elementary symmetric function $e_k \in \mathbb{F}[V_B]^{\Sigma_{|B|}}$ into the k-th orbit Chern class of B.

We say that $\mathbb{F}[V]^G$ satisfies the **weak splitting principle** if there are a finite number of orbits whose orbit Chern classes generate $\mathbb{F}[V]^G$. If a single orbit suffices then we say that $\mathbb{F}[V]^G$ satisfies the **splitting principle**. Practically by definition we have:

PROPOSITION 3.1.1 : *If V is a finite dimensional representation of the finite group G then $\mathbb{F}[V]^G$ satisfies the weak splitting principle if and only if there exists orbits $B_1, \dots, B_k \subset V$ such that the induced map*

$$\bigotimes_{i=1}^{k} \mathbb{F}[V_{B_i}]^{\Sigma_{|B_i|}} \longrightarrow \mathbb{F}[V]^G$$

is surjective. $\square$

Topologists should recognize the significance of this when they think of $\mathbb{F}[t_1, \dots, t_k]^{\Sigma_k}$ as $H^*(BU(k); \mathbb{F})$ after doubling degrees.

The orbit polynomial, though neither in this language nor generality, was known to the 19-th century invariant theorists and provided them with a bridge between invariant theory and Galois theory. (See e.g. Weber [257] Band II § 58–§ 60, § 123–§ 130. As noted by Noether [170] p. 91 f., Weber's argument in § 58 has a gap (which I hope to have fixed!). See also Klein [124], [125] lecture IX.)

LEMMA 3.1.2 : *Suppose $\varrho : G \hookrightarrow \text{GL}(n, \mathbb{F})$ is a representation of a finite group G and $B \subset V^*$ is an orbit. Then the orbit polynomial*

$$\varphi_B(X) = \prod_{b \in B}(X + b) \in FF(\mathbb{F}[V]^G)[X]$$

is irreducible.

Here as usual $FF(-)$ denotes the field of fractions functor.

PROOF : Let $f(X)$ be a divisor of $\varphi_B(X)$ and $X + b$ a factor of $f(X)$. Since $f(X)$ is G-invariant $(X + gb) \mid gf(X) = f(X)$ for all $g \in G$, so $\varphi_B(X) \mid f(X)$. $\square$

THEOREM 3.1.3 : *Let $\varrho : G \hookrightarrow \mathrm{GL}(n, \mathbb{F})$ be an irreducible representation of a finite group G. If $B \subset V^*$ is an orbit then $FF(\mathbb{F}[V]) > FF(\mathbb{F}[V])^G = FF(\mathbb{F}[V]^G)$ is the splitting field of $\varphi_B(X) \in FF(\mathbb{F}[V]^G)[X]$.*

PROOF : Since ϱ is irreducible and $\mathrm{Span}_{\mathbb{F}}(B) \subset V^*$ is G-invariant it follows that B must contain a basis for V^*. (Otherwise $\mathrm{Span}_{\mathbb{F}}(B)$ is a proper invariant subspace of V^* contrary to irreducibility.) Therefore

$$FF(\mathbb{F}[V]^G)(B) \supseteq FF(\mathbb{F}[V])$$

and, the reverse inclusion being trivial, the result follows. $\square$

LEMMA 3.1.4 : *Let $\varrho : G \hookrightarrow \mathrm{GL}(n, \mathbb{F})$ be a representation of a finite group G. If $\mathbb{F}$ is large enough there is always an orbit with $|G|$ points.*

PROOF : Suppose the representation ϱ be defined over $\mathbb{L} \subseteq \mathbb{F}$. Let $\mathbb{K} > \mathbb{L}$ be an extension of degree n and let $b_1, \ldots, b_n \in \mathbb{K}$ be linearly independent over $\mathbb{L}$. Consider the vector $b = (b_1, \ldots, b_n) \in \tilde{V} = \mathbb{K} \otimes_{\mathbb{F}} V$ where $V = \mathbb{L}^n$. We claim that the isotropy group of b is trivial. For if $g \in G$ and $\varrho(g) = (a_{i,j}) \in \mathrm{GL}(n, \mathbb{L}) \subset \mathrm{GL}(n, \mathbb{K})$ then $\varrho(g) \cdot b = b$ if and only if

$$\sum_{j=1}^{n} a_{i,j} b_j = b_i \qquad i = 1, \ldots, n$$

or

$$\sum_{j=1}^{n} (a_{i,j} - \delta_{i,j}) b_j = 0 \qquad i = 1, \ldots, n \,.$$

By the linear independence of $b_1, \ldots, b_n \in \mathbb{K}$ over $\mathbb{L}$ we get

$$a_{i,j} = \begin{cases} 1 & i = j \\ 0 & \text{otherwise} \end{cases}$$

and since ϱ is faithful $g = 1$. Thus the orbit of b contains $|G|$ elements. $\square$

DEFINITION : *Let $\varrho : G \hookrightarrow \mathrm{GL}(n, \mathbb{F})$ be a representation of a finite group G and $B \subset V^*$ an orbit. The subalgebra of $\mathbb{F}[V]$ generated by the orbit Chern classes $c_1(B), \ldots, c_{|B|}(B)$ will be denoted by $\mathbb{S}_B^*$ and called the* **characteristic subalgebra** *of B. Note that $\mathbb{S}_B^* \subset \mathbb{F}[V]^G$.*

LEMMA 3.1.5 : *Let $\varrho : G \hookrightarrow \mathrm{GL}(n, \mathbb{F})$ be a representation of a finite group G and $B \subset V^*$ an orbit with $|G|$ elements. For $h \in G$ and $b \in B$ set*

$$\psi_h(X) = \varphi_B(X) \sum_{g \in G} \frac{ghb}{X + gb} \,.$$

Then $\psi_h(X) \in \mathbb{S}_B^*[X]$, i.e. the coefficients of $\psi_h(X)$ belong to the characteristic subalgebra of B.

PROOF : It is clear that $\psi_h(X)$ is a polynomial of degree $|G| - 1$. We have the equalities

$$\frac{ghb}{X + gb} = \frac{ghb}{X}\left(\frac{1}{1 - \frac{-gb}{X}}\right)$$

$$= \frac{ghb}{X}\sum_{i=0}^{\infty}(-1)^i\left(\frac{gb}{X}\right)^i$$

$$= \sum_{i=0}^{\infty}(-1)^i\frac{ghb(gb)^i}{X^{i+1}}$$

and therefore

$$\sum_{g \in G}\frac{ghb}{X + gb} = \sum_{g \in G}\sum_{i=0}^{\infty}(-1)^i\frac{ghb(gb)^i}{X^{i+1}}$$

$$= \sum_{i=0}^{\infty}\sum_{g \in G}(-1)^i\frac{ghb(gb)^i}{X^{i+1}}$$

$$= \sum_{i=0}^{\infty}\frac{s_i(b,\ h)}{X^{i+1}}$$

where

$$s_i(b,\ h) = \sum_{g \in G}(-1)^i ghb(gb)^i \qquad i = 0,\ 1,\ldots,\ .$$

The function $s_i(b,\ h)$, regarded as a function of the elements of the orbit B, is invariant with respect to permutation, i.e. is symmetric in the elements of B. Hence by 1.1.1 $s_i(b,\ h) \in \mathbb{S}_B^*$. If

$$\psi_h(X) = \sum_{i=0}^{|G|-1} A_i X^i$$

the identity

$$\psi_h(X) = \varphi_B(X)\sum_{i=0}^{\infty}\frac{s_i(b,\ h)}{X^{i+1}}$$

yields upon equating coefficients

$$A_i = \sum_{j=0}^{|G|-(1+i)} c_{|B|-(i+j+1)}(B)s_j(b,h) \in \mathbb{S}_B^*$$

and the result is established. □

THEOREM 3.1.6 : *Suppose $\varrho : G \hookrightarrow \mathrm{GL}(n, \mathbb{F})$ is an irreducible representation of a finite group G and $B \subset V^*$ is an orbit with $|G|$ elements. Then every element $f \in \mathbb{F}[V]$ may be written as a rational function of $c_1(B), \ldots, c_{|G|}(B)$ and one fixed element $b \in B$.*

PROOF : Since ϱ is irreducible it follows that B contains a basis for V^*. Hence it suffices to prove the assertion for all $b' \in B$. Let $b' = hb$, $h \in G$. By the preceding lemma

$$\psi_h(X) = \varphi_B(X) \sum_{g \in G} \frac{ghb}{X + gb} \in \mathbb{S}_B^*[X] \, .$$

Direct computation yields

$$\frac{\psi_h(-b)}{\varphi_B'(-b)} = hb + \frac{1}{\varphi_B'(-b)} \sum_{\substack{g \in G \\ g \neq 1}} \frac{\varphi_B(-b)ghb}{-b + gb} = hb$$

because $-b + gb \neq 0$ for $g \neq 1$, but $\varphi_B(-b) = 0$. Finally

$$\varphi_B'(X) = \sum_{i=1}^{|G|} ic_i(|B| - i)X^{i-1} \in \mathbb{S}_B^*[X]$$

so the identity

$$hb = \frac{\psi_h(-b)}{\varphi_B'(-b)}$$

expresses $b' = hb$ as a rational function of b and the orbit Chern classes $c_1(B), \ldots, c_{|G|}(B)$. $\square$

This result has a nice Galois-theoretic interpretation for which we require a (not so common any longer) definition.

DEFINITION : *If $\mathbb{L} < \mathbb{K}$ is a finite Galois extension, then an irreducible polynomial $g(X) \in \mathbb{L}[X]$ is called a **Galois resolvent** for $\mathbb{K} > \mathbb{L}$ if adjoining a single root of $g(X)$ to $\mathbb{L}$ yields $\mathbb{K}$, i.e. there is a root $\theta \in \mathbb{K}$ of $g(X)$ such that $\mathbb{K} = \mathbb{L}(\theta)$.*

Note that this says that $g(X)$ is a Galois resolvent for $\mathbb{K} > \mathbb{L}$ if it is irreducible, $\mathbb{K}$ is a splitting field of $g(X)$ over $\mathbb{L}$, and one of the roots of $g(X)$ is a primitive element. The preceding theorem may then be rephrased as follows.

THEOREM 3.1.7 : *Suppose $\varrho : G \hookrightarrow \mathrm{GL}(n, \mathbb{F})$ is an irreducible representation of a finite group G and $B \subset V^*$ is an orbit with $|G|$ elements. Then $\varphi_B(X) \in \mathbb{F}[V]^G[X]$ is a Galois resolvent for the field extension $FF(\mathbb{F}[V]) > FF(\mathbb{F}[V]^G)$. Hence the Galois group of $\varphi_B(X) \in FF(\mathbb{F}[V]^G)[X]$ is G and $FF(\mathbb{F}[V]) = \mathbb{F}(V)$ is its splitting field.* $\square$

Suppose $\varrho : G \hookrightarrow \mathrm{GL}(n, \mathbb{F})$ is an irreducible representation of a finite group G and $B \subset V^*$ is an orbit with $|G|$ elements. If we think of the elements of B as formal variables over $\mathbb{F}$, i.e. $\mathbb{F}[B]$ should be thought of as a polynomial algebra on the formal variables $b \in B$, then the polynomial

$$\varphi_B(X) = \prod(X + b) \in FF(\mathbb{F}[B])[X] = \mathbb{F}(B)[X]$$

is referred to as the general polynomial of degree $d = |B|$ and its Galois group over $\mathbb{F}(B)$ is known to be Σ_d. By specializing the variables to an orbit of the G-action in V^* we convert the problem of finding the solutions of $\varphi_B(X) = 0$ into a problem of invariant theory. To obtain such a specialization we must impose conditions on the coefficients of the general polynomial. This is the viewpoint adopted by Klein [124] in studying certain equations of degree 5, 6, and 7.

This circle of ideas has the following consequence for invariant theory. (Compare Weber [257] Band II § 58 and the comment of Noether loc. cit.)

THEOREM 3.1.8 (H. Weber): *Let $\varrho : G \hookrightarrow \mathrm{GL}(n, \mathbb{F})$ be an irreducible representation of a finite group G and $B \subset V^*$ an orbit with $|G|$ elements. Assume $|G|$ is prime to the characteristic of $\mathbb{F}$. Then every element $f \in \mathbb{F}[V]^G$ may be expressed as a rational function of the orbit Chern classes $c_1(B), \ldots, c_{|G|}(B)$ of B.*

PROOF :   By 3.1.6 we have a representation

$$f = \frac{\sum_{i=0}^{r} p_i b^i}{\sum_{j=0}^{s} q_j b^j} \qquad p_i, \ q_j \in \mathbb{S}_B^*.$$

By multiplying numerator and denominator by

$$\prod_{\substack{g \in G \\ g \neq 1}} \left(\sum_{j=0}^{s} q_j (gb)^j \right)$$

we may suppose that the denominator is G-invariant. Cross multiplying shows that we may assume the numerator to be G-invariant as well. If $h = \sum_{i=0}^{m} h_i b^i$, $h_i \in \mathbb{S}_B^*$, is G-invariant, then averaging over G yields

$$|G| \cdot h = \sum_{i=0}^{m} \sum_{g \in G} h_i (gb)^i = \sum_{i=0}^{m} \left(\sum_{g \in G} (gb)^i \right) h_i.$$

The power sum functions

$$\sum_{g \in G} (gb)^i \qquad i = 1, 2, \ldots, \ .$$

may be expressed as polynomials in the orbit Chern classes $c_1(B), \ldots, c_{|G|}(B)$ by the recursive formula of Newton (see section 1.1). Hence

$$f = \frac{|G| \sum_{i=0}^{r} p_i b^i}{|G| \sum_{j=0}^{s} q_j b^j} \qquad p_i, \ q_j \in \mathbf{S}_B^*.$$

has both numerator and denominator in $\mathbf{S}_B^*$. $\square$

To quote Weber on this point *"Ob diese Darstellung freilich durch ganze Funktionen möglich ist, würde bei diesem Beweis unentschieden bleiben."* We take up this point after the next corollary.

COROLLARY 3.1.9 : *Let* $\varrho : G \hookrightarrow GL(n, \mathbb{F})$ *be an irreducible representation of a finite group* G *and* $B \subset V^*$ *an orbit with* $|G|$ *elements. Assume* $|G|$ *is prime to the characteristic of* $\mathbb{F}$. *Then* $\mathbb{F}[V]^G$ *is finite over* $\mathbf{S}_B^*$, *i.e.* $\mathbb{F}[V]^G$ *is a finitely generated module over* $\mathbb{F}[B]^{\Sigma_{|B|}}$ *with respect to the module structure induced by the natural map* $\mathbb{F}[V_B]^{\Sigma_{|B|}} \longrightarrow \mathbb{F}[V]^G$. $\square$

We turn next to the question of the integrality of the representation 3.1.8. The following result is by no means optimal. There is need for more work when the field has small characteristic compared to the order of the group.

THEOREM 3.1.10 (L. Smith and R.E. Stong): *Let* $\varrho : G \hookrightarrow GL(n, \mathbb{F})$ *be a representation of a finite group* G *over a field* $\mathbb{F}$. *Suppose either the field* $\mathbb{F}$ *is of characteristic zero or that the order of* G *is less than the characteristic of* $\mathbb{F}$. *Then* $\mathbb{F}[V]^G$ *is generated by orbit Chern classes. If* b *is the size of the largest orbit of* G *acting on* V^* *then* $\mathbb{F}[V]^G$ *is generated by classes of degree at most* b.

PROOF : The hypotheses on $\mathbb{F}$ is equivalent to the invertibility of all the integers $1, 2, \ldots, |G|$ in $\mathbb{F}$. So by Noether's theorem 2.4.3, $\mathbb{F}[V]^G$ is generated by forms of degree at most $|G|$. If $f \in \mathbb{F}[V]^G$ is a polynomial of degree at most $|G|$ then the formula

$$(-1)^j j! \, u_1 \ldots u_j = \sum_{I \subseteq \{1, \ldots, n\}} (-1)^{|I|} \left(\sum_{i \in I} u_i \right)^j$$

used in the proof of 2.4.1 shows that we may write f in the form

$$f = \sum_{\ell \in V^*} \ell^d$$

where $d = \deg(f)$ and the sum extends over a finite set of linear polynomials $\ell \in V^*$.

The transfer gives us a surjective map (see section 2.4) $\pi^G : \mathbb{F}[V] \longrightarrow \mathbb{F}[V]^G$. Therefore we may suppose that $\mathbb{F}[V]^G$ is generated by elements $\pi^G(f)$ of degree at most $|G|$. If $f \in \mathbb{F}[V]$ has degree $d \leq |G|$, then f may be expressed as a sum of d-th powers, so we find

$$\pi^G(f) = \frac{1}{|G|} \sum_{\ell \in V^*} \sum_{g \in G} (g\ell)^d .$$

Therefore it will suffice to show that

$$\sum_{g \in G} (g\ell)^d$$

may be written as a polynomial in orbit Chern classes for any linear polynomial $\ell \in V^*$. This is seen as follows.

Let $\ell \in V^*$ be a linear polynomial and G_ℓ the isotropy group of ℓ. Denote by $[\ell]$ the orbit of ℓ. The map

$$G \longrightarrow [\ell] \qquad g \mapsto g\ell$$

induces an isomorphism of G-sets

$$G/G_\ell \overset{\simeq}{\longrightarrow} [\ell] .$$

Therefore

$$\sum_{g \in G} (g\ell)^r = |G : G_\ell| \sum_{h \in [\ell]} (h)^r .$$

Notice that $\sum_{h \in [\ell]} (h)^r$ is the r-th power sum polynomial, which is a symmetric polynomial of the elements of the orbit $[\ell]$ and hence may be written as a polynomial in the elementary symmetric polynomials $e_1, \ldots, e_{|G:G_\ell|}$ of the elements of the orbit, which by definition are the orbit Chern classes of the orbit $[\ell]$. $\square$

Section 3.2 contains a number of examples using orbit Chern classes to compute rings of invariants. To close this section, we present[1] some examples that show the limitations of this method. In particular, the assumption, in theorem 3.1.10, that the characteristic of $\mathbb{F}$ is zero or larger than the order of $|G|$ cannot be relaxed to the assumption that $|G|$ is prime to the characteristic of $\mathbb{F}$. The first example illustrates several techniques for constructing invariants when ordinary orbit Chern classes are insufficient. It is clear however that in finite characteristic, there is need for other efficient algorithms for constructing invariants.

---

[1] I am indebted to Karen Parshall for bringing these examples to my attention.

EXAMPLE 1 (L. E. Dickson [69]): Consider the subgroup of GL(2, $\mathbb{F}_3$) generated by the matrices

$$A = \begin{bmatrix} 0 & 1 \\ -1 & 0 \end{bmatrix}, \quad B = \begin{bmatrix} -1 & 1 \\ 1 & 1 \end{bmatrix} \in \text{GL}(2, \ \mathbb{F}_3).$$

Set

$$C = AB = \begin{bmatrix} 1 & 1 \\ 1 & -1 \end{bmatrix}.$$

One readily checks that

$$A^2 = B^2 = C^2 = -I,$$

where I is the identity matrix. Hence the subgroup of GL(2, $\mathbb{F}_3$) generated by A and B is isomorphic to the quaternion group Q_8 of order 8. The order of Q_8 is prime to the characteristic of $\mathbb{F}_3$. The elements of Q_8 are $\pm I$, $\pm A$, $\pm B$, $\pm C$. Inspection of these matrices shows that every nonzero vector in $\mathbb{F}_3^2$ occurs exactly once as a first column, so that Q_8 acts transitively on $\mathbb{F}_3^2$. Thus the orbits of Q_8 acting on V^* are $\{0\}$ and $V^* \smallsetminus \{0\}$ and the only Chern classes are therefore (see section 5.6 example 4 for a detailed computation)

$$\frac{xy^9 - x^9y}{xy^3 - x^3y}, \quad (xy^3 - x^3y)^2$$

where $\{x, \ y\}$ is the dual of the canonical basis of $\mathbb{F}_3^2$. These polynomials are of degree 6 and 8. However, as we will presently see (or as direct computation shows) $x^4 + y^4$ is invariant and therefore $\mathbb{F}_3[x,y]^{Q_8}$ cannot be generated by orbit Chern classes alone.

To compute the invariants in this example we employ a generalization of the orbit Chern classes. Let V^* denote the dual of $\mathbb{F}_3^2$. The action of Q_8 on $S^2(V^*)$ decomposes $S^2(V^*)$ into orbits, and just as for the orbits of Q_8 on V^*, the elementary symmetric functions of the elements of an orbit $B \subset S^2(V^*)$ are invariants of Q_8 called generalized orbit Chern classes. The action of Q_8 on $S^2(V^*)$ is given by $S^2(\varrho^*) : Q_8 \longrightarrow \text{GL}(S^2(V^*))$ and [2] has kernel $\mathbb{Z}/2$ generated by $-I$. Therefore the action of Q_8 on $S^2(V^*)$ is given by:

$$
\begin{array}{lll}
A(x^2) = y^2, & B(x^2) = x^2 - 2xy + y^2, & C(x^2) = x^2 + 2xy + y^2 \\
A(xy) = -xy, & B(xy) = y^2 - x^2, & C(xy) = x^2 - y^2 \\
A(y^2) = x^2, & B(y^2) = x^2 + 2xy + y^2, & C(y^2) = x^2 - 2xy + y^2.
\end{array}
$$

Using these formulas we compute the orbits of the three polynomials

$$x^2 + y^2, \ x^2 + xy - y^2, \ x^2 - xy - y^2$$

---

[2] This is easy to see: $-I(uv) = (-I)u(-I)v = (-u)(-v) = uv$.

and find

$$[x^2 + y^2] = \{\pm(x^2 + y^2)\}$$
$$[x^2 + xy - y^2] = \{\pm(x^2 + xy - y^2)\}$$
$$[x^2 - xy - y^2] = \{\pm(x^2 - xy - y^2)\}.$$

Taking the second Chern class of each orbit gives us the invariants

$$(x^2 + y^2)^2, \ (x^2 + xy - y^2)^2, \ (x^2 - xy - y^2)^2 \in \mathbb{F}_3[x, y]^{Q_8}.$$

The sum of these invariants is zero so they span a 2-dimensional subspace for which a suitable basis is

$$\varphi = (x^2 + y^2)^2$$
$$\psi = ((x^2 + xy - y^2)^2 - (x^2 - xy - y^2)^2) = (x^3 y - xy^3).$$

The Jacobian of φ and ψ provide us with a third invariant

$$\vartheta = x^6 + x^4 y^2 + x^2 y^4 + y^6$$

and together φ, ψ and ϑ are a complete fundamental system[3] of invariants for $\mathbb{F}_3[x, y]^{Q_8}$ (see section 4.3 example 1 for a justification of this claim and a further discussion of this example).

The method of constructing invariants with generalized orbit Chern classes is not restricted to using orbits in $S^2(V^*)$. For any representation $\varrho : G \hookrightarrow \mathrm{GL}(n, \mathbb{F})$ of a finite group G and $d \in \mathbb{N}$, we may define the **generalized Chern classes** $c_i(B)$ of an orbit $B \subset S^d(V^*)$ by the formula

$$c(B) = \prod_{b \in B}(X + b) = \sum_{i+j=|B|} c_i(B) X^j \in \mathbb{F}[V][X].$$

As with the Chern classes of orbits $B \subset V^* = S^1(V^*)$ one sees that $c_i(B) \in \mathbb{F}[V]^G$. The class $c_i(B)$ has degree di. This construction of invariants is only interesting if we can find a complete fundamental system of invariants among a small set of orbits for small values of d.

If $f \in \mathbb{F}[V]^G$ then $c_1([f]) = f$, where $[f]$ is the orbit of f, so $\mathbb{F}[V]^G$ is always generated in a trivial way by generalized orbit Chern classes. If $|G|$ is less than the characteristic of $\mathbb{F}$ then $\mathbb{F}[V]^G$ is generated by generalized Chern classes of orbits $B \subset S^d(V^*)$ where d ranges between 1 and $|G|$. This too is trivial, being a consequence of Noether's theorem 2.4.2. However, sharp estimates on the number and degree of a minimal set of such orbits are lacking, both in the modular and nonmodular cases.

---

[3] Dickson seems to have overlooked the invariant of degree 6 : strange since it is the Jacobian of two of the invariants of degree 4.

The preceding example is one of a family of examples that may be constructed using subgroups $G_{p^{2nm}-1}$ of GL(n, $\mathbb{F}_{p^m}$) of order $p^{2nm} - 1$ acting transitively on $\mathbb{F}_{p^m}^n \smallsetminus \{0\}$ originally discovered by L. E. Dickson [69] [70]. The full general linear group GL(n, $\mathbb{F}_{p^m}$) has order $(p^{nm} - 1)(p^{nm} - p^m) \cdots (p^{nm} - p^{(n-1)m})$ so the subgroup $G_{p^{2nm}-1} \subset$ GL(n, $\mathbb{F}_{p^m}$) is proper but has the same orbit structure on $V = \mathbb{F}_{p^m}^n$ as the full general linear group. Therefore $G_{p^{2nm}-1}$ and GL(n, $\mathbb{F}_{p^m}$) have the same Chern classes, hence by the remark following proposition 1.2.4 their ring of invariants cannot both be generated by orbit Chern classes. In [68] (see also section 8.1) Dickson showed that $\mathbb{F}_{p^m}[V]^{\text{GL}(n,\ \mathbb{F}_{p^m})}$ is indeed generated by Chern classes and therefore $\mathbb{F}_{p^m}[V]^{G_{p^{2nm}-1}}$ cannot be generated by orbit Chern classes, even though the order of $G_{p^{2nm}-1}$ is prime to p, the characteristic of the Galois field $\mathbb{F}_{p^m}$.

For the sake of concreteness here is a family of cases when $n = 2$ and $m = 1$. For a description of the groups in the general case see [70] (theorem on page 16). The industrious reader can try to work out a complete fundamental system of invariants for these groups.

EXAMPLE 2 : Let p be an odd prime, $p \equiv 3 \mod 4$. Then -1 is not a quadratic residue modulo p. Consider the set of matrices

$$G_{p^2-1} = \left\{ \begin{bmatrix} a & -b \\ b & a \end{bmatrix} \in \text{GL}(2, \mathbb{F}_p) \mid a^2 + b^2 \neq 0 \right\} .$$

For $p = 3$, G_8 is the quaternion group of the preceding example. The identities

$$\begin{bmatrix} a & -b \\ b & a \end{bmatrix}^{-1} = \frac{1}{a^2 + b^2} \begin{bmatrix} a & b \\ -b & a \end{bmatrix}$$

$$\begin{bmatrix} a & -b \\ b & a \end{bmatrix} \begin{bmatrix} c & -d \\ d & c \end{bmatrix} = \begin{bmatrix} ac - bd & -(ad + bc) \\ ad + bc & ac - bd \end{bmatrix}$$

$$(ac - bd)^2 + (ad+bc)^2 = (a^2 + b^2)(c^2 + d^2)$$

show that $G_{p^2-1} < $ GL(2, $\mathbb{F}_p$) is a subgroup. The equation

$$x^2 + y^2 = 0$$

has only the trivial solution $x = 0 = y$ in $\mathbb{F}_p$. For if $x = a$, $y = b$ were a nontrivial solution, then without loss of generality we could suppose $b \neq 0$, whence $-1 = (\frac{a}{b})^2$, contrary to the assumption that -1 is not a quadratic residue modulo p. Hence every pair $(a, b) \neq (0, 0)$ occurs exactly once as a first column of one of the matrices in G_{p^2-1}. Therefore the elements of G_{p^2-1} are in bijective correspondence with the nonzero vectors $\begin{bmatrix} a \\ b \end{bmatrix} \in \mathbb{F}_p^2$, a bijection being given by

$$T \mapsto T\left(\begin{bmatrix} 1 \\ 0 \end{bmatrix} \right) \qquad \forall\, T \in G_{p^2-1},$$

so $|G_{p^2-1}| = p^2 - 1$ and G_{p^2-1} acts transitively on $\tilde{V} = \mathbb{F}_p^2 \smallsetminus \{0\}$. Hence the only invariants in $\mathbb{F}_p[x, y]^{G_{p^2-1}}$ that arise as Chern classes are the Dickson polynomials (see section 5.6 example 4)

$$c_{p^2-1}(\tilde{V}) = (xy^p - x^p y)^{p-1}$$

$$c_{p^2-p}(\tilde{V}) = \frac{xy^{p^2} - x^{p^2}y}{xy^p - x^p y} \, .$$

A direct computation shows that

$$T(x^2 + y^2) = (a^2 + b^2)(x^2 + y^2) \quad \text{for } T = \begin{bmatrix} a & -b \\ b & a \end{bmatrix} \in G_{p^2-1}$$

so $x^2 + y^2$ is a det relative invariant, i.e. $x^2 + y^2 \in \mathbb{F}_p[x, y]^{G_{p^2-1}}_{\det}$, from which it follows that $(x^2 + y^2)^{p-1} \in \mathbb{F}_p[x, y]^{G_{p^2-1}}$. Since $p^2 - p$ is the degree of the Chern class of least degree and $p^2 - p > 2(p - 1)$ the invariant $(x^2 + y^2)^{p-1} \in \mathbb{F}_p[x, y]^{G_{p^2-1}}$ cannot be expressed as a polynomial in Chern classes.

3.2   Examples

We collect a number of examples to illustrate the ideas developed so far.

EXAMPLE 1 : Let G be the symmetric group Σ_n acting on $V = \mathbb{F}^n$ by permuting a basis $x_1, \ldots, x_n$. Then $\{x_1, \ldots, x_n\}$ is an orbit whose orbit Chern classes are the elementary symmetric functions in $x_1, \ldots, x_n$, which generate $\mathbb{F}[V]^{\Sigma_n}$.

Let us suppose for the rest of this example that $\mathbb{F} = \mathbb{R}$, the real numbers. Then the action of Σ_n preserves the standard inner product on $V = \mathbb{R}^n$. The line spanned by the vector $x_1 + \cdots + x_n$ is left fixed by the action and the orthogonal complement $W = \mathrm{Span}(x_1 + \cdots + x_n)^{\perp}$ is invariant under the action. Let $\Delta^n \subset \mathbb{R}^n$ be the convex hull of the unit basis vectors and the zero vector. Then Δ^n is an n-dimensional simplex: for example, for $n = 3$ it is a tetrahedron. Let $\pi : V \longrightarrow W$ be the orthogonal projection. The image of the $(n-1)$-dimensional face of Δ^n opposite the origin under π is an $(n-1)$-dimensional simplex Δ^{n-1} centered at the origin in W. For example, for $n = 3$ the image is a triangle centered at the origin of $W \cong \mathbb{R}^2$. The restriction of the action of Σ_n from $\mathbb{R}^n = V$ to W presents Σ_n as the full group of symmetries of Δ_{n-1}. The inclusion $W \hookrightarrow V$ induces a map

$$\mathbb{R}[e_1, \ldots, e_n] \cong \mathbb{R}[V]^{\Sigma_n} \longrightarrow \mathbb{R}[W]^{\Sigma_n}$$

which one easily sees is surjective (the projection π even provides a splitting). Since the only polynomial functions on V that vanish on W are in the ideal spanned by e_1 we see that $\mathbb{R}[W]^{\Sigma_n} \cong \mathbb{R}[e_2, \dots, e_n]$ with the isomorphism induced by the inclusion $W \hookrightarrow V$. The vertices of Δ_{n-1} are an orbit of the Σ_n-action. Under this isomorphism the fundamental invariants $e_2, \dots, e_n$ are the Chern classes $c_2, \dots, c_n$ in the vertices of Δ_{n-1}, the first Chern class in the vertices being zero. In particular when $n = 4$ we have described the invariants of the symmetry group of a regular tetrahedron. By restricting attention to the alternating subgroup we obtain the rotational symmetries of the $(n-1)$-dimensional simplex, and 1.3.2 describes the invariants for us.

EXAMPLE 2 : Consider the representation of $\mathbb{Z}/4$ defined by the matrix

$$T = \begin{bmatrix} 0 & -1 \\ 1 & 0 \end{bmatrix}.$$

If $\{x,\ y\}$ is the dual of the canonical basis of $\mathbb{F}^2$, $\mathbb{F}$ not of characteristic 2, then the matrix T defines a faithful representation of $\mathbb{Z}/4$ on the dual space given by

$$T(x) = y \qquad T(y) = -x$$

where we think of $T \in \mathbb{Z}/4$ as the generator. By Noether's theorem[4] the ring of invariants $\mathbb{F}[x,\ y]^{\mathbb{Z}/4}$ is generated by polynomials of degree at most 4. If $v \in V^*$ then the orbit of v consists of $\{v,\ T(v),\ -v,\ -T(v)\}$, so setting $u = T(v)$ we compute

$$c_i([v]) = \begin{cases} 1 & \text{for } i = 0 \\ -(u^2 + v^2) & \text{for } i = 2 \\ u^2 v^2 & \text{for } i = 4 \\ 0 & \text{otherwise.} \end{cases}$$

So for $v = x + a \cdot y$ we obtain

$$c_0 = 1$$
$$c_2 = -(1 + a^2)(x^2 + y^2)$$
$$c_4 = (1 - 4a^2 + a^4)x^2 y^2 + a^2(x^4 + y^4) + 2(a^3 - a)(x^3 y - xy^3)$$

so choosing successively $a = 0$ and $a = 2$ we find the invariants

$$\varrho_1 = x^2 + y^2$$
$$\varrho_2 = x^2 y^2$$
$$\varrho_3 = x^3 \cdot y - x \cdot y^3$$

---

[4] Characteristic 3 presents a problem here, since $3 < 4 = |\mathbb{Z}/4|$, and Noether's bound does not a priori apply. The final result can be justified by appealing to Molien's theorem (see chapter 4 section 4.3), or the fact that T defines a signed permutation representation (see chapter 4 section 4.2), to compute the Poincaré series (see chapter 4) of the invariants and compare it against the Poincaré series of the subring generated by $\varrho_1,\ \varrho_2,\ \varrho_3 \in \mathbb{F}[x,\ y]$.

Direct computation shows that any invariant of degree less than or equal to four may be written as linear combination of ϱ_1 and ϱ_2, or a multiple of ϱ_3, or a linear combination of ϱ_1^2, $\varrho_1\varrho_2$ and ϱ_2^2. Therefore ϱ_1, ϱ_2, ϱ_3 generate $\mathbb{F}[x, y]^{\mathbb{Z}/4}$. These polynomials satisfy one relation, namely:

$$\varrho_3^2 = \varrho_1^2\varrho_2 - 4\varrho_2^2$$

so $\mathbb{F}[x, y]^{\mathbb{Z}/4} \cong \mathbb{F}[u, v, w]/(w^2 - u^2v + 4v^2)$.

EXAMPLE 3 : Consider the dihedral group D_8 of order 8. This group has a real representation of dimension 2 as the group of symmetries of a square with vertices $\pm E_1$, $\pm E_2$ where E_1, E_2 are an orthonormal basis for $\mathbb{R}^2$.

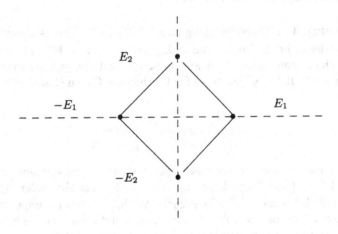

3.2.1: Square

The action on the dual vector space (with basis $\{x, y\}$ dual to E_1, E_2) is generated by the matrices

$$D = \begin{bmatrix} 0 & -1 \\ 1 & 0 \end{bmatrix} \qquad S = \begin{bmatrix} -1 & 0 \\ 0 & 1 \end{bmatrix}$$

where D is a rotation through $90°$ and S a reflection in the y-axis. By 3.1.10, $\mathbb{R}[x, y]^{D_8}$ is generated by orbit Chern classes. Since all orbits contain exactly 4 elements, $\mathbb{R}[x, y]^{D_8}$ is generated by forms of degree at most 4. Each orbit is invariant under the involution $v \mapsto -v$, so the Chern classes, being symmetric functions of the elements of the orbit, are even functions, and hence there are no odd degree invariants. The action being orthogonal the norm must be preserved, so $x^2 + y^2$ is an invariant. The orbit of x is $\{\pm x, \pm y\}$ and has as top Chern class $x^2 y^2$. It is not too hard to see that these elements are a basis for the invariants of degree 2 and 4 respectively, and hence that

$$\mathbb{R}[x, y]^{D_8} = \mathbb{R}[x^2 + y^2, x^2 y^2].$$

This kind of analysis can be applied to the representation of the dihedral group of order $2k$ as the group of symmetries of a regular k-gon in $GL(2, \mathbb{R})$. One quickly obtains as before a quadratic invariant $x^2 + y^2$ and an invariant of degree k which is algebraically independent of $x^2 + y^2$, **but** it is not apparent whether or not these generate the ring of invariants. In fact they do and this may be seen by applying 4.3.2 or 5.5.5.

EXAMPLE 4 : Consider the group $GL(2, \mathbb{F}_2)$. This is a non-abelian group of order 6, hence isomorphic to Σ_3, acting on $V = \mathbb{F}_2^2$. There are two orbits for the action namely, $\{0\}$ and $V \smallsetminus \{0\}$ and analogously for the dual vector space V^*. If $\{x, y\}$ is a basis for V^* then the Chern classes of $V^* \smallsetminus \{0\}$ are

$$c_i = \begin{cases} 0 & \text{for } i = 1 \\ x^2 + xy + y^2 & \text{for } i = 2 \\ xy^2 + x^2 y & \text{for } i = 3 \end{cases}$$

as they are the elementary symmetric polynomials in the elements x, y and $x + y$ of $V^* \smallsetminus \{0\}$. The classes c_1, c_2 are algebraically independent, so $\mathbb{F}_2[V]^G \supset \mathbb{F}_2[x^2 + xy + y^2, x^2 y + xy^2]$. Without further computations or further theory (see for example 5.5.5, 8.1.5) it is not clear that we have found all the invariants. In fact we have, as may be verified by 5.5.5.

EXAMPLE 5 : Fix a prime p and let m be a divisor of $p - 1$. The dihedral group $D_{2m} := \mathbb{Z}/m \rtimes \mathbb{Z}/2$ of order $2m$ has a faithful representation of dimension 2 over $\mathbb{F}_p$ given by the matrices

$$\begin{bmatrix} \theta & 0 \\ 0 & \theta^{-1} \end{bmatrix}, \begin{bmatrix} 0 & 1 \\ 1 & 0 \end{bmatrix} \in GL(2, \mathbb{F}_p)$$

where $\theta \in \mathbb{F}_p$ is a primitive m-th root of unity. Let $\{u, v\}$ denote a basis for $V = \mathbb{F}_p^2$ with respect to which the generators of D_{2m} have the above form. It is not hard to see that

$$\mathbb{F}_p[V]^{D_{2m}} \simeq \mathbb{F}_p[\varrho_1, \varrho_2]$$

where

$$\varrho_1 = uv, \quad \varrho_2 = u^m + v^m$$

are possible choices of polynomial generators. Let B be the orbit of $u + v$. The orbit polynomial of B is

$$\varphi_B(X) = \prod_{i=1}^{m}(X + \theta^i u + \theta^{-i} v).$$

To compute this polynomial we develop a certain identity for the polynomial

$$\alpha(u,\, v) := (u + v)^m - (u^m + v^m) \in \mathbb{F}_p[u,\, v].$$

To this end note that $\alpha(u,\, v)$ is invariant with respect to the involution that interchanges u and v. Therefore it is possible to write $\alpha(u,\, v)$ as a polynomial in the elementary symmetric functions $e_1 = u+v$ and $e_2 = uv$. Taking account of homogeneity we see that

$$\alpha(u,\, v) = \sum_{i_1 + 2i_2 = m} a_{i_1 i_2} e_1^{i_1} e_2^{i_2},$$

where $a_{i_1 i_2} \in \mathbb{F}_p$. Since $\alpha(u,\, v)$ does not contain the terms u^m, v^m it follows that $i_1 < m$, so we may rewrite this formula in the form

$$\alpha(u,\, v) = \sum_{j=1}^{[m/2]} b_j e_1^{m-2j} e_2^j,$$

where $b_j \in \mathbb{F}_p$ and $[m/2]$ denotes the integral part of $m/2$. A simple computation shows that $b_1 = m \not\equiv 0 \bmod p$. Computing further we obtain

$$\alpha(u,\, v) = (u + v)^m - \varrho_2 = \sum_{j=1}^{[m/2]} b_j e_1^{m-2j} e_2^j$$

$$= \sum_{j=1}^{[m/2]} b_j (u + v)^{m-2j} (uv)^j = \sum_{j=1}^{[m/2]} b_j \varrho_1^j (u + v)^{m-2j}$$

which yields the identity

(∗) $$(u + v)^m - \sum_{j=1}^{[m/2]} b_j \varrho_1^j (u + v)^{m-2j} - \varrho_2 = 0.$$

Introduce the polynomial

$$h(X) = X^m - \left(\sum b_j \varrho_1^j X^{m-2j} \right) - \varrho_2 \in \mathbb{F}_p[u,\, v]^{D_{2m}}[X].$$

The identity (∗) shows $h(u+v) = 0$. The coefficients of $h(X)$ are invariant with respect to the action of D_{2m}, so D_{2m} acts on the roots of $h(X)$ in $\mathbb{F}_p[u,\, v]$. Hence $h(X)$ is zero on the elements of the orbit $D_{2m} \cdot (u+v) = \{\theta^i u + \theta^{-i} v\}$.

But the degree of $h(X)$ is $m = |D_{2m} \cdot (u + v)|$ and $h(X)$ is monic, hence $h(X) = \varphi_{D_{2m} \cdot (u+v)}(X)$, the orbit polynomial of $D_{2m} \cdot (u + v)$. From this we read off the Chern classes of the orbit of $u + v$, in particular we have:

$$
c_{2i}(D_{2m} \cdot (u + v)) = \begin{cases} -b_i \varrho_1^i & \text{for } 1 \le i \le [\frac{m-1}{2}] \\ b_{[m/2]} \varrho_1^{[m/2]} - \varrho_2 & \text{for } i = [\frac{m}{2}] \text{ and } m \text{ odd} \\ -\varrho_2 & \text{for } i = [\frac{m}{2}] \text{ and } m \text{ even} \\ 0 & \text{otherwise.} \end{cases}
$$

Since $b_1 = m \not\equiv 0 \bmod p$ it follows that $c_2(D_{2m} \cdot (u+v))$ and $c_m(D_{2m} \cdot (u+v))$ generate the ring of invariants, so $\mathbb{F}_p[V]^{D_{2m}}$ satisfies the splitting principle.

Similarly if m is a divisor of $p + 1$ and $\gamma = \theta + \theta^{-1}$, where θ is a primitive mth root of unity in $\mathbb{F}_{p^2}$, then Galois theory shows $\gamma \in \mathbb{F}_p$. The matrices

$$
\begin{bmatrix} -1 & 0 \\ \gamma & 1 \end{bmatrix}, \begin{bmatrix} 0 & 1 \\ 1 & 0 \end{bmatrix} \in \mathrm{GL}(2, \mathbb{F}_p)
$$

define a faithful representation of D_{2m}. An argument similar to the preceding shows that the ring of invariants is again generated by Chern classes of a single orbit. (For a discussion of the invariant theory of $D_{2(p+1)}$ see section 5.6 example 6 and section 10.3 example 2.)

3.3   Another Finiteness Proof

We begin by describing another finiteness proof of E. Noether [170] which leads to an algorithm, easily implemented on a computer, for finding a complete fundamental system of invariants for the ring of invariants of a finite group in the non-modular case. The starting point for this version of the finiteness theorem is the fundamental theorem on symmetric polynomials 1.1.1

$$
\mathbb{F}[x_1, \ldots x_m]^{\Sigma_m} = \mathbb{F}[e_1, \ldots, e_m]
$$

where Σ_m acts on V by permuting the basis $\{x_1, \ldots, x_m\}$. Denote this representation of Σ_m by $\tau : \Sigma_m \hookrightarrow \mathrm{GL}(m, \mathbb{F})$. For each integer k there is the representation

$$
\tau_k : \Sigma_m \hookrightarrow \mathrm{GL}(km, \mathbb{F})
$$

which in matrix form is

$$
\tau_k(s) = \begin{bmatrix} \tau(s) & 0 & \cdots \\ 0 & \tau(s) & \cdots \\ \vdots & \vdots & \vdots \\ 0 & \cdots & \tau(s) \end{bmatrix} \qquad s \in \Sigma_m .
$$

One way to think of this is by replacing each (scalar) variable $x_1, \ldots, x_m$ by a vector variable of dimension k yielding the representation τ_k. If we regard the vector space $\mathbb{F}^{km}$ as the space of $k \times m$ matrices then via τ_k the symmetric group Σ_m acts by permuting the columns of the matrices. Let us write $\{x_{i,j} \mid i = 1, \ldots, k, \; j = 1, \ldots, m\}$ for the basis of $\mathbb{F}^{km}$ where $x_{i,j}$ is the matrix with a 1 as (i, j)-entry and 0 elsewhere. In this way the representation τ_k yields an action of the symmetric group Σ_m on the polynomial algebra $\mathbb{F}[x_{i,j} \mid i = 1, \ldots, k, \; j = 1, \ldots, m]$. Naturally we ask: what is the structure of $\mathbb{F}[x_{i,j} \mid i = 1, \ldots, k, \; j = 1, \ldots, m]^{\Sigma_m}$? The answer to this question is what Weyl calls the First Main Theorem of invariant theory for the symmetric group Σ_m. The fact that this theorem does not play as central a role in this book as in Weyl's book [259] is a sign of how our point of view differs from his. Still, Weyl's summary (loc. cit. pp. 275-276) of Noether's proof of Hilbert's finiteness theorem using the First Main Theorem for the symmetric group has served as an inspiration for more than one modern paper [221], [23].

In order to describe the invariants of Σ_m acting as above on $\mathbb{F}^{km}$, introduce the identity

$$\prod_{j=1}^{m}(1 + u_1 x_{1,j} + \cdots + u_k x_{k,j}) = \sum_{B=(b_1, \ldots, b_k)} \sigma_B(x_{i,j})(u_1^{b_1} \cdots u_k^{b_k})$$

defining $\sigma_B \in \mathbb{F}[x_{i,j} \mid 1 \leq i \leq k, \; 1 \leq j \leq m]$, $B = (b_1, \ldots, b_k)$ where $0 \leq b_1, \ldots, b_k \leq m$, $b_1 + \cdots + b_k \leq m$. (One should think of the preceding identity as being in $\mathbb{F}[x_{i,j} \mid 1 \leq i \leq k, \; 1 \leq j \leq m][u_1, \ldots, u_k]$.) The function σ_B is the B-th elementary symmetric function in the (vector) variables $\{x_{i,j} \mid 1 \leq i \leq k, \; 1 \leq j \leq m\}$. Weyl ([259] p.37) refers to these polynomials as the **polarized elementary symmetric polynomials**.

The following table lists the polarized elementary symmetric polynomials when the dimension of V is $2(= k)$ and $m = 4$. It was generated by a small computer program written by Nelson Killius. In reading the table note that the second index j of $x_{i,j}$ denotes which vector variable is involved and the first index i which component of that vector variable.

THEOREM 3.3.1 (First Main Theorem of Invariant Theory for Σ_m): *Let m a positive integer and $\mathbb{F}$ a field of characteristic p, with $p = 0$ or $p > m$. Then with the preceding notations the polynomials*

$$\{\sigma_B \mid B = (b_1, \ldots, b_k), \; 0 \leq b_1, \ldots, b_k \leq m, \text{ and } \sum b_i \leq m\}$$

are a system of generators for $\mathbb{F}[x_{i,j} \mid i = 1, \ldots, k, \; j = 1, \ldots, m]^{\Sigma_m}$ as an algebra over $\mathbb{F}$.

We postpone the proof of the First Main Theorem for Σ_m to section 3.4. Granted this theorem we proceed to recycle Noether's arguments.

Let us suppose that $\varrho : G \hookrightarrow \mathrm{GL}(V)$ is a finite group represented in an n-dimensional vector space V over the field $\mathbb{F}$ of characteristic p. **We make**

$\sigma_{(0,0)} = 1$

$\sigma_{(1,0)} = x_{11} + x_{12} + x_{13} + x_{14}$

$\sigma_{(0,1)} = x_{21} + x_{22} + x_{23} + x_{24}$

$\sigma_{(2,0)} = x_{11}x_{12} + x_{11}x_{13} + x_{11}x_{14} + x_{12}x_{13} + x_{12}x_{14} + x_{13}x_{14}$

$\sigma_{(0,2)} = x_{21}x_{22} + x_{21}x_{23} + x_{21}x_{24} + x_{22}x_{23} + x_{22}x_{24} + x_{23}x_{24}$

$\sigma_{(1,1)} = x_{11}x_{22} + x_{11}x_{23} + x_{11}x_{24} + x_{12}x_{21} + x_{12}x_{23} + x_{12}x_{24}$
$\qquad + x_{13}x_{21} + x_{13}x_{22} + x_{13}x_{24} + x_{14}x_{21} + x_{14}x_{22} + x_{14}x_{23}$

$\sigma_{(3,0)} = x_{11}x_{12}x_{13} + x_{11}x_{12}x_{14} + x_{11}x_{13}x_{14} + x_{12}x_{13}x_{14}$

$\sigma_{(0,3)} = x_{21}x_{22}x_{23} + x_{21}x_{22}x_{24} + x_{21}x_{23}x_{24} + x_{22}x_{23}x_{24}$

$\sigma_{(2,1)} = x_{11}x_{12}x_{23} + x_{11}x_{12}x_{24} + x_{11}x_{13}x_{22} + x_{11}x_{13}x_{24}$
$\qquad + x_{11}x_{14}x_{22} + x_{11}x_{14}x_{23} + x_{12}x_{13}x_{21} + x_{12}x_{13}x_{24}$
$\qquad + x_{12}x_{14}x_{21} + x_{12}x_{14}x_{23} + x_{13}x_{14}x_{21} + x_{13}x_{14}x_{22}$

$\sigma_{(1,2)} = x_{11}x_{22}x_{23} + x_{11}x_{22}x_{24} + x_{11}x_{23}x_{24} + x_{12}x_{21}x_{23}$
$\qquad + x_{12}x_{21}x_{24} + x_{12}x_{23}x_{24} + x_{13}x_{21}x_{22} + x_{13}x_{21}x_{24}$
$\qquad + x_{13}x_{22}x_{24} + x_{14}x_{21}x_{22} + x_{14}x_{21}x_{23} + x_{14}x_{22}x_{23}$

$\sigma_{(4,0)} = x_{11}x_{12}x_{13}x_{14}$

$\sigma_{(0,4)} = x_{21}x_{22}x_{23}x_{24}$

$\sigma_{(3,1)} = x_{11}x_{12}x_{13}x_{24} + x_{11}x_{12}x_{14}x_{23} + x_{11}x_{13}x_{14}x_{22} + x_{12}x_{13}x_{14}x_{21}$

$\sigma_{(1,3)} = x_{11}x_{22}x_{23}x_{24} + x_{12}x_{21}x_{23}x_{24} + x_{13}x_{21}x_{22}x_{24} + x_{14}x_{21}x_{22}x_{23}$

$\sigma_{(2,2)} = x_{11}x_{12}x_{23}x_{24} + x_{11}x_{13}x_{22}x_{24} + x_{11}x_{14}x_{22}x_{23}$
$\qquad + x_{12}x_{13}x_{21}x_{24} + x_{12}x_{14}x_{21}x_{23} + x_{13}x_{14}x_{21}x_{22}$

Table 3.3.1 : The polarized elementary symmetric functions
$\dim_{\mathbb{F}}(V) = k = 2, m = 4$

the standing assumption for the rest of this discussion that $p \nmid |G|$, so $|G|$ is invertible in $\mathbb{F}$. Let $f \in \mathbb{F}[V]^G$ be an invariant polynomial and write

$$f = \sum_I a_I x^I$$

where $I = (i_1, \ldots, i_n)$, $x_1, \ldots, x_n$ is a basis for V^*, $a_I \in \mathbb{F}$, and $x^I = x_1^{i_1} \cdots x_n^{i_n}$. Then

$$f = \pi^G f = \frac{1}{|G|} \sum_{g \in G} gf$$

$$= \frac{1}{|G|} \sum_{I, \, g \in G} a_I g(x^I)$$

$$= \sum_I a_I \left(\frac{1}{|G|} \sum_{g \in G} (gx_1)^{i_1} \cdots (gx_n)^{i_n} \right).$$

Let us examine

$$\frac{1}{|G|} \sum_{g \in G} (gx_1)^{i_1} \cdots (gx_n)^{i_n}.$$

For $\{1 = g_1, \ldots, g_d\}$ introduce new variables

$$x_{i,j} = g_j x_i \qquad \text{with } i = 1, \ldots, n \text{ and } j = 1, \ldots, d$$

where for convenience in what follows we have set $|G| = d$. In this notation we are looking at

$$(*) \qquad\qquad \varphi_I = \frac{1}{d} \sum_{j=1}^d x_{1,j}^{i_1} \cdots x_{n,j}^{i_n}$$

which we may regard as a polynomial in the nd variables $\{x_{i,j}\}$. We regard $\mathbb{F}^{nd}$ as the vector space of $n \times d$ matrices on which the symmetric group Σ_d acts by permuting the columns. The effect of $s \in \Sigma_d$ on the sum $(*)$ is to change the order of the summation, hence the expression for φ_I as an polynomial in $\mathbb{F}[x_{i,j} \mid 1 \le i \le n, \, 1 \le j \le d]$ is an invariant of Σ_d acting as above.

Define a homomorphism of algebras

$$\eta : \mathbb{F}[x_{i,j} \mid 1 \le i \le n, \, 1 \le j \le d] \longrightarrow \mathbb{F}[x_1, \ldots, x_n]$$

by requiring $x_{i,j} \mapsto g_j x_i$ for $i = 1, \ldots, n$ and $j = 1, \ldots, d$. For a polynomial $f \in \mathbb{F}[x_1, \ldots, x_n]^G$ we have the formula

$$f = \sum_I a_I \left(\frac{1}{|G|} \sum_{g \in G} (gx_1)^{i_1} \cdots (gx_n)^{i_n} \right)$$

derived above. Introduce the polynomial

$$\varphi(f) = \sum a_I \varphi_I \in \mathbb{F}[x_{i,j} \mid 1 \le i \le n, \, 1 \le j \le d]$$

where φ_I is defined as in formula (*). Unraveling the definitions we see that $\varphi(f) \in \mathbb{F}[x_{i,j} \mid 1 \le i \le n,\ 1 \le j \le d\,]^{\Sigma_d}$ and $\eta(\varphi(f)) = f$.

Let us reformulate this in a more coordinate-free way. Our group G is acting on the vector space V. Introduce the vector space

$$V(G) := \mathrm{Fun}(G,\ V)$$

of functions from G to V. This is a vector space of dimension dn. There is the *diagonal* map

$$\Delta : V \longrightarrow V(G)$$

given by the formula

$$\Delta(v)(g) := g \cdot v$$

for $v \in V$, and $g \in G$.

Think of the symmetric group Σ_d as the group of permutations of the underlying set of G. We may regard G as a subgroup of Σ_d by letting G act on itself by left translation. The formula

$$(s \cdot \lambda)(g) := \lambda(s(g))$$

defines an action of Σ_d on $V(G)$ which restricts to give an action of G on $V(G)$ for which Δ is equivariant.

The algebra homomorphism Δ induces a ring homomorphism ($\mathbb{F}[-]$ is a contravariant functor)

$$\Delta^* : \mathbb{F}[V(G)] \longrightarrow \mathbb{F}[V]\,.$$

Since Δ is G-equivariant, $\Delta^*(\mathbb{F}[V(G)]^G) \subset \mathbb{F}[V]^G$, and as $\mathbb{F}[V(G)]^{\Sigma_d} \subset \mathbb{F}[V(G)]^G$ restriction induces a homomorphism of algebras

$$\eta := \Delta^*\big|_{\mathbb{F}[V(G)]^{\Sigma_d}} : \mathbb{F}[V(G)]^{\Sigma_d} \longrightarrow \mathbb{F}[V]^G$$

which we call the **Noether map**.

THEOREM 3.3.2 (E. Noether): *Let $\varrho : G \hookrightarrow \mathrm{GL}(V)$ be a representation of a finite group G of order d in the n-dimensional vector space V over $\mathbb{F}$. Assume that d is invertible in $\mathbb{F}$. Then the Noether map*

$$\eta : \mathbb{F}[V(G)]^{\Sigma_d} \longrightarrow \mathbb{F}[V]^G$$

is an epimorphism.

PROOF : Choose a basis $\{x_1, \ldots, x_n\}$ for V^*, the dual of V. As in the preceding discussion this leads to a basis for $V(G)^*$. Unraveling the definitions we see that the map η of the coordinate encumbered discussion is the Noether map. Hence for any $f \in \mathbb{F}[V]^G$ we have $\eta(\varphi(f)) = f$ and the result follows. $\square$

The First Main Theorem for the symmetric group provides a set of generators for $\mathbb{F}[x_{i,j} \mid 1 \le i \le n,\ 1 \le j \le d\,]^{\Sigma_d}$. This leads to the following:

COROLLARY 3.3.3 : *Let $\varrho : G \hookrightarrow \mathrm{GL}(V)$ be a representation of a finite group G of order d in the n-dimensional vector space V over a field $\mathbb{F}$ of characteristic p. Assume that $p = 0$ or $p > d$. Then $\mathbb{F}[V]^G$ is generated by $\binom{n+|G|}{|G|}$, or fewer, elements of degree at most $d = |G|$.*

PROOF :   The Noether map η is surjective, and by the First Main Theorem 3.4.1 for the symmetric group the polynomials

$$\{\eta(\sigma_B) \mid B = (b_1, \ldots, b_n) \text{ with } 0 \leq b_1, \ldots, b_n \leq d, \ \sum b_i \leq d\}$$

generate the image. The number of such polynomials is the number of sequences $B = (b_1, \ldots, b_n)$ with

$$b_i \geq 0 \text{ for } i = 1, \ldots, n \qquad \sum b_i \leq d = |G|$$

which is $\binom{n+|G|}{|G|}$. The degree of such a polynomial is $\sum b_i$ which is at most $d = |G|$. $\square$

The proof of the preceding corollary provides us with an algorithm to compute a generating set of invariants. Namely, choose a basis $z_1, \ldots, z_n$ for V^* and order the elements of G, say $1 = g_1, \ldots, g_d$. In the polarized elementary symmetric polynomials σ_I in the nd variables $x_{i,j}$, where $I = (i_1, \ldots, i_n)$ $0 \leq i_j \leq d$, and $i_1 + \cdots + i_n \leq d$, make the substitution $x_{i,j} = g_j z_i$. The resulting $\binom{n+|G|}{|G|}$ polynomials generate $\mathbb{F}[V]^G$.

EXAMPLE 1   : Consider the representation ϱ of $\mathbb{Z}/4$ defined by the matrix

$$T = \begin{bmatrix} 0 & -1 \\ 1 & 0 \end{bmatrix}.$$

If $\{x, y\}$ is the dual of the canonical basis of $\mathbb{F}^2$, $\mathbb{F}$ not of characteristic 2, then the matrix T defines a faithful representation of $\mathbb{Z}/4$ on the dual space given by

$$T(x) = y \qquad T(y) = -x$$

where we think of $T \in \mathbb{Z}/4$ as the generator. We can apply the above algorithm to this example. The table 3.3.2 of polarized symmetric functions in 2 vector variables of dimension 4 provides us with a complete list of the necessary polarized elementary symmetric functions. If we order the group elements $\{I, T, T^2, T^3\}$, the matrix

$$\begin{bmatrix} x & y & -x & -y \\ y & -x & -y & x \end{bmatrix}$$

provides us with a substitution scheme to follow: the entries in the first row being $x_{1,j}$ and those in the second row $x_{2,j}$ where $j = 1, \ldots, 4$. To quote

Nelson Killius: *"Nun kann ich (und der Computer noch viel besser) die erzeugenden Invarianten durch Einsetzen der Werte der Matrix, die sich aus der Darstellung ϱ ergibt, direkt angeben:"*

$$\eta(\sigma_{(0,0)}) = 1$$
$$\eta(\sigma_{(1,0)}) = 0$$
$$\eta(\sigma_{(0,1)}) = 0$$
$$\eta(\sigma_{(2,0)}) = -(x^2 + y^2)$$
$$\eta(\sigma_{(0,2)}) = -(x^2 + y^2)$$
$$\eta(\sigma_{(1,1)}) = 0$$
$$\eta(\sigma_{(3,0)}) = 0$$
$$\eta(\sigma_{(0,3)}) = 0$$
$$\eta(\sigma_{(2,1)}) = 0$$
$$\eta(\sigma_{(1,2)}) = 0$$
$$\eta(\sigma_{(4,0)}) = x^2 y^2$$
$$\eta(\sigma_{(0,4)}) = x^2 y^2$$
$$\eta(\sigma_{(3,1)}) = 2(xy^3 - x^3 y)$$
$$\eta(\sigma_{(1,3)}) = 2(x^3 y - xy^3)$$
$$\eta(\sigma_{(2,2)}) = x^4 + y^4 - 4x^2 y^2$$

Apparently there is a great deal of redundancy in the algorithm. In the next chapter we will reexamine this example with other methods and confirm that (compare section 3.2 example 2)

$$f_1 = x^2 + y^2$$
$$f_2 = x^2 y^2$$
$$f_3 = x^3 \cdot y - x \cdot y^3$$

are a set of generators for $\mathbb{F}[V]^{\mathbb{Z}/4}$.

3.4   The First Main Theorem of Invariant Theory for Σ_m

Let $\varrho : G \hookrightarrow \mathrm{GL}(n, \mathbb{F})$ be a representation of a finite group. There are many ways to construct further representations of G from ϱ. For example, if $k \in \mathbb{N}$ is a positive integer, one has the representation

$$\overset{\longleftarrow k \longrightarrow}{\varrho \oplus \cdots \oplus \varrho} = \underset{k}{\oplus} \varrho : G \hookrightarrow \mathrm{GL}(nk, \mathbb{F}),$$

which is the k-fold direct sum of ϱ with itself. In matrix notation

$$\underset{k}{\oplus}\, \varrho(g) = \begin{bmatrix} \varrho(g) & 0 \dots & 0 \\ 0 & \ddots & 0 \\ 0 & \dots & \varrho(g) \end{bmatrix} \qquad \forall\, g \in G\,.$$

If $V = \mathbb{F}^n$, and we identify $\mathbb{F}^{nk}$ with $\underset{k}{\oplus} V = \overset{\xleftarrow{\quad k \quad}}{V \oplus \cdots \oplus V}$ then the action of G on $\mathbb{F}^{nk}$ is given by

$$\underset{k}{\oplus}\, \varrho(g)(v_1, \dots, v_k) = ((\varrho(g)(v_1), \dots, \varrho(g)(v_k))$$
$$\forall\, g \in G, \quad (v_1, \dots, v_k) \in \underset{k}{\oplus} V\,.$$

A natural question to ask is: given a complete fundamental system of invariants for $\mathbb{F}[V]^G$ can one construct a complete fundamental system of invariants for $\mathbb{F}[\underset{k}{\oplus} V]^G$? In this section we will examine this problem for the tautological representation of the symmetric group.

By definition of the representation $\underset{k}{\oplus} \varrho$, the maps

$$V \xrightarrow{\;\Delta\;} V \oplus \dots \oplus V \xrightarrow{\;\alpha\;} V$$

given by

$$\Delta(v) = (v, \dots, v) \qquad v \in V$$
$$\alpha(v_1,\, v_2, \dots, v_k) = v_1 + v_2 + \cdots + v_k \qquad v_i \in V\ i = 1, \dots, k\,,$$

are G-equivariant, and hence induce maps of algebras Δ^* and α^* that yield a commutative diagram

$$\begin{array}{ccccc} \mathbb{F}[V] & \xleftarrow{\;\Delta^*\;} & \mathbb{F}[V \oplus \cdots \oplus V] & \xleftarrow{\;\alpha^*\;} & \mathbb{F}[V] \\ \cup & & \cup & & \cup \\ \mathbb{F}[V]^G & \xleftarrow{\;\Delta^*\;} & \mathbb{F}[V \oplus \cdots \oplus V]^G & \xleftarrow{\;\alpha^*\;} & \mathbb{F}[V]^G \end{array}$$

which will provide us with a means to construct invariants in $\mathbb{F}[\underset{k}{\oplus} V]^G$ from invariants in $\mathbb{F}[V]^G$. The maps Δ^* and α^* have the following simple formulas:

$$\Delta^*(f)(v) = f(v, \dots, v) \qquad v \in V, \quad f \in \mathbb{F}[\underset{k}{\oplus} V]$$
$$\alpha^*(h)(v_1, \dots, v_k) = h(v_1 + \cdots + v_k) \qquad v_i \in V,\ i = 1, \dots, k,\quad h \in \mathbb{F}[V]\,.$$

The symmetric group Σ_k acts on $\underset{k}{\oplus} V$ by permuting the factors and hence also on $\mathbb{F}[\underset{k}{\oplus} V]$. If we regard Σ_k as acting on V trivially (i.e. $\sigma(v) = v$, $\forall$

$\sigma \in \Sigma_k$, $v \in V$) then the maps α and Δ are also Σ_k-equivariant. The composition $\alpha\Delta : V \longrightarrow V$ is multiplication by k and hence $\Delta^* \alpha^*(f) = k^d f$ where $d = \deg(f)$.

A polynomial $h \in \mathbb{F}[\underset{k}{\oplus} V]$ is said to be **multihomogeneous of multidegree** $(d_1, \ldots, d_k)$ if

$$h(v_1, \ldots, v_{i-1}, \lambda v_i, v_{i+1}, \ldots, v_k) = \lambda^{d_i} h(v_1, \ldots, v_i, \ldots, v_k)$$
$$\forall \lambda \in \mathbb{F}, \quad (v_1, \ldots, v_k) \in \underset{k}{\oplus} V.$$

If $x_1, \ldots, x_n$ is a basis for V^* and we write $x_{i,1}, \ldots, x_{i,n}$ for these vectors regarded as a basis for the i-th factor of $\underset{k}{\oplus} V$, then the monomials of degree d in $x_{i,1}, \ldots, x_{i,n}$ form a basis for the multihomogeneous polynomials of multidegree $(0, \ldots, d, \ldots, 0)$, where the d is in the i-th place. Therefore the monomials

$$x_{1,1}^{d_{1,1}} \cdots x_{1,n}^{d_{1,n}} \cdot x_{2,1}^{d_{2,1}} \cdots x_{2,n}^{d_{2,n}} \cdots x_{k,1}^{d_{k,1}} \cdots x_{k,n}^{d_{k,n}}$$

are a basis for the space of multihomogeneous polynomials of multidegree $(d_1, \ldots, d_k)$, where $d_i = d_{i,1} + \cdots d_{i,n}$. The collection of all these monomials with $d_1 + \cdots + d_k = d$ forms a basis for the homogeneous polynomials in $\mathbb{F}[\underset{k}{\oplus} V]$ of degree d. So sorting out the monomials according to their multihomogeneous degrees allows us to write any homogeneous polynomial $h \in \mathbb{F}[\underset{k}{\oplus} V]$ as a sum of multihomogeneous polynomials, with multidegrees $(d_1, \ldots, d_k)$ satisfying $d_1 + \cdots + d_k = d$. In more abstract language[5] this just corresponds to the isomorphism

$$S^d(\underset{k}{\oplus} V^*) = \underset{d_1 + \cdots + d_k = d}{\oplus} S^{d_1}(V^*) \otimes \cdots \otimes S^{d_k}(V^*).$$

If we write

$$h = \sum_{d_1 + \cdots + d_k = d} h_{d_1, \ldots, d_k}$$

where $h_{d_1, \ldots, d_k}$ is multihomogeneous of multidegree $(d_1, \ldots, d_k)$, then $h_{d_1, \ldots, d_k}$ is called the **component of h of (multi)degree** $(d_1, \ldots, d_k)$.

Suppose $f \in \mathbb{F}[V]$ and

$$\alpha^*(f) = \sum_{d_1 + \cdots + d_k = \deg(f)} \alpha^*(f)_{d_1, \ldots, d_k}$$

---

[5] In the language of gradings (see chapters 4 and 9) we are regarding $\mathbb{F}[\underset{k}{\oplus} V]$ as **multigraded** where the i-th factor of V^* in $\mathbb{F}[\underset{k}{\oplus} V]$ has the multigrading $(\delta_{1,i}, \ldots, \delta_{k,i}) = (0, \ldots, 0, 1, 0, \ldots, 0)$.

is the decomposition of $\alpha^*(f)$ into multihomogeneous components, then $\alpha^*(f)_{d_1,\ldots,d_k} \in \mathbb{F}[\underset{k}{\oplus} V]$ is called the $(d_1,\ldots,d_k)$-**polarization** of f and will be denoted by $\mathrm{Pol}_{(d_1,\ldots,d_k)}(f)$. The polarization operators are linear maps

$$\mathrm{Pol}_{(d_1,\ldots,d_k)} : S^d(V^*) \longrightarrow S^{d_1}(V^*) \otimes \cdots \otimes S^{d_k}(V^*) \qquad d_1 + \cdots + d_k = d$$

commuting with the G-action.

EXAMPLE 1 : Consider the polynomial $x_1^3 + x_1 x_2^2 \in \mathbb{F}[x_1, x_2]$. The polarization of this polynomial from two scalar variables x_1 and x_2 to two vector variables of dimension 3, viz. $x_{1,1}$, $x_{1,2}$, $x_{2,1}$, $x_{2,2}$, $x_{3,1}$, $x_{3,2}$, is given by computing

$$\alpha^* : \mathbb{F}[x_1, x_2] \longrightarrow \mathbb{F}[x_{1,1}, x_{1,2}] \otimes \mathbb{F}[x_{2,1}, x_{2,2}] \otimes \mathbb{F}[x_{3,1}, x_{3,2}]$$
$$\alpha^*(x_j) = x_{1,j} + x_{2,j} + x_{3,j}, \; j = 1, \, 2.$$

We find

$$\alpha^*(x_1^3 + x_1 x_2^2) =$$

$$x_{1,1}^3 + x_{1,1} x_{1,2}^2 + 3 x_{1,1}^2 x_{2,1} + x_{1,2}^2 x_{2,1} + \mathbf{3\, x_{1,1}\, x_{2,1}^2} + x_{2,1}^3 +$$

$$2\, x_{1,1}\, x_{1,2}\, x_{2,2} + \mathbf{2\, x_{1,2}\, x_{2,1}\, x_{2,2}} + \mathbf{x_{1,1}\, x_{2,2}^2} + x_{2,1}\, x_{2,2}^2 +$$

$$3\, x_{1,1}^2 x_{3,1} + x_{1,2}^2 x_{3,1} + 6\, x_{1,1}\, x_{2,1}\, x_{3,1} + 3\, x_{2,1}^2 x_{3,1} +$$

$$2\, x_{1,2}\, x_{2,2}\, x_{3,1} + x_{2,2}^2 x_{3,1} + 3\, x_{1,1}\, x_{3,1}^2 + 3\, x_{2,1}\, x_{3,1}^2 +$$

$$x_{3,1}^3 + 2\, x_{1,1}\, x_{1,2}\, x_{3,2} + 2\, x_{1,2}\, x_{2,1}\, x_{3,2} + 2\, x_{1,1}\, x_{2,2}\, x_{3,2} +$$

$$2\, x_{2,1}\, x_{2,2}\, x_{3,2} + 2\, x_{1,2}\, x_{3,1}\, x_{3,2} + 2\, x_{2,2}\, x_{3,1}\, x_{3,2} + x_{1,1}\, x_{3,2}^2 +$$

$$x_{2,1}\, x_{3,2}^2 + x_{3,1}\, x_{3,2}^2$$

where the multihomogeneous component of degree $(1, 2, 0)$ have been printed in boldface. Thus $\mathrm{Pol}_{(1, 2, 0)}(x_1^3 + x_1 x_2^2) = 3\, x_{1,1}\, x_{2,1}^2 + 2\, x_{1,2}\, x_{2,1}\, x_{2,2} + x_{1,1}\, x_{2,2}^2$.

If $h \in \mathbb{F}[\underset{k}{\oplus} V]$ is multihomogeneous of multidegree $(d_1,\ldots,d_k)$ and $g \in G$, then gh is again multihomogeneous of multidegree $(d_1,\ldots,d_k)$. Therefore, for $h \in \mathbb{F}[\underset{k}{\oplus} V]^G$ each of its multihomogeneous components is invariant. If $f \in \mathbb{F}[V]^G$ then $\alpha^*(f) \in \mathbb{F}[\underset{k}{\oplus} V]^G$ so $\mathrm{Pol}_{(d_1,\ldots,d_k)} \in \mathbb{F}[\underset{k}{\oplus} V]^G$ for all $(d_1,\ldots,d_k)$ with $d_1 + \cdots + d_k = \deg(f)$. The polarizations of f do not in general lie in $\mathrm{Im}(\alpha^*)$, and so provide us with new invariants in $\mathbb{F}[\underset{k}{\oplus} V]^G$.

The polarized elementary symmetric polynomials $\sigma_{d_1,\ldots,d_k}$ described in section 3.3 are the polarizations $\mathrm{Pol}_{(d_1,\ldots,d_k)}(e_i)$ of the elementary symmetric polynomials $e_1,\ldots,e_n \in \mathbb{F}[x_1,\ldots,x_n]^{\Sigma_n}$.

THEOREM 3.4.1 (First Main Theorem of Invariant Theory for Σ_n):
Let $n, k \in \mathbb{N}$ be positive integers and $\tau : \Sigma_n \hookrightarrow \mathrm{GL}(n, \mathbb{F})$ the tautological
representation of Σ_n as the permutation group on the canonical basis of $V :=$
$\mathbb{F}^n$. If $n!$ is invertible in $\mathbb{F}$ then the polarized elementary symmetric functions

$$\{\sigma_B \mid B = (b_1, \ldots, b_k) \, , \, b_1, \ldots, b_k \in \mathbb{N}_0 \text{ and } \sum b_i \leq n\}$$

are a complete fundamental system of invariants for

$$\mathbb{F}[\underset{k}{\oplus} V]^{\Sigma_n} = \mathbb{F}[x_{i,j} \mid i = 1, \ldots, k, \, j = 1, \ldots, n]^{\Sigma_n} .$$

PROOF : The proof is by induction on n. For $n = 1$, Σ_1 is the trivial
group and the polarizations of the polynomial $e_1 = x_1$ are $x_1, x_2, \ldots, x_k$,
which are a basis for $\underset{k}{\oplus} V$, where $V = \mathbb{F}$ has dimension 1 as a vector space
over $\mathbb{F}$. The conclusion is therefore trivially valid.

Assume the result established for all $0 < n' < n$. Let $x_1, \ldots, x_n$ be a basis
for V^* and set $\tilde{V}^* = \mathrm{Span}\{x_1, \ldots, x_{n-1}\}$. Let $e_1, \ldots, e_n$ be the elementary
symmetric polynomials in $x_1, \ldots, x_n$ and set

$$\tilde{e}_i(x_1, \ldots, x_{n-1}) = \begin{cases} e_i(x_1, \ldots, x_{n-1}, 0) & \text{for } i = 1, \ldots, n-1 \\ 0 & \text{for } i = n. \end{cases}$$

The polynomials $\tilde{e}_1, \ldots, \tilde{e}_{n-1}$ are the elementary symmetric polynomials in
the variables $x_1, \ldots, x_{n-1}$. We have the following additional formulae relating
$e_1, \ldots, e_n$ to $\tilde{e}_1, \ldots, \tilde{e}_{n-1}$:

$$e_1 = \tilde{e}_1 + x_n$$
$$e_2 = \tilde{e}_2 + x_n \tilde{e}_1$$

($\spadesuit$) $\qquad\qquad\qquad \vdots$

$$e_{n-1} = \tilde{e}_{n-1} + x_n \tilde{e}_{n-2}$$
$$e_n = x_n \tilde{e}_{n-1} .$$

To verify this, note that by the definitions

$$\sum_{i+j=n} e_i t^j = \prod_{i=1}^n (t + x_i) = \left(\prod_{i=1}^{n-1} (1 + x_i t) \right) (1 + x_n t)$$

$$= \left(\sum_{i+j=n-1} \tilde{e}_i t^j \right)(1 + x_n t)$$

$$= \sum_{i+j=n} (\tilde{e}_i + x_n \tilde{e}_{i-1}) \, t^j \in \mathbb{F}[x_1, \ldots, x_n][t],$$

from which the desired formulae are obtained by equating coefficients of powers of t.

The formulae ($\spadesuit$) may be solved recursively for $\tilde{e}_1, \ldots, \tilde{e}_{n-1}$ to give

$$\tilde{e}_1 = e_1 - x_n$$
$$\tilde{e}_2 = e_2 - x_n \tilde{e}_1 = e_2 - x_n(e_1 - x_n)$$

($\clubsuit$)
$$\vdots$$

$$\tilde{e}_{n-1} = e_n - x_n \tilde{e}_{n-1} = \cdots.$$

In other words, there are formulae

$$\tilde{e}_i = F_i(e_1, \ldots, e_i, x_n) = \sum_{j=1}^{i} F_{i,j}(e_1, \ldots, e_i) x_n^j$$

for $i = 1, \ldots, n - 1$. Polarizing these formulae give

($\diamondsuit$)
$$\mathrm{Pol}_D(\tilde{e}_i) = \sum H_{D,A} \, x^A$$

where $D = (d_1, \ldots, d_k)$, $i = d_1 + \cdots + d_k$, $A = (a_1, \ldots, a_n)$, $x^A = x_{1,n}^{a_1} \cdots x_{k,n}^{a_n}$ and $H_{D,A}$ is a polynomial in the polarized symmetric polynomials of $e_1, \ldots, e_n$. Of course $\mathrm{Pol}_D(\tilde{e}_i) = \tilde{\sigma}_D$ is a polarized elementary symmetric polynomial of $n - 1$ variables.

Let $f \in \mathbb{F}[\underset{k}{\oplus} V]^{\Sigma_n}$ and write f in the form

($\heartsuit$)
$$f = \sum_A f_A x^A$$

where $f_A \in \mathbb{F}[\underset{k}{\oplus} \tilde{V}]$. One sees directly that in fact $f_A \in \mathbb{F}[\underset{k}{\oplus} \tilde{V}]^{\Sigma_{n-1}}$. By the induction hypothesis[6] f_A can be written[7] as a polynomial in the polarized elementary symmetric polynomials $\tilde{\sigma}_D = \mathrm{Pol}_D(\tilde{e}_i)$, $D = (d_1, \ldots, d_{n-1})$, where $d_1 + \cdots + d_{n-1} = i$, for $i = 1, \ldots, n - 1$. Let us indicate this by writting $f_A = L_A(\tilde{\sigma}_D)$. Substituting for $\tilde{\sigma}_D$ from the formula ($\diamondsuit$) in the formula ($\heartsuit$) leads to

$$f = \sum_A f_A x^A = \sum_{A,D} L_A(\tilde{\sigma}_D) \, x^A$$
$$= \sum_A K_A \, x^A$$

---

[6] In this step we need $n - 1 \in \mathbb{F}^{\times}$. It is here that we use the hypothesis that $n!$ is prime to the characteristic of $\mathbb{F}$.

[7] Since the polarized elementary symmetric polynomials are not algebraically independent for $n > 1$ this way of writing f_A is not unique.

where K_A is a polynomial in the polarized elementary symmetric polynomials σ_B. The polynomials f and K_A are invariant with respect to Σ_n. Therefore averaging the preceding formula over the subgroup of Σ_n generated by the cyclic permutation of $x_1, \ldots, x_n$ (this is possible because n is prime to the characteristic of $\mathbb{F}$) yields

$(\star)$
$$f = \frac{1}{n} \sum_{j=1}^{n} \sum_{A} K_A \, x_{1,j}^{a_1} \cdots x_{k,j}^{a_k} \,.$$

Polarizing Newton's formula

$$x_1^a + \cdots + x_n^a = s_a(e_1, \ldots, e_n)$$

that expresses the power sum symmetric polynomial in terms of the elementary symmetric polynomials leads to a formula expressing

$$\sum_{j=1}^{n} x_{1,j}^{a_1} \cdots x_{k,j}^{a_k}$$

in terms of the polarized symmetric polynomials of $e_1, \ldots, e_n$. Substituting this formula in $(\star)$, the result is an expression for f as a polynomial in the polarized elementary symmetric polynomials σ_B of $e_1, \ldots, e_n$. $\square$

The First Main Theorem for the symmetric group shows that for $n!$ prime to the characteristic of $\mathbb{F}$ one obtains a complete fundamental system of invariants for $\mathbb{F}[\bigoplus_k V]^{\Sigma_n}$, where Σ_n acts on $\bigoplus_k V$ by $\bigoplus_k \tau$, and $\tau : \Sigma_n \hookrightarrow \mathrm{GL}(n, \, \mathbb{F})$ is the tautological representation, by polarizing a complete fundamental system of invariants for $\mathbb{F}[V]^{\Sigma_n}$. To what extent this result remains true for other finite groups and other representations $\varrho : G \hookrightarrow \mathrm{GL}(n, \, \mathbb{F})$ is largely uncharted territory. For a discussion of polarization in connectton with Coxeter groups see [253]. The case $G = \mathrm{GL}(n, \, \mathbb{F}_p)$ and its tautological representation is studied in [142] and [185].

Even for the symmetric group Σ_n and its tautological representation there are many questions that remain unanswered, such as a description of all the relations, or syzygies (see chapter 6), between the polarized elementary symmetric polynomials. (For a discussion of the case $k = 2$ see [32] chapter XIX.) Nevertheless there is the following special case of a theorem of Weyl [259] chapter II § 4, § 5 and Theorem 2.5.A.

THEOREM 3.4.2 (H. Weyl): *Suppose that* $\varrho : G \hookrightarrow \mathrm{GL}(n, \, \mathbb{F})$ *is a representation of a finite group* G, *where* $\mathbb{F}$ *is a field of characteristic zero. If* $k \in \mathbb{N}$ *is an integer with* $k > n$ *then a complete fundamental system of invariants for* $\mathbb{F}[\bigoplus_k V]^G$ *(where* G *acts via* $\bigoplus_k \varrho$*) is obtained by polarizing a* complete fundamental system of invariants for $\mathbb{F}[\underset{n}{\oplus} V]^G$ (where G acts via the representation $\underset{n}{\oplus} \varrho$).

The polarizing operators used by Weyl arise from a linear map

$$\underset{k}{\oplus} V \longrightarrow \underset{n}{\oplus} V$$

built out of vector space addition in a manner similar to the way $\mathrm{Pol}_{(d_1, \ldots, d_k)}(f)$ was defined. For a proof and discussion of Weyl's theorem see [259] and [194].

In Example 2 in § 2.4 we computed the invariants of the threefold direct sum of the tautological representation of the symmetric group Σ_2 over $\mathbb{F}_2$ using the basis of the Jordan normal form. We saw that the invariants were not generated by polynomials of degree 1 and 2. Hence the First Main Theorem fails in the modular case. In [45] Campbell, Hughes and Pollack describe a set of algebra generators for the ring of vector invariants of the symmetric group in the modular case.

Chapter 4
Poincaré Series

Let $G \hookrightarrow \mathrm{GL}(V)$ be a linear representation of a finite group G. We would like to compute the dimension of the homogeneous component $\mathbb{F}[V]_j^G$ consisting of invariant polynomials of degree j. These numbers may be conveniently arranged in a generating function called the Poincaré series.

DEFINITION : *A* **graded vector space** *over* $\mathbb{F}$ *is a family of vector spaces* $M = \{M_i \mid i \in \mathbb{Z}\}$. *If* $M_i = 0$ *for all* $i < 0$ *we call* M *a* **positively graded** *vector space over* $\mathbb{F}$. *If* $\dim_{\mathbb{F}} M_i$ *is finite for each* i *we say* M *is of* **finite type**. *If* M *and* N *are graded vector spaces over* $\mathbb{F}$ *a* **morphism of graded vector spaces of degree** d, $f : M \longrightarrow N$, *is a sequence of linear transformations* $\{f_i : M_i \longrightarrow N_{i+d} \mid i \in \mathbb{Z}\}$. *If* $d = 0$, *we speak simply of a morphism of graded vector spaces.*

REMARK : A **graded module** M over a ring A is defined analogously. The elements of M_k are said to have **degree** or **grading** k and of course likewise for a graded vector space.

If $M = \{M_i\}$ is a graded vector space then we call M_i the **component** of degree i. The usual concepts of linear algebra can be applied to graded vector spaces in the obvious way, that is by working componentwise when necessary.

DEFINITION : *Let* M *be a positively graded vector space over* $\mathbb{F}$ *of finite type. The* **Poincaré series** *of* M *is the formal power series*

$$P(M, t) = \sum_{j=0}^{\infty} \dim_{\mathbb{F}}(M_j) t^j .$$

The Poincaré series $P(M, t)$ has a finite value at $t = 1$ if and only if the series contains only finitely many terms, in which case $\oplus M_i$ is a finite dimensional vector space. In this case we speak of $\dim_{\mathbb{F}}(M)$ as $P(M, 1)$ and say that M is **totally finite**.

For graded vector spaces M', M'', their direct sum is defined componentwise: that is $(M' \oplus M'')_i = M_i' \oplus M_i''$. The tensor product of M' and M'' is the

73

graded vector space whose component of degree n is $\bigoplus_{i+j=n} M_i' \otimes M_j''$.

The Poincaré series of a graded vector space is *additive* and *multiplicative* in the sense that for an exact sequence

$$0 \longrightarrow M' \longrightarrow M \longrightarrow M'' \longrightarrow 0$$

of graded vector spaces of finite type

$$P(M,\, t) = P(M',\, t) + P(M'',\, t)\,,$$

and for M', M'' of finite type

$$P(M' \otimes_{\mathbb{F}} M'',\, t) = P(M',\, t) \cdot P(M'',\, t)\,.$$

Both assertions are easily verified.

In general, there is not very much interesting one can say about the Poincaré series of unstructured vector spaces. For finitely generated graded modules over Noetherian rings, however, there is the remarkable fact that the Poincaré series is a rational function, which is a consequence of a general result (the Hilbert-Serre theorem) about finitely generated modules over graded finitely generated algebras. The Poincaré series of a graded connected commutative Noetherian algebra (see section 4.1 for the definitions) has a number of properties reflecting the structure of the algebra. The papers [234] and [225] may be consulted for further information. The special form the series takes for a ring of invariants is elucidated by the formula of Molien.

4.1 The Hilbert–Serre Theorem

The first result concerning the Poincaré series of the polynomial invariants of a finite group is a special case of a more general theorem for finitely generated graded modules over graded Noetherian rings. We begin by defining the concepts involved.

DEFINITION : *A* **graded algebra** *A over the field $\mathbb{F}$ is a graded vector space together with morphisms $\eta : \mathbb{F} \longrightarrow A$ and $\mu : A \otimes A \longrightarrow A$ such that the diagrams*

$$\mathbb{F} \otimes A$$
$$\eta \otimes 1 \downarrow \qquad \searrow^{\cong}$$
$$A \otimes A \xrightarrow{\ \mu\ } A$$
$$1 \otimes \eta \uparrow \qquad \nearrow_{\cong}$$
$$A \otimes \mathbb{F}$$

$$\begin{array}{ccc} A \otimes A \otimes A & \xrightarrow{\ \mu \otimes 1\ } & A \otimes A \\ {\scriptstyle 1 \otimes \mu} \big\downarrow & & \big\downarrow {\scriptstyle \mu} \\ A \otimes A & \xrightarrow{\quad \mu \quad} & A \end{array}$$

are *commutative*.

Here we regard $\mathbb{F}$ as a graded vector space over itself concentrated in degree 0, i.e. all components apart from the zero component are the zero vector space, and the zero component is $\mathbb{F}$. The first diagram expresses the fact that $\eta(1) \in A$ is a two sided unit, and the second diagram that the associative law holds. The distributive laws are a consequence of μ being a linear map. If A is positively graded and η is an isomorphism on the component of degree 0 we say that A is **connected**.

The usual notions of ring theory, such as ideals, modules, finite generation, etc. may be carried over to graded algebras by working componentwise. See for example [136]. There are several standard ways to pass back and forth between graded and ungraded modules, rings, etc. In one direction we can replace a graded object A by its **totalization** defined to be $\mathrm{Tot}(A) = \oplus A_i$, thereby throwing away the grading. In the other direction we can regard a non-graded object B as graded by **concentrating** it in degree 0, i.e. by declaring $B_0 = B$ and $B_i = 0$ for $i \neq 0$. We often use the same symbol for a graded object and its ungraded totalization, as well as for an ungraded object, and that object graded by being concentrated in degree 0: be warned.

If A is a graded polynomial algebra on a single generator x of degree d then

$$A = \left\{ A_i \;\middle|\; \begin{array}{ll} A_i = 0 & \text{for } i < 0 \text{ or } d \nmid i \\ A_i = \mathbb{F} \cdot x^j & \text{for } i = dj \end{array} \right\},$$

so the Poincaré series of A is given by

$$P(A,\, t) = \sum_{i=0}^{\infty} t^{id} = \frac{1}{1 - t^d}.$$

More generally, for a polynomial algebra over $\mathbb{F}$ generated by elements $x_1, \ldots, x_n$ of degrees $d_1, \ldots, d_n$ the Poincaré series is given by

$$P(A,\, t) = \prod_{i=1}^{n} \frac{1}{1 - t^{d_i}}.$$

This is a rational function with a pole at $t = 1$ of order n. The theorem of Hilbert and Serre generalizes this.

THEOREM 4.1.1 (D. Hilbert, J. -P. Serre): *Suppose that A is a graded connected commutative $\mathbb{F}$-algebra finitely generated over $\mathbb{F}$ by homogeneous elements $x_1, \ldots, x_s$ in positive degrees $k_1, \ldots, k_s$. Suppose that M is a finitely generated graded A-module. Then the Poincaré series $P(M, t)$ is of the form*

$$\frac{f(t)}{\prod_{j=1}^{s}(1 - t^{k_j})}$$

where $f(t)$ is a polynomial in t with integer coefficients.

PROOF : The proof is by induction on s. If $s = 0$ then M is just a finite dimensional vector space over $\mathbb{F}$ so $P(M, t)$ is a polynomial. So suppose $s > 0$. Let $\lambda : M \longrightarrow M$ denote left multiplication by the element $x_s \in A$. We have exact sequences

$$0 \longrightarrow \ker(\lambda) \longrightarrow M \longrightarrow M/\ker(\lambda) \longrightarrow 0$$

$$0 \longrightarrow \operatorname{Im}(\lambda) \longrightarrow M \longrightarrow \operatorname{coker}(\lambda) \longrightarrow 0,$$

yielding

(4.1.1) $P(M, t) = P(\ker(\lambda), t) + P(M/\ker(\lambda), t)$

(4.1.2) $P(M, t) = P(\operatorname{Im}(\lambda), t) + P(\operatorname{coker}(\lambda), t).$

The map λ is a morphism of degree $\deg(x_s)$, i.e., $\lambda = \{\lambda_j : M_j \longrightarrow M_{j+\deg(x_s)}\}$ where M_i denotes the homogeneous component of M of degree (or grading) i. So

$$(M/\ker(\lambda))_j \cong (\operatorname{Im}(\lambda))_{j+\deg(x_s)},$$

hence

$$P(M/\ker(\lambda), t) = t^{-\deg(x_s)} P(\operatorname{Im}(\lambda), t)$$

and substituting in the equations (4.1.1), (4.1.2), we obtain

$$P(M, t) = P(\ker(\lambda), t) + t^{-\deg(x_s)} P(\operatorname{Im}(\lambda), t)$$

$$P(M, t) = P(\operatorname{Im}(\lambda), t) + P(M/\operatorname{Im}(\lambda), t).$$

Multiplying the first of these equations by $t^{\deg(x_s)}$ and subtracting from the second equation yields

$$P(M, t) = \frac{-t^{\deg(x_s)}}{1 - t^{\deg(x_s)}} P(\ker(\lambda), t) + \frac{1}{1 - t^{\deg(x_s)}} P(M/\operatorname{Im}(\lambda), t).$$

The modules $\ker(\lambda)$ and $M/\operatorname{Im}(\lambda)$ both have trivial x_s-action, and are thus finitely generated over the subalgebra of A generated by $x_1, \ldots, x_{s-1}$. By the inductive hypothesis the right hand side of the preceding equation is a rational function of the stated form. $\square$

EXAMPLE 1 (R. P. Stanley): Consider the two matrices

$$\begin{bmatrix} -1 & 0 & 0 \\ 0 & -1 & 0 \\ 0 & 0 & 1 \end{bmatrix} \ , \ \begin{bmatrix} 1 & 0 & 0 \\ 0 & 1 & 0 \\ 0 & 0 & i \end{bmatrix} \ \in \mathrm{GL}(3, \ \mathbb{C}).$$

Together they generate a subgroup isomorphic to $\mathbb{Z}/2 \times \mathbb{Z}/4$. Since the group contains $-I$ the invariants are all of even degree. The induced action on $\mathbb{C}[x, \ y, \ z]$ preserves monomials, and one easily sees that

$$x^2, \ xy, \ y^2, \ z^4$$

are a minimal set of algebra generators for the invariants $\mathbb{C}[x, \ y, \ z]^{\mathbb{Z}/2 \times \mathbb{Z}/4}$. The invariants contain $\mathbb{C}[x^2, \ y^2, \ z^4]$ as a subalgebra and $\mathbb{C}[x, \ y, \ z]^{\mathbb{Z}/2 \times \mathbb{Z}/4}$ is a free $\mathbb{C}[x^2, \ y^2, \ z^4]$-module on the two generators 1, xy. (This is a special case of theorem 6.7.8.) In other words

$$\mathbb{C}[x, \ y, \ z]^{\mathbb{Z}/2 \times \mathbb{Z}/4} = \mathbb{C}[x^2, \ y^2, \ z^4] \oplus \mathbb{C}[x^2, \ y^2, \ z^4] \cdot (xy)$$

as $\mathbb{C}[x^2, \ y^2, \ z^4]$-module. Thus we obtain for the Poincaré series

$$P(\mathbb{C}[x, \ y, \ z]^{\mathbb{Z}/2 \times \mathbb{Z}/4}, \ t) = \frac{1}{(1 - t^2)^2(1 - t^4)} + \frac{t^2}{(1 - t^2)^2(1 - t^4)}$$

$$= \frac{1}{(1 - t^2)^3} \ .$$

Therefore the invariants $\mathbb{C}[x, \ y, \ z]^{\mathbb{Z}/2 \times \mathbb{Z}/4}$ have the same Poincaré series as a polynomial algebra $\mathbb{C}[f_1, \ f_2, \ f_3]$ on three generators f_1, f_2, f_3 all of degree 2. **But** $\mathbb{C}[x, \ y, \ z]^{\mathbb{Z}/2 \times \mathbb{Z}/4}$ is not isomorphic to such a polynomial algebra. Even worse, the three matrices

$$\begin{bmatrix} -1 & 0 & 0 \\ 0 & 1 & 0 \\ 0 & 0 & 1 \end{bmatrix} \ , \ \begin{bmatrix} 1 & 0 & 0 \\ 0 & -1 & 0 \\ 0 & 0 & 1 \end{bmatrix} \ , \ \begin{bmatrix} 1 & 0 & 0 \\ 0 & 1 & 0 \\ 0 & 0 & -1 \end{bmatrix} \ \in \mathrm{GL}(3, \ \mathbb{C})$$

define an action of $(\mathbb{Z}/2)^3$ on $\mathbb{C}^3$ and the ring of invariants is $\mathbb{C}[x^2, \ y^2, \ z^2]$, which is a polynomial algebra, and also has Poincaré series $\frac{1}{(1 - t^2)^3}$. Thus, the form of the Poincaré series $P(\mathbb{F}[V]^G, \ t)$ can be an aid in studying the algebra structure of $\mathbb{F}[V]^G$ but it does not determine it.

If we think of $P(M, \ t)$ as a function of a complex variable then in this situation, the following proposition shows that we may interpret the order of the pole at $t = 1$ as telling us about the polynomial rate of growth of the dimension of the graded pieces.

PROPOSITION 4.1.2 : *Suppose that*

$$P(t) = \frac{f(t)}{\prod_{j=1}^{s}(1 - t^{k_j})} = \sum_{r \geq 0} a_r t^r$$

where $f(t)$ is a polynomial with integer coefficients and the a_r are non-negative integers. Let $d(M)$ be the order of the pole of $P(t)$ at $t = 1$. Then

    (i) *there exists a constant $\kappa > 0$ such that $a_n \leq \kappa \cdot n^{d(M)-1}$ for $n > 0$, but*

    (ii) *there does not exist a constant $\kappa > 0$ such that $a_n \leq \kappa \cdot n^{d(M)-2}$ for $n > 0$.*

PROOF : The hypothesis and conclusion remain unaltered if we replace $P(t)$ by $P(t) \cdot (1 + t + \cdots + t^{k_j-1})$, and so without loss of generality we may assume each $k_j = 1$. So we may suppose $P(t) = f(t)/(1-t)^{d(M)}$ with $f(1) \neq 0$. Let $f(t) = \alpha_m t^m + \cdots + \alpha_0$. We have

$$a_n = \alpha_0 \binom{n + d(M) - 1}{d(M) - 1} + \alpha_1 \binom{n + d(M) - 2}{d(M) - 1} + \cdots + \alpha_m \binom{n + d(M) - m - 1}{d(M) - 1}.$$

The condition $f(1) \neq 0$ implies that $\alpha_0 + \cdots + \alpha_m \neq 0$, so this expression is a polynomial of degree exactly $d(M) - 1$ in n. $\square$

NOTATION : We write $d(M)$ for the order of the pole at $t = 1$ of $P(M, t)$.

LEMMA 4.1.3 : *Suppose that $A \supseteq B$ is a finite extension of graded algebras, each of which is finitely generated as an algebra over $\mathbb{F}$ by homogeneous elements of positive degrees. Then $d(A) = d(B)$.*

PROOF : As a B-module, A is finitely generated, so it is a quotient of a finitely generated free graded B-module M. Then $P(M, t) = f(t)P(B, t)$ for some polynomial $f(t)$ with $f(1) > 0$. Thus as power series[1] we have

$$P(B, t) \leq P(A, t) \leq f(t)P(B, t),$$

which proves the lemma. $\square$

PROPOSITION 4.1.4 : $d(\mathbb{F}[V]^G) = \dim_{\mathbb{F}}(V)$.

PROOF : By Theorem 2.3.1, $\mathbb{F}[V]$ is a finite extension of $\mathbb{F}[V]^G$, so by the lemma, it suffices to look at the pole at $t = 1$ of

$$P(\mathbb{F}[V], t) = 1/(1 - t)^n,$$

where $n = \dim_{\mathbb{F}}(V)$. $\square$

---

[1] I.e. the stated inequalities hold coefficient for coefficient

4.2   Permutation Representations

Suppose that G is a finite group and X is a finite G-set. If $\mathbb{F}$ is a field we denote by V_X the vector space $\operatorname{Fun}(X,\ \mathbb{F})$ of all functions from X to $\mathbb{F}$. The group G acts on V_X by

$$(g \cdot \varphi)(x) = \varphi(g^{-1}x) \qquad \forall\, x \in X \text{ and } g \in G$$

so we obtain the linear representation

$$\varrho_X : G \longrightarrow \operatorname{GL}(m,\ \mathbb{F})$$

where $m = |X|$ is the cardinality of X. In this section we investigate the ring of invariants of $\mathbb{F}[V_X]^G$.

The vector space V_X has a basis $\{\delta_x \mid x \in X\}$ consisting of the delta functions defined by

$$(*) \qquad\qquad \delta_{x'}(x'') = \begin{cases} 1 & \text{for } x' = x'' \\ 0 & \text{otherwise.} \end{cases}$$

We may identify $X \subset V_X^*$ with the dual basis as follows. An element $x \in X$ defines a linear functional by evaluation at x, namely

$$x : V_X \longrightarrow \mathbb{F} \qquad x(\varphi) = \varphi(x)\,.$$

This defines a map

$$e : X \longrightarrow V_X^*\,.$$

If x', $x'' \in X$ then $e(x') = e(x'')$ if and only if

$$\varphi(x') = \varphi(x'') \qquad \forall\, \varphi \in V_X$$

since $\operatorname{Fun}(X,\ \mathbb{F})$ separates points of X. (A field $\mathbb{F}$ must have at least 2 distinct elements, $0 \neq 1$.) In particular, if $e(x') = e(x'')$ then

$$1 = \delta_{x'}(x') = \delta_{x'}(x'')$$

implies that $x' = x''$. Finally $(*)$ shows that X and $\{\delta_x \mid x \in X\}$ are dual bases. Thus the dual representation

$$\varrho_X^* : G \longrightarrow \operatorname{GL}(m,\ \mathbb{F})$$

is the usual permutation representation of G on X made into a linear representation over $\mathbb{F}$. If the action of G on X is **effective**, i.e. $gx = x\ \forall\, x \in X$ if and only if $g = 1 \in G$, then the representations ϱ_X and ϱ_X^* are faithful. We will assume that this is the case in the sequel.

For a finite G-set X the action of G on X extends to V_X and hence also to

$$\mathbb{F}[V_X] = \mathbb{F}[x_1, \ldots, x_m]$$

where it sends monomials to monomials. The monomials of degree n are a basis for $\mathbb{F}[V_X]_n = S^n(V_X^*)$ and hence $S^n(V_X^*)$ is the linear representation associated to the permutation representation of G on the monomial basis. Therefore we have from 1.3.3:

LEMMA 4.2.1 : *Let X be a finite G-set. Then $S^n(V_X^*)^G$ has a basis consisting of the orbit sums of the G-action on the monomial basis.* □

To obtain more information about the Poincaré series of $\mathbb{F}[X]^G$ we must study the action of G on the monomial basis. To this end we introduce some new concepts.

For a set X denote by $\underset{n}{\times} X$ the n-fold cartesian product of X with itself. The symmetric group Σ_n acts on $\underset{n}{\times} X$ by permutation of the factors. If X is a G-set then $\underset{n}{\times} X$ is made into a G-set by setting

$$g \cdot (x_1, \ldots, x_n) = (gx_1, \ldots, gx_n).$$

The Σ_n and G actions commute with each other, i.e.

$$\sigma g(x_1, \ldots, x_n) = g\sigma(x_1, \ldots, x_n) \qquad \forall\, g \in G,\ \sigma \in \Sigma_n,\ (x_1, \ldots, x_n) \in \underset{n}{\times} X.$$

The n-**fold symmetric product** of X, $\mathbb{SP}^n(X)$, is defined to be the orbit space of $\underset{n}{\times} X$ by the Σ_n-action, i.e.

$$\mathbb{SP}^n(X) = \underset{n}{\times} X / \Sigma_n.$$

The elements of $\mathbb{SP}^n(X)$ may be thought of as unordered n tuples of elements $[x_1, \ldots, x_n]$ of X. Since the G and Σ_n-actions on $\underset{n}{\times} X$ commute the action of G on $\underset{n}{\times} X$ passes down to $\mathbb{SP}^n(X)$, i.e.

$$\alpha : G \times \mathbb{SP}^n(X) \longrightarrow \mathbb{SP}^n(X)$$

$$g \cdot [x_1, \ldots, x_n] \mapsto [gx_1, \ldots, gx_n] \qquad \forall\, g \in G,\quad [x_1, \ldots, x_n] \in \mathbb{SP}^n(X)$$

defines an action of G on $\mathbb{SP}^n(X)$ and the quotient map

$$\underset{n}{\times} X \xrightarrow{\ q\ } \mathbb{SP}^n(X)$$

is a G-map. The following lemma explains our interest in the functor $\mathbb{SP}^n(-)$.

LEMMA 4.2.2 : *Let X be a finite set. Then the map*

$$M : \mathbb{SP}^n(X) \longrightarrow S^n(V_X^*)$$

defined by

$$M([x_1, \ldots, x_n]) = x_1 \cdots x_n$$

is a bijection between the elements of $\mathbb{SP}^n(X)$ and the monomial basis of $S^n(V_X)$. If G is a finite group and X is a G-set then M is a G-map.

PROOF : This is clear, since to a monomial $x_1^{e_1} \cdots x_m^{e_m}$ of degree n we may associate the unordered n tuple $[x_1, \ldots, x_1, \ldots, x_m, \ldots, x_m]$ where each x_i is repeated e_i times. This correspondence is an inverse to M. $\square$

The opening discussion of section 1.2 and Lemma 4.2.2 allows us to count the number of elements of $\mathbb{SP}^n(X)$, namely:

LEMMA 4.2.3 : Let X be a finite set, then

$$|\mathbb{SP}^n(X)| = \binom{|X| + n - 1}{n}. \quad \square$$

EXAMPLE 1 : Let $X = \{x, y, z\}$ be a set with 3 elements. Then the first three symmetric powers of X are

$$\mathbb{SP}^1(X) = \{x, y, z\} = X$$

$$\mathbb{SP}^2(X) = \left\{ \begin{matrix} [x, x], [y, y], [z, z] \\ [x, y], [y, z], [z, x] \end{matrix} \right\}$$

$$\mathbb{SP}^3(X) = \left\{ \begin{matrix} [x, x, x], [y, y, y], [z, z, z] \\ [x, x, y], [y, y, x], [z, z, x] \\ [x, x, z], [y, y, z], [z, z, y] \\ [x, y, z] \end{matrix} \right\}.$$

DEFINITION : Let X be a finite G-set, G a finite group. Define

$$A(n, X, G) = |\mathbb{SP}^n(X)/G|,$$

i.e. $A(n, X, G)$ is the number of G-orbits on $\mathbb{SP}^n(X)$.

From 1.3.3 and 4.2.2 we obtain:

PROPOSITION 4.2.4 : Let G be a finite group and X be a finite G-set. Then the Poincaré series of the ring of invariants $\mathbb{F}[X]^G$ is independent of $\mathbb{F}$ and is given by

$$P(\mathbb{F}[X]^G, t) = \sum_{n=0}^{\infty} A(n, X, G)t^n. \quad \square$$

To study the numbers $A(n, X, G)$ it is convenient to introduce another way to visualize the elements of $\mathbb{SP}^n(X)$ which is suggested by 4.2.2.

DEFINITION : *A multiset of size* n *consists of a pair* (Y, μ) *where* Y *is a set and* $\mu : Y \longrightarrow \mathbb{N}$ *is a function, such that* $\sum_{y \in Y} \mu(y) = n$. *The set* Y *is called the* **underlying set** *of the multiset,* $\mu(y)$ *the* **multiplicity** *of the element* $y \in Y$, *and* μ *the* **multiplicity function** *of the multiset.*

Since multiplicities are strictly positive the underlying set of a multiset of size n has cardinality at most n.

To a point $[x] \in \mathbb{SP}^n(X)$ we associate a multiset of size n as follows. The point $[x]$ is an unordered n tuple $[x_1, \ldots, x_n]$ of elements of X. For the underlying set of the multiset $E_{[x]}$ we take $\{x_1, \ldots, x_n\}$. (N.b. When dealing with sets repetitions may be struck from the list of elements, i.e. $\{x, x\} = \{x\}$. In other words $E_{[x]} = \{z_1, \ldots, z_m\}$ where $z_1, \ldots, z_m$ are the distinct elements of $[x] = [x_1, \ldots, x_n]$.) The multiplicity function $\mu_{[x]}$ is defined by setting $\mu_{[x]}(z)$ equal to the number of occurrences of z in the unordered n tuple $[x_1, \ldots, x_n]$. This procedure associates a multiset $(E_{[x]}, \mu_{[x]})$, with underlying set a subset of X, to every point $[x]$ of $\mathbb{SP}^n(X)$.

Conversely, if (Y, μ) is a multiset of size n whose underlying set is a subset of X, we may associate to (Y, μ) a point $[x] \in \mathbb{SP}^n(X)$ by taking the unordered n-tuple of points consisting of the elements of Y, each repeated as often as its multiplicity.

Thus the points of $\mathbb{SP}^n(X)$ may be viewed as the multisets of size n whose underlying set is a subset of X.

In the language of multisets the symmetric square, i.e. $\mathbb{SP}^2(X)$, of $X = \{x, y, z\}$ may be described by the table which follows. Each row of the table is a point of $\mathbb{SP}^2(\{x, y, z\})$ viewed as a multiset of size 2.

UNDERLYING SET	$\mu(x)$	$\mu(y)$	$\mu(z)$
$\{x\}$	2		
$\{y\}$		2	
$\{z\}$			2
$\{x, y\}$	1	1	
$\{x, z\}$	1		1
$\{y, z\}$		1	1

Table 4.2.1 : $\mathbb{SP}^2(\{x, y, z\})$

DEFINITION : *If (Y, μ) is a multiset of size n, then the* **type** *of (Y, μ) is the $|Y|$-tuple of numbers $\{\mu(y) \mid y \in Y\}$ arranged in decreasing order.*

The type of (Y, μ) is a partition of n with parts the numbers $\{\mu(y) \mid y \in Y\}$.

LEMMA 4.2.5 : *Let G be a finite group and X a finite G-set. Then for every $[x] \in \mathbb{SP}^n(X)$ and $g \in G$ the multisets $[x]$ and $g[x]$ have the same type.*

PROOF : Suppose $[x] = [x_1, \ldots, x_n]$. Since the order does not matter we may write $[x]$ in the form

$$[\, \underset{\leftarrow \mu(y_1) \rightarrow}{y_1, \ldots, y_1} \, , \, \underset{\leftarrow \mu(y_2) \rightarrow}{y_2, \ldots, y_2} \, , \ldots, \, \underset{\leftarrow \mu(y_s) \rightarrow}{y_1, \ldots, y_s} \,]$$

where $\mu(y_1) \geq \mu(y_2) \geq \cdots \geq \mu(y_s)$ is the type of $[x]$. By definition of the G-action

$$g \cdot [x] = [\, \underset{\leftarrow \mu(y_1) \rightarrow}{gy_1, \ldots, gy_1} \, , \, \underset{\leftarrow \mu(y_2) \rightarrow}{gy_2, \ldots, gy_2} \, , \ldots, \, \underset{\leftarrow \mu(y_s) \rightarrow}{gy_s, \ldots, gy_s} \,]$$

and the type has not changed. $\square$

The proof of the preceding lemma shows that the action of G on $\mathbb{SP}^n(X)$ sends a point y in the multiset $[x] \in \mathbb{SP}^n(X)$ of multiplicity k to a point of the same multiplicity. Classifying multisets by their type can be of use in determining the orbit structure of $\mathbb{SP}^n(X)$. Since the type of a multiset of size n is just a partition of n we can use standard partition theory notation to indicate the type. (See for example [188] chapter 6.)

EXAMPLE 2 : Let $G = \mathbb{Z}/3$ acting on $\{x, y, z\}$ by cyclic permutation. Let us compute the orbits of the action on $\mathbb{SP}^1(X)$, $\mathbb{SP}^2(X)$ and $\mathbb{SP}^3(X)$ (refer to example 1). $X = \mathbb{SP}^1(X)$ consists of a single $\mathbb{Z}/3$ orbit because the action is transitive. $\mathbb{SP}^2(X)$ consists of two $\mathbb{Z}/3$ orbits corresponding to the two types of multisets of size 2. They are:

$$\{[x, x], [y, y], [z, z]\} \quad \text{of type } 2^1,$$
$$\{[x, y], [y, z], [z, x]\} \quad \text{of type } 1^2,$$

where we have used one of the standard partition theory notations [188] to indicate the types. (2^1 is the partition 2 of 2 and 1^2 the partition 1, 1 of 2.) The three types of multisets of size 3 are 3^1, $2^1 1^1$ and 1^3 and to each of these there corresponds $\mathbb{Z}/3$ orbits. These are:

$$\{[x, y, z]\} \qquad \text{of type } 1^3$$
$$\{[x, x, y], [y, y, z], [z, z, x]\} \quad \text{of type } 2^1 1^1$$
$$\{[x, x, z], [y, y, x], [z, z, y]\} \quad \text{of type } 2^1 1^1$$
$$\{[x, x, x], [y, y, y], [z, z, z]\} \quad \text{of type } 3^1$$

so that there are four orbits of the $\mathbb{Z}/3$ action on $\mathbb{SP}^3(X)$.

The orbit sums are listed in the following table:

deg	ORBIT SUM	NOTATION
1	$x + y + z$	e_1
2	$x^2 + y^2 + z^2$	s_2
2	$xy + yz + zx$	e_2
3	xyz	e_3
3	$x^2y + y^2z + z^2x$	f
3	$x^2z + y^2x + z^2y$	h
3	$x^3 + y^3 + z^3$	s_3

Table 4.2.2 : Orbit sum polynomials for $\mathbb{Z}/3$ cyclically permuting $\{x,\ y,\ z\}$

By Noether's theorem 2.4.2, $\mathbb{F}[x,\ y,\ z]^{\mathbb{Z}/3}$ is generated by elements of degree at most 3 provided 3! is invertible in $\mathbb{F}$. From this table we see that the polynomials e_1, e_2, e_3 and f (or h) may be chosen as generators for $\mathbb{F}[x,\ y,\ z]^{\mathbb{Z}/3}$. This should be no surprise as $\mathbb{Z}/3$ acting cyclically on $\{x,\ y,\ z\}$ is just the alternating group A_3 in its tautological permutation representation (the invariants were computed in section 1.3) and so these elements generate in characteristic 2 and 3 also.

If G is a finite group and X a finite G-set Garsia and Stanton ([89] theorem 7.4, see also theorem 6.2.9 for a more direct proof) have shown that $\mathbb{Q}[X]^G$ is generated by the orbit sums of degree less than or equal to $\binom{|X|}{2} = \frac{|X|(|X|-1)}{2}$. This bound is generally better than $|G|$ which is provided by Noether's theorem 2.4.2. Moreover the result of Garsia and Stanton is sharp, because (see section 1.3.2)

$$\mathbb{Q}[x_1,\ldots,x_n]^{A_n} = \mathbb{Q}[e_1,\ldots,e_n,\ \Delta]/(\Delta^2 - D(e_1,\ldots,e_n))$$

where $e_1,\ldots,e_n$ are the elementary symmetric functions, Δ the discriminant, which has degree $n(n-1)/2$, and $D(e_1,\ldots,e_n)$ the polynomial expressing Δ^2 as a polynomial in the elementary symmetric polynomials. The bound of Garsia and Stanton on the degrees of generating invariants has been shown to hold in arbitrary characteristic by M. Göbel [93].

If X is a finite G-set, then X decomposes into a disjoint union

$$X = X^G \sqcup X_1 \sqcup \cdots \sqcup X_s$$

where $X_1, \ldots, X_s$ are the nontrivial orbits of G on X. If the isotropy group of $x \in X$ is denoted by G_x then points x', x'' in the same orbit have conjugate isotropy groups. Let $G_1, \ldots, G_s$ be a set of representatives of the isotropy groups of the orbits $X_1, \ldots, X_s$, then

$$|X| = |X^G| + \sum_{i=1}^{s} |G : G_i|$$

is the **class equation** which counts the elements of X in a way appropriate to the G-action.

PROPOSITION 4.2.6 : *Let the finite group G act transitively on the set* $\{x_1, \ldots, x_m\} = X$. *Then*

$$\mathbb{SP}^n(X)^G = \begin{cases} \varnothing & n \not\equiv 0 \bmod m \\ [\underbrace{x_1,, \ldots, x_m, \ldots, x_1, \ldots, x_m}_{k}] & n = km. \end{cases}$$

PROOF : The only thing clear at the outset is that the indicated point is indeed a fixed point.

If $[x] \in \mathbb{SP}^n(X)^G$ then the underlying set of $[x]$ must contain all the elements $x_1, \ldots, x_m$ because G acts transitively on X. Moreover invariance and transitivity imply each pair of elements $x_{i'}$, $x_{i''}$ must have the same multiplicity. This is possible if and only if n is a multiple of m and $[x]$ is the indicated element. □

If G is cyclic of prime order then the only subgroups of G are G itself and the trivial subgroup. This greatly simplifies the orbit structure of $\mathbb{SP}^n(X)$ as the following result shows.

COROLLARY 4.2.7 : *Let the cyclic group $\mathbb{Z}/p$ of prime order act by cyclic permutation on the set $\{x_1, \ldots, x_p\} = X$. Then*

$$|\mathbb{SP}^n(X)/\mathbb{Z}/p| = \begin{cases} \frac{1}{p}\binom{p+n-1}{n} & n \not\equiv 0 \bmod p \\ 1 + \frac{1}{p}\left[\binom{p+n-1}{n} - 1\right] & n \equiv 0 \bmod p. \end{cases}$$

PROOF : Since $\mathbb{Z}/p$ has prime order the only possible orbits are fixed points and orbits with p elements. The result then follows from 4.2.3, 4.2.6 and the class equation. □

We are now in a position to compute the Poincaré series of the ring of invariants of the regular representation of a cyclic group of prime order.

PROPOSITION 4.2.8 : *Let* $\varrho : Z/p \hookrightarrow \mathrm{GL}(p, \mathbb{F})$ *denote the regular representation of the cyclic group of prime order* p. *Then the Poincaré series of the ring of invariants* $\mathbb{F}[x_1, \ldots, x_p]^{\mathbb{Z}/p}$ *is given by*

$$P(\mathbb{F}[x_1, \ldots, x_p]^{\mathbb{Z}/p}, \ t) = \frac{1}{p} \left[\frac{1}{(1-t)^p} + \frac{p-1}{(1-t^p)} \right]$$

$$= \frac{1}{p} \left[\frac{(1 + t + \cdots + t^{p-1})^p + (p-1)(1-t^p)^{p-1}}{(1-t^p)^p} \right].$$

PROOF : The regular representation is the permutation representation associated to the action of $\mathbb{Z}/p$ on itself given by left translation. In other words ϱ is the linear representation associated to the action of $\mathbb{Z}/p$ on the set $X = \{x_1, \ldots, x_p\}$ by cyclic permutation. By 4.2.7 and the definitions we have

$$A(n, \ X, \ \mathbb{Z}/p) = \begin{cases} \frac{1}{p} \binom{p+n-1}{n} & n \not\equiv 0 \bmod p \\ 1 + \frac{1}{p} \left[\binom{p+n-1}{n} - 1 \right] & n \equiv 0 \bmod p. \end{cases}$$

By 4.2.4 we obtain

$$P(\mathbb{F}[x_1, \ldots, x_p]^{\mathbb{Z}/p}, \ t) = \sum_{n=0}^{\infty} A(n, \ X, \ \mathbb{Z}/p) \, t^n$$

$$= \frac{1}{p} \left[\sum_{n=0}^{\infty} \binom{p+n-1}{n} t^n + (p-1) \sum_{k=0}^{\infty} t^{kp} \right]$$

$$= \frac{1}{p} \left[\frac{1}{(1-t)^p} + \frac{p-1}{1-t^p} \right]$$

and the result follows from formal manipulations. $\square$

4.3   Molien's Theorem

If $\mathbb{F}$ is a field of characteristic zero and $\varrho : G \hookrightarrow \mathrm{GL}(n, \mathbb{F})$ a representation of the finite group G then the map $\pi^G = \frac{1}{|G|} \sum_{g \in G} g$ is a projection operator on $\mathbb{F}[V]$ with image $\mathbb{F}[V]^G$. So the dimension of $\mathbb{F}[V]_j^G$ is equal to the trace of the matrix representing π^G on $\mathbb{F}[V]_j$ and we have

$$P(\mathbb{F}[V]^G, t) = \frac{1}{|G|} \sum_{g \in G} \sum_{j=0}^{\infty} \mathrm{tr}(g, \mathbb{F}[V]_j) t^j.$$

The following formula for the trace actually works independently of the characteristic of $\mathbb{F}$.

PROPOSITION 4.3.1   (Trace Formula): Let $\varrho : G \hookrightarrow \mathrm{GL}(n, \mathbb{F})$ be a representation of a finite group G over a field $\mathbb{F}$. If $g \in G$, then the trace of the action of G on $\mathbb{F}[V]$ is given by $\displaystyle\sum_{j=0}^{\infty} \mathrm{tr}(g, \ \mathbb{F}[V]_j) t^j = \dfrac{1}{\det(1 - g^{-1}t)}.$

REMARK :   Apart from a factor of $\det(g^{-1})$ the terms in the sum on the right hand side are the reciprocals of the characteristic polynomials of the elements $\varrho(g)$ for $g \in G$. One should think of the right hand side of this formula as a power series in t with coefficients certain sums of determinants of the action of $g^{-1} \in G$ on V. Note that $\mathrm{Mat}_n(\mathbb{F})[t] \cong \mathrm{Mat}_n(\mathbb{F}[t])$, so det makes sense on $\mathrm{Mat}_n(\mathbb{F})[t]$.

PROOF :   Extending the field does not affect either side of this equation, so we may assume that $\mathbb{F}$ is algebraically closed. Then g is upper triangularizable on V (if $\mathbb{F}$ has characteristic zero, it is even diagonalizable), say with eigenvalues $\lambda_1, \ldots, \lambda_n$. The eigenvalues on V^* are $\lambda_1^{-1}, \ldots, \lambda_n^{-1}$, so the eigenvalues on $\mathbb{F}[V]_j$ are the products of j not necessarily distinct λ_i^{-1}'s. So using a basis in which g is upper triangular we have

$$\sum_{j=0}^{\infty} \mathrm{tr}(g, \ \mathbb{F}[V]_j) t^j = \left(\sum_{j=0}^{\infty} \lambda_1^{-j} t^j\right) \cdots \left(\sum_{j=0}^{\infty} \lambda_n^{-j} t^j\right) = \prod_{i=1}^{n} \frac{1}{1 - \lambda_i^{-1}t} = \frac{1}{\det(1 - g^{-1}t)},$$

which proves the proposition.   □

THEOREM 4.3.2   (T. Molien): Let $\varrho : G \hookrightarrow \mathrm{GL}(n, \ \mathbb{F})$ be a representation of a finite group G over a field $\mathbb{F}$ of characteristic zero. Then the Poincaré series of the ring of invariants is given by

$$P(\mathbb{F}[V]^G, \ t) = \frac{1}{|G|} \sum_{g \in G} \frac{1}{\det(1 - g^{-1}t)} = \frac{1}{|G|} \sum_{g \in G} \frac{1}{\det(1 - g\,t)}.$$

PROOF :   This follows from 4.3.1 and the formula immediately preceding it, since summing over $g^{-1} \in G$ is the same as summing over $g \in G$.   □

REMARK :   In nonzero characteristic coprime to $|G|$, Molien's theorem still holds, as long as one uses a "Brauer lift" of the trace and determinant, obtained by lifting eigenvalues to characteristic zero. See [207] § 12 for a nice discussion with examples. Even in non-coprime characteristic, Molien's formula gives information. It no longer calculates the Poincaré series of the invariants, but rather the Poincaré series for the multiplicity of the trivial module as a composition factor.

If $\varrho : G \hookrightarrow \mathrm{SL}(n, \ \mathbb{F})$ is a representation into the special linear group then the Poincaré series of $\mathbb{F}[V]^G$ has a very special symmetry property used by R. P. Stanley in [234].

COROLLARY 4.3.3 (R. P. Stanley): Let $\varrho : G \hookrightarrow \mathrm{SL}(n, \ \mathbb{F})$ be a representation of a finite group over a field whose characteristic is prime to the order of G. (N. b. $\det(\varrho(g)) = 1 \ \forall \ g \in G$.) Then

$$P(\mathbb{F}[V]^G, \ \frac{1}{t}) = (-1)^n t^n P(\mathbb{F}[V]^G, \ t).$$

PROOF : By Molien's theorem we have

$$P(\mathbb{F}[V]^G, \ \frac{1}{t}) = \frac{1}{|G|} \sum_{g \in G} \frac{1}{\det(I - g^{-1}t^{-1})}$$

$$= \frac{t^n}{|G|} \sum_{g \in G} \frac{1}{\det(tI - g^{-1})} = \frac{t^n}{|G|} \sum_{g \in G} \frac{1}{\det(gt - I)\det(g^{-1})}$$

$$= \frac{(-1)^n t^n}{|G|} \sum_{g \in G} \frac{1}{\det(I - gt)} = (-1)^n t^n P(\mathbb{F}[V]^G, \ t)$$

since $\det(g) = 1$ for all $g \in G$. $\square$

In the preceding section we found a formula for the Poincaré series of the regular representation of the cyclic group of prime order over an arbitrary field. The formula is independent of the field. If the field $\mathbb{F}$ is the complex numbers then we can check our work by applying the theorem of Molien. To do so, notice that over $\mathbb{C}$ the regular representation of the cyclic group of order p is equivalent to the representation given by the matrix

$$\begin{bmatrix} 1 & 0 & \cdots & 0 \\ 0 & \lambda & \cdots & 0 \\ \vdots & \vdots & \vdots & \vdots \\ 0 & \cdots & 0 & \lambda^{p-1} \end{bmatrix} \in \mathrm{GL}(p, \ \mathbb{C}) \qquad \lambda = \exp(2\pi i/p).$$

The formula of Molien then gives

$$P(\mathbb{F}[x_1, \ldots, x_p]^{\mathbb{Z}/p}, \ t) = \frac{1}{p} \sum_{\zeta \in \mathbb{Z}/p} \frac{1}{(1 - \zeta^0 t)(1 - \zeta^1 t) \cdots (1 - \zeta^{p-1}t)}.$$

For $\zeta \neq 1$ we have an equality of sets

$$\{\zeta^0, \ \zeta^1, \ldots, \ \zeta^{p-1}\} = \{1, \ \lambda, \ldots, \ \lambda^{p-1}\}$$

so

$$(1 - \zeta^0 t)(1 - \zeta^1 t) \cdots (1 - \zeta^{p-1}t) = (1 - t)(1 - \lambda t) \cdots (1 - \lambda^{p-1}t)$$

$$= 1 - t^p$$

hence the formula given by Molien's theorem reduces to

$$\frac{1}{p}\left[\frac{p-1}{1-t^p}+\frac{1}{(1-t)^p}\right]$$

which checks with the combinatorial formula obtained from 4.2.8.

For a permutation representation we may interpret the trace in another way. If the finite group G acts by permutations on the finite set X we may choose as basis for $\mathbb{F}[X]_n = S^n(V_X^*)$ the set $\mathbb{SP}^n(X)$. The group G acts on $\mathbb{F}[X]_n$ by permuting this basis. Hence for each $g \in G$ the matrix of g with respect to the basis $\mathbb{SP}^n(X)$ has exactly one nonzero entry, a 1, in each row and each column. So the number of 1's on the diagonal is nothing but the number of fixed points of the action of g on $\mathbb{SP}^n(X)$. Hence the trace formula gives us:

PROPOSITION 4.3.4 : Let G be a finite group acting on the finite set X and $\mathbb{F}$ a field. Then

$$P(\mathbb{F}[X]^G,\ t) = \frac{1}{|G|}\sum_{n=0}^{\infty}\sum_{g\in G}|\mathbb{SP}^n(X)^g|\,t^n\ .$$

(N.b. This formula is independent of the field $\mathbb{F}$.) $\square$

REMARK : This proposition could also be obtained from 4.2.4 directly by using the formula

$$|X/G| = \frac{1}{|G|}\sum_{g\in G}|X^g|$$

which is valid for any finite G-set X. If we replace X by $\mathbb{SP}^n(X)$ in this formula we see

$$A(n,\ X,\ G) = \frac{1}{|G|}\sum_{g\in G}|\mathbb{SP}^n(X)^g|$$

for $n = 1,\ 2,\ldots$, and substituting in 4.2.4 yields the preceding proposition.

The proof of Molien's theorem may be modified to yield a formula for relative invariants.

THEOREM 4.3.5 : Let $\varrho : G \hookrightarrow \mathrm{GL}(n,\ \mathbb{F})$ be a representation of a finite group G over a field $\mathbb{F}$ of characteristic zero. If S is a simple $\mathbb{F}(G)$-module, then

$$P(\mathbb{F}[V]_S^G,\ t) = \frac{\dim_\mathbb{F}(S)}{|G|}\sum_{g\in G}\frac{\mathrm{tr}(g^{-1},\ S)}{\det(1 - g^{-1}t)}\ .$$

PROOF : By character theory, we have

$$P(\mathbb{F}[V]_S^G, \ t) = \sum_{j=0}^{\infty} \dim_{\mathbb{F}}(\mathbb{F}[V]_S^G)_j t^j$$

$$= \sum_{j=0}^{\infty} \left(\frac{\dim_{\mathbb{F}}(S)}{|G| \sum_{g \in G} \text{tr}(g^{-1}, \ S) \text{tr}(g, \mathbb{F}[V]_j)} \right) t^j$$

$$= \frac{\dim_{\mathbb{F}}(S)}{|G|} \sum_{g \in G} \left(\text{tr}(g^{-1}, \ S) \sum_{j=0}^{\infty} \text{tr}(g, \ \mathbb{F}[V]_j) t^j \right)$$

$$= \frac{\dim_{\mathbb{F}}(S)}{|G|} \sum_{g \in G} \frac{\text{tr}(g^{-1}, \ S)}{\det(1 - g^{-1}t)} \ .$$

In the last step, we have used Proposition 4.3.1. $\square$

EXAMPLE 1 : Let $G = Q_8$ be the quaternion group of order 8 with representation $\varrho : G \hookrightarrow GL(2, \ \mathbb{C})$ given by the matrices in table 4.3.1.

g	$\varrho(g)$	$\det(1 - g^{-1}t)$	g	$\varrho(g)$	$\det(1 - g^{-1}t)$
1	$\begin{bmatrix} 1 & 0 \\ 0 & 1 \end{bmatrix}$	$(1 - t)^2$	-1	$\begin{bmatrix} -1 & 0 \\ 0 & -1 \end{bmatrix}$	$(1 + t)^2$
i	$\begin{bmatrix} i & 0 \\ 0 & -i \end{bmatrix}$	$1 + t^2$	$-i$	$\begin{bmatrix} -i & 0 \\ 0 & i \end{bmatrix}$	$1 + t^2$
j	$\begin{bmatrix} 0 & 1 \\ -1 & 0 \end{bmatrix}$	$1 + t^2$	$-j$	$\begin{bmatrix} 0 & -1 \\ 1 & 0 \end{bmatrix}$	$1 + t^2$
k	$\begin{bmatrix} 0 & i \\ i & 0 \end{bmatrix}$	$1 + t^2$	$-k$	$\begin{bmatrix} 0 & -i \\ -i & 0 \end{bmatrix}$	$1 + t^2$

Table 4.3.1 : The quaternion subgroup Q_8 in $GL(2, \ \mathbb{C})$

Molien's theorem then gives

$$P(\mathbb{C}[V]^{Q_8}, t) = \frac{1}{8} \left(\frac{1}{(1-t)^2} + \frac{1}{(1+t)^2} + \frac{6}{1+t^2} \right) = \frac{1 + t^6}{(1 - t^4)^2}$$

$$= 1 + 2t^4 + t^6 + \cdots .$$

It follows that there are two algebra generators in degree four and one in degree six, and possibly more. The form of the Poincaré series suggests that there is a polynomial subring on two generators in degree four, and that $\mathbb{C}[V]^{Q_8}$ is a free module over this polynomial subring, on two generators, one of degree 0 and one of degree 6. To verify this we write down some invariants by computing Chern classes of the orbits of x and $x+y$. The orbit of x consists of the eight points $\pm x$, $\pm y$, $\pm ix$, $\pm iy$ so the total Chern class[2] of the orbit of x is

$$c([x]) = (1-x)(1+x)(1-ix)(1+ix)(1-y)(1+y)(1-iy)(1+iy)$$
$$= 1 - (x^4 + y^4) + x^4 y^4$$

from which we obtain the invariant $\alpha = x^4 + y^4 = -c_4([x])$ of degree 4. The orbit of $x+y$ is

$$[x+y] = \{\pm(x+y),\ \pm(x-y),\ \pm i(x+y),\ \pm i(x-y)\}$$

and hence the total Chern class of this orbit is

$$c([x+y]) =$$
$$(1+(x+y))(1-(x+y))(1-i(x+y))(1+i(x+y))(1-(x-y))(1+(x-y))(1-i(x-y))(1+i(x-y))$$
$$= 1 - 2(x^4 + 6x^2 y^2 + y^4) + (x^2 - y^2)^4 .$$

From this we see that $-c_4([x+y]) = x^4 + 6x^2 y^2 + y^4$ is an invariant and hence so is

$$\beta = \frac{1}{6}(c_4([x]) - c_4([x+y])) = x^2 y^2 .$$

The Jacobian of α and β

$$\begin{bmatrix} \frac{\partial \alpha}{\partial x} & \frac{\partial \alpha}{\partial y} \\ \frac{\partial \beta}{\partial x} & \frac{\partial \beta}{\partial y} \end{bmatrix} = \det \begin{bmatrix} 4x^3 & 4y^3 \\ 2xy^2 & 2x^2 y \end{bmatrix} = 8(x^5 y - xy^5)$$

is an invariant by 1.5.6, since $\varrho(Q_8) < \mathrm{SL}(2,\ \mathbb{C})$. Thus

$$\left. \begin{aligned} \alpha &= x^4 + y^4 \\ \beta &= x^2 y^2 \\ \gamma &= x^5 y - xy^5 \end{aligned} \right\} \in \mathbb{C}[x,\ y]^{Q_8} .$$

The polynomials α and β are algebraically independent and the elements of $\mathbb{C}[\alpha,\ \beta]$ all have degrees congruent to 0 modulo 4. The elements of $\mathbb{C}[\alpha,\ \beta] \cdot \gamma$ have degrees congruent to 2 modulo 4, so

$$\mathbb{C}[\alpha,\ \beta] \cap \mathbb{C}[\alpha,\ \beta] \cdot \gamma = 0$$

and hence

$$\mathbb{C}[\alpha, \beta] \oplus \mathbb{C}[\alpha, \beta] \cdot \gamma \subseteq \mathbb{C}[V]^{Q_8} .$$

---

[2] I.e formal sum of the Chern classes, which is **nonhomogeneous**.

A comparison of Poincaré series shows that we have equality. The only relation between α, β and γ is the one expressing γ^2 as a polynomial in α and β, namely

$$\gamma^2 = \alpha^2\beta - 4\beta^3 \,,$$

and we arrive at the presentation

$$\mathbb{C}[V]^{Q_8} \simeq \mathbb{C}[\alpha,\beta,\gamma]/(\gamma^2 - \alpha^2\beta + 4\beta^3) \,.$$

Molien's theorem does not directly apply to the representation[3] of the quaternion group, $\varrho : Q_8 \hookrightarrow GL(2, \mathbb{F}_3)$, studied in section 3.3 example 2. To apply Molien's theorem to a representation over a finite field $\mathbb{F}_q$ with q elements we need to choose a Brauer lift (see e.g. [202] § 18, [207] § 12), i.e. an identification of the multiplicative group $\mathbb{F}_q^\times$ with the group of $(q-1)$-st roots of unity in $\mathbb{C}$ to give the proper interpretation to the determinants in Molien's formula. For $\mathbb{F}_3$ however this presents no difficulties. As an aid to computation we use table 4.3.2. Molien's theorem then gives, just as in the case of the preceding representation $Q_8 \hookrightarrow GL(2, \mathbb{C})$,

$$P(\mathbb{F}_3[V]^{Q_8}, t) = \frac{1}{8}\left(\frac{1}{(1-t)^2} + \frac{1}{(1+t)^2} + \frac{6}{1+t^2}\right) = \frac{1+t^6}{(1-t^4)^2}$$
$$= 1 + 2t^4 + t^6 + \cdots \,.$$

In view of the discussion of $\mathbb{C}[x,\ y]^{Q_8}$ this justifies our conclusion in section 3.1 example 1 that

$$\left.\begin{array}{l} \varphi = (x^2 + y^2)^2 \\ \psi = xy^3 - x^3y \\ \vartheta = x^6 + x^4y^2 + x^2y^4 + y^6 \end{array}\right\} \in \mathbb{F}_3[V]^{Q_8}$$

is a complete fundamental system of invariants.

We precede the next example with an observation to ease the computation of the Poincaré series of a ring of invariants by means of Molien's theorem. Let $\mathbb{F}$ be a field of characteristic zero and $\varrho : G \hookrightarrow GL(n, \mathbb{F})$ a representation of

---

[3] Alternatively one could write down a lift of the mod 3 representation to characteristic zero. For example the matrices

$$\tilde{A} = \begin{bmatrix} 0 & 1 \\ -1 & 0 \end{bmatrix}, \quad \tilde{B} = \frac{1}{\sqrt{-2}}\begin{bmatrix} -1 & 1 \\ 1 & 1 \end{bmatrix} \in GL(2,\ \mathbb{Z}_{(3)}(\sqrt{-2}))\,,$$

where $\mathbb{Z}_{(3)}$ denotes the subring of the rationals whose denominator is prime to 3, generate a subgroup isomorphic to Q_8. Under the map $\mathbb{Z}_{(3)}(\sqrt{-2}) \longrightarrow \mathbb{F}_3$ given by $a + b\sqrt{-2} \mapsto a + b$ the subgroup $Q_8 < GL(2,\ \mathbb{Z}_{(3)}(\sqrt{-2}))$ maps isomorphically onto the subgroup $Q_8 < GL(2,\ \mathbb{F}_3)$. Character theory says that the representation $Q_8 < GL(2,\ \mathbb{Z}_{(3)}(\sqrt{-2}))$ is conjugate in $GL(2,\ \mathbb{C})$ to the subgroup $Q_8 < GL(2,\ \mathbb{C})$, etc.

g	$\varrho(g)$	$\det(1 - g^{-1}t)$	g	$\varrho(g)$	$\det(1 - g^{-1}t)$
I	$\begin{bmatrix} 1 & 0 \\ 0 & 1 \end{bmatrix}$	$(1 - t)^2$	$-I$	$\begin{bmatrix} -1 & 0 \\ 0 & -1 \end{bmatrix}$	$(1 + t)^2$
A	$\begin{bmatrix} 0 & 1 \\ -1 & 0 \end{bmatrix}$	$1 + t^2$	$-A$	$\begin{bmatrix} 0 & -1 \\ 1 & 0 \end{bmatrix}$	$1 + t^2$
B	$\begin{bmatrix} -1 & 1 \\ 1 & 1 \end{bmatrix}$	$1 + t^2$	$-B$	$\begin{bmatrix} 1 & -1 \\ -1 & -1 \end{bmatrix}$	$1 + t^2$
C	$\begin{bmatrix} 1 & 1 \\ 1 & -1 \end{bmatrix}$	$1 + t^2$	$-C$	$\begin{bmatrix} -1 & -1 \\ -1 & 1 \end{bmatrix}$	$1 + t^2$

Table 4.3.2: The quaternion subgroup Q_8 in $GL(2, \mathbb{F}_3)$

a finite group G. If $g \in G$ then $\det(1 - gt)$ does not change if we think of the entries of the matrix of g as belonging to some field extension of $\mathbb{F}$. (We used this in the course of proving Molien's theorem.) If over an extension field of $\mathbb{F}$ the matrix of g is diagonalizable with eigenvalues $\lambda_1(g), \ldots, \lambda_n(g)$, then

$$\det(1 - gt) = \prod_{i=1}^{n} (1 - \lambda_i(g)t).$$

The following example is based in part on an example of Stanley ([235] example 5.1).

EXAMPLE 2 : Consider the dihedral group D_{2k} of order $2k$ represented in $GL(2, \mathbb{R})$ as the group of symmetries of a regular k-gon centered at the origin. In this representation the group D_{2k} is generated by the matrices

$$D = \begin{bmatrix} \cos \frac{2\pi}{k} & -\sin \frac{2\pi}{k} \\ \sin \frac{2\pi}{k} & \cos \frac{2\pi}{k} \end{bmatrix} \qquad S = \begin{bmatrix} 1 & 0 \\ 0 & -1 \end{bmatrix},$$

where D is a rotation through $2\pi/k$ radians and S is a reflection in an axis. Thus the elements of D_{2k} are the identity, the $k - 1$ rotations D^i, $i = 1, \ldots, k - 1$, and the k reflections SD^i, $0 = 1, \ldots, k - 1$. Over the complex numbers the rotation D is diagonalizable with complex conjugate eigenvalues $\lambda = \exp(2\pi i/k)$, $\bar{\lambda} = \lambda^{-1}$. Thus D^i is diagonalizable with eigenvalues λ^i, λ^{-i} and hence

$$\det(1 - D^i t) = (1 - \lambda^i t)(1 - \lambda^{-i} t).$$

The reflections SD^i are all diagonalizable with eigenvalues 1, -1, so

$$\det(1 - SD^i t) = (1 - t)(1 + t) = 1 - t^2.$$

By Molien's theorem 4.3.2 we obtain

$$(\star) \qquad P(\mathbb{R}[x,\,y]^{D_{2k}},\,t) = \frac{1}{2k}\Big\{\frac{k}{1-t^2} + \sum_{i=0}^{k-1}\frac{1}{(1-\lambda^i t)(1-\lambda^{-i}t)}\Big\}.$$

To make use of this we need to evaluate the second sum. We follow Stanley (loc. cit.) and observe that by Molien's theorem

$$\frac{1}{k}\sum_{i=0}^{k-1}\frac{1}{(1-\lambda^i t)(1-\lambda^{-i}t)} = P(\mathbb{C}[x,\,y]^{\mathbb{Z}/k},\,t)$$

where $\mathbb{Z}/k \hookrightarrow D_{2k}$ is the cyclic subgroup generated by D and we have complexified the original representation. The action of $\mathbb{Z}/k$ over $\mathbb{C}$ is diagonalizable, given by the matrix

$$\begin{bmatrix} \lambda & 0 \\ 0 & \lambda^{-1} \end{bmatrix}$$

and therefore the action of $\mathbb{Z}/k$ on $\mathbb{C}[x,\,y]$ sends monomials to monomials, and $\mathbb{C}[x,y]^{\mathbb{Z}/k}$ has a $\mathbb{C}$-basis consisting of monomials $x^{ak}y^{bk}(xy)^c$; a, b, $c \in \mathbb{N}_0$, $0 \le c \le k-1$. Hence

$$\mathbb{C}[x,\,y]^{\mathbb{Z}/k} = \bigoplus_{i=0}^{k-1} \mathbb{C}[x^k,\,y^k]\cdot(xy)^i.$$

The Poincaré series of $\mathbb{C}[x^k,\,y^k]\cdot(xy)^i$ is

$$\frac{t^{2i}}{(1-t^k)^2}$$

so by the additivity property of Poincaré series

$$P(\mathbb{C}[x,\,y]^{\mathbb{Z}/k},\,t) = \frac{1+t^2+\cdots+t^{2k-2}}{(1-t^k)^2}.$$

From this we obtain

$$\sum_{i=0}^{k-1}\frac{1}{(1-\lambda^i t)(1-\lambda^{-i}t)} = k\Big[\frac{1+t^2+\cdots+t^{2k-2}}{(1-t^k)^2}\Big]$$

which by substituting in $(\star)$ yields

$$\begin{aligned}
P(\mathbb{R}[x,\,y]^{D_{2k}},\,t) &= \frac{1}{2k}\Big\{k\Big[\frac{1+t^2+\cdots+t^{2k-2}}{(1-t^k)^2}\Big] + k\frac{1}{1-t^2}\Big\} \\
&= \frac{1}{2}\Big[\frac{1-t^{2k}+(1-t^k)^2}{(1-t^k)^2(1-t^2)}\Big] \\
&= \frac{1}{2}\Big[\frac{(1-t^k)(1+t^k)+(1-t^k)^2}{(1-t^k)^2(1-t^2)}\Big] \\
&= \frac{1}{(1-t^k)(1-t^2)}
\end{aligned}$$

and we see $\mathbb{R}[x, y]^{D_{2k}}$ has the Poincaré series of a polynomial algebra on two generators, one of degree 2 and one of degree k.

The action of D_{2k} on $\mathbb{R}^2$ is orthogonal and hence preserves the square of the norm, giving the invariant $q = x^2 + y^2 \in \mathbb{R}[x, y]^{D_{2k}}$. The vertices of the regular k-gon form an orbit and have the coordinates

$$\left(\cos(\frac{2\pi i}{k}), \sin(\frac{2\pi i}{k})\right) \qquad i = 0, \ldots, k-1.$$

The k-th Chern class of this orbit is

$$h = \prod_{i=0}^{k-1} \left((\cos(\frac{2\pi i}{k}))x + (\sin(\frac{2\pi i}{k}))y \right).$$

If $k \not\equiv 0 \bmod 4$ then the coefficient of x^k in this polynomial is nonzero and one sees that q and h are algebraically independent. If $k \equiv 0 \bmod 4$ then the coefficient of $y^2 x^{k-2}$ is likewise nonzero and q and h are algebraically independent in this case also. Therefore $\mathbb{R}[q, h] \subseteq \mathbb{R}[x, y]^{D_{2k}}$. Since

$$P(\mathbb{R}[x, y]^{D_{2k}}, t) = \frac{1}{(1-t^2)(1-t^k)} = P(\mathbb{R}[q, h], t)$$

it follows $\mathbb{R}[x, y]^{D_{2k}} = \mathbb{R}[q, h]$. (For another discussion of this example see section 5.5. The case $k = 4$ is discussed at the end of the next section.)

The use of invariant theory to derive the formula

$$(\spadesuit) \qquad \frac{1}{k} \sum_{i=0}^{k-1} \frac{1}{(1-\lambda^i t)(1-\lambda^{-i} t)} = \frac{1 + t^2 + \cdots + t^{2k-2}}{(1-t^k)^2} \qquad \lambda = \exp(2\pi i/k)$$

is a fruitful one: see [235], [228], [217] and the example of the generalized quaternion groups in section 6.7.

For the sake of future reference we extract the essential points about the invariants of the cyclic group $\mathbb{Z}/k$ contained in the preceding example.

EXAMPLE 3 : Consider the representation $\varrho : \mathbb{Z}/k \hookrightarrow \mathrm{GL}(2, \mathbb{C})$ given by the matrix

$$T = \begin{bmatrix} \lambda & 0 \\ 0 & \lambda^{-1} \end{bmatrix}$$

where $\lambda = \exp(2\pi i/k)$. For $k = 4$ and $\mathbb{F} = \mathbb{C}$ this is conjugate to the representation of section 3.3 example 1. In the preceding example we saw that

$$\mathbb{C}[x, y]^{\mathbb{Z}/k} = \bigoplus_{i=0}^{k-1} \mathbb{C}[x^k, y^k] \cdot (xy)^i,$$

and

$$P(\mathbb{C}[x,\ y]^{\mathbb{Z}/k},\ t) = \frac{1 + t^2 + \cdots + t^{2k-2}}{(1 - t^k)^2}\ .$$

From this we obtain easily that $\mathbb{C}[x,\ y]^{\mathbb{Z}/k}$ is generated as an algebra by the polynomials $f' = x^k,\ f = xy,\ f'' = y^k$ and

$$\mathbb{C}[x,\ y]^{\mathbb{Z}/k} = \mathbb{C}[f'\ f,\ f'']/(f^k - f'f'')$$

EXAMPLE 4 : Consider the permutation representation τ of $\mathbb{Z}/2$ on $V = \mathbb{F}^2$ given by the matrix

$$T = \begin{bmatrix} 0 & 1 \\ 1 & 0 \end{bmatrix} \in \mathrm{GL}(2,\ \mathbb{F})\,.$$

For each integer $k \in \mathbb{N}$ we have the representation $k\tau$ (see section 1.5 example 2). Since the representations $k\tau$ are permutation representations, proposition 4.2.4 implies that the Poincaré series $P(\mathbb{F}[\underset{k}{\oplus} V]^{\mathbb{Z}/2},\ t)$ is independent of the field $\mathbb{F}$. Therefore to compute the Poincaré series we may suppose that the field $\mathbb{F}$ has characteristic different from 2. The matrix representing the non-trivial element of $\mathbb{Z}/2$ acting on $\underset{k}{\oplus} V$ is

$$T_k = \begin{bmatrix} T & 0 & \cdots & 0 \\ 0 & T & \cdots & 0 \\ \vdots & \vdots & \vdots & \vdots \\ 0 & \cdots & 0 & T \end{bmatrix} \in \mathrm{GL}(2k,\ \mathbb{F})$$

and $\det(I - tT_k) = (1 - t^2)^k$. Therefore Molien's theorem gives after substitution and a short computation

$$\begin{aligned}
P(\mathbb{F}[\underset{k}{\oplus} V]^{\mathbb{Z}/2},\ t) &= \frac{1}{2}\left[\frac{1}{\det(I - tI)} + \frac{1}{\det(I - tT_k)} \right] \\
&= \frac{1}{2}\left[\frac{1}{(1 - t)^{2k}} + \frac{1}{(1 - t)^k(1 + t)^k} \right] \\
&= \frac{1}{2}\left[\frac{(1 + t)^k + (1 - t)^k}{(1 - t)^k(1 - t^2)^k} \right]\,.
\end{aligned}$$

For example, when $k = 2$ we find

$$P(\mathbb{F}[V \oplus V]^{\mathbb{Z}/2},\ t) = \frac{1}{2}\left[\frac{(1 + t)^2 + (1 - t)^2}{(1 - t)^2(1 - t^2)^2} \right] = \frac{1 + t^2}{(1 - t)^2(1 - t^2)^2}$$

which justifies the remark in section 1.5 example 2 that $\mathbb{F}[V \oplus V]^{\mathbb{Z}/2}$ is generated by the five polynomials $x' + y',\ x'y',\ x'' + y'',\ x''y'',\ x'x'' + y'y''$. For, the Poincaré series of the subalgebra A^* of $\mathbb{F}[x', y', x'', y'']$ generated by these five elements is isomorphic to the direct sum

$$\mathbb{F}[x'+y', x'y', x''+y'', x''y''] \oplus \mathbb{F}[x'+y', x'y', x''+y'', x''y''] \cdot (x'x''+y'y''),$$

as may be seen directly, or by applying theorem 6.7.8. Therefore the Poincaré series of this subalgebra is also

$$P(A^*, t) = \frac{1+t^2}{(1-t)^2(1-t^2)^2}$$

and hence $A^* = \mathbb{F}[x', y', x'', y'']^{\mathbb{Z}/4}$.

EXAMPLE 5 (T. Molien): Consider the group $\mathcal{I}$ of rotations of a regular icosahedron embedded in $\mathbb{R}^3$ centered at the origin. (See for example [97] chapter 2.) The icosahedron has 20 faces, each one an equilateral triangle. It has 12 vertices and 30 edges. The group $\mathcal{I} \subset \mathsf{SO}(3) \subset \mathrm{GL}(3, \mathbb{R})$ has order 60 and is isomorphic to the alternating group A_5. Every element T of $\mathcal{I}$ is a rotation, so has a polar axis L_T by Euler's theorem ([97] theorem 2.3.1) and an angle of rotation θ_T in the plane orthogonal to the polar axis. The points of intersection of L_T with the unit sphere of $\mathbb{R}^3$ are called the **poles** of T. The polar axes of the icosahedron consist of the 6 lines joining opposite vertices, 15 lines joining midpoints of opposite edges and 10 lines joining the centers of opposite faces. Thus there are $6 + 15 + 10 = 31$ polar axes and 62 poles. The rotations about the lines joining opposite vertices are through an angle $2\pi k/5$ where $k = 1, 2, 3, 4$, those about polar axes joining midpoints of opposite edges through an angle π and those about a polar axis through the centers of opposite sides are rotations through an angle $2\pi k/3$ where $k = 1, 2$. Since

$$1 + 6 \cdot 4 + 15 \cdot 1 + 10 \cdot 2 = 60$$

this accounts for all the elements of $\mathcal{I}$. The following table will allow us to compute the Poincaré series $P(\mathbb{R}[x, y, z]^{\mathcal{I}}, t)$ with the aid of Molien's theorem. In the table $\zeta = \exp(2\pi i/3)$ and $\omega = \exp(2\pi i/5)$.

Substituting from this table into the formula of Molien gives for the Poincaré series:

$$P(\mathbb{R}[x, y, z]^{\mathcal{I}}, t) =$$

$$\frac{1}{60} \left[\frac{1}{(1-t)^3} + \frac{15}{(1-t)(1-t^2)} + \frac{20}{1-t^3} + 6 \sum_{i=1}^{4} \frac{1}{(1-\omega^i t)(1-\omega^{-i} t)(1-t)} \right].$$

The sum[4]

$$\sum_{i=1}^{4} \frac{1}{(1-\omega^i t)(1-\omega^{-i} t)(1-t)} = \frac{1}{1-t} \sum_{i=1}^{4} \frac{1}{(1-\omega^i t)(1-\omega^{-i} t)}$$

---

[4] Molien, by the way, gives no clue how he made this calculation.

ORDER OF ROTATION	DIAGONAL FORM OF THE MATRIX	det(1 $-$ gt)	NUMBER OF TERMS
1	$\begin{bmatrix} 1 & 0 & 0 \\ 0 & 1 & 0 \\ 0 & 0 & 1 \end{bmatrix}$	$(1 - t)^3$	1
2	$\begin{bmatrix} 0 & 1 & 0 \\ 1 & 0 & 0 \\ 0 & 0 & 1 \end{bmatrix}$	$(1 - t)(1 - t^2)$	15
3	$\begin{bmatrix} \zeta^{\pm 1} & 0 & 0 \\ 0 & \zeta^{\mp 1} & 0 \\ 0 & 0 & 1 \end{bmatrix}$	$1 - t^3$	20
5	$\begin{bmatrix} \omega^i & 0 & 0 \\ 0 & \omega^{-i} & 0 \\ 0 & 0 & 1 \end{bmatrix}$ $i = 1, 2, 3, 4$	$(1 - (\omega^i + \omega^{-i})t + t^2)(1 - t)$ $i = 1, 2, 3, 4$	24

Table 4.3.3: The icosahedral group $\mathcal{I}$.

may be evaluated with the aid of (♠) (see example 2). We find

$$\sum_{i=1}^{4} \frac{1}{(1 - \omega^i t)(1 - \omega^{-i} t)} = \sum_{i=0}^{4} \frac{1}{(1 - \omega^i t)(1 - \omega^{-i} t)} - \frac{1}{(1 - t)^2}$$

$$= 5 \frac{1 + t^2 + t^4 + t^6 + t^8}{(1 - t^5)^2} - \frac{1}{(1 - t)^2}.$$

Combining these results and simplifying gives

$$P(\mathbb{R}[x, \ y, \ z]^{\mathcal{I}}, \ t) = \frac{1}{60} \left[\frac{15}{(1-t)(1-t^2)} + \frac{20}{1-t^3} + 30 \frac{1+t^2+t^4+t^6}{(1-t^4+t^8)^2(1-t)} - \frac{6}{(1-t)^3} \right]$$

$$= \frac{1 + t^{15}}{(1 - t^2)(1 - t^6)(1 - t^{10})}.$$

This formula suggests that $\mathbb{R}[x, \ y, \ z]^{\mathcal{I}}$ contains a polynomial subalgebra generated by three polynomials f, h, k of degrees 2, 6, 10, and a polynomial r of degree 15, such that

$$\mathbb{R}[x, \ y, \ z]^{\mathcal{I}} = \mathbb{R}[f, \ h, \ k] \oplus \mathbb{R}[f, \ h, \ k] \cdot r.$$

This is indeed the case, and suitable polynomials f, h, k and r may be found as follows.

The group $\mathcal{I}$ acts by rotations of $\mathbb{R}^3$ so preserves the inner product, and hence for suitably chosen x, y, z the quadratic polynomial $f = x^2 + y^2 + z^2$ is invariant. The set of polar axes $\mathcal{L}$ of $\mathcal{I}$ is permuted by the action of $\mathcal{I}$ on $\mathbb{R}^3$. Each polar axis is represented by a linear functional $\ell : \mathbb{R}^3 \longrightarrow \mathbb{R}$ whose kernel is the plane through the origin orthogonal to the polar axis. The action of $\mathcal{I}$ on $\mathcal{L}$ separates $\mathcal{L}$ into orbits. Two polar axes ℓ' and ℓ'' belong to the same orbit if and only if the corresponding angles of rotation $\theta_{\ell'}$ and $\theta_{\ell''}$ are the same. Thus there are three orbits $\mathcal{A}$, $\mathcal{B}$, $\mathcal{C} \subset \mathcal{L}$ of the $\mathcal{I}$ action on $\mathcal{L}$ consisting of 6, 10 and 15 axes corresponding to the rotations of order 5, 3 and 2 respectively. The top Chern class of each orbit provides polynomials $h = c_6(\mathcal{A})$, $k = c_{10}(\mathcal{B})$ and $r = c_{15}(\mathcal{C})$ suitable to the above description of $\mathbb{R}[x, y, z]^{\mathcal{I}}$.

The group $\mathcal{I}$ is the rotation subgroup of the real crystallographic group (see chapter 7) $\mathbf{I}_3$ which is group number 23 in the Shephard-Todd list (see section 7.3). The invariants of $\mathbf{I}_3$ are a polynomial algebra on the elements f, h and k and $\mathbb{R}[x, y, z]^{\mathcal{I}}$ is the degree two extension of $\mathbb{R}[x, y, z]^{\mathbf{I}_3} = \mathbb{R}[f, h, k]$ by a square root of an invariant R of degree 30 (see section 7.6 corollary 7.6.8).

4.4   Reflecting Hyperplanes

Let $\mathbb{F}$ be a field of characteristic zero and V a faithful representation of G of dimension n. Then the formula of Molien gives

$$P(\mathbb{F}[V]^G, t) = \frac{1}{|G|} \sum_{g \in G} \frac{1}{\det(1 - g^{-1}t)}$$

$$= \frac{1}{|G|} \left\{ \frac{1}{\det(1 - t)} + \sum_{g \neq 1 \in G} \frac{1}{\det(1 - g^{-1}t)} \right\}.$$

None of the group elements $g \neq 1 \in G$ can have 1 as an eigenvalue of order n and therefore

$$\sum_{g \neq 1 \in G} \frac{1}{\det(1 - g^{-1}t)}$$

has a pole of order at most $n - 1$ at $t = 1$. Thus the value at $t = 1$ of the holomorphic function $(1 - t)^n P(\mathbb{F}[V]^G, t)$ is equal to $1/|G|$.

We can obtain further information about the coefficients in the Laurent expansion of $P(\mathbb{F}[V]^G, t)$ by the same method. Namely, once the contribution from the identity element has been subtracted, the remaining expression has a pole of order at most $n - 1$ at $t = 1$. An element $g \in G$ contributes to this pole if and only if $\dim_{\mathbb{F}}(V^g) = n - 1$. An element $1 \neq g \in G$ with this property is called a **pseudoreflection**. So a pseudoreflection[5] has eigenvalue 1 with multiplicity $n - 1$, and one other eigenvalue, say λ. A special case is when $\lambda = -1$, in which case we speak of a **real** reflection or just a reflection. Let $s(G)$ denote the set of pseudoreflections in G. The Poincaré series may then be written in the form

$$\frac{1}{|G|} \left\{ \frac{1}{(1-t)^n} + \sum_{g \neq 1 \in s(G)} \frac{1}{\det(1 - g^{-1}t)} + \cdots \right\}.$$

For a pseudoreflection we have

$$\frac{1}{\det(1 - g^{-1}t)} = \frac{1}{(1-t)^{n-1}(1 - \lambda^{-1}t)}$$

and after multiplying by $(1 - t)^{n-1}$, the value at $t = 1$ is $1/(1 - \lambda^{-1})$. Next, we observe that if g is a pseudoreflection so is g^{-1}, and $g = g^{-1}$ precisely when g is a real reflection. The identity

$$\frac{1}{1 - \lambda^{-1}} + \frac{1}{1 - \lambda} = \frac{(1 - \lambda) + (1 - \lambda^{-1})}{1 - \lambda^{-1} - \lambda + 1} = 1.$$

shows that if we pair together each non-real pseudoreflection with its inverse, the total contribution from the non-real pseudoreflections is equal to one half of their number. Real reflections have $\lambda = -1$, so that their contribution is also one half their number. Thus we have shown:

PROPOSITION 4.4.1 : *Let* $\mathbb{F}$ *be a field of characteristic zero and* $\varrho :$ $G \hookrightarrow \mathrm{GL}(n, \mathbb{F})$ *a finite dimensional representation of* G *on* $V = \mathbb{F}^n$, *then the Laurent expansion of the Poincaré series* $P(\mathbb{F}[V]^G, t)$ *begins as follows:*

$$P(\mathbb{F}[V]^G, t) = \frac{\frac{1}{|G|}}{(1-t)^n} + \frac{\frac{|s(G)|}{2 \cdot |G|}}{(1-t)^{n-1}} + \cdots$$

where $|s(G)|$ *is the number of pseudoreflections of* G *in the representation* ϱ. $\square$

The connection expressed in 4.4.1 between the coefficients of the Poincaré series of the ring of invariants and the structure of the representation $\varrho :$ $G \hookrightarrow \mathrm{GL}(n, \mathbb{F})$ leads to a number of unexpected conclusions. For example, the following result will be of use in section 7.4 (see (7.4.2)).

---

[5] See section 7.1 for a discussion of pseudoreflections.

PROPOSITION 4.4.2 : *Let $\varrho : G \hookrightarrow \mathrm{GL}(n,\ \mathbb{F})$ be a representation of a finite group G over a field $\mathbb{F}$ of characteristic prime to the order of G. Suppose $f_1, \ldots, f_n \in \mathbb{F}[V]^G$, $\deg(f_i) = d_i$ for $i = 1, \ldots, n$, satisfy*

(i) $\displaystyle\prod_{i=1}^{n} d_i = |G|$,

(ii) $f_1, \ldots, f_n$ *are algebraically independent.*

Then $|s(G)| \geq \displaystyle\sum_{i=1}^{n}(d_i - 1)$.

PROOF : By 4.4.1 we have

$$P(\mathbb{F}[V]^G, t) = \frac{\frac{1}{|G|}}{(1-t)^n} + \frac{\frac{|s(G)|}{2 \cdot |G|}}{(1-t)^{n-1}} + \cdots$$

and by direct computation (see the discussion following the proof)

$$P(\mathbb{F}[f_1, \ldots, f_n], t) = \frac{\frac{1}{d_1 \cdots d_n}}{(1-t)^n} + \frac{\frac{\sum_{i=1}^{n}(d_i - 1)}{2 \cdot d_1 \cdots d_n}}{(1-t)^{n-1}} + \cdots.$$

We are going to treat t as a complex variable and do some analysis with these functions.

Both $P(\mathbb{F}[V]^G,\ t)$ and $P(\mathbb{F}[f_1, \ldots, f_n],\ t)$ have a pole of order n at $t = 1$, and by (i)

$$\lim_{t \to 1}(1 - t)^n P(\mathbb{F}[V]^G,\ t) = \frac{1}{|G|} = \frac{1}{d_1 \cdots d_n} = \lim_{t \to 1}(1 - t)^n P(\mathbb{F}[f_1, \ldots, f_n],\ t).$$

Therefore the difference

$$\Delta(t) = P(\mathbb{F}[V]^G,\ t) - P(\mathbb{F}[f_1, \ldots, f_n],\ t)$$

has a pole of order $n - 1$ at $t = 1$. Since $\mathbb{F}[f_1, \ldots, f_n] \subseteq \mathbb{F}[V]^G$ the Poincaré series of $\mathbb{F}[V]^G$ dominates coefficient for coefficient the Poincaré series of $\mathbb{F}[f_1, \ldots, f_n]$. Hence $\Delta(t)$ is a power series with non-negative integral coefficients and therefore takes on non-negative real values for t real and near 1. The same is therefore true of the function $(1-t)^{n-1}\Delta(t)$ which is holomorphic at $t = 1$. Evaluating at $t = 1$ gives

$$0 \leq \lim_{t \to 1}\left[(1 - t)^{n-1}\Delta(t)\right] = \left[(1 - t)^{n-1}\Delta(t)\right]\Big|_{t=1}$$

$$= \frac{|s(G)|}{2|G|} - \frac{\sum_{i=1}^{n}(d_i - 1)}{2 \cdot d_1 \cdots d_n} = \frac{1}{2|G|}\left(|s(G)| - \sum_{i=1}^{n}(d_i - 1)\right)$$

as desired. □

We can derive from proposition 4.4.1 some rather remarkable facts about the case of a faithful representation $G \hookrightarrow GL(n, \mathbb{F})$ for which the ring of invariants $\mathbb{F}[V]^G$ is itself again a polynomial algebra. For suppose

$$\mathbb{F}[V]^G \cong \mathbb{F}[f_1, \ldots, f_n] \quad \deg(f_i) = d_i, \quad i = 1, \ldots, n.$$

Then

$$P(\mathbb{F}[V]^G, t) = \left(\frac{1}{1 - t^{d_1}}\right) \cdots \left(\frac{1}{1 - t^{d_n}}\right).$$

Let us expand this in a Laurent series about $t = 1$. First note

$$\prod_{i=1}^{n} \frac{1}{1 - t^{d_i}} = \left(\frac{1}{1 - t}\right)^n \prod_{i=1}^{n} \frac{1}{1 + t + t^2 + \cdots + t^{d_i - 1}}$$

and

$$\prod_{i=1}^{n} \frac{1}{1 + t + t^2 + \cdots + t^{d_i - 1}} \bigg|_{t=1} = \prod_{i=1}^{n} \frac{1}{d_i}$$

and therefore

$$P(\mathbb{F}[V]^G, t) = \frac{\frac{1}{d_1 \cdots d_n}}{(1 - t)^n} + \cdots.$$

Equating the leading coefficient of this computation of the Laurent series with the one given by 4.4.1 we obtain $d_1 \cdots d_n = |G|$.

The next step is to compute the coefficient of $1/(1 - t)^{n-1}$. To this end note the identity

$$\frac{1}{1 - t^d} - \frac{1/d}{1 - t} = \frac{1}{d}\left[\frac{d}{1 - t^d} - \frac{1}{1 - t}\right]$$

$$= \frac{1}{d}\left[\frac{d - (1 + t + \cdots + t^{d-1})}{1 - t^d}\right]$$

$$= \frac{1}{d}\left[\frac{1 - 1}{1 - t^d} + \frac{1 - t}{1 - t^d} + \cdots + \frac{1 - t^{d-1}}{1 - t^d}\right]$$

$$= \frac{1}{d}\left[0 + \frac{1}{1 + t + \cdots + t^{d-1}} + \cdots + \frac{1 + t + \cdots + t^{d-2}}{1 + t + \cdots + t^{d-1}}\right]$$

so evaluating at $t = 1$ yields

$$\left[\frac{1}{1 - t^d} - \frac{1/d}{1 - t}\right]_{t=1} = \frac{d - 1}{2d}$$

from which we conclude

$$\prod_{i=1}^{n} \frac{1}{1-t^{d_i}} = \frac{\frac{1}{d_1 \cdots d_n}}{(1-t)^n} + \frac{\frac{1}{2 \cdot d_1 \cdots d_n} \sum_{i=1}^{n}(d_i - 1)}{(1-t)^{n-1}} + \cdots .$$

Equating this expression for $P(\mathbb{F}[V]^G, t)$ with 4.4.1, the one derived from Molien's theorem, we obtain:

PROPOSITION 4.4.3 : *Suppose that $\mathbb{F}$ is a field of characteristic zero and $G \hookrightarrow \mathrm{GL}(n, \mathbb{F})$ is a faithful representation of a finite group G. If*

$$\mathbb{F}[V]^G \cong \mathbb{F}[f_1, \ldots, f_n] \quad \deg(f_i) = d_i, \quad i = 1, \ldots, n,$$

then

$$|G| = d_1 \cdots d_n$$

$$|s(G)| = \sum_{i=1}^{n}(d_i - 1)$$

where $|s(G)|$ is the number of pseudoreflections in G. □

The question of when the ring of invariants of a finite group acting on a polynomial algebra is again a polynomial algebra has been only partially answered, namely in the non-modular case (i.e. when $|G| \in \mathbb{F}$ is a unit) [206], [75] and the purely modular case [161] when G is a p-group and the field $\mathbb{F}$ is the Galois field with p elements. We will examine pseudoreflection groups in more detail in chapters 7, 8 and 10.

EXAMPLE 1 : Consider the subgroup of $\mathrm{GL}(n, \mathbb{R})$ generated by the two matrices

$$\frac{1}{\sqrt{2}}\begin{bmatrix} 1 & -2 \\ 2 & 1 \end{bmatrix} \qquad \begin{bmatrix} 1 & 0 \\ 0 & -1 \end{bmatrix}.$$

Both of these are reflections and their product is a rotation through an angle of $2\pi/8$ so the group is D_{16}, the dihedral group of order 16 represented as the full symmetry group of a regular octagon in the plane. The group contains 8 reflections and 7 rotations (we don't count the identity as a rotation). From the discussion of example 2 in section 4.3 we know that $\mathbb{R}[x, y]^{D_{16}}$ has the Poincaré series of a polynomial algebra on two generators f_1 of degree 2 and f_2 of degree 8. Explicit generators may be found by noting that the eight vertices of the regular octagon

$$\pm(1, \ 0), \ \pm(0, \ 1), \ \pm(1\sqrt{2}, 1/\sqrt{2}), \ \pm(-1/\sqrt{2}, \ 1/\sqrt{2})$$

are an orbit E of the action. The Chern classes

$$c_2(E) = x^2 + y^2$$

$$c_8(E) = x^2 y^2 (x^2 - y^2)^2$$

are algebraically independent and may be chosen as f_1, f_2. Thus

$$\mathbb{R}[x,\ y]^{D_{16}} = \mathbb{R}[x^2 + y^2,\ x^2 y^2 (x^2 - y^2)^2].$$

Note $8 = 1 + 7 = (\deg(f_1) - 1) + (\deg(f_2) - 1)$ as predicted by theorem 4.4.3.

Chapter 5
Dimension Theoretic Properties of Rings of Invariants

In the previous chapters we have learned some of the basic finiteness properties of rings of polynomial invariants of finite groups: the finiteness of the number of generators and bounds for their number and degrees. In addition to the number of generators there are other measures of finiteness. For example, in a Noetherian ring A every ascending chain of ideals must eventually stabilize: but for a given ring A is there an upper bound on the length of such a sequence? For arbitrary ideals the answer must be no, as for $x \in A$ one has for each integer k the chain of ideals

$$(x^{2^k}) \subset (x^{2^{k-1}}) \subset \cdots \subset (x^2) \subset (x) \subset A$$

so if x is not nilpotent this yields an ascending chain of arbitrary length. By restricting our attention to prime ideals however we receive a useful invariant called the **Krull dimension** of A. This is the first of the further finiteness properties we investigate.

5.1 Review of some Graded Commutative Algebra

We begin by reviewing the ideal theory of graded commutative Noetherian rings. For further details we refer to [21] and [266], Volume II, especially pages 149 - 154.

We will be concerned with a finitely generated, connected, commutative positively graded[1] algebra over a field $\mathbb{F}$. If A is such an algebra then choosing a finite set of generators $a_1, \ldots, a_n$ for A we receive a surjective map

$$\mathbb{F}[x_1, \ldots, x_n] \xrightarrow{\alpha} A \qquad \deg(x_i) = \deg(a_i) \quad \text{for } i = 1, \ldots, n$$

by requiring $\alpha(x_i) = a_i$ for $i = 1, \ldots, n$. Since a graded polynomial algebra is Noetherian it follows that A is Noetherian.

---

[1] Throughout this chapter, unless explicitly stated to the contrary the word graded means positively graded.

An ideal $\mathfrak{m}$ is **maximal** if it is not contained in any proper ideal of A. Since A is connected and positively graded it has a unique maximal ideal consisting of all the elements of A of strictly positive degree. An ideal I of A is **prime** if whenever $a, b \in A$ and $a \cdot b \in I$ then at least one of the elements a or b belongs to I. An ideal I of A is a **primary** ideal if whenever $a, b \in A$ with $a \cdot b \in I$, and a not in I, then some positive power of b belongs to I. The **radical** of an ideal I denoted by $\sqrt{I}$, is the ideal of all $a \in A$ such that some positive power of a lies in I. If $\mathfrak{q}$ is primary then $\sqrt{\mathfrak{q}}$ is prime.

There are simple characterizations of the structure of ideals in terms of the structure of the ring of quotients by the ideal. For example:
 — An ideal $\mathfrak{m} \subset A$ is maximal if and only if $A/\mathfrak{m}$ is a field.
 — An ideal $\mathfrak{p} \subset A$ is prime if and only if $A/\mathfrak{p}$ is an integral domain.
 — An ideal $\mathfrak{q} \subset A$ is a primary ideal if and only if every zero divisor in $A/\mathfrak{q}$ is nilpotent.
If I is any ideal of A then the Lasker-Noether theorem [266], volume I chapter IV § 4, says there are a finite number of primary ideals $\mathfrak{q}_1, \ldots, \mathfrak{q}_n$ such that
 — $I = \mathfrak{q}_1 \cap \cdots \cap \mathfrak{q}_n$
 — no $\mathfrak{q}_i$ contains $\bigcap_{i \neq j} \mathfrak{q}_j$
 — if $j \neq i$ then $\sqrt{\mathfrak{q}_i} \neq \sqrt{\mathfrak{q}_j}$.
Such a representation of I as the intersection of primary ideals is called an **irredundant primary decomposition** of I. If

$$\mathfrak{q}'_1 \cap \cdots \cap \mathfrak{q}'_{n'} = I = \mathfrak{q}''_1 \cap \cdots \cap \mathfrak{q}''_{n''}$$

are two irredundant primary decompositions of I then $n' = n''$ and, after reordering, $\sqrt{\mathfrak{q}'_i} = \sqrt{\mathfrak{q}''_i}$. The prime ideals $\mathfrak{p}_i = \sqrt{\mathfrak{q}_i}$ are thus independent of the choice of irredundant primary decomposition of I and are called the **associated primes** of I. A minimal associated prime[2] of I is called an **isolated prime** of I. The isolated primes of I are the minimal prime ideals among those that include I, and $\sqrt{I}$ is the intersection of the isolated primes of I. The associated primes of an ideal I in A are uniquely determined by I as follows: $\mathfrak{p}$ is an associated prime of I if and only if $\mathfrak{p}$ is the radical of the annihilator ideal $\mathrm{Ann}_A(x)$ of some element $x \in A/I$.

We also recall some basic facts about ring extensions. (A good reference for information about integral ring extensions is [266] volume 1 chapter V.)

DEFINITION : *Let $A \supset B$ be an extension of graded connected commutative algebras over a field. We say that A **is finitely generated over** B if there are elements $x_1, \ldots, x_n \in A$ such that every element of A is expressible as a polynomial in $x_1, \ldots, x_n$ with coefficients in B. We say A **is finite over** B if A is finitely generated as a B-module. The extension $A \supset B$ is called*

---

[2] I.e., an associated prime of I containing no other associated prime of I.

integral if every element of A is the root of a monic polynomial with coefficients in B. B is called **integrally closed in** A if every element of A that is integral over B belongs to B. An integral domain D that is integrally closed in its own field of fractions is called **normal**.

PROPOSITION 5.1.1 : Let A be finitely generated over B. Then A is finite over B if and only if A is integral over B.

PROOF : Let A be finite over B and $x_1, \ldots, x_n \in A$ a set of B-module generators for A. If $x \in A$ then we may write

$$x \cdot x_i = \sum_{j=1}^{n} b_{i,j} x_j \qquad i = 1, \ldots, n$$

which we may choose to rewrite in the matrix form

$$(b - xI) \begin{bmatrix} x_1 \\ \vdots \\ x_n \end{bmatrix} = \begin{bmatrix} 0 \\ \vdots \\ 0 \end{bmatrix},$$

where $b = (b_{i,j})$ and I is the $n \times n$ identity matrix. Let C be the matrix obtained from $(b - xI)$ by replacing the j-th column with a column of zeros. Then Cramer's rule (see page 157 footnote number 3) says

$$\det(b - xI)x_j = \det(C) = 0,$$

so $\det(b - xI)$ must annihilate $x_1, \ldots, x_n$ and hence also $1 \in A$, which implies $\det(b - xI) = 0$, and expanding out the determinant gives a monic polynomial equation with x as a root.

Conversely suppose $A \supset B$ is an integral extension and choose generators $x_1, \ldots, x_n$ for A as an algebra over B. Each of these generators is integral over B so there are integers $k_1, \ldots, k_n$ and elements $b_{i,j} \in B$, $i = 1, \ldots, n$, $j = 0, \ldots, k_n$ such that

$$x_i^{k_i+1} = \sum_{j=0}^{k_i} b_{i,j} x_i^j \qquad i = 1, \ldots, n.$$

Hence any polynomial in $x_1, \ldots, x_n$ may be expressed as a polynomial of degree at most k_i in the each x_i, so the monomials $x_1^{e_1} \cdots x_n^{e_n}$ where $0 \leq e_i \leq k_i$ and $i = 1, \ldots, n$ generate A as a B-module. $\square$

We made no use of the grading in the proof, so this result is also valid for ungraded commutative rings.

5.2 Krull Dimension

The basic finiteness theorem of Hilbert-Noether 2.3.1 says that the ring of invariants of a finite group is finitely generated and the ambient polynomial algebra is a finite integral extension of the ring of invariants. We exploit this relationship to relate the ideal theory of the ring of invariants to that of the ambient polynomial ring.

We begin with a definition motivated in part by the needs of algebraic geometry to assign a *dimension* to abstract algebraic varieties and a *multiplicity* to their intersection. (See for example [171].)

DEFINITION : *The **Krull dimension** of a graded connected commutative algebra over a field $\mathbb{F}$ is the maximum length k of a chain of proper inclusions of prime ideals $\mathfrak{p}_0 \subset \mathfrak{p}_1 \subset \cdots \subset \mathfrak{p}_k$, or ∞ if there are such chains of unbounded length. If $\mathfrak{p}$ is a prime ideal of A the **height** of $\mathfrak{p}$ is the largest integer h such that there exists a chain of proper inclusions of prime ideals $\mathfrak{p}_0 \subset \mathfrak{p}_1 \subset \cdots \subset \mathfrak{p}_h = \mathfrak{p}$, the **depth** of $\mathfrak{p}$ is the largest integer d such that there exists a chain of proper inclusions of prime ideals $\mathfrak{p}_0 \supset \mathfrak{p}_1 \supset \cdots \supset \mathfrak{p}_d = \mathfrak{p}$.*

We write $\dim(A)$ with no subscript to denote the Krull dimension of A, and $\mathrm{ht}(\mathfrak{p})$ for the height of a prime ideal $\mathfrak{p}$ and $\mathrm{dp}(\mathfrak{p})$ for its depth. To avoid confusion with the notation for the dimension of a vector space we always add the field as a subscript, viz. $\dim_{\mathbb{F}}(V)$, to denote the dimension of the vector V space over the field $\mathbb{F}$.

The height and depth are measures of where in the lattice of prime ideals ordered under inclusions the ideal $\mathfrak{p}$ is located relative to the bottom and the top of the lattice. It may be helpful to recall that (0) is a prime ideal in A if and only if A is an integral domain. A prime ideal is called **minimal** if it is proper and contains no other prime ideal. Thus (0) is a minimal prime only in an integral domain, otherwise the minimal primes are the isolated primes of (0) and are minimal elements of the lattice of prime ideals.

If A is a graded connected algebra then the **augmentation ideal** of A, denoted by $\bar{A}$ or IA, is the ideal of elements of strictly positive degree. The ideal $\bar{A}$ is the unique maximal *graded* ideal in A, and in this sense A is a local ring. The height of $\bar{A}$ is the Krull dimension of A and $\bar{A}$ is at the top of the lattice of prime ideals of A ordered under inclusion.

Our first goal is to compute the Krull dimension of polynomial algebras and use this to obtain an upper bound for the Krull dimension of an arbitrary graded commutative finitely generated $\mathbb{F}$-algebra.

LEMMA 5.2.1 : *If A is an integral domain and $I \subset A[x]$ a prime ideal in the graded polynomial algebra over A, then $I = (J)$ or (J, x), where $J = A \cap I$.*

PROOF : Consider the quotient ring A/J, which is an integral domain since $J \subset A$ is a prime ideal. Let $FF(A/J)$ denote the field of fractions of A/J and consider $I/J \subset FF(A/J)[x]$. Since $FF(A/J)$ is a field, the ideal $I/J \subset FF(A/J)[x]$ is principal. Since it is also a graded prime ideal in a graded polynomial lagebra it is either (0) or (x) and the result follows. $\square$

PROPOSITION 5.2.2 : *The Krull dimension of* $\mathbb{F}[V]$ *is equal to* $\dim_{\mathbb{F}}(V)$.

PROOF : If $\dim_{\mathbb{F}}(V) = 0$ there is nothing to prove, so we may proceed by induction over $\dim_{\mathbb{F}}(V)$. Suppose $x_1, \ldots, x_n$ is a basis for V^* and

$$\mathfrak{p}_0 \subset \mathfrak{p}_1 \subset \cdots \subset \mathfrak{p}_d \subset \mathbb{F}[x_1, \ldots, x_n]$$

is an ascending chain of proper inclusions of prime ideals. Set $\mathfrak{p}_i' = \mathfrak{p}_i \cap \mathbb{F}[x_1, \ldots, x_{n-1}]$, $i = 1, \ldots, d$. If

$$\mathfrak{p}_0' \subset \mathfrak{p}_1' \subset \cdots \subset \mathfrak{p}_d' \subset \mathbb{F}[x_1, \ldots, x_{n-1}]$$

is a proper chain then by the induction hypothesis $d \leq n-1$ and we are done. If the chain is not proper, then there is a smallest integer j such that $\mathfrak{p}_j' = \mathfrak{p}_{j+1}' = \cdots = \mathfrak{p}_d'$. By the preceding lemma $\mathfrak{p}_j = (\mathfrak{p}_j')$ or $(\mathfrak{p}_j', x)$, $\mathfrak{p}_{j+1} = (\mathfrak{p}_{j+1}')$ or $(\mathfrak{p}_{j+1}', x), \ldots, \mathfrak{p}_d = (\mathfrak{p}_d')$ or $(\mathfrak{p}_d', x)$. Hence there are at most two distinct ideals in this list. Since $j \leq n-1$ by induction, $\dim(\mathbb{F}[x_1, \ldots, x_n]) \leq n$. The ascending chain of prime ideals

$$0 \subset (x_1) \subset (x_1, x_2) \subset \cdots \subset (x_1, \ldots, x_n)$$

has length n, so $\dim(\mathbb{F}[x_1, \ldots, x_n]) \geq n$ and this completes the proof. $\square$

DEFINITION : *If* A *is a graded connected algebra over a field* $\mathbb{F}$ *and* M *is a graded* A-*module, then* $QM = M/(\bar{A} \cdot M)$ *is the module of* **indecomposable** *elements of* M. *The indecomposable elements of the algebra* A, *denoted by* QA, *are by definition* $Q(\bar{A})$, *where* $\bar{A}$ *is regarded as a module over* A.

If A is a connected algebra over $\mathbb{F}$ then $A/\bar{A} \cong \mathbb{F}$ and it is easy to see that $QM \cong \mathbb{F} \otimes_A M$. Hence $Q(M)$ is just a graded vector space over $\mathbb{F}$.

PROPOSITION 5.2.3 (Nakayama's Lemma (graded analog)): *If* A *is a connected commutative algebra over a field* $\mathbb{F}$ *and* M *is a positively graded* A-*module, then* $M = 0$ *if and only if* $QM = 0$.

PROOF : If $M = 0$ then $QM = 0$. Suppose on the other hand that $QM = 0$. Since M is positively graded $M_i = 0$ for $i < 0$, so we may suppose inductively that we have shown that $M_i = 0$ for $i < s$. If $x \in M_s$ then since $QM = M/(\bar{A} \cdot M) = 0$ it follows that $x = \sum a_r \cdot x_r$ with $a_j \in \bar{A}$. Hence $\deg(a_j) > 0$ and therefore $\deg(x_j) < s$, so $x_j = 0$ and hence $x = 0$. $\square$

COROLLARY 5.2.4 : *If A is a graded connected commutative algebra and M', M'' are positively graded A-modules, then an A-module morphism $f : M' \longrightarrow M''$ is surjective if and only if $Qf : QM' \longrightarrow QM''$ is surjective.*

PROOF : This follows immediately from the fact that the functor $\mathbb{F} \otimes_A -$ sends epimorphisms to epimorphisms and 5.2.3. $\square$

COROLLARY 5.2.5 : *If A and M are as in the preceding proposition and $x_1, \ldots, x_n \in M$ then $x_1, \ldots, x_n$ generate M if and only if they project in QM to a spanning set.*

PROOF : Form a free A-module F generated by classes z_i of degree $\deg(x_i)$, $i = 1, \ldots, n$, and define an A-module map $\zeta : F \longrightarrow M$ by requiring that $z_i \mapsto x_i$ for $i = 1, \ldots, n$. Now apply the preceding corollary. $\square$

Thus a basis for QM lifts to an A-module generating set for M, and M cannot be generated by fewer than $\dim_{\mathbb{F}}(QM)$ elements. Likewise, a basis for QA lifts to a minimal generating set for A as an algebra.

PROPOSITION 5.2.6 : *If A is a finitely generated commutative graded algebra over a field then $\dim(A) \leq \dim_{\mathbb{F}}(QA)$.*

PROOF : Choose a basis for QA, say $a_1, \ldots, a_n$ and introduce the polynomial algebra $\mathbb{F}[x_1, \ldots, x_n]$, where $\deg(x_i) = \deg(a_i)$ for $i = 1, \ldots, n$. There is the surjective map $\varphi : \mathbb{F}[x_1, \ldots, x_n] \longrightarrow A$ defined by requiring that $x_i \mapsto a_i$ for $i = 1, \ldots, n$. If

$$\mathfrak{p}_0 \subset \mathfrak{p}_1 \subset \cdots \subset \mathfrak{p}_m$$

is an ascending chain of proper inclusions of prime ideals in A then the pre-images

$$\varphi^{-1}(\mathfrak{p}_0) \subset \varphi^{-1}(\mathfrak{p}_1) \subset \cdots \subset \varphi^{-1}(\mathfrak{p}_m)$$

form a proper ascending chain of prime ideals in $\mathbb{F}[x_1, \ldots, x_n]$. Hence by 5.2.2, $m \leq n$. $\square$

If $\varrho : G \hookrightarrow \mathrm{GL}(n, \ F)$ is a representation of a finite group G then the ring of coinvariants $\mathbb{F}[V]_G$ is also the module of $\mathbb{F}[V]^G$-indecomposables of $\mathbb{F}[V]$. If $\sigma : \mathbb{F}[V]_G \longrightarrow \mathbb{F}[V]$ is a vector space splitting to the quotient map $\mathbb{F}[V] \longrightarrow \mathbb{F}[V]_G$ we obtain an $\mathbb{F}[V]^G$-module map

$$\mathbb{F}[V]^G \otimes \mathbb{F}[V]_G \longrightarrow \mathbb{F}[V]$$

extending the map $f \otimes [h] \mapsto f \cdot \sigma([h])$. By 5.2.4 this is an epimorphism. Therefore passing to quotient fields we see

$$\dim_{\mathbb{F}(V)^G}(\mathbb{F}(V)) \leq \dim_{\mathbb{F}}(\mathrm{Tot}(\mathbb{F}[V]_G)).$$

By 1.2.4 $\mathbb{F}(V) \supseteq \mathbb{F}(V)^G$ is Galois with Galois group G and hence

$$\dim_{\mathbb{F}(V)^G}(\mathbb{F}(V)) = |G|.$$

Combining these two computations gives:

PROPOSITION 5.2.7 : *Let $\varrho : G \hookrightarrow \mathrm{GL}(n, \mathbb{F})$ be a representation of a finite group G. Then $\dim_{\mathbb{F}}(\mathrm{Tot}(\mathbb{F}[V]_G)) \geq |G|$.* $\square$

In other words, the minimum number of generators of $\mathbb{F}[V]$ as an $\mathbb{F}[V]^G$ module is $|G|$. The following example shows that strict inequality can occur.

EXAMPLE 1 : Consider the representation $\varrho : \mathbb{Z}/2 \hookrightarrow \mathrm{GL}(4, \mathbb{F})$ of section 1.5 example 2. Then

$$\mathbb{F}[x', \, y', \, x'', \, y'']_{\mathbb{Z}/2} = \frac{\mathbb{F}[x', \, y', \, x'', \, y'']}{(x' + y', \, x'y', \, x'' + y'', \, x''y'', \, x'x'' + y'y'')}$$

If $\mathbb{F}$ is a field of characteristic not equal to 2 then the following picture describes $\mathbb{F}[x', \, y', \, x'', \, y'']_{\mathbb{Z}/2}$, where u is the residue class of x' and v the residue class of x''.

$$u \bullet \qquad\qquad\qquad\qquad \bullet\, v$$
$$\bullet\, 1$$

$$\mathbb{F}[x', y', \, x'', \, y'']_{\mathbb{Z}/2}$$

Hence $\dim(\mathrm{Tot}(\mathbb{F}[V]_{\mathbb{Z}/2})) = 3 > 2 = |\mathbb{Z}/2|$.

5.3   Noether Normalization

The theorem of Hilbert-Serre 4.1.1 gives the general form of the Poincaré series of a graded connected finitely generated algebra over $\mathbb{F}$ and suggests that such an algebra might contain a polynomial subalgebra over which it is finitely generated as a module. This is the content of the Noether Normalization Theorem. We require a pair of lemmas for the proof.

LEMMA 5.3.1 : *If an ideal I is contained in a finite union of prime ideals $\mathfrak{p}_1, \ldots, \mathfrak{p}_n$ then I is contained in some $\mathfrak{p}_i$.*

PROOF : Without loss of generality we may suppose that $n > 1$ and I is not contained in the union of fewer than n of the ideals $\mathfrak{p}_1, \ldots, \mathfrak{p}_n$. Choose $y \in I$ with $y \notin \mathfrak{p}_1 \cup \cdots \cup \mathfrak{p}_{n-1}$. Then $y \in \mathfrak{p}_n$. Choose $z \in I$ with $z \notin \mathfrak{p}_n$, and $t_i \in \mathfrak{p}_i$ with $t_i \notin \mathfrak{p}_n$ for each i between 1 and $n-1$. Then $y + zt_1 \cdots t_{n-1}$ is in I but is not in $\mathfrak{p}_1 \cup \cdots \cup \mathfrak{p}_n$. This contradiction proves the lemma. $\square$

REMARK : In the above proof, we only used the fact that $\mathfrak{p}_n$ was prime, and even this is only necessary if $n > 2$. So in fact it suffices to assume that all except possibly two of the $\mathfrak{p}_i$ are prime.

LEMMA 5.3.2 : *Let A be a finitely generated graded commutative algebra over a field $\mathbb{F}$ of Krull dimension d and*

$$\mathfrak{p}_0 \subset \mathfrak{p}_1 \subset \cdots \subset \mathfrak{p}_d$$

a chain of prime ideals in A of length d. Then for $i = 1, \ldots, d$ there exist elements $a_i \in \mathfrak{p}_i \smallsetminus \mathfrak{p}_{i-1}$ with $\mathfrak{p}_i$ an isolated prime of $(a_0, \ldots, a_i)$, where $a_0 = 0$.

PROOF : $\mathfrak{p}_0$ is an isolated prime of (0) and by Lasker-Noether $\mathfrak{p}_1$ cannot be. Therefore by 5.3.1 $\mathfrak{p}_1$ is not included in the union of the isolated primes of (0) and we choose a_1 to be an element of $\mathfrak{p}_1$ which is not in any isolated prime of (0). Passing down to $A/\mathfrak{p}_0$ yields the conclusion by induction. $\square$

THEOREM 5.3.3 (Noether Normalization Theorem): *Let A be a finitely generated graded connected algebra over a field $\mathbb{F}$. Then the following integers are equal:*
  (i) $\dim(A)$,
  (ii) *the smallest integer r such that there exist r elements $a_1, \ldots, a_r \in A$ with $A/(a_1, \ldots, a_r)$ totally finite,*
  (iii) *the largest integer s such that there exist s algebraically independent elements in A,*
  (iv) $d(A)$, *the order of the pole of $P(A, t)$ at $t = 1$.*

PROOF : If $\dim(A) = 0$ then $\bar{A}$ is the only prime ideal in A so $\bar{A} = \sqrt{0}$ and every element of A is nilpotent. In particular a finite system of generators for A is nilpotent so A is totally finite and the result holds.

We proceed by induction over the Krull dimension of A. Let $d = \dim(A)$ and apply the preceding lemma to a chain of prime ideals

$$\mathfrak{p}_0 \subset \mathfrak{p}_1 \subset \cdots \subset \mathfrak{p}_d$$

of maximal length. We receive elements $a_1, \ldots, a_d$ with $\mathfrak{p}_d = \bar{A}$ an isolated prime of $(a_1, \ldots, a_d)$. If $\mathfrak{p}$ is any other isolated prime of $(a_1, \ldots, a_d)$ then $\mathrm{ht}(\mathfrak{p}) \geq d$ so $\mathfrak{p} = \bar{A}$. Hence $\bar{A}$ is the only isolated prime of $(a_1, \ldots, a_d)$ so $\sqrt{(a_1, \ldots, a_d)} = \bar{A}$ and every element of A is nilpotent mod $(a_1, \ldots, a_d)$. In particular a finite system of generators for A is nilpotent mod $(a_1, \ldots, a_d)$ so $A/(a_1, \ldots, a_d)$ is totally finite. Therefore $r \leq d$.

We next show that $a_1, \ldots, a_d$ are algebraically independent. To do this let $B := A/\mathfrak{p}_0$ and $b_i = a_i + \mathfrak{p}_0 \in B$. It is clearly enough to show that $b_1, \ldots, b_d$ in B are algebraically independent: pass down from B to $C := A/\mathfrak{p}_1$ and set $c_i = a_i + \mathfrak{p}_1 \in C$. The ring C has Krull dimension $d - 1$ and $C/(c_2, \ldots, c_d)$ is

totally finite so by the induction hypothesis $c_2, \ldots, c_d$ are algebraically independent in C. Thus the elements $b_2, \ldots, b_d \in B$ are algebraically independent.
Suppose that $p(z_1, \ldots, z_d) \in \mathbb{F}[z_1, \ldots, z_d]$ is chosen of minimal degree so that
$p(b_1, b_2, \ldots, b_d) = 0$ is an algebraic relation between $b_1, \ldots, b_d \in B$. We may
write this relation in the form

$$0 = p(b_1, \ldots, b_d) = b_1 \cdot q(b_1, \ldots, b_d) + r(b_2, \ldots, b_d).$$

Passing down to C gives

$$0 = r(c_2, \ldots, c_d)$$

and since $c_2, \ldots, c_d$ are algebraically independent this shows $r(\ldots)$ is identically zero. B is an integral domain and $b_1 \neq 0 \in B$ so $q(b_1, \ldots, b_d) = 0$
contrary to the choice of $p(z_1, \ldots, z_d)$ of minimal degree, and hence $b_1, \ldots, b_d$
are algebraically independent. Hence $s \geq d$.

At this stage we know that $a_1, \ldots, a_d \in A$ generate a polynomial subalgebra,
i.e., a subalgebra isomorphic to $\mathbb{F}[z_1, \ldots, z_d]$ where $\deg(z_i) = \deg(a_i)$ for
$i = 1, \ldots, d$, over which A is finite. By 4.1.3 and 4.1.4 we therefore have
$d(A) = d$. This same pair of results shows that the number s defined by (iii)
is $d(A) = \dim(A)$. The coefficients of the Poincaré series of A have polynomial
growth of exponent $d(A) - 1$, and those of a polynomial algebra $\mathbb{F}[w_1, \ldots, w_s]$
with exponent $s - 1$, so if $\mathbb{F}[w_1, \ldots, w_s] \subset A$ the coefficients of the Poincaré
series of A dominate those of $\mathbb{F}[w_1, \ldots, w_s]$, and hence $d(A) \geq s$.

Similarly we see that the integer defined in (ii) is also $d(A)$: for if A is finite
over $a_1, \ldots, a_r$ then there is a surjective map

$$\mathbb{F}[z_1, \ldots, z_r] \otimes A/(a_1, \ldots, a_r) \longrightarrow A$$

so the coefficients of the Poincaré series of $\mathbb{F}[z_1, \ldots, z_r] \otimes A/(a_1, \ldots, a_r)$ dominate term by term those of A and 4.1.3 implies $r \geq d(A)$. $\square$

COROLLARY 5.3.4 : *If $A \supseteq B$ is a finite extension of finitely generated
graded connected commutative algebras over a field $\mathbb{F}$ then $\dim(A) = \dim(B)$.*

PROOF : Finiteness is transitive, so this follows from (ii) of 5.3.3. $\square$

COROLLARY 5.3.5 : *Let $\varrho : G \hookrightarrow \mathrm{GL}(n, \mathbb{F})$ be a representation of a
finite group G. Then $\dim(\mathbb{F}[V]^G) = n = \dim_{\mathbb{F}}(V)$.*

PROOF : This follows from the preceding corollary, 4.1.4, and 5.2.2.
$\square$

PROPOSITION 5.3.6 : *Let A be a finitely generated graded commutative
algebra of Krull dimension d over the field $\mathbb{F}$. Then any set of elements
$a_1, \ldots, a_d \in A$ with $A/(a_1, \ldots, a_d)$ totally finite is algebraically independent.*

PROOF : We recycle a portion of the argument used to prove 5.3.3. If $d = 0$ there is nothing to prove. Let

$$\mathfrak{p}_0 \subset \mathfrak{p}_1 \subset \cdots \subset \mathfrak{p}_d$$

be a proper chain of prime ideals of maximal length in A. Set $B = A/\mathfrak{p}_0$. It suffices to show that $b_i = a_i + \mathfrak{p}_0 \in B$ are algebraically independent. Set $C = B/(b_1)$. Using 5.3.3 (ii) we see that $\dim(C) = d - 1$ so passing down to C we see by induction that the elements $c_i = b_i + (b_1)$ for $i = 2, \ldots, d$ are algebraically independent in C, and hence $b_2, \ldots, b_n$ are algebraically independent in B. The proof may be completed as in 5.3.3. $\quad\square$

Thus every finitely generated graded commutative algebra over a field is finitely generated over a polynomial subalgebra. In this context the following definition has proven of use.

DEFINITION : *Let A be a finitely generated graded commutative algebra over the field $\mathbb{F}$ of Krull dimension d. A **graded (or homogeneous) system of parameters** for A is any set of d elements $a_1, \ldots, a_d \in A$ such that $A/(a_1, \ldots, a_d)$ is totally finite.*

By the Noether Normalization Theorem a system of parameters always exists and is algebraically independent. By contrast $x, xy \in \mathbb{F}[x, y]$ are algebraically independent, but $\mathbb{F}[x, y]/(x, xy) \cong \mathbb{F}[y]$ is not totally finite, so $x, xy \in \mathbb{F}[x, y]$ is not a system of parameters. Hence the converse of proposition 5.3.6 does not hold. The following criteria is often useful in verifying that $f_1, \ldots, f_n \in \mathbb{F}[z_1, \ldots, z_n]$ is a system of parameters.

PROPOSITION 5.3.7 : $f_1, \ldots, f_n \in \mathbb{F}[z_1, \ldots, z_n]$ *are a system of parameters if and only if for every field extension $\bar{\mathbb{F}} \supset \mathbb{F}$ the variety*

$$V(f_1, \ldots, f_n; \bar{\mathbb{F}}) = \{(x_1, \ldots, x_n) \in \bar{\mathbb{F}}^n \mid f_i(x_1, \ldots, x_n) = 0 \text{ for } i = 1, \ldots, n\}$$

consists of the point $(0, \ldots, 0)$ alone.

PROOF : Suppose that $f_1, \ldots, f_n \in \mathbb{F}[z_1, \ldots, z_n]$ is a system of parameters. Then $\mathbb{F}[z_1, \ldots, z_n]/(f_1, \ldots, f_n)$ is finite dimensional. Hence there exist integers $m_i \in \mathbb{N}$, such that $z_i^{m_i} \in (f_1, \ldots, f_n)$ for $i = 1, \ldots, n$. If $x = (x_1, \ldots, x_n) \in \bar{\mathbb{F}}^n$ belongs to $V(f_1, \ldots, f_n; \bar{\mathbb{F}})$ then

$$0 = z_i^{m_i}(x) = x_i^{m_i} \quad i = 1, \ldots, n$$

so $x = (0, \ldots, 0)$.

Conversely, let $\bar{\mathbb{F}}$ be the algebraic closure of $\mathbb{F}$ and suppose $V(f_1, \ldots, f_n; \bar{\mathbb{F}}) = \{0\}$. Then the Hilbert Nullstellensatz [102] (see also [252] § 79 or [14] chapter 7 exercise 14) implies

$$\sqrt{(f_1, \ldots, f_n)} = (z_1, \ldots, z_n).$$

Hence for each i, $1 \leq i \leq n$, there exists $m_i \in \mathbb{N}$ such that $z_i^{m_i} \in (f_1, \ldots, f_n)$. The quotient algebra $\mathbb{F}[z_1, \ldots, z_n]/(f_1, \ldots, f_n)$ is generated by the residue classes of $z_1, \ldots, z_n$ which are nilpotent and therefore the totalization of the algebra $\mathbb{F}[z_1, \ldots, z_n]/(f_1, \ldots, f_n)$ is finite dimensional. $\square$

EXAMPLE 1 : Consider the representation of $\mathbb{Z}/4$ defined by the matrix

$$T = \begin{bmatrix} 0 & -1 \\ 1 & 0 \end{bmatrix}.$$

If $\{x, y\}$ is the dual of the canonical basis of $\mathbb{F}^2$, $\mathbb{F}$ not of characteristic 2, then in example 2 of section 3.2 we saw that

$$\varrho_1 = x^2 + y^2$$
$$\varrho_2 = x^2 y^2$$
$$\varrho_3 = x^3 \cdot y - x \cdot y^3$$

generate $\mathbb{F}[x, y]^{\mathbb{Z}/4}$. These polynomials satisfy one relation, namely:

$$\varrho_3^2 = \varrho_1^2 \varrho_2 - 4\varrho_2^2$$

so $\mathbb{F}[x, y]^{\mathbb{Z}/4} \cong \mathbb{F}[u, v, w]/(w^2 - u^2 v + 4v^2)$. The system of equations

$$x^2 + y^2 = 0$$
$$x^2 y^2 = 0$$

has only the trivial solution $(0, 0)$ in any field extension $\bar{\mathbb{F}}$ of $\mathbb{F}$ so by 5.3.7 are a system of parameters. In addition the invariants ϱ_1 and ϱ_2 are algebraically independent and moreover

$$\mathbb{F}[\varrho_1, \varrho_2] \oplus \mathbb{F}[\varrho_1, \varrho_2] \cdot \varrho_3 \subset \mathbb{F}[x, y]^{\mathbb{Z}/4}.$$

The Poincaré series of $\mathbb{F}[x, y]^{\mathbb{Z}/4}$ may be computed with the aid of table 5.3.1 and Molien's theorem.

We find

$$P(\mathbb{F}[V]^{\mathbb{Z}/4}, t) = \frac{1}{4}\left[\frac{1}{(1-t)^2} + \frac{1}{(1+t)^2} + \frac{2}{(1+t^2)}\right]$$

$$= \frac{1+t^4}{(1-t^2)(1-t^4)} = \frac{1}{(1-t^2)(1-t^4)} + \frac{t^4}{(1-t^2)(1-t^4)}$$

$$= P(\mathbb{F}[\varrho_1, \varrho_2] \oplus \mathbb{F}[\varrho_1, \varrho_2] \cdot \varrho_3, t)$$

and hence

$$\mathbb{F}[\varrho_1, \varrho_2] \oplus \mathbb{F}[\varrho_1, \varrho_2] \cdot \varrho_3 \cong \mathbb{F}[x, y]^{\mathbb{Z}/4}.$$

g	$\varrho(g)$	$\det(1 - g^{-1}t)$
1	$\begin{bmatrix} 1 & 0 \\ 0 & 1 \end{bmatrix}$	$(1 - t)^2$
T	$\begin{bmatrix} 0 & -1 \\ 1 & 0 \end{bmatrix}$	$1 + t^2$
T^2	$\begin{bmatrix} -1 & 0 \\ 0 & -1 \end{bmatrix}$	$(1 + t)^2$
T^3	$\begin{bmatrix} 0 & 1 \\ -1 & 0 \end{bmatrix}$	$1 + t^2$

Table 5.3.1: The group $\mathbb{Z}/4$

Thus $\mathbb{F}[x, y]^{\mathbb{Z}/4}$ is a free $\mathbb{F}[\varrho_1, \varrho_2]$-module. This is a special case of a theorem (6.7.8) due to Eagon and Hochster.

The following proposition provides a way to find a system of parameters for rings of invariants in many cases (see also [183], [235] proposition 3.4 and [220] proposition 2.2).

PROPOSITION 5.3.8 : If $\varrho : G \hookrightarrow \mathrm{GL}(n, \mathbb{F})$ is an irreducible representation of a finite group G with $G \in \mathbb{F}^\times$. If $B \subset V^*$ is an orbit containing $|G|$ elements, then the Chern classes $\{c_i(G) \mid i = 1, \ldots, |B|\}$ contain a system of parameters for $\mathbb{F}[V]^G$.

PROOF : This is an immediate consequence of 3.1.9 and the definitions. □

The Noether Normalization Theorem is the stepping stone to a number of further properties of the Krull dimension.

COROLLARY 5.3.9 : If A and B are finitely generated commutative graded algebras over a field $\mathbb{F}$, then $\dim(A \otimes_\mathbb{F} B) = \dim(A) + \dim(B)$.

PROOF : Choose minimal sets of algebraically independent elements $a_1, \ldots, a_n \in A$ and $b_1, \ldots, b_m \in B$ such that the quotient algebras $A/(a_1, \ldots, a_n)$ and $B/(b_1, \ldots, b_m)$ are totally finite. Then the elements

$$a_1 \otimes 1, \ldots, a_n \otimes 1, 1 \otimes b_1, \ldots, 1 \otimes b_m \in A \otimes_\mathbb{F} B$$

are algebraically independent,

$$A \otimes_{\mathbb{F}} B/(a_1 \otimes 1, \ldots, a_n \otimes 1, \ 1 \otimes b_1, \ldots, 1 \otimes b_m)$$

is totally finite and thus the result follows. $\square$

PROPOSITION 5.3.10 : *Let A be a finitely generated graded commutative algebra over the field $\mathbb{F}$ of Krull dimension d and $a_1, \ldots, a_j$ elements of A. Then $\dim(A/(a_1, \ldots, a_j)) \geq d - j$ and equality holds if and only if $a_1, \ldots, a_j$ is a subset of a system of parameters.*

PROOF : Choose a system of parameters $b_1, \ldots, b_n$ for $A/(a_1, \ldots, a_j)$ and lift them to elements $c_1, \ldots, c_n \in A$. Then $A/(a_1, \ldots, a_j, c_1, \ldots c_n)$ is totally finite so $n + j \geq d$ and if we have equality then by 5.3.3 (ii) they are a system of parameters. $\square$

We will return to systems of parameters when we study Cohen-Macaulay rings in chapter 6.

5.4   Going Up and Down: Localization

It follows from the Noether Normalization Theorem 5.3.3 and 5.3.4 that the Krull dimension of a graded commutative algebra B over a field $\mathbb{F}$ does not change when we pass to a finite extension $A \supseteq B$ of B. It is therefore natural to ask if there is some direct relationship between the chains of prime ideals in A and B.

To handle this problem we pass from graded algebras to local rings. For a graded connected algebra A over a field $\mathbb{F}$ there is the ungraded ring $\mathrm{Tot}(A) = \oplus_{i=0}^{\infty} A_i$, and if $I \in A$ is a graded ideal then $\mathrm{Tot}(I) \subset \mathrm{Tot}(A)$ is an ideal in the usual sense. An element of $\mathrm{Tot}(A)$ is called **homogeneous** if it lies in some $A_i \subset \mathrm{Tot}(A)$, in which case the **degree** of the element is said to be i. If $J \subset \mathrm{Tot}(A)$ is an ideal in the usual sense then we obtain a graded ideal $I(J)$ in A by defining $I(J)_i = \{a \in J \mid a \text{ is homogeneous of degree } i\}$.

These constructions are by no means bijective. One readily sees that $I = I(\mathrm{Tot}(I))$, but there may be many ideals J in $\mathrm{Tot}(A)$ with $I(J) = I$ for a graded ideal I in A. For example, if $J = (x - 1) \subset \mathrm{Tot}(\mathbb{F}[x])$, then J contains no nonzero homogeneous elements, so $I(J) = (0) \subset A$. $\mathrm{Tot}(A)$ contains many more ideals than A but every ideal of A comes from some ideal of $\mathrm{Tot}(A)$. Despite this we will see shortly that A and $\mathrm{Tot}(A)$ have the same Krull dimension.

The notation $\mathrm{Tot}(A)$ is cumbersome, and we often leave it to the context to clarify if A denotes A as graded connected commutative algebra over a field or $\mathrm{Tot}(A)$ as a ring.

For the remainder of this section we will be working with non-graded rings. The Krull dimension for such a ring is defined as in the graded case as the length of the longest chain of prime ideals in the ring.

Given a prime ideal $\mathfrak{p}$ of A there is a bijective correspondence between chains of prime ideals

$$\mathfrak{p} = \mathfrak{p}_0 \subset \mathfrak{p}_1 \subset \cdots \subset \mathfrak{p}_n$$

starting at $\mathfrak{p}$ and arbitrary chains of prime ideals (starting at (0)) in the quotient ring $A/\mathfrak{p}$. This allows us to reduce questions involving $\mathrm{dp}(\mathfrak{p})$ to questions about $\dim(A/\mathfrak{p})$. The analogous effect for chains of prime ideals

$$\mathfrak{p}_0 \subset \cdots \subset \mathfrak{p}_{n-1} \subset \mathfrak{p}_n = \mathfrak{p}$$

ending at $\mathfrak{p}$ is achieved by localizing with respect to $\mathfrak{p}$, which is a way of turning the prime ideal $\mathfrak{p}$ into a maximal ideal.

DEFINITION : *If A is a ring and $\mathfrak{p}$ is a prime ideal in A the* **localization** *of A with respect to $\mathfrak{p}$, denoted by $A_\mathfrak{p}$, is the ring of formal fractions*

$$A_\mathfrak{p} = \left\{ \frac{a}{b} \mid a \in A,\ b \in A \smallsetminus \mathfrak{p} \right\},$$

two such fractions $\frac{a'}{b'}$ and $\frac{a''}{b''}$ being considered as equal if and only if there is an element $c \in A \smallsetminus \mathfrak{p}$ such that

$$(*) \qquad\qquad (a'b'' - a''b')c = 0 \in A.$$

Addition and multiplication of fractions is defined as usual.

Note that for an integral domain the relation $(*)$ simplifies to

$$a'b'' - a''b' = 0 \in A$$

as no element $c \in A \smallsetminus \mathfrak{p}$ is a zero divisor.

DEFINITION : *If A is a ring, $\mathfrak{p}$ is a prime ideal in A, and M is an A-module we set $M_\mathfrak{p} = A_\mathfrak{p} \otimes_A M$ and call $M_\mathfrak{p}$ the* **localization** *of M with respect to $\mathfrak{p}$.*

For an A-module M the kernel of the map $M \longrightarrow M_\mathfrak{p}$ consists of the elements which are annihilated by some element of A not in $\mathfrak{p}$. A short exact sequence

$$0 \longrightarrow M' \longrightarrow M \longrightarrow M'' \longrightarrow 0$$

of A-modules gives rise to a short exact sequence

$$0 \longrightarrow (M')_\mathfrak{p} \longrightarrow (M)_\mathfrak{p} \longrightarrow (M'')_\mathfrak{p} \longrightarrow 0$$

of $A_\mathfrak{p}$-modules, since checking if an element of M' is annihilated by an element of A not in $\mathfrak{p}$ does not depend on whether we view it as an element of M' or M, and $A_\mathfrak{p} \otimes_A - = -_\mathfrak{p}$ preserves epimorphisms.

If I is a proper ideal in $A_{\mathfrak{p}}$ then $I \cap A$ is an ideal in A contained in $\mathfrak{p}$ (because if it contains an element outside $\mathfrak{p}$ then it contains an invertible element and hence the identity element). Conversely, if J is an ideal in A contained in $\mathfrak{p}$ then $J_{\mathfrak{p}}$ is an ideal in $A_{\mathfrak{p}}$. These processes set up a one to one correspondence between proper ideals in $A_{\mathfrak{p}}$ and ideals in A contained in $\mathfrak{p}$. In particular, $\mathfrak{p}_{\mathfrak{p}}$ is the unique maximal ideal of $A_{\mathfrak{p}}$. A commutative ring with a unique maximal ideal is called a **local ring**. Localization with respect to $\mathfrak{p}$ provides a bijective correspondence between chains of prime ideals

$$\mathfrak{p}_0 \subset \cdots \subset \mathfrak{p}_{n-1} \subset \mathfrak{p}_n = \mathfrak{p}$$

ending at $\mathfrak{p}$ and chains of prime ideals in $A_{\mathfrak{p}}$, which end at the maximal ideal $\mathfrak{p}_{\mathfrak{p}}$.

For a representation of a finite group $G \hookrightarrow \mathrm{GL}(V)$ we have seen that $\mathbb{F}[V]$ is a finite extension of the finitely generated $\mathbb{F}$-algebra $\mathbb{F}[V]^G$. We can use this to describe the relationship between the prime ideals of $\mathbb{F}[V]$ and $\mathbb{F}[V]^G$. By the preceding discussion we see that the relation of the maximal ideals is the key, so we begin with this.

LEMMA 5.4.1 : *Suppose that $A' \supseteq A''$ is a finite integral extension of commutative rings. If $\mathfrak{m}'$ is a maximal ideal of A' then $\mathfrak{m}' \cap A''$ is a maximal ideal of A''. Conversely, if $\mathfrak{m}''$ is a maximal ideal of A'' then there is a maximal ideal $\mathfrak{m}'$ of A' with $\mathfrak{m}' \cap A'' = \mathfrak{m}''$. Moreover, any prime ideal $\mathfrak{p}'$ of A' with $\mathfrak{p}' \cap A'' = \mathfrak{m}''$ is maximal in A'.*

PROOF : If $\mathfrak{m}'$ is maximal in A' then $A'/\mathfrak{m}'$ is a field that is integral over $A''/\mathfrak{m}' \cap A''$. If $x \in A''/\mathfrak{m}' \cap A''$ then $x^{-1} \in A'/\mathfrak{m}'$ satisfies some monic equation over $A''/\mathfrak{m}' \cap A''$, say $x^{-n} + b_{n-1}x^{-n+1} + \cdots + b_0 = 0$. But then $x^{-1} = -b_{n-1} - \cdots - b_0 x^{n-1} \in A''/\mathfrak{m}' \cap A''$, so $A''/\mathfrak{m}' \cap A''$ is a field and hence $\mathfrak{m}' \cap A''$ is a maximal ideal in A''.

Conversely, if $\mathfrak{m}'' \subset A''$ is a maximal ideal consider $J = \mathfrak{m}'' \cdot A'$. This is an ideal in A'. If it is proper we may choose any maximal ideal $\mathfrak{m}'$ in A' such that $J \subset \mathfrak{m}'$ and $\mathfrak{m}' \cap A'' = \mathfrak{m}''$. So we need only show that $\mathfrak{m}'' \cdot A' \neq A'$. Suppose the contrary. Choose a set of generators $a_1', \ldots, a_n'$ for A' over A''. Since we are assuming $\mathfrak{m}'' \cdot A' = A'$ there are elements $c_{i,j} \in \mathfrak{m}''$ such that

$$a_i' = \sum_j c_{i,j} a_j'$$

which we choose to rewrite in matrix form as $(I - C)(a') = 0$. Cramer's rule (see page 157 footnote number 3) says that $\det(I - C) \in \mathrm{Ann}_{A'}(a_1', \ldots, a_n') = 0$, and so $\det(I - C) = 0$. On the other hand $C \equiv 0 \bmod \mathfrak{m}''$, so $\det(I - C) \equiv 1 \bmod \mathfrak{m}''$ which is a contradiction.

Finally if $\mathfrak{p}' \subset A'$ is prime and $\mathfrak{p}' \cap A'' = \mathfrak{m}''$ then $A'/\mathfrak{p}' \supset A''/\mathfrak{m}''$ is a finite extension, and $A''/\mathfrak{m}''$ is a field. If $a \in A'/\mathfrak{p}'$ then there are $b_0'', \ldots, b_{n-1}'' \in$

$A''/\mathfrak{m}''$ such that

$$a^n + b''_{n-1}a^{n-1} + \cdots + b''_0 = 0.$$

Choosing n minimal and noting that $A'/\mathfrak{p}'$ is an integral domain we may assume that $b''_0 \neq 0$, so rewriting the previous equation in the form

$$a(a^{n-1} + b''_{n-1}a^{n-2} + \cdots + b''_1) = -b''_0$$

we see that

$$a^{-1} = -\frac{a^{n-1} + b''_{n-1}a^{n-2} + \cdots + b''_1}{b''_0}$$

and hence $A'/\mathfrak{p}'$ is a field, whence $\mathfrak{p}'$ is maximal. $\square$

THEOREM 5.4.2 : *Suppose that $A' \supseteq A''$ is a finite integral extension of commutative rings.*

   (i) **(Lying over)** *If $\mathfrak{p}''$ is a prime ideal of A'' then there is a prime ideal $\mathfrak{p}'$ of A' with $\mathfrak{p}' \cap A'' = \mathfrak{p}''$. There are no strict inclusions between such prime ideals $\mathfrak{p}'$. In this situation we say that $\mathfrak{p}'$ lies over $\mathfrak{p}''$.*

   (ii) **(Going up)** *If $\mathfrak{p}''_1 \supset \mathfrak{p}''_0$ are prime ideals in A'' and $\mathfrak{p}'_0$ is a prime ideal in A' lying over $\mathfrak{p}''_0$ then there is a prime ideal $\mathfrak{p}'_1$ in A' lying over $\mathfrak{p}''_1$ with $\mathfrak{p}'_1 \supset \mathfrak{p}'_0$.*

   (iii) *If $\mathfrak{p}' \subset A'$ lies over $\mathfrak{p}'' \subset A''$ then one is maximal if and only if the other is.*

   PROOF : This follows from 5.4.1 by localizing. $\square$

COROLLARY 5.4.3 : *Let $A' \supseteq A''$ be a finite integral extension of commutative rings. If*

$$\mathfrak{p}'_0 \subset \cdots \subset \mathfrak{p}'_r \subset A'$$

is a proper chain of prime ideals in A' and $\mathfrak{p}''_i = \mathfrak{p}'_i \cap A''$ then

$$\mathfrak{p}''_0 \subset \cdots \subset \mathfrak{p}''_r \subset A''$$

is a proper chain of prime ideals in A''. Conversely, if

$$\mathfrak{p}''_0 \subset \cdots \subset \mathfrak{p}''_s \subset A''$$

is a proper chain of prime ideals in A'' and $\mathfrak{p}'_0 \subset A'$ lies over $\mathfrak{p}''_0$ then there is a proper chain of prime ideals in A'

$$\mathfrak{p}'_0 \subset \cdots \subset \mathfrak{p}'_s \subset A'$$

with $\mathfrak{p}'_i$ lying over $\mathfrak{p}''_i$ for $i = 1, \ldots, s$. Hence the Krull dimensions of A' and A'' are equal.

PROOF : The first assertion follows from 5.4.2 part (i). To prove the second assertion proceed inductively applying 5.4.2 (ii) to the successive quotient rings $A'/\mathfrak{p}'_i \supset A''/\mathfrak{p}''_i$. $\square$

The correspondence provided by 5.4.3 yields another proof of:

COROLLARY 5.4.4 : *Suppose that G is a finite group, $\mathbb{F}$ a field and V a finite dimensional $\mathbb{F}(G)$-module. Then the Krull dimension of $\mathbb{F}[V]^G$ is equal to $\dim_{\mathbb{F}}(V)$.*

PROOF : By 2.3.2 $\mathbb{F}[V] \supseteq \mathbb{F}[V]^G$ is a finite integral extension so we may apply the previous corollary. □

The extension $\mathbb{F}[V] \supseteq \mathbb{F}[V]^G$ is not only integral, but also the corresponding extension of fields of fractions is Galois by 1.2.4. We can exploit such extra structure to improve on 5.4.2.

THEOREM 5.4.5 : *Suppose that $A' \supseteq A''$ is a finite extension of commutative integral domains, A'' is integrally closed, and the corresponding extension $\mathbb{L}' \supseteq \mathbb{L}''$ of fields of fractions is* **normal** *(i.e., an irreducible polynomial over $\mathbb{L}''$ with a root in $\mathbb{L}'$ splits completely in $\mathbb{L}'$, but is not necessarily separable).*

(i) (**Transitivity**) *The Galois group $G = \mathrm{Gal}(\mathbb{L}' \mid \mathbb{L}'')$ acts transitively on the prime ideals $\mathfrak{p}'$ of A' lying over a given prime ideal $\mathfrak{p}''$ of A''.*

(ii) (**Going down**) *If $\mathfrak{p}_0'' \subset \mathfrak{p}_1''$ are prime ideals in A'' and $\mathfrak{p}_1'$ is a prime ideal of A' lying over $\mathfrak{p}_1''$ then there is a prime ideal $\mathfrak{p}_0'$ in A' lying over $\mathfrak{p}_0''$, with $\mathfrak{p}_0' \subset \mathfrak{p}_1'$.*

PROOF : Since the extension $A' \supseteq A''$ is finite the corresponding extension of fields of fractions is finite, and hence so is the Galois group. We begin the proof of (i) with the case where $\mathbb{L}' \supseteq \mathbb{L}''$ is separable, so that $\mathbb{L}'' = \mathbb{L}'^G$ and $A'' = A'^G$ (by integral closure). Suppose that $\mathfrak{p}_1'$ and $\mathfrak{p}_2'$ are prime ideals in A' lying over $\mathfrak{p}''$, and suppose that $\mathfrak{p}_1'$ and $\mathfrak{p}_2'$ are not G-conjugate. By 5.4.2 (i), no conjugate of $\mathfrak{p}_1'$ contains $\mathfrak{p}_2'$. So by the preceding lemma we may choose an element $x \in \mathfrak{p}_2'$ such that for all $g \in G$, $x \notin g\mathfrak{p}_1'$. Then $\prod_{g \in G} g(x)$ is an element of $A'^G = A''$ lying in $\mathfrak{p}_2'$ but not in $\mathfrak{p}_1'$, which contradicts the hypothesis that $\mathfrak{p}_1' \cap A'' = \mathfrak{p}'' = \mathfrak{p}_2' \cap A''$.

If $\mathbb{L}' \supseteq \mathbb{L}''$ is purely inseparable then the only prime ideal of A' lying over $\mathfrak{p}''$ is $\{x \in A' \mid x^{p^n} \in \mathfrak{p}''$ for some $n \geq 0\}$. Finally, every extension is a composition of a purely inseparable and a separable extension, so the statement is proved.

We turn to (ii). Let $\mathbb{L}' \supseteq \mathbb{L}''$ be the fields of fractions of $A' \supseteq A''$, and let $\mathbb{L}$ be a finite normal extension of $\mathbb{L}''$ containing $\mathbb{L}'$. Let A be the integral closure of A'' in $\mathbb{L}$, so that A is a finite extension of A', and choose primes $\mathfrak{p}''$ and $\mathfrak{p}'$ in A lying over $\mathfrak{p}_0''$ in A'' and $\mathfrak{p}_1'$ in A' respectively. By 5.4.2 (ii), we can find $\mathfrak{p} \supset \mathfrak{p}''$ with $\mathfrak{p}$ lying over $\mathfrak{p}_1'$. By (i), for some $g \in \mathrm{Gal}(\mathbb{L} \mid \mathbb{L}')$ we have $g(\mathfrak{p}) = \mathfrak{p}'$. Set $\mathfrak{p}_0' = g(\mathfrak{p}'') \cap A'$. Then $\mathfrak{p}_0' \subset g(\mathfrak{p}) \cap A' = \mathfrak{p}' \cap A' = \mathfrak{p}_1'$ and $\mathfrak{p}_0'$ lies over $\mathfrak{p}_0''$. □

REMARK : According to Hilbert's weak Nullstellensatz (see Matsumura [141], theorem 5.3 or van der Waerden [252] volume II chapter X1 § 79), if $\mathbb{F}$ is algebraically closed then the maximal ideals in $\mathbb{F}[V]$ are in natural one–one correspondence with the points of V. So if we denote by max $(—)$ the set of maximal ideals in $—$ we have $V \cong \max(\mathbb{F}[V])$. By 5.4.2 part (i) the inclusion $i : \mathbb{F}[V]^G \hookrightarrow \mathbb{F}[V]$ gives rise to a *surjective* map

$$i^* : \max(\mathbb{F}[V]) \longrightarrow \max(\mathbb{F}[V]^G),$$

and by 5.4.5 part (i),

$$\max(\mathbb{F}[V]^G) \cong V/G.$$

This relationship is the *Ansatz* for one of the geometric approachs to invariant theory. See for example [51] and [126].

We close this section with a proof of the promised fact that for a graded connected commutative algebra A over a field $\mathbb{F}$, $\dim(A) = \dim(\text{Tot}(A))$.

PROPOSITION 5.4.6 : *Let A be a a graded connected commutative algebra over a field $\mathbb{F}$. Then $\dim(A) = \dim(\text{Tot}(A))$.*

PROOF : By 5.3.3 there is a polynomial subalgebra $\mathbb{F}[z_1, \ldots, z_d] \subset A$, $d = \dim(A)$, over which A is finite and hence integral by 2.3.2. Therefore $\text{Tot}(A)$ is finite and integral over $\text{Tot}(\mathbb{F}[z_1, \ldots, z_d])$ and by 5.4.3 $\dim(\text{Tot}(A)) = d$ since the computation of the Krull dimension of a polynomial algebra 5.2.2 made no use of the grading. $\square$

5.5 Degree

In Section 5.3 we saw that if A is a a graded connected commutative algebra over a field $\mathbb{F}$ the Krull dimension of A is equal to the order of the pole of the rational function $P(A, t)$ at $t = 1$. If the Krull dimension is n, then the value of the rational function $(1 - t)^n P(A, t)$ at $t = 1$ is a nonzero rational number, called the **degree** of A [225], written $\deg(A)$. If $A = \mathbb{F}[V]^G$ where $\mathbb{F}$ has characteristic prime to $|G|$, then as a consequence of Molien's theorem 4.3.2 $\deg(\mathbb{F}[V]^G) = \frac{1}{|G|}$ (see section 4.4). Our goal in this section is to remove the restriction that $|G|$ be relatively prime to the characteristic of $\mathbb{F}$.

If $B \supseteq A$ is a finite extension of graded integral domains, each a graded connected commutative algebra over a field $\mathbb{F}$, we shall see that the ratio of the degrees is equal to the degree of the corresponding field extension. The fact 1.2.4 that the field extension $\mathbb{F}(V) \mid \mathbb{F}(V)^G$ is a Galois extension will lead to our goal. But first, we need to interpret this properly.

We define the **graded field of fractions** of A to be $\mathbb{L} = \{\mathbb{L}_j \mid -\infty < j < \infty\}$ where $\mathbb{L}_j$ consists of the fractions x/y with $x \in A_{m+j}$, $y \in A_m$ with

$m \geq \max\{0, -j\}$ and $y \neq 0$. These fractions are added and multiplied in the usual way. The result is a **graded field** in the sense that every nonzero homogeneous element has a homogeneous inverse. Thus $\mathbb{L}_0$ is a field, and as long as A has nonzero elements of positive degree, it is easy to see that $\mathbb{L}$ may be identified with the Laurent polynomials $\mathbb{L}_0[X, X^{-1}]$, with X corresponding to some nonzero element of minimal nonzero degree in A. The **full** field of fractions of A (i.e. the field of fractions of $\mathrm{Tot}(A)$) is of course $\mathbb{L}_0(X)$.

If $\mathbb{K} \supseteq \mathbb{L}$ are graded fields, with $\mathbb{K}$ finitely generated as an $\mathbb{L}$-module, then $\mathbb{K}$ is clearly free as an $\mathbb{L}$-module, and we write $|\mathbb{K} : \mathbb{L}|$ for the number of generators. This is easily seen to be equal to the degrees of the (ungraded) field extensions $|\mathbb{K}_0 : \mathbb{L}_0|$ and $|\mathbb{K}_0(X) : \mathbb{L}_0(X)|$.

    LEMMA 5.5.1 : *Suppose that $B \supseteq A$ are (graded) integral domains with B integral over A, and with (graded) fields of fractions $\mathbb{K} \supseteq \mathbb{L}$ a finite extension. Then there is a basis of $\mathbb{K}$ over $\mathbb{L}$ consisting of (homogeneous) elements of B. The A-submodule of B generated by this basis is free.*

    PROOF : Since B is integral over A, every $b \in B$ satisfies some equation of the form

$$b^r + a_{r-1}b^{r-1} + \cdots + a_0 = 0$$

with $a_i \in A$, and since B is an integral domain we may suppose that $a_0 \neq 0$. Thus there is an element of B, namely $\tilde{b} = -b^{r-1} - a_{r-1}b^{r-2} - \cdots - a_1$, with the property that $\tilde{b}b = a_0$ is a nonzero element of A.

Let $|\mathbb{K} : \mathbb{L}| = m$, and choose elements $y_1, \ldots, y_m$ forming a basis of $\mathbb{K}_0$ as a vector space over $\mathbb{L}_0$. If $y_i = b'_i/b_i$ with $b_i, b'_i \in B$, then choosing $\tilde{b}_i$ as above, we see that the elements $x_i = \tilde{b}_i b_i y_i = \tilde{b}_i b'_i \in B$ also form a basis of $\mathbb{K}$ over $\mathbb{L}$. These elements are linearly independent in B regarded as an A-module (since they are linearly independent over the field of fractions), so the A-submodule of B generated by them is free. If $B \supseteq A$ and $\mathbb{K} \supseteq \mathbb{L}$ are graded, the above argument may be performed with homogeneous elements. $\square$

    PROPOSITION 5.5.2 : *Suppose that $B \supseteq A$ is a finite extension of finitely generated connected graded integral domains over $\mathbb{F}$, and that $\mathbb{K} \supseteq \mathbb{L}$ are their graded fields of fractions. Then*

$$\deg(B) = |\mathbb{K} : \mathbb{L}| \deg(A).$$

    PROOF : Apply the lemma to obtain a basis $x_1, \ldots, x_m$ of $\mathbb{K}$ over $\mathbb{L}$ consisting of homogeneous elements of B generating a free A-submodule $Ax_1 \oplus \cdots \oplus Ax_m$ of B of rank m. Hence $P(A, t)\left(\sum_{i=1}^m t^{\deg(x_i)}\right) \leq P(B, t)$.

Since B is finitely generated as an A-module we may choose an element $a \in A$, with inverse a^{-1} in $\mathbb{L}$, such that $B \subseteq A[a^{-1}]x_1 \oplus \cdots \oplus A[a^{-1}]x_m$. This implies

$P(B, t) \leq P(A, t)t^{-deg(a)} \left(\sum_{i=1}^{m} t^{\deg(x_i)} \right)$. Therefore we have

$$P(A, t) \left(\sum_{i=1}^{m} t^{\deg(x_i)} \right) \leq P(B, t) \leq P(A, t)t^{-deg(a)} \left(\sum_{i=1}^{m} t^{\deg(x_i)} \right)$$

and multiplying this inequality by $(t-1)^{\deg(a)}$ and putting $t = 1$ we get

$$\deg(A) \cdot m \leq \deg(B) \leq m \cdot \deg(A)$$

completing the proof. $\square$

THEOREM 5.5.3 : *Suppose that G is a finite group and V is an n-dimensional faithful representation of G over a field $\mathbb{F}$. Then $\deg(\mathbb{F}[V]^G) = \frac{1}{|G|}$. Hence the Laurent expansion of the Poincaré series of $\mathbb{F}[V]^G$ about $t = 1$ begins*

$$P(\mathbb{F}[V]^G, t) = \frac{\frac{1}{|G|}}{(1-t)^n} + \cdots.$$

PROOF : $\mathbb{F}[V]$ is a finite extension of $\mathbb{F}[V]^G$, so we may apply proposition 5.5.2. At the level of (ungraded) fields of fractions, $\mathbb{F}(V)$ is a Galois extension of $\mathbb{F}(V)^G$ with Galois group G. So we conclude from proposition 5.5.2 that

$$\deg(\mathbb{F}[V]) = |G| \cdot \deg(\mathbb{F}[V]^G).$$

Since $P(\mathbb{F}[V], t) = 1/(1-t)^n$, $\deg \mathbb{F}[V] = 1$, and the theorem is proved. $\square$

We illustrate the results obtained so far by an example. We first record yet another pair of results related to when the ring of invariants is a polynomial algebra (see also [120]).

COROLLARY 5.5.4 : *Suppose that $\varrho : G \hookrightarrow GL(n, \mathbb{F})$ is a faithful representation of a finite group G over a field $\mathbb{F}$. If $\mathbb{F}[V]^G = \mathbb{F}[f_1, \ldots, f_n]$, where $\deg(f_i) = d_i$ for $i = 1, \ldots, n$, then $d_1 d_2 \cdots d_n = |G|$.*

PROOF : By 5.5.3 $\deg(\mathbb{F}[V]^G) = \frac{1}{|G|}$ and by direct computation

$$\deg(\mathbb{F}[f_1, \ldots, f_n]) = (1-t)^n \prod_{i=1}^{n} \frac{1}{1 - t^{d_i}} \Big|_{t=1}$$

$$= \prod_{i=1}^{n} \frac{1}{1 + t + t^2 + \cdots + t^{d_i-1}} \Big|_{t=1} = \frac{1}{d_1 d_2 \cdots d_n},$$

so $\frac{1}{|G|} = \frac{1}{d_1 d_2 \cdots d_n}$ and the result follows. $\square$

PROPOSITION 5.5.5 : *Suppose* $G \hookrightarrow \mathrm{GL}(V)$ *is a finite dimensional representation of a finite group* G *and* $\mathbb{F}[V]^G$ *contains elements* $f_1, \ldots, f_n$, $n = \dim_{\mathbb{F}}(V)$, *such that* $\deg(f_1) \cdots \deg(f_n) = |G|$. *If* $f_1, \ldots, f_n$ *are a system of parameters then* $\mathbb{F}[V]^G \cong \mathbb{F}[f_1, \ldots, f_n]$.

PROOF : Since finiteness is transitive it is immaterial whether we assume $f_1, \ldots, f_n$ is a system of parameters for $\mathbb{F}[V]^G$ or $\mathbb{F}[V]$, as the two conditions are equivalent. If $\mathbb{F}[V]^G$ is finite over $f_1, \ldots, f_n$ then $f_1, \ldots, f_n$ are algebraically independent by 5.3.6.

If A denotes the subalgebra of $\mathbb{F}[V]^G$ generated by $f_1, \ldots, f_n$ then $A \cong \mathbb{F}[f_1, \ldots, f_n]$ and hence

$$P(A, t) = \prod_{i=1}^{n} \left(\frac{1}{1 - t^{d_i}} \right) = \frac{\frac{1}{d_1 \cdots d_n}}{(1 - t)^n} + \cdots$$

where $d_i = \deg(f_i)$ $i = 1, \ldots, n$ and so $\deg(A) = d_1 \cdots d_n$. Also by 5.3.3 the Poincaré series $P(\mathbb{F}[V]^G, t)$ has a pole of order n at $t = 1$. By 5.5.3 $\deg(\mathbb{F}[V]^G) = \frac{1}{|G|}$. Hence by 5.5.3 A and $\mathbb{F}[V]^G$ have the same field of fractions $\mathbb{K}$. Since $\mathbb{F}[V]^G$ is finite over A the extension $A \subset \mathbb{F}[V]^G$ is integral. However $A \cong \mathbb{F}[f_1, \ldots, f_n]$ is integrally closed in its field of fractions. Moreover, $\mathbb{F}[V]^G \subset \mathbb{K}$, so we have $\mathbb{F}[V]^G \subset A \cong \mathbb{F}[f_1, \ldots, f_n]$ and the result follows. $\square$

EXAMPLE 1 (L. Flatto [85]): As a first example to illustrate the use of 5.5.5 we reconsider (see section 4.3 example 2) the dihedral group D_{2k} of order $2k$ represented in $\mathrm{GL}(2, \mathbb{R})$ as the group of symmetries of a regular k-gon centered at the origin. The group D_{2k} in this representation is generated by the matrices

$$D = \begin{bmatrix} \cos \frac{2\pi}{k} & -\sin \frac{2\pi}{k} \\ \sin \frac{2\pi}{k} & \cos \frac{2\pi}{k} \end{bmatrix} \qquad S = \begin{bmatrix} 1 & 0 \\ 0 & -1 \end{bmatrix}$$

where D is a rotation through $2\pi/k$ radians and S a reflection in an axis. To compute some elements in $\mathbb{R}[x, y]^{D_{2k}}$ we note first of all that as the action is orthogonal, it preserves the norm and hence $f_1 = x^2 + y^2$ is an invariant quadratic polynomial. We next introduce the complex variable $z = x + iy$ and note that with respect to z the action of D_{2k} is given by

$$S(z) = \bar{z} \qquad D(z) = \exp \frac{2\pi i}{k} z .$$

Thus $f_2 = \mathfrak{Re}(z^k)$, where $\mathfrak{Re}(\text{---})$ denotes the real part of ---, is a polynomial in x and y that is invariant under the action of D_{2k}. The system of equations

$$x^2 + y^2 = 0$$
$$\mathfrak{Re}(z^k) = 0$$

have only the trivial solution $(0, 0)$ in any extension field and hence by 5.3.7 f_1, f_2 is a system of parameters. The classes f_1, f_2 are algebraically independent as may be seen by applying lemma 5.6.1 (see the next section) to their Jacobian determinant which is

$$\det \begin{bmatrix} \frac{\partial f_1}{\partial x} & \frac{\partial f_2}{\partial x} \\ \frac{\partial f_1}{\partial y} & \frac{\partial f_2}{\partial y} \end{bmatrix} = -2k\Im\mathfrak{m}(z^k) \neq 0,$$

where $\Im\mathfrak{m}(\text{---})$ denotes the imaginary part of ---. Moreover $\deg(f_1)\deg(f_2) = 2k = |D_{2k}|$ and hence by 5.5.5

$$\mathbb{R}[x, y]^{D_{2k}} = \mathbb{R}[f_1, f_2].$$

EXAMPLE 2 : Consider the group $\mathbb{Z}/2 \int \Sigma_n$ of signed permutations acting on $\mathbb{F}^n = V$, where $\mathbb{F}$ is a field of characteristic different from 2, by letting Σ_n permute a basis $\{t_1, \ldots, t_n\}$ and $\mathbb{Z}/2 \times \cdots \times \mathbb{Z}/2$ act on $x = a_1 t_1 + \cdots + a_n t_n$ by changing the signs of the coefficients $a_1, \ldots, a_n$. Then

$$\mathbb{F}[V]^{\mathbb{Z}/2 \int \Sigma_n} = \mathbb{F}[e_1(t_1^2,, \ldots, t_n^2), \ldots, e_n(t_1^2, \ldots, t_n^2)]$$

where $e_1, \ldots, e_n$ are the elementary symmetric polynomials. To see this we note:

  (i) $e_1(t_1^2, \ldots, t_n^2), \ldots, e_n(t_1^2, \ldots, t_n^2) \in \mathbb{F}[V]^{\mathbb{Z}/2 \int \Sigma_n}$,
  (ii) these polynomials are a system of parameters,
  (iii) $\deg e_i(t_1^2, \ldots, t_n^2) = 2i$, and therefore
  (iv) $\displaystyle\prod_{i=1}^{n} \deg(e_i(t_1^2, \ldots, t_n^2)) = 2^n n! = |\mathbb{Z}/2 \int \Sigma_n|$,

so the formula for the invariants follows from 5.5.5.

EXAMPLE 3 (C. Peterson): Let us return to example 2 of section 3.2 where we examined $\varrho : \mathbb{Z}/4 \hookrightarrow GL(2, \mathbb{F})$ generated by the matrix

$$T = \begin{bmatrix} 0 & -1 \\ 1 & 0 \end{bmatrix}.$$

Here $\mathbb{F}$ is a field of characteristic different from 2. If $\{x, y\}$ is the dual of the canonical basis of $\mathbb{F}^2$ then in example 2 of 3.2 we saw that

$$f_1 = x^2 + y^2$$
$$f_2 = x^2 y^2$$
$$f_3 = x^3 \cdot y - x \cdot y^3$$

generate $\mathbb{F}[x, y]^{\mathbb{Z}/4}$ and satisfy one relation

$$f_3^2 = f_1^2 f_2 - 4f_2^2.$$

The polynomials f_1, f_2 are a system of parameters and $\mathbb{F}[x, y]^{\mathbb{Z}/4}$ is a free module over $\mathbb{F}[f_1, f_2]$ on the generators $\{1, f_3\}$, i.e.

$$\mathbb{F}[x, y]^{\mathbb{Z}/4} = \mathbb{F}[f_1, f_2] \oplus \mathbb{F}[f_1, f_2] \cdot f_3 .$$

The matrix T is a signed permutation matrix so there is a canonical inclusion $i : \mathbb{Z}/4 \hookrightarrow \mathbb{Z}/2 \int \Sigma_2$. By the preceding example

$$\mathbb{F}[V]^{\mathbb{Z}/2 \int \Sigma_2} = \mathbb{F}[f_1, f_2]$$

and the inclusion i induces the inclusion

$$\mathbb{F}[V]^{\mathbb{Z}/2 \int \Sigma_2} = \mathbb{F}[f_1, f_2] \hookrightarrow \mathbb{F}[f_1, f_2, f_3]/(f_3^2 - f_1^2 f_2 + 4 f_2^2) = \mathbb{F}[V]^{\mathbb{Z}/4} .$$

Therefore $\mathbb{F}[x, y]^{\mathbb{Z}/4}$ is a free module over $\mathbb{F}[x, y]^{\mathbb{Z}/2 \int \Sigma_2}$ on the generators 1 and f_3. Moreover f_3 is a det relative invariant of $\mathbb{Z}/2 \int \Sigma_2$. Thus

$$\mathbb{F}[x, y]_{\det}^{\mathbb{Z}/2 \int \Sigma_2} = \mathbb{F}[f_1, f_2] \cdot f_3 ,$$

yielding precisely the same relationship between the invariants of $\mathbb{Z}/4$ and $\mathbb{Z}/2 \int \Sigma_2$ that we found for the alternating invariants related to the symmetric invariants in section 1.3.

The following result will be of use in chapter 8 and in many examples.

PROPOSITION 5.5.6 (H. Nakajima and R.E. Stong): *Let G be a finite group and $\varrho : G \hookrightarrow \mathrm{GL}(n, \mathbb{F})$ a representation. Suppose there is a chain of G-invariant subspaces*

$$V^* = V_n^* \supset V_{n-1}^* \supset \cdots \supset V_0^* = \{0\}$$

with $\dim_{\mathbb{F}}(V_i^*) = i$ and vectors $z_i \in V_i^* \setminus V_{i-1}^*$ with G-orbits $[z_i]$ for $i = 1, \ldots, n$ satisfying the following condition:

$$\prod_{i=1}^{n} \|[z_i]\| = |G| .$$

Then $\mathbb{F}[V]^G = \mathbb{F}[f_1, \ldots, f_n]$ where

$$f_i = c_{d_i}([z_i]), \quad d_i = \|[z_i]\| \quad \text{for } i = 1, \ldots, n$$

are the top Chern classes of the orbits $[z_1], \ldots, [z_n]$.

PROOF : The polynomial f_i has degree d_i so we may apply 5.5.5 to obtain the conclusion, provided that $f_1, \ldots, f_n$ is a system of parameters. Therefore it suffices to show that $\mathbb{F}[V]$ is integral over the subalgebra $A_0 \subseteq \mathbb{F}[V]^G$ generated by $f_1, \ldots, f_n$. Let $A_i \subseteq \mathbb{F}[V]$ be the subalgebra generated by $f_1, \ldots, f_n$ together with $z_1, \ldots, z_i$. Thus we have

$$A_0 \subseteq A_1 \subseteq \cdots \subseteq A_n = \mathbb{F}[V].$$

Every element of $[z_i]$ is of the form $\lambda z_i + u$ where $\lambda \in \mathbb{F}$ is nonzero and $u \in V_{i-1}^*$. Therefore

$$f_i = \prod_{z \in [v_i]} z = \sum_{j=1}^{d_i} \varphi_{i,j}(z_1, \ldots, z_{i-1}) z_i^j$$

with $\varphi_{i,d_i}(z_1, \ldots, z_{i-1})$ a nonzero field element. Hence z_i is a root of the monic polynomial

$$\frac{1}{\varphi_{i,d_i}(z_1, \ldots, z_{i-1})} \left[f_i - \sum_{j=1}^{d_i} \varphi_{i,j}(z_1, \ldots, z_{i-1}) X^j \right] \in A_{i-1}[X].$$

Therefore A_i is integral over A_{i-1} for $i = 1, \ldots, n$ and hence by transitivity of integrality $\mathbb{F}[V] = A_n$ is integral over A_0. $\square$

5.6   Examples : GL(2, $\mathbb{F}_p$) and its Subgroups

In section 3.2 we considered some of the subgroups of GL(2, $\mathbb{F}_p$) and their invariants. Here we consider some further subgroups. The group GL(2, $\mathbb{F}_p$) has order $(p^2 - 1)(p^2 - p)$ so the largest power of p dividing the order of a subgroup is the first power.

EXAMPLE 1 (C.W. Wilkerson): Let p be an odd prime and consider the representation of the dihedral group of order $2p$, $\varrho : D_{2p} \hookrightarrow \mathrm{GL}(2, \mathbb{F}_p)$, where $\mathbb{F}_p$ is the Galois field of p elements, given by the matrices

$$S = \begin{bmatrix} 1 & 1 \\ 0 & 1 \end{bmatrix} \qquad T = \begin{bmatrix} -1 & 0 \\ 0 & 1 \end{bmatrix}.$$

S has order p and T has order 2. The action of D_{2p} on $\mathbb{F}[x, y]$ is given by the transposed matrices so

$$S(x) = x + y \qquad T(x) = -x$$
$$S(y) = y \qquad T(y) = y.$$

The class y is clearly invariant. The orbit of x is

$$x, \; x + y, \ldots, \; x + (p-1)y, \; -x, \; -x + y, \ldots, \; -x + (p-1)y$$

so the top Chern class of the orbit of x is[3]

$$\prod_{k=0}^{p-1}(ky + x)\prod_{\ell=0}^{p-1}(\ell y - x) = -x^2 \prod_{k=1}^{p-1}(ky + x)\prod_{\ell=1}^{p-1}(\ell y - x)$$

$$= -x^2 \prod_{k=1}^{p-1}(ky + x)\prod_{\ell=1}^{p-1}(-\ell y - x)$$

$$= -x^2(-1)^{p-1}\left[\prod_{k=1}^{p-1}(ky + x)\right]^2$$

$$= -x^2(x^{p-1} - y^{p-1})^2 = (xy^{p-1} - x^p)^2$$

which is an invariant of degree $2p$. The classes y and $(xy^{p-1} - x^p)^2$ are a system of parameters and the product of their degrees is the order of D_{2p} so by 5.5.5 (alternatively we could apply 5.5.6) we obtain

$$\mathbb{F}_p[x, \; y]^{D_{2p}} \cong \mathbb{F}_p[y, \; (xy^{p-1} - x^p)^2].$$

The dual representation ϱ^* is given by the transposed matrices, so the action of D_{2p} on $\mathbb{F}[x, y]$ is given by the matrices S and T and

$$S(x) = x \qquad T(x) = -x$$
$$S(y) = x + y \qquad T(y) = y.$$

This time the orbit of x gives us the invariant x^2 of degree 2 as top Chern class and the orbit of y has as top Chern class

$$\prod_{k=0}^{p-1}(kx + y) = y(y^{p-1} - x^{p-1})$$

of degree p. Again, since x and y are integral over $\mathbb{F}[x^2, \; y(y^{p-1} - x^{p-1})]$, 5.5.5 applies to yield

$$\mathbb{F}_p[x, \; y]^{D_{2p}} \cong \mathbb{F}[x^2, \; y(y^{p-1} - x^{p-1})].$$

This example illustrates several points. First, dual (transposed) representations can have very different invariants, both of which are polynomial algebras. This cannot happen in the non-modular case where we may compute

---

[3] We make use of the formula $t^{p-1} - s^{p-1} = \prod_{\lambda \in \mathbb{F}_p^\times} (t - \lambda s) \in \mathbb{F}_p[t, s]$ at several points in

the following computations.

the Poincaré series of the invariants with Molien's formula 4.3.2. The determinants that appear in this formula are the same for a representation and its dual. If the invariants are polynomial the degrees of the generators may be read off from the Poincaré series. Second, note how the *theory* helped to make the computations: the Chern classes of orbits gave us invariants, which proposition 5.5.5 said were enough to generate the ring of invariants. Finally, notice that for the representation ϱ^* every nonidentity element of D_{2p} fixes the subspace spanned by the vector $\begin{bmatrix} 0 \\ 1 \end{bmatrix} \in \mathbb{F}_p^2$ and therefore every nonidentity element is a pseudoreflection, so $|s_{\varrho^*}(D_{2p})| = 2p - 1$. However the degrees of the polynomial generators of $\mathbb{F}_p[x, y]^{D_{2p}}$ are $d_1 = 2$ and $d_2 = p$ and $(d_1 - 1) + (d_2 - 1) = 1 + p - 1 = p \neq |s_{\varrho^*}(D_{2p})| = 2p - 1$. Therefore the formula in proposition 4.4.3 for $|s(G)|$ need not hold when the order of the group G is divisible by the characteristic of the ground field $\mathbb{F}$. For a formula that holds in the modular case see [163].

The method used to compute the invariants in the preceding example is by no means an isolated instance. We illustrate this by examining the invariants of some other subgroups of $\mathrm{GL}(2, \mathbb{F}_p)$. The analysis of the following example is based on ideas of Tamagawa [246] (see also chapter 8).

EXAMPLE 2 : Consider the (parabolic) subgroup $P < \mathrm{GL}(2, \mathbb{F}_p)$ consisting of the lower triangular matrices. This group has order $p(p - 1)^2$ and acts on $\mathbb{F}[x, y]$ via the upper triangular matrices, so that in this notation the line L spanned by x is an invariant subset as is its complement $V^* \setminus L$. If we denote by $\tilde{L}$ the set of nonzero elements in L, then the top Chern class of $\tilde{L}$ is given by

$$c_{p-1}(\tilde{L}) = \prod_{\lambda \in \mathbb{F}^\times} \lambda x = x^{p-1}.$$

The orbit $V^* \setminus L$ contains $p^2 - p$ elements. These in turn lie on p lines $L_0, \ldots, L_{p-1}$ and we may choose

$$L_i = \mathrm{Span}_{\mathbb{F}_p}\{y + ix\} \qquad 0 \leq i \leq p - 1.$$

So the top Chern class of the orbit $V^* \setminus L$ is

$$c_{p^2-p} = \prod_{v \in V^* \setminus L} v = \prod_{i=0}^{p-1} \prod_{v \in L_i} v = \prod_{i=0}^{p-1} \prod_{\lambda \in \mathbb{F}^\times} \lambda(y + ix)$$

$$= \prod_{i=0}^{p-1} (y + ix)^{p-1} = \left[\prod_{i=0}^{p-1} (y + ix) \right]^{p-1}$$

$$= \left[y(y^{p-1} - x^{p-1}) \right]^{p-1} = (y^p - yx^{p-1})^{p-1}.$$

The classes $c_{p-1}(\tilde{L})$, $c_{p^2-p}(V^* \smallsetminus L)$ are algebraically independent, a system of parameters and the product of their degrees is $(p-1)(p^2-p) = |P|$, so 5.5.5 applies and we conclude that

$$\mathbb{F}[x, y]^P = \mathbb{F}[x^{p-1}, (y^p - yx^{p-1})^{p-1}].$$

EXAMPLE 3 : Consider the subgroup of GL(2, $\mathbb{F}_p$) generated by the matrix

$$S = \begin{bmatrix} 1 & 0 \\ 1 & 1 \end{bmatrix} \in \text{GL}(2, \mathbb{F}_p).$$

The matrix S has order p and generates a p-Sylow subgroup $\text{Syl}_p\text{GL}(2, \mathbb{F}_p)$ of GL(2, $\mathbb{F}_p$). The action of $\text{Syl}_p\text{GL}(2, \mathbb{F}_p)$ on the standard basis of the dual vector space is given by:

$$S(x) = x \qquad S(y) = x + y.$$

Thus x is invariant and the orbit of y is

$$y, \; x+y, \; 2x+y, \ldots, \; (p-1)x+y$$

so the top Chern class of the orbit of y is $y(y^{p-1} - x^{p-1})$. The classes $x, y(y^{p-1} - x^{p-1})$ are a system of parameters for $\mathbb{F}[x, y]$ and the product of their degrees is $p = |\text{Syl}_p\text{GL}(2, \mathbb{F}_p)|$. Hence by 5.5.5 we conclude that

$$\mathbb{F}[x, y]^{\text{Syl}_p\text{GL}(2, \mathbb{F}_p)} = \mathbb{F}[x, y(y^{p-1} - x^{p-1})].$$

We next consider the full finite general linear group. The invariants of this group were first computed by L. E. Dickson [68]. In chapter 8 we will consider the groups GL(n, $\mathbb{F}_p$) in more detail using ideas of Tamagawa.

EXAMPLE 4 (L. E. Dickson): Consider the tautological representation of GL(2, $\mathbb{F}_p$) on $V = \mathbb{F}_p^2$. Let $\{x, y\}$ be a basis for the dual vector space V^*. The only orbits of GL(2, $\mathbb{F}_p$) on V^* are $\{0\}$ and $\tilde{V} = V^* \smallsetminus \{0\}$. To compute the Chern classes of the orbit $\tilde{V}^*$ it will be convenient to consider

$$\varphi_{V^*}(t) = t\varphi_{\tilde{V}^*}(t) = \prod_{v \in V^*} (t + v).$$

(Recall that $\varphi_B(t)$ was defined for any G-invariant set B and that the coefficients of the powers of t in the expansion are invariants.) Let $L = \text{Span}_{\mathbb{F}_p}\{x\} \subset V^*$ and note that

$$\varphi_L(s) = \prod_{w \in L} (s + w) = s^p - x^{p-1}s$$

is $\mathbb{F}_p$-linear in s. Thus

$$\varphi_{V^*}(t) = \prod_{v \in V^*} (t + v) = \prod_{\lambda, \mu \in \mathbb{F}_p} (t + \lambda x + \mu y)$$

$$= \prod_{\mu \in \mathbb{F}_p} \prod_{\lambda \in \mathbb{F}_p} (t + \lambda x + \mu y)$$

$$= \prod_{\mu \in \mathbb{F}_p} \varphi_L(t + \mu y) = \prod_{\mu \in \mathbb{F}_p} (\varphi_L(t) + \mu \varphi_L(y))$$

$$= \varphi_L(t)^p - \varphi_L(y)^{p-1} \varphi_L(t) = (t^p - x^{p-1}t)^p - (y^p - x^{p-1}y)^{p-1}(t^p - x^{p-1}t)$$

$$= t^{p^2} - (x^{p(p-1)} + y^{p-1}(y^{p-1} - x^{p-1})^{p-1})t^p - (xy^p - x^p y)^{p-1}t$$

and hence

$$c_{p^2-1}(\tilde{V}^*) = (xy^p - x^p y)^{p-1}$$

$$c_{p^2-p}(\tilde{V}^*) = (x^{p(p-1)} + y^{p-1}(y^{p-1} - x^{p-1})^{p-1})$$

$$= \frac{xy^{p^2} - x^{p^2} y}{xy^p - x^p y}.$$

We claim that these polynomials are algebraically independent. To see this we make use of the following lemma.

LEMMA 5.6.1 : If $f_1, \ldots, f_n \in \mathbb{F}[z_1, \ldots, z_n]$ and

$$\det \begin{bmatrix} \frac{\partial f_1}{\partial z_1} & \cdots & \frac{\partial f_n}{\partial z_1} \\ \vdots & \vdots & \vdots \\ \frac{\partial f_1}{\partial z_n} & \cdots & \frac{\partial f_n}{\partial z_n} \end{bmatrix} \neq 0$$

then $f_1, \ldots, f_n$ are algebraically independent.

PROOF : Suppose $f_1, \ldots, f_n$ are algebraically dependent and that $r(u_1, \ldots, u_n) \in \mathbb{F}[u_1, \ldots, u_n]$ is a polynomial of minimal degree such that

$$r(f_1, \ldots, f_n) = 0.$$

Applying the chain rule gives

$$\frac{\partial r}{\partial f_1} \frac{\partial f_1}{\partial z_i} + \cdots + \frac{\partial r}{\partial f_n} \frac{\partial f_n}{\partial z_i} = 0 \qquad i = 1, \ldots, n.$$

If we rewrite this as a matrix equation

$$\begin{bmatrix} \frac{\partial f_1}{\partial z_1} & \cdots & \frac{\partial f_n}{\partial z_1} \\ \vdots & \vdots & \vdots \\ \frac{\partial f_1}{\partial z_n} & \cdots & \frac{\partial f_n}{\partial z_n} \end{bmatrix} \cdot \begin{bmatrix} \frac{\partial r}{\partial f_1} \\ \vdots \\ \frac{\partial r}{\partial f_n} \end{bmatrix} = \begin{bmatrix} 0 \\ \vdots \\ 0 \end{bmatrix}$$

then

$$\frac{\partial r}{\partial f_1} = \cdots = \frac{\partial r}{\partial f_n} = 0$$

or

$$\det \begin{bmatrix} \frac{\partial f_1}{\partial z_1} & \cdots & \frac{\partial f_n}{\partial z_1} \\ \vdots & \vdots & \vdots \\ \frac{\partial f_1}{\partial z_n} & \cdots & \frac{\partial f_n}{\partial z_n} \end{bmatrix} = 0.$$

If the first of these conditions were to hold then in characteristic zero we would have $r(u_1, \ldots, u_n) = 0$ contrary to $r \neq 0$, and in characteristic p that $r(u_1, \ldots, u_n) = s(u_1, \ldots, u_n)^p$ contrary to r being of minimal degree. $\quad\square$

REMARK :   The converse of this lemma can fail in characteristic $p \neq 0$. For example, if $\mathbb{F}_p$ is the Galois field with p elements then x^p, $y^p \in \mathbb{F}_p[x, y]$ are algebraically independent but their Jacobian matrix is identically zero.

The verification that

$$\det \begin{bmatrix} \frac{\partial c_{p^2-1}(\tilde{V})}{\partial x} & \frac{\partial c_{p^2-p}(\tilde{V})}{\partial x} \\ \frac{\partial c_{p^2-1}(\tilde{V})}{\partial y} & \frac{\partial c_{p^2-p}(\tilde{V})}{\partial y} \end{bmatrix} = x^{p^2+1}y^{2p} - x^{p+1}y^{p^2+p} + x^{p^2+p}y^{p+1} - x^{2p}y^{p^2+1},$$

which is nonzero, is left as an exercise. Thus $c_{p^2-1}(\tilde{V})$, $c_{p^2-p}(\tilde{V})$ are algebraically independent. The product of their degrees is $(p^2 - 1)(p^2 - p) = |GL(2, \mathbb{F}_p)|$. Finally we use 5.3.7 to show they are a system of parameters. To do so we must show that the only solution to the equations

$$xy^p - x^p y = 0$$

$$\frac{xy^{p^2} - x^{p^2} y}{xy^p - x^p y} = 0$$

over the algebraic closure $\bar{\mathbb{F}}_p$ of $\mathbb{F}_p$ is the trivial solution. To this end note

$$xy^p - x^p y = xy(y^{p-1} - x^{p-1})$$

$$\frac{xy^{p^2} - x^{p^2} y}{xy^p - x^p y} = \frac{y^{p^2-1} - x^{p^2-1}}{y^{p-1} - x^{p-1}}$$

$$= \frac{(y^{p-1})^{p+1} - (x^{p-1})^{p+1}}{y^{p-1} - x^{p-1}}$$

$$= (y^{p-1})^p + (y^{p-1})^{p-1}(x^{p-1}) + \cdots + (y^{p-1})(x^{p-1})^{p-1} + (x^{p-1})^p$$

so after the substitution $u = y^{p-1}$, $v = x^{p-1}$ the preceding equations are equivalent to the system

$$\begin{aligned} u - v &= 0 \\ u^p + u^{p-1}v + \cdots + uv^{p-1} + v^p &= 0. \end{aligned}$$

The first of these equations implies $u = v$ and substituting in the second then yields $u = (p + 1)u = 0$, so indeed the only solution is the trivial solution $(0, 0)$ and c_{p^2-1}, c_{p^2-p} are a system of parameters. From 5.5.5 we conclude that

$$\mathbf{D}^*(2) := \mathbb{F}_p[x, y]^{\mathrm{GL}(2, \mathbb{F}_p)} = \mathbb{F}_p[c_{p^2-1}(\tilde{V}), c_{p^2-p}(\tilde{V})].$$

Given this result it is an easy matter to work out the invariants of the subgroups lying between $\mathrm{SL}(2, \mathbb{F}_p)$ and $\mathrm{GL}(2, \mathbb{F}_p)$:

EXAMPLE 5 (L.E. Dickson): Let p be an odd prime, k a divisor of $p-1$ and set

$$\mathrm{SL}_k(2, \mathbb{F}_p) = \left\{ T \in \mathrm{GL}(2, \mathbb{F}_p) \mid \det(T)^{\frac{p-1}{k}} = 1 \right\}.$$

For $k = 1$ we have $\mathrm{SL}_1(2, \mathbb{F}_p) = \mathrm{GL}(2, \mathbb{F}_p)$ and for $k = p - 1$ we have $\mathrm{SL}_{p-1}(2, \mathbb{F}_p) = \mathrm{SL}(2, \mathbb{F}_p)$. We claim that

$$\mathbb{F}_p[x, y]^{\mathrm{SL}_k(2, \mathbb{F}_p)} = \mathbb{F}_p\left[(xy^p - x^p y)^{\frac{p-1}{k}}, \ \frac{xy^{p^2} - x^{p^2} y}{xy^p - x^p y} \right].$$

To this end note that

$$\mathbb{F}[x, y]^{\mathrm{GL}(2, \mathbb{F}_p)} \subset \mathbb{F}[x, y]^{\mathrm{SL}_k(2, \mathbb{F}_p)}$$

so

$$(xy^p - x^p y)^{p-1}$$

$$\frac{xy^{p^2} - x^{p^2} y}{xy^p - x^p y}$$

are both $\mathrm{SL}_k(2, \mathbb{F}_p)$-invariant. Next note that $(xy^p - x^p y)^{p-1}$ is a k-th power (since k is a divisor of $p - 1$) and for any $T \in \mathrm{GL}(2, \mathbb{F}_p)$

$$\left(T(xy^p - x^p y)^{\frac{p-1}{k}} \right)^k = T((xy^p - x^p y)^{p-1}) = (xy^p - x^p y)^{p-1}$$

so

$$\left(T(xy^p - x^p y)^{\frac{p-1}{k}} \right) = \lambda(T) \cdot (xy^p - x^p y)^{\frac{p-1}{k}}$$

where $\lambda : \mathrm{GL}(2, \mathbb{F}_p) \longrightarrow \mathbb{F}_p^{\times}$ is a homomorphism and $\lambda(T)^k = 1$. Thus $\lambda = \det^{k\ell}$, $(k, \ell) = 1$, $(xy^p - x^p y)^{\frac{p-1}{k}}$ is a relative invariant for $\det^{k\ell}$ and $(xy^p - x^p y)^{\frac{p-1}{k}} \in \mathbb{F}[x, y]^{\mathrm{SL}_k(2, \mathbb{F}_p)}$. The classes

$$(xy^p - x^p y)^{\frac{p-1}{k}}$$

$$\frac{xy^{p^2} - x^{p^2} y}{xy^p - x^p y}$$

have degrees $\frac{p^2-1}{k}$ and $p^2 - p$ whose product is the order of $SL_k(2, \mathbb{F}_p)$. By 5.5.5 we conclude that

$$\mathbb{F}_p[x, y]^{SL_k(2, \mathbb{F}_p)} = \mathbb{F}_p[c_{p^2-p}(\tilde{V}), c_{p^2-1}(\tilde{V})^{\frac{1}{k}}]$$

as claimed.

At this point one could also compute the modules of relative invariants for the linear representations $\det^k$ of $GL(2, \mathbb{F}_p)$.

EXAMPLE 6 : Let p be an odd prime. We begin by describing a representation of the dihedral group $\varrho : D_{2(p+1)} \hookrightarrow GL(2, \mathbb{F}_p)$. To this end we first construct $\mathbb{F}_{p^2}$, the Galois field with p^2 elements, as a vector space of dimension two over $\mathbb{F}_p$.

Choose $\lambda \in \mathbb{F}_p$ so that $-\lambda$ is not a quadratic residue modulo p. Then the polynomial $x^2 + \lambda \in \mathbb{F}_p[x]$ is irreducible, and hence the principal ideal $(x^2 + \lambda) \subset \mathbb{F}_p[x]$ is prime. Therefore the quotient ring $\mathbb{F}_p[x]/(x^2 + \lambda)$ is an integral domain. It contains only finitely many elements, namely p^2, since $\{1, x\}$ is a basis for $\mathbb{F}_p[x]/(x^2 + \lambda)$ over $\mathbb{F}_p$. A finite integral domain is a field, so $\mathbb{F}_p[x]/(x^2 + \lambda) \cong \mathbb{F}_{p^2}$.

The group of units of $\mathbb{F}_{p^2}$ is cyclic of order $p^2 - 1$. Let $\xi \in \mathbb{F}_{p^2}^{\times}$ be an element of order $p+1$. Write $\xi = a + bx$, a, $b \in \mathbb{F}_p$. Consider the linear transformation

$$T : \mathbb{F}_p \oplus \mathbb{F}_p \longrightarrow \mathbb{F}_p \oplus \mathbb{F}_p$$

given by left multiplication by ξ, where we identify $\mathbb{F}_p \oplus \mathbb{F}_p$ with $\mathbb{F}_{p^2}$ using $\{1, x\}$ as a basis for $\mathbb{F}_{p^2}$ over $\mathbb{F}_p$. Then

$$T(1) = \xi \cdot 1 = a + bx$$
$$T(x) = \xi \cdot x = (a + bx)x = ax + bx^2$$
$$= ax - \lambda b = -\lambda b + ax,$$

so the matrix of T with respect to the basis $\{1, x\}$ is

$$T = \begin{bmatrix} a & -\lambda b \\ b & a \end{bmatrix}.$$

By construction this matrix has order $p + 1$, hence

$$1 = \det(T^{p+1}) = (\det(T))^{p+1} = (\det(T))^p \det(T) = (\det(T))^2,$$

so $\det(T) = \pm 1$. Since the sign plays no role in the discussion we suppose the matrix T has $\det(T) = 1 = a^2 + \lambda b^2$. The formula for the inverse of a 2×2 matrix, [210] Proposition 11.3, then gives

$$T^{-1} = \begin{bmatrix} a & \lambda b \\ -b & a \end{bmatrix} \in GL(2, \mathbb{F}_p).$$

To simplify the remainder of the discussion we suppose $p \equiv 3 \bmod 4$, so -1 is not a quadratic residue in $\mathbb{F}_p$. Therefore we may choose $\lambda = 1$ when constructing $\mathbb{F}_{p^2}$ from $\mathbb{F}_p$, so

$$T = \begin{bmatrix} a & -b \\ b & a \end{bmatrix}$$

$$T^{-1} = \begin{bmatrix} a & b \\ -b & a \end{bmatrix} = T^{\mathrm{tr}}.$$

Consider the matrix

$$S = \begin{bmatrix} 0 & 1 \\ 1 & 0 \end{bmatrix} \in \mathrm{GL}(2, \mathbb{F}_p).$$

A short computation shows that $S^2 = 1$ and $STS = T^{\mathrm{tr}} = T^{-1}$ so S and T generate a dihedral subgroup $D_{2(p+1)}$ of $\mathrm{GL}(2, \mathbb{F}_p)$ of order $2(p+1)$. This is the representation $\varrho : D_{2(p+1)} \hookrightarrow \mathrm{GL}(2, \mathbb{F}_p)$ whose invariants we wish to study.

Since $|D_{2(p+1)}| = 2(p+1) \not\equiv 0 \bmod (p)$ we may use Molien's theorem 4.3.2 to compute the Poincaré series of $\mathbb{F}_p[x, y]^{D_{2(p+1)}}$. Without loss of generality we may pass to an algebraic closure $\bar{\mathbb{F}}_p$ of $\mathbb{F}_p$ where the matrix T diagonalizes to

$$\begin{bmatrix} \xi & 0 \\ 0 & \xi^{-1} \end{bmatrix} \qquad \xi^{p+1} = 1 \in \bar{\mathbb{F}}_p.$$

The computation then proceeds as in section 4.3 Example 2 and we find

$$P(\mathbb{F}_p[x, y]^{D_{2(p+1)}}, \ t) = \frac{1}{(1 - t^2)(1 - t^{p+1})}.$$

This suggests that $\mathbb{F}[x, y]^{D_{2(p+1)}}$ is a polynomial algebra on two generators, one of degree 2 and one of degree $p + 1$. If we pass from $\mathbb{F}_p$ to the extension field $\mathbb{F}_{p^2}$ then this may be verified as follows.

The minimal polynomial of T is $t^{p+1} - 1$. Since $\mathbb{F}_{p^2}$ contains ξ, which is a $(p+1)$-st root of unity, we may find a basis $\{E_1, E_2\}$ for $V = \mathbb{F}_{p^2} \oplus \mathbb{F}_{p^2}$ of eigenvectors of T. With respect to this basis T is represented by the diagonal matrix

$$\begin{bmatrix} \xi & 0 \\ 0 & \xi^{-1} \end{bmatrix} \in \mathrm{GL}(2, \mathbb{F}_{p^2}).$$

Let the matrix of S with respect to the basis $\{E_1, E_2\}$ be

$$\begin{bmatrix} \alpha & \beta \\ \gamma & \delta \end{bmatrix} \in \mathrm{GL}(2, \mathbb{F}_{p^2}).$$

Then

$$\begin{bmatrix} \alpha & \beta \\ \gamma & \delta \end{bmatrix} \begin{bmatrix} \xi & 0 \\ 0 & \xi^{-1} \end{bmatrix} = \begin{bmatrix} \xi^{-1} & 0 \\ 0 & \xi \end{bmatrix} \begin{bmatrix} \alpha & \beta \\ \gamma & \delta \end{bmatrix}$$

$$\begin{bmatrix} \alpha & \beta \\ \gamma & \delta \end{bmatrix} \begin{bmatrix} \alpha & \beta \\ \gamma & \delta \end{bmatrix} = \begin{bmatrix} 1 & 0 \\ 0 & 1 \end{bmatrix}.$$

Multiplying out and equating coefficients allows us to conclude

$$\alpha = 0 = \delta, \quad \beta = e = \gamma^{-1},$$

so S is represented by a matrix

$$\begin{bmatrix} 0 & e^{-1} \\ e & 0 \end{bmatrix} \in \mathrm{GL}(2, \ \mathbb{F}_{p^2})$$

for some $0 \neq e \in \mathbb{F}_{p^2}$. The vectors $\{E_1, \ eE_2\}$ are also an eigenbasis for V with respect to T and in addition

$$S(E_1) = eE_2, \quad S(eE_2) = E_1.$$

So with respect to the basis $\{E_1, \ eE_2\}$ for V the generators $T, \ S \in D_{2(p+1)}$ are represented by the matrices

$$T = \begin{bmatrix} \xi & 0 \\ 0 & \xi^{-1} \end{bmatrix}, \quad S = \begin{bmatrix} 0 & 1 \\ 1 & 0 \end{bmatrix} \in \mathrm{GL}(2, \ \mathbb{F}_{p^2}).$$

If $\{x, \ y\}$ is the dual basis then one readily sees that the polynomials

$$f = xy, \ h = x^{p+1} + y^{p+1} \in \mathbb{F}_{p^2}[x, \ y]$$

are invariant, a system of parameters and algebraically independent. Moreover

$$\deg(f) \cdot \deg(h) = 2(p+1) = |D_{2(p+1)}|,$$

so by 5.5.5 we conclude

$$\mathbb{F}_{p^2}[x, \ y]^{D_{2(p+1)}} = \mathbb{F}_{p^2}[xy, \ x^{p+1} + y^{p+1}].$$

Since xy and $x^{p+1} + y^{p+1}$ become invariant after extending scalars from $\mathbb{F}_p$ to $\mathbb{F}_{p^2}$ they are already invariant to begin with, and hence

$$\mathbb{F}_p[x, \ y]^{D_{2(p+1)}} = \mathbb{F}_p[xy, \ x^{p+1} + y^{p+1}].$$

The quadratic form xy is equivalent over $\mathbb{F}_p$ to the quadratic form $u^2 - v^2$ through the change of variables $x = u + v$ and $y = u - v$, so the group $D_{2(p+1)}$ is the orthogonal group $O_2^-(\mathbb{F}_p)$ in the notations of Dickson [71]. For another discussion of this example see section 10.4 example 2.

Chapter 6
Homological Properties of Invariants

Let G be a finite group acting on the finite dimensional vector space V over $\mathbb{F}$. Then $\mathbb{F}[V]^G$ is finitely generated as an algebra over $\mathbb{F}$ by 2.3.1. We may therefore write $\mathbb{F}[V]^G \cong \mathbb{F}[f_1, \ldots, f_d]/J$ where J is the ideal of relations for this presentation of $\mathbb{F}[V]^G$ as a quotient of a polynomial algebra. The ideal J in $\mathbb{F}[f_1, \ldots, f_d]$ is finitely generated since $\mathbb{F}[f_1, \ldots, f_d]$ is a Noetherian ring. Thus there is a finitely generated free $\mathbb{F}[f_1, \ldots, f_d]$-module F' and an epimorphism $\partial : F' \longrightarrow J$. The kernel J' of the map ∂ is a finitely generated $\mathbb{F}[f_1, \ldots, f_d]$-module, and so there exists a finitely generated free $\mathbb{F}[f_1, \ldots, f_d]$-module F'' and an epimorphism $\partial' : F'' \longrightarrow J'$. Proceeding in this way we obtain short exact sequences

$$0 \longrightarrow J \longrightarrow \mathbb{F}[f_1, \ldots, f_d] \longrightarrow \mathbb{F}[V]^G \longrightarrow 0$$

$$0 \longrightarrow J' \longrightarrow F' \longrightarrow J \longrightarrow 0$$

$$0 \longrightarrow J'' \longrightarrow F'' \longrightarrow J' \longrightarrow 0$$

$$\vdots \qquad \vdots \qquad \vdots$$

$$0 \longrightarrow J^{(n)} \longrightarrow F^{(n)} \longrightarrow J^{(n-1)} \longrightarrow 0$$

which may be spliced together to form a free resolution (see [136] or [173])

$$(\mathcal{F}) \qquad \cdots \longrightarrow F^{(n)} \longrightarrow \cdots \longrightarrow F' \longrightarrow \mathbb{F}[f_1, \ldots, f_d] \longrightarrow \mathbb{F}[V]^G \longrightarrow 0$$

of $\mathbb{F}[V]^G$ as an $\mathbb{F}[f_1, \ldots, f_d]$-module. The successive modules $J^{(n)}$, which appear first here and then there, are called the **syzygies** (that is about what the word 'syzygy' means with reference to the phases of the moon) of $\mathbb{F}[V]^G$. Of course they depend on the choice of the generators $f_1, \ldots, f_d \in \mathbb{F}[V]^G$. It is natural to ask if there is some choice of generators for which the process stops, i.e. for which the free resolution $\mathcal{F}$ has finite length. Moreover is there anything invariant to be gotten out of the syzygies, that is, something independent of the choice of the generators. This is the subject of Hilbert's Syzygy Theorem and one of the main sources of the origins of homological algebra.

EXAMPLE : Consider the quaternion group (see section 4.3 example 1) in its natural representation in GL(2, $\mathbb{C}$). The ring of invariants in $\mathbb{C}[x, y]$ is generated by

$$\alpha = x^4 + y^4$$
$$\beta = x^2 y^2$$
$$\gamma = x^5 y - x y^5$$

which satisfy the single relation

$$\gamma^2 = \alpha^2 \beta - 4\beta^3 .$$

So in this context,

$$0 \longrightarrow (\gamma^2 - \alpha^2\beta + 4\beta^3) \longrightarrow \mathbb{C}[\alpha, \beta, \gamma] \longrightarrow \mathbb{C}[x, y]^G \longrightarrow 0$$

defines the first syzygy sequence. The second syzygy sequence is

$$0 \longrightarrow 0 \longrightarrow \mathbb{C}[\alpha, \beta, \gamma] \cdot \xi \longrightarrow (\gamma^2 - \alpha^2\beta + 4\beta^3) \longrightarrow 0,$$

since the ideal $(\gamma^2 - \alpha^2\beta + 2\beta^3)$ is a principal ideal, so isomorphic to a free module over $\mathbb{C}[\alpha, \beta, \gamma]$ on a single generator ξ of degree 12. There are no more syzygies and the sequence

$$0 \longrightarrow \mathbb{C}[\alpha, \beta, \gamma] \cdot \xi \longrightarrow \mathbb{C}[\alpha, \beta, \gamma] \longrightarrow \mathbb{C}[x, y]^G \longrightarrow 0$$

is then a free resolution of $\mathbb{C}[x, y]^G$ as a $\mathbb{C}[\alpha, \beta, \gamma]$-module.

In this chapter we will assume a certain amount of familiarity with the rudiments of homological algebra (chain complexes, derived functors of $\otimes$ and Hom) as found for example in [54], chapters V and VI, or [136], chapters I-III and V, or [173], chapters I - VII.

6.1   Flat, Projective and Free Modules

Let A be a graded connected commutative algebra over a field $\mathbb{F}$ and M an A-module. M is called **flat** if the functor $- \otimes_A M$ is exact. M is called **projective** if every exact sequence of the form

$$0 \longrightarrow L \longrightarrow N \longrightarrow M \longrightarrow 0$$

splits or, equivalently, the functor $\mathrm{Hom}_A(M, -)$ is exact. M is called **free** if there is a graded vector space V over $\mathbb{F}$ and an isomorphism $M \cong A \otimes_\mathbb{F} V$ as A-modules. Over a general ring (i.e. not graded and connected) flat, projective and free are different concepts, but in our case we have:

PROPOSITION 6.1.1 : *Let A be a graded connected commutative algebra over a field $\mathbb{F}$ and M a positively graded A-module. Then the following are equivalent:*

(i) *M is a free A-module*

(ii) *M is a projective A-module*

(iii) *M is a flat A-module*

(iv) $\operatorname{Tor}_1^A(\mathbb{F}, M) = 0$.

PROOF : Clearly (i) $\Rightarrow$ (ii) $\Rightarrow$ (iii) $\Rightarrow$ (iv) so it will suffice to show (iv) $\Rightarrow$ (i). Let $q : M \longrightarrow QM \cong \mathbb{F} \otimes_A M$ denote the canonical projection. Choose a splitting $\sigma : QM \longrightarrow M$ as graded vector spaces and introduce the map of A-modules $\varphi : A \otimes_{\mathbb{F}} QM \longrightarrow M$ defined by $\varphi(a \otimes v) = a \cdot \sigma(v)$. Clearly $Q(A \otimes_{\mathbb{F}} QM) \cong QM$ and $Q\varphi$ is an isomorphism. By the corollary to Nakayama's Lemma 5.2.4, φ is an epimorphism and there is the exact sequence

$$0 \longrightarrow K \longrightarrow A \otimes_{\mathbb{F}} QM \longrightarrow M \longrightarrow 0.$$

Applying the functor $\mathbb{F} \otimes_A -$ to this sequence yields by virtue of (iv) the short exact sequence

$$0 \longrightarrow QK \longrightarrow A \otimes_{\mathbb{F}} QM \overset{\cong}{\longrightarrow} QM \longrightarrow 0$$

so $QK = 0$ and hence $K = 0$ by Nakayama's Lemma 5.2.3. $\quad\square$

Note that as a consequence of (iv) $\Rightarrow$ (i) we see that

$$\operatorname{Tor}_1^A(\mathbb{F}, M) = 0 \Rightarrow \operatorname{Tor}_i^A(\mathbb{F}, M) = 0 \quad \forall\, i > 0.$$

This type of result is true in more generality and could be obtained from the theory of satellites ([54] chapter III) and dimension shifting. Instead we offer a direct proof. (Compare this to the behavior of $\operatorname{Ext}_A^n(\mathbb{F}, A)$ described in section 6.6, particularly corollary 6.6.3.)

PROPOSITION 6.1.2 : *Let A be a graded connected commutative algebra over a field $\mathbb{F}$ and M a positively graded A-module. If for some $n \in \mathbb{N}$*

$$\operatorname{Tor}_n^A(\mathbb{F}, M) = 0$$

then

$$\operatorname{Tor}_m^A(\mathbb{F}, M) = 0 \quad \forall\, m \geq n.$$

PROOF : We proceed by induction on n. The case $n = 0$ is Nakayama's Lemma and the case $n = 1$ is 6.1.1. Let $\sigma : QM \longrightarrow M$ be a splitting of the canonical projection $q : M \longrightarrow QM = \mathbb{F} \otimes_A M$, and define as before $\varphi : A \otimes_{\mathbb{F}} QM \longrightarrow M$ by $\varphi(a \otimes v) = a \cdot \sigma(v)$. Then as in the proof of 6.1.1 the map φ is an epimorphism of A-modules and so we have an exact sequence

$$0 \longrightarrow N \longrightarrow A \otimes_{\mathbb{F}} QM \overset{\varphi}{\longrightarrow} M \longrightarrow 0.$$

The module $A \otimes_{\mathbb{F}} QM$ is free, and since we may suppose $n \geq 2$ we find in the long exact sequence

$$\cdots \to \mathrm{Tor}_n^A(\mathbb{F}, A \otimes_{\mathbb{F}} QM) \to \mathrm{Tor}_n^A(\mathbb{F}, M) \to \mathrm{Tor}_{n-1}^A(\mathbb{F}, N) \to \mathrm{Tor}_{n-1}^A(\mathbb{F}, A \otimes_{\mathbb{F}} QM) \to \cdots$$

that the terms

$$\mathrm{Tor}_n^A(\mathbb{F}, A \otimes_{\mathbb{F}} QM)$$
$$\mathrm{Tor}_{n-1}^A(\mathbb{F}, A \otimes_{\mathbb{F}} QM)$$

both vanish. Thus N satisfies the hypothesis of the proposition with n replaced by $n - 1$ so by induction we are done. $\square$

6.2 Regular Sequences : The Koszul Complex

The proof of the Syzygy Theorem, although it was conceived for rings of invariants, is best placed in the more general context of homological algebra over a graded polynomial algebra. This viewpoint is due to H. Cartan. For a polynomial algebra $\mathbb{F}[z_1, \ldots, z_n]$ (n.b. we do not require that $\deg(z_i) = 1$, only that $\deg(z_i) > 0$) our goal in this section is to construct for each module M over $\mathbb{F}[z_1, \ldots, z_n]$ a functorial (and minimal) resolution of M as $\mathbb{F}[z_1, \ldots, z_n]$-module.

DEFINITION : *Let A be a graded connected commutative algebra over a field $\mathbb{F}$. A sequence $a_1, \ldots, a_n \in \bar{A}$ is called a* **regular sequence** *if a_1 is not a zero divisor in A and a_i is not a zero divisor in $A/(a_1, \ldots, a_{i-1})$ for $i = 2, \ldots, n$. (N.b. $\bar{A}$ is the augmentation ideal of A.)*

EXAMPLE 1 : The elementary symmetric polynomials $e_1, \ldots, e_n \in \mathbb{F}[z_1, \ldots, z_n]$ are a regular sequence. We verify this by induction on n. The case $n = 1$ is trivial, so we may suppose that $n > 1$ and $e_1, \ldots, e_{n-1} \in \mathbb{F}[x_1, \ldots, x_{n-1}]$ is a regular sequence. Define algebra homomorphisms

$$\varrho_j : \mathbb{F}[z_1, \ldots, z_n] \longrightarrow \mathbb{F}[z_1, \ldots, \hat{z}_j, \ldots, z_n] \qquad j = 1, \ldots, n$$

by dividing out the ideal $(z_j) \subset \mathbb{F}[z_1, \ldots, z_n]$, where the variable under the $\hat{\ }$ is omitted. The elementary symmetric polynomials satisfy

$$\varrho_j(e_i(z_1, \ldots, z_n)) = \begin{cases} e_i(z_1, \ldots, \hat{z}_j, \ldots, z_n) & \text{for } i \neq n \\ 0 & \text{for } i = n \end{cases}$$

and hence $\varrho_j(e_1), \ldots, \varrho_j(e_{n-1}) \in \mathbb{F}[z_1, \ldots, \hat{z}_j, \ldots, z_n]$ is a regular sequence for $j = 1, \ldots, n$ by the induction hypothesis.

We claim that $e_1, \ldots, e_{n-1} \in \mathbb{F}[z_1, \ldots, z_n]$ is a regular sequence. For if e_m is a zero divisor modulo $e_1, \ldots, e_{m-1}$ for some $m \leq n - 1$, then we may choose a relation of minimal degree of the form

$$f_1 e_1 + f_2 e_2 + \cdots + f_m e_m = 0 \in \mathbb{F}[z_1, \ldots, z_n]$$

with $f_m \neq 0$. If we apply ϱ_n to this equation we obtain

$$\varrho_n(f_1)\varrho_n(e_1) + \varrho_n(f_2)\varrho_n(e_2) + \cdots + \varrho_n(f_m)\varrho_n(e_m) = 0 \in \mathbb{F}[z_1, \ldots, z_{n-1}].$$

Since $\varrho_n(e_1), \ldots, \varrho_n(e_m) \in \mathbb{F}[z_1, \ldots, z_{n-1}]$ is a regular sequence we conclude $\varrho_n(f_m) = 0$, $\varrho_n(f_{m-1}) = 0, \ldots, \varrho_n(f_1) = 0$ from successive application of the regularity condition. Hence z_n divides f_i in $\mathbb{F}[z_1, \ldots, z_n]$ for $i = 1, \ldots, n-1$ and setting $h_i = f_i/z_n$ for $i = 1, \ldots, n-1$ we obtain

$$h_1 e_1 + h_2 e_2 + \cdots + h_m e_m = 0 \in \mathbb{F}[z_1, \ldots, z_n].$$

This is a relation of lower degree than the one we started with, which is a contradiction to how we chose the original relation.

Finally we must show that e_n is not a zero divisor in the quotient algebra $\mathbb{F}[z_1, \ldots, z_n]/(e_1, \ldots, e_{n-1})$. To this end suppose we have an equation

$$(*) \qquad f e_n = f_1 e_1 + \cdots + f_{n-1} e_{n-1} \in \mathbb{F}[z_1, \ldots, z_n].$$

If we apply ϱ_j to this equation we obtain

$$0 = \varrho_j(f_1) e_1(z_1, \ldots, \hat{z}_j, \ldots, z_n) + \cdots + \varrho_j(f_{n-1}) e_{n-1}(z_1, \ldots, \hat{z}_j, \ldots, z_n)$$

in $\mathbb{F}[z_1, \ldots, \hat{z}_j, \ldots, z_n]$. Since

$$e_1(z_1, \ldots, \hat{z}_j, \ldots, z_n), \ldots, e_{n-1}(z_1, \ldots, \hat{z}_j, \ldots, z_n) \in \mathbb{F}[z_1, \ldots, \hat{z}_j, \ldots, z_n]$$

is a regular sequence it follows that $\varrho_j(f_1) = \cdots = \varrho_j(f_{n-1}) = 0$. Hence $f_1, \ldots, f_{n-1}$ are divisible by z_j for $j = 1, \ldots, n$ in $\mathbb{F}[z_1, \ldots, z_n]$ and therefore by the product $z_1 \cdots z_n = e_n$. Dividing equation $(*)$ by e_n we obtain

$$f = \frac{f_1}{e_n} e_1 + \cdots + \frac{f_{n-1}}{e_n} e_{n-1} \in \mathbb{F}[z_1, \ldots, z_n]$$

so $f \in (e_1, \ldots, e_{n-1})$ and e_n is not a zero divisor modulo $(e_1, \ldots, e_{n-1})$ as was to be shown. $\quad \square$

PROPOSITION 6.2.1 : *If A is a graded connected commutative algebra over a field $\mathbb{F}$ and $a_1, \ldots, a_n \in A$ a regular sequence, then $a_1, \ldots, a_n$ are algebraically independent.*

PROOF : We recycle yet again a portion of the proof of the Noether Normalization Theorem 5.3.3. Introduce the polynomial algebra $\mathbb{F}[z_1, \ldots, z_n]$, where $\deg(z_i) = \deg(a_i)$ for $i = 1, \ldots, n$. By requiring $\varphi(z_i) = a_i$ for $i = 1, \ldots, n$ we obtain a map of algebras $\varphi : \mathbb{F}[z_1, \ldots, z_n] \longrightarrow A$. The elements $a_1, \ldots, a_n$ are algebraically independent if and only if φ is monic.

To show φ is a monomorphism we proceed by induction on n. For $n = 1$ the element $a_1 \in A$ is not a zero divisor so φ is monic. Suppose the result proven for all sequences of length $n-1$ in all graded connected commutative algebras over $\mathbb{F}$. Let $a_1, \ldots, a_n \in A$ be a regular sequence of length n. Evidently the residue classes $\bar{a}_2, \ldots, \bar{a}_n \in A/(a_1)$ are a regular sequence, and hence by the induction hypothesis are algebraically independent in $A/(a_1)$. Suppose that $f(z_1, \ldots, z_n) \in \ker(\varphi)$ has minimal degree. We write

$$f(z_1, \ldots, z_n) = z_1 q(z_1, \ldots, z_n) + r(z_2, \ldots, z_n).$$

Then $r(\bar{a}_2, \ldots, \bar{a}_n) = 0$ implies that r is identically zero, and hence

$$0 = f(a_1, \ldots, a_n) = a_1 \cdot q(a_1, \ldots, a_n) \in A.$$

But $a_1 \in A$ is not a zero divisor and hence

$$q(a_1, \ldots, a_n) = 0$$

contrary to the choice of $f(z_1, \ldots, z_n)$ of minimal degree. $\square$

EXAMPLE 2 : The converse of the preceding proposition is false: x, $xy \in \mathbb{F}[x, y]$ are algebraically independent but not a regular sequence in $\mathbb{F}[x, y]$. However, because they are algebraically independent, they are a regular sequence in the subalgebra $A = \mathbb{F}[x, xy] \subset \mathbb{F}[x, y]$. Therefore regularity does not behave well under extension.

Nor does regularity behave well under restriction. Consider the subalgebra $A \subset \mathbb{F}[x, y]$ generated by

$$x^2, \ x^2 y, \ y^2, \ y^3 \in \mathbb{F}[x, y].$$

The elements x^2, $y^2 \in A$ regarded as elements of $\mathbb{F}[x, y]$ are a regular sequence. However

$$y^2(x^2 y) = x^2 y^3 = 0 \in A/(x^2)$$

and, since $x^2 y$ does not belong to the ideal generated by x^2 in A, it follows that y^2 is a zero divisor in $A/(x^2)$. Hence x^2, y^2 is not a regular sequence in A.

COROLLARY 6.2.2 : Let A be a graded connected commutative algebra over a field $\mathbb{F}$ and $a_1, \ldots, a_n \in A$ a regular sequence. Then $n \leq \dim(A)$.

PROOF : This follows from the Noether Normalization Theorem 5.3.3 and the preceding proposition. $\square$

Algebras for which the length of the longest regular sequence they contain is equal to their Krull dimension are called **Cohen-Macaulay algebras** and will be examined in more detail in section 6.7.

NOTATION : *If V is a graded vector space over $\mathbb{F}$ we denote by $E[V]$ the exterior algebra on the dual vector space V^*. If $u_1, \ldots, u_n$ is a basis for V^*, then $E[V] = E[u_1, \ldots, u_n]$ has as a basis the elements*

$$u_{i_1} \cdots u_{i_m} \qquad i_1 < i_2 < \cdots < i_m$$

of degree $\deg(u_{i_1}) + \cdots + \deg(u_{i_m})$. Multiplication is by juxtaposition and subject to the rules

$$u_i u_j = \begin{cases} -u_j u_i & \text{for } i \neq j \\ 0 & \text{for } i = j. \end{cases}$$

The elements of $E[V]$ may be thought of as alternating multilinear functions from $\mathrm{Tot}(V)$ to $\bar{\mathbb{F}}$, where $\bar{\mathbb{F}}$ is the algebraic closure of $\mathbb{F}$. The homogeneous component of $E[V]$ of degree k is denoted by $\Lambda^k(V^*)$ or $\Lambda^k(u_1, \ldots, u_n)$ and is isomorphic to the k-th exterior power of V^*.

DEFINITION : *Let A be a graded connected commutative algebra over a field $\mathbb{F}$ and $a_1, \ldots, a_n \in A$. The **Koszul complex** of A with respect to $a_1, \ldots, a_n \in A$ is the differential graded commutative algebra*

$$\mathcal{K} = \mathcal{K}(a_1, \ldots, a_n) = A \otimes E[sa_1, \ldots, sa_n]$$

where[1] $\deg(sa_i) = 1 + \deg(a_i)$ and the differential ∂ is defined by requiring

$$\partial|_A = 0$$

$$\partial(sa_i) = a_i \qquad \text{for } i = 1, \ldots, n$$

$$\partial(x \cdot y) = \partial(x)y + (-1)^{\deg(x)} x \partial(y) \qquad \forall \, x, y \in \mathcal{K}.$$

The problem of syzygies asks for a finite free resolution of modules over the polynomial algebra $\mathbb{F}[z_1, \ldots, z_n]$. The Koszul complex provides a solution to this problem by **bigrading** it. In $\mathcal{K}_A(a_1, \ldots, a_n) = A \otimes E[sa_1, \ldots, sa_n]$ we introduce a bigrading by demanding

$$\mathrm{bideg}(a \otimes sa_{i_1} sa_{i_2} \cdots sa_{i_k}) = (k, \ \deg(a) + \deg(a_{i_1}) + \cdots + \deg(a_{i_k})).$$

where $a \in A$. The first component of the bidegree is called the **homological degree** or **resolution degree**, and the second component the **internal degree**. The homogeneous component of homological degree k is isomorphic to $A \otimes \Lambda^k(sa_1, \ldots, sa_n)$. The differential in the Koszul complex lowers the homological degree by 1, and so we may arrange the Koszul complex in the form of a sequence of graded A-modules

$$0 \longleftarrow \mathcal{K}_{0,*} \xleftarrow{\partial_1} \mathcal{K}_{1,*} \xleftarrow{\partial_2} \cdots \xleftarrow{\partial_n} \mathcal{K}_{n,*} \longleftarrow 0$$

where $\mathcal{K}_{i,*} \cong A \otimes \Lambda^i(V^*)$, $\Lambda^i(V^*)$ being the i-th exterior power of the vector space V^* dual to the vector space with basis $sa_1, \ldots, sa_n$. The homology of the Koszul complex is the homology of this chain complex of A-modules.

---

[1] The s is meant to suggest **suspension** or **shift**.

THEOREM 6.2.3 : *Let A be a graded connected commutative algebra over a field $\mathbb{F}$ and $a_1, \ldots, a_n \in A$. The Koszul complex $\{\mathcal{K}(a_1, \ldots, a_n), \partial\}$ is acyclic if and only if $a_1, \ldots, a_n$ is a regular sequence.*

PROOF : It is straightforward (but a tedious manipulation of signs) to verify that $\partial \cdot \partial = 0$ so $\mathcal{K}$ is a complex. For convenience set

$$\mathcal{K}(i) = \begin{cases} A & \text{for } i = 0 \\ \mathcal{K}(a_1, \ldots, a_i) & \text{for } i > 0 \end{cases}.$$

Note that $\mathcal{K}(i-1) \hookrightarrow \mathcal{K}(i)$ is a subcomplex, and moreover

$$\mathcal{K}(i-1) \xrightarrow{\cong} \mathcal{K}(i)/\mathcal{K}(i-1).$$

The isomorphism, which is of degree $1 + \deg(a_i)$, is given by

$$x \mapsto x \cdot sa_i \qquad x \in \mathcal{K}(i-1).$$

We therefore receive an exact homology sequence

$$\cdots \longrightarrow H_j(\mathcal{K}(i-1)) \longrightarrow H_j(\mathcal{K}(i)) \longrightarrow H_j(\mathcal{K}(i-1)) \longrightarrow H_{j-1}(\mathcal{K}(i-1)) \longrightarrow \cdots$$

Using the induction hypothesis reduces this sequence to

$$0 \longrightarrow H_1(\mathcal{K}(i)) \longrightarrow A/(a_1, \ldots, a_{i-1}) \xrightarrow{\cdot a_i} A/(a_1, \ldots, a_{i-1}) \longrightarrow A/(a_1, \ldots, a_i) \longrightarrow 0.$$

Since multiplication by a_j on $A/(a_1, \ldots, a_{j-1})$ is a monomorphism for $j = 1, \ldots, i$ if and only if $a_1, \ldots, a_i$ is a regular sequence, the result follows. $\square$

If we regard $\mathcal{K}$ as a chain complex of graded A-modules augmented by the map

$$\mathcal{K}(a_1, \ldots, a_n) \xrightarrow{1 \otimes \epsilon} A \xrightarrow{q} A/(a_1, \ldots, a_n)$$

where ϵ denotes the augmentation of $E[V]$ and q the quotient map we obtain:

THEOREM 6.2.4 (J. -L. Koszul): *Let A be a graded connected commutative algebra over a field $\mathbb{F}$ and $a_1, \ldots, a_n \in A$ a regular sequence. Then the Koszul complex $\mathcal{K}_A(a_1, \ldots, a_n)$ provides a free acyclic resolution of $A/(a_1, \ldots, a_n)$ as an A-module.* $\square$

A number of special cases come readily to hand.

COROLLARY 6.2.5 : *The Koszul complex*

$$\mathcal{K} = \mathbb{F}[z_1, \ldots, z_n] \otimes_{\mathbb{F}} E[sz_1, \ldots, sz_n]$$

with differential

$$\partial(z_i) = 0, \qquad \partial(sz_i) = z_i \qquad i = 1, \ldots, n$$

is acyclic and provides a free acyclic resolution of $\mathbb{F}$ as $\mathbb{F}[z_1, \ldots, z_n]$-module.

PROOF : $z_1, \ldots, z_n \in \mathbb{F}[z_1, \ldots, z_n]$ is a regular sequence. $\square$

COROLLARY 6.2.6 : *The Koszul complex*
$$\mathcal{L} = \mathbb{F}[z_1, \ldots, z_n] \otimes_{\mathbb{F}} E[sz_1, \ldots, sz_n] \otimes_{\mathbb{F}} \mathbb{F}[z_1, \ldots, z_n]$$
with differential
$$\partial(1 \otimes z_i) = 0, \qquad \partial(z_i \otimes 1) = 0, \qquad \partial(sz_i) = z_i \otimes 1 - 1 \otimes z_i$$
is acyclic and provides a free resolution of $\mathbb{F}[z_1, \ldots, z_n]$ *as an* $\mathbb{F}[z_1, \ldots, z_n]$-*bimodule, i.e. as* $\mathbb{F}[z_1, \ldots, z_n] \otimes \mathbb{F}[z_1, \ldots, z_n]$-*module.*

PROOF : The sequence
$$z_1 \otimes 1 - 1 \otimes z_1, \ldots, z_n \otimes 1 - 1 \otimes z_n \in \mathbb{F}[z_1, \ldots, z_n] \otimes \mathbb{F}[z_1, \ldots, z_n]$$
is a regular sequence and the multiplication map
$$\mu : \mathbb{F}[z_1, \ldots, z_n] \otimes \mathbb{F}[z_1, \ldots, z_n] \longrightarrow \mathbb{F}[z_1, \ldots, z_n]$$
induces an isomorphism
$$\frac{\mathbb{F}[z_1, \ldots, z_n] \otimes \mathbb{F}[z_1, \ldots, z_n]}{(z_1 \otimes 1 - 1 \otimes z_1, \ldots, z_n \otimes 1 - 1 \otimes z_n)} \cong \mathbb{F}[z_1, \ldots, z_n]$$
of bimodules. We may bigrade $\mathcal{L}$ by setting
$$\mathrm{bideg}(f' \otimes sz_{i_1} \cdots sz_{i_k} \otimes f'') = (k, \ \deg(f') + \deg(z_{i_1}) + \cdots + \deg(z_{i_k}) + \deg(f''))$$
and then rearrange $\mathcal{L}$ into the required resolution as with the Koszul complex $\mathcal{K}$. $\square$

The complex $\mathcal{K}$ is often referred to simply as <u>the</u> Koszul complex for $\mathbb{F}[z_1, \ldots, z_n]$ and [24] $\mathcal{L}$ as <u>the</u> two sided Koszul complex for $\mathbb{F}[z_1, \ldots, z_n]$.

COROLLARY 6.2.7 : *Let M be a module over* $\mathbb{F}[z_1, \ldots, z_n]$. *Then the complex*
$$\mathcal{L}(M) = \mathcal{L} \otimes_{\mathbb{F}[z_1, \ldots, z_n]} M$$
is a free resolution of M as an $\mathbb{F}[z_1, \ldots, z_n]$-*module.*

PROOF :   The complex $\mathcal{L}$ is a complex of free right $\mathbb{F}[z_1, \ldots, z_n]$-modules and $H_*(\mathcal{L})$ is a free $\mathbb{F}[z_1, \ldots, z_n]$-module as well. The complex $\mathcal{L}$ is acyclic and its homology is isomorphic to
$$\frac{\mathbb{F}[z_1, \ldots, z_n] \otimes \mathbb{F}[z_1, \ldots, z_n]}{(z_1 \otimes 1 - 1 \otimes z_1, \ldots, z_n \otimes 1 - 1 \otimes z_n)} \cong \mathbb{F}[z_1, \ldots, z_n].$$
We may apply the Künneth theorem [136] V.10.1 and 6.2.6 to conclude
$$H_*(\mathcal{L}(M)) \cong H_*(\mathcal{L}) \otimes_{\mathbb{F}[z_1, \ldots, z_n]} M \cong M .$$

We bigrade the complex $\mathcal{L}(M)$ by giving the elements $x \in M$ bidegree $(0, \deg(x))$ and in this way obtain the desired resolution from 6.2.6. $\square$

COROLLARY 6.2.8 : *Suppose that A is a graded connected commutative algebra over a field $\mathbb{F}$ and $a_1, \ldots, a_n \in A$. Regard A as a module over the polynomial algebra $\mathbb{F}[\bar{a}_1, \ldots, \bar{a}_n]$, where $\deg(\bar{a}_i) = \deg(a_i)$ for $i = 1, \ldots, n$, by*

$$f(\bar{a}_1, \ldots, \bar{a}_n) \cdot a = f(a_1, \ldots, a_n) \cdot a \qquad \forall \, f \in \mathbb{F}[\bar{a}_1, \ldots, \bar{a}_n] \qquad a \in A.$$

Then A is a free $\mathbb{F}[\bar{a}_1, \ldots, \bar{a}_n]$-module if and only if $a_1, \ldots, a_n$ is a regular sequence.

PROOF : By 6.1.1 it suffices to show that $\operatorname{Tor}_1^{\mathbb{F}[\bar{a}_1, \ldots, \bar{a}_n]}(\mathbb{F}, \, A) = 0$. The Koszul complex

$$\mathcal{K} = \mathbb{F}[\bar{a}_1, \ldots, \bar{a}_n] \otimes_{\mathbb{F}} E[s\bar{a}_1, \ldots, s\bar{a}_n]$$

of 6.2.5 is a free resolution of $\mathbb{F}$ as $\mathbb{F}[\bar{a}_1, \ldots, \bar{a}_n]$-module and

$$\operatorname{Tor}_*^{\mathbb{F}[\bar{a}_1, \ldots, \bar{a}_n]}(\mathbb{F}, \, A) = H_*(\mathcal{K} \otimes_{\mathbb{F}[\bar{a}_1, \ldots, \bar{a}_n]} A) \cong H_*(\mathcal{K}_A(a_1, \ldots, a_n))$$

which by Koszul's theorem is acyclic if and only if $a_1, \ldots, a_n$ is a regular sequence. $\square$

The preceding result provides us with the tool needed to prove[2] the theorem of Garsia and Stanton [89], which improves Noether's upper bound 2.4.2 for the degrees of algebra generators of the ring of invariants of a permutation group.

THEOREM 6.2.9 (A. K. Garsia and D. Stanton): *Let G be a finite group and X a finite G-set. If $\mathbb{F}$ is a field of characteristic prime to the order of G then $\mathbb{F}[X]^G$ is generated by the orbit sums of degree less than or equal to* $\min \left\{ |X|, \, \binom{|X|}{2} = \frac{|X|(|X|-1)}{2} \right\}$.

PROOF : Let $|X| = n$ and write $\varrho : G \hookrightarrow \Sigma_n$ for the permutation representation corresponding to the action of G on X. We have the following relations between the rings of invariants

$$\mathbb{F}[e_1, \ldots, e_n] = \mathbb{F}[X]^{\Sigma_n} \subseteq \mathbb{F}[X]^G \subseteq \mathbb{F}[X].$$

The projection operator (see section 2.4)

$$\pi^G : \mathbb{F}[X] \longrightarrow \mathbb{F}[X]^G$$

is a $\mathbb{F}[X]^{\Sigma_n}$-module homomorphism and splits the inclusion $\mathbb{F}[X]^G \subseteq \mathbb{F}[X]$. Hence we obtain a monomorphism

$$\mathbb{F} \otimes_{\mathbb{F}[X]^{\Sigma_n}} \mathbb{F}[X]^G \hookrightarrow \mathbb{F} \otimes_{\mathbb{F}[X]^{\Sigma_n}} \mathbb{F}[X] = \mathbb{F}[X]_{\Sigma_n}.$$

---

[2] This proof is based on a conversation with Hanspeter Kraft. See also [194] proposition 9.2 (where $\binom{n-1}{2}$ should be replaced by $\binom{n}{2}$).

The elementary symmetric polynomials $e_1, \ldots, e_n \in \mathbb{F}[X]^{\Sigma_n}$ are a regular sequence, so by 6.2.8 $\mathbb{F}[X]$ is a free $\mathbb{F}[e_1, \ldots, e_n]$-module and we have an isomorphism $\mathbb{F}[X] \cong \mathbb{F}[X]^{\Sigma_n} \otimes \mathbb{F}[X]_{\Sigma_n}$ as $\mathbb{F}[X]^{\Sigma_n}$-modules. Hence the Poincaré series of $\mathbb{F}[X]_{\Sigma_n}$ is given by

$$P(\mathbb{F}[X]_{\Sigma_n}, \, t) = \frac{P(\mathbb{F}[X], \, t)}{P(\mathbb{F}[X]^{\Sigma_n}, \, t)} = \frac{\prod_{i=1}^n (1 - t^i)}{(1 - t)^n}$$

$$= \prod_{i=1}^n (1 + t + \cdots + t^{i-1}),$$

which is a polynomial of degree $1 + 2 + \cdots + (n-1) = \frac{n(n-1)}{2}$. It follows that $\mathbb{F}[X]_{\Sigma_n}$ is zero in degrees greater than $\frac{n(n-1)}{2}$, and since $\mathbb{F} \otimes_{\mathbb{F}[X]^{\Sigma_n}} \mathbb{F}[X]^G$ is a submodule of $\mathbb{F}[V]_{\Sigma_n}$ the same holds for $\mathbb{F} \otimes_{\mathbb{F}[X]^{\Sigma_n}} \mathbb{F}[X]^G$. Therefore by Nakayama's lemma 5.2.3 $\mathbb{F}[X]^G$ is generated as a $\mathbb{F}[X]^{\Sigma_n}$-module by forms of degree less than or equal to $\frac{n(n-1)}{2}$, and hence a fortiori, $\mathbb{F}[X]^G$ is generated as an algebra by polynomials of degree at most $\frac{n(n-1)}{2}$. $\quad\square$

This bound is generally better than $|G|$, which is what Noether's theorem 2.4.2 provides, and has been extended to an arbitrary ring of coefficients by M. Göbel [93]. Moreover the bound $\frac{n(n-1)}{2}$ for transitive permutation representations of degree n is sharp because (see section 1.3) the ring of invariants of the alternating group A_n in its tautological representation is generated by the elementary symmetric functions $e_1, \ldots, e_n$ and Δ, the discriminant, which has degree $n(n-1)/2$. If the action of G on X is not transitive, then X decomposes into a disjoint union $A \sqcup B$ of G-invariant subsets. Let $|A| = a$, $|B| = b$, and $|X| = n = a + b$. The operation of G on X determines representations

$$\varrho_A : G \hookrightarrow \Sigma_a$$
$$\varrho_B : G \hookrightarrow \Sigma_b$$
$$\varrho : G \hookrightarrow \Sigma_n$$

with $\varrho = \varrho_A \oplus \varrho_B$. We have inclusions

$$\mathbb{F}[A]^{\Sigma_a} \otimes \mathbb{F}[B]^{\Sigma_b} = \mathbb{F}[X]^{\Sigma_a \times \Sigma_b} \subseteq \mathbb{F}[X]^G \subseteq \mathbb{F}[X].$$

As in the proof of 6.2.9 $\mathbb{F}[X]$ is a free $\mathbb{F}[A]^{\Sigma_a} \otimes \mathbb{F}[B]^{\Sigma_b}$-module and if the characteristic of $\mathbb{F}$ is prime to the order of G the projection $\pi^G : \mathbb{F}[X] \longrightarrow \mathbb{F}[X]^G$ is defined and splits the inclusion $\mathbb{F}[X]^G \subseteq \mathbb{F}[X]$ as $\mathbb{F}[A]^{\Sigma_a} \otimes \mathbb{F}[B]^{\Sigma_b}$-module. Therefore the inclusion $\mathbb{F}[X]^G \subseteq \mathbb{F}[X]$ induces a monomorphism

$$\mathbb{F} \otimes_{\mathbb{F}[A]^{\Sigma_a} \otimes \mathbb{F}[B]^{\Sigma_b}} \mathbb{F}[X]^G \hookrightarrow \mathbb{F}[X]_{\Sigma_a \times \Sigma_b}.$$

By 1.5.3
$$\mathbb{F}[X]_{\Sigma_a \times \Sigma_b} \cong \mathbb{F}[A]_{\Sigma_a} \otimes \mathbb{F}[B]_{\Sigma_b} .$$

The Poincaré series of $\mathbb{F}[A]_{\Sigma_a}$ and $\mathbb{F}[B]_{\Sigma_b}$ have degrees $a(a-1)/2 = \binom{a}{2}$ and $b(b-1)/2 = \binom{b}{2}$ respectively. Therefore the Poincaré series of $\mathbb{F}[X]_{\Sigma_a \times \Sigma_b}$ has degree $\binom{a}{2} + \binom{b}{2}$. Hence, as in the proof of 6.2.9, we conclude that $\mathbb{F}[X]^G$ is generated as an algebra by invariants of degree at most $\binom{a}{2} + \binom{b}{2}$. Note

$$\binom{a}{2} + \binom{b}{2} = \binom{a+b}{2} - ab = \binom{n}{2} - ab$$

so this improves the bound of theorem 6.2.9. Since any finite G-set X decomposes into a finite disjoint union of orbits, we obtain by induction the following refinement of the bound of Garsia and Stanton.

THEOREM 6.2.10 : *Let G be a finite group and X a finite G-set with orbit decomposition $X = X_1 \sqcup \cdots \sqcup X_m$, and set $n_i = |X_i|$ for $i = 1, \ldots, m$. If $\mathbb{F}$ is a field of characteristic prime to the order of G then $\mathbb{F}[X]^G$ is generated by the orbit sums of degree less than or equal to $\binom{n_1}{2} + \binom{n_2}{2} + \cdots + \binom{n_m}{2}$.* □

6.3   Hilbert's Syzygy Theorem

The Koszul complex provides us with a solution to the problem of syzygies. To wit:

THEOREM 6.3.1   (Hilbert's Syzygy Theorem): *A graded module M over the graded polynomial algebra $\mathbb{F}[z_1, \ldots, z_n]$ has a finite free resolution of length at most n, i.e. there is a finite free resolution*

$$(\mathcal{F}) \qquad 0 \longrightarrow F_m \longrightarrow \cdots \longrightarrow F_1 \longrightarrow F_0 \longrightarrow M \longrightarrow 0$$

of M as an $\mathbb{F}[z_1, \ldots, z_n]$-module with $m \leq n$.

PROOF : Choose $\mathcal{F} = \mathcal{L}(M)$. □

COROLLARY 6.3.2 : *Let $G \hookrightarrow \mathrm{GL}(V)$ be a finite dimensional representation of a finite group G and $f_1, \ldots, f_s \in \mathbb{F}[V]^G$ a system of algebra generators. Then $\mathbb{F}[V]^G$ has at most s syzygies.*

PROOF : With respect to the surjective homomorphism of $\mathbb{F}[z_1, \ldots, z_s]$-modules

$$\mathbb{F}[z_1, \ldots, z_s] \longrightarrow \mathbb{F}[V]^G \qquad z_i \mapsto f_i,$$

$F[V]^G$ admits a finite free resolution of length at most s. □

DEFINITION :   Let A be a graded connected commutative algebra over a field $\mathbb{F}$ and M an A-module. If

$$(\mathcal{F}) \qquad \cdots \longrightarrow F_s \longrightarrow \cdots \longrightarrow F_1 \longrightarrow F_0 \longrightarrow M \longrightarrow 0$$

is a free resolution of M as an A-module we say $\mathcal{F}$ **has finite length** n if $F_n \neq 0$ and $F_s = 0$ for $s > n$. If M admits a free resolution of finite length then the minimal such length is called the **projective dimension** or **homological dimension** of M, and is denoted by proj–dim$_A(M)$ or hom–dim$_A(M)$. The algebra A is said to have **finite global (projective) dimension** d if $\sup_{M \in \mathbf{MOD}/A}\{\text{proj–dim}_A(M)\}$ is finite and equal to d (MOD/A denotes the category of graded modules over A). The global dimension of A is denoted by gl–dim(A).

As usual when the suprema are not finite we set hom–dim$_A(M) = \infty$, respectively gl–dim$(A) = \infty$.

It is not our purpose in this chapter to present a course in homological algebra, but we do need a few facts about projective dimension.

PROPOSITION 6.3.3 :   Let A be a graded connected commutative algebra over a field $\mathbb{F}$, M an A-module and $n \in \mathbb{N}$. Then the following are equivalent:

(i) M has projective dimension less than or equal to n.

(ii) $\text{Ext}_A^{n+1}(M, \text{—}) = 0$.

(iii) $\text{Ext}_A^n(M, \text{—})$ is right exact

(iv) Given an exact sequence

$$0 \longrightarrow K_n \longrightarrow P_{n-1} \longrightarrow \cdots \longrightarrow P_0 \longrightarrow M \longrightarrow 0$$

with P_i projective for $0 \leq i < n$, the module K_n is projective.

PROOF :   The implications (i) $\Rightarrow$ (ii) $\Rightarrow$ (iii) are clear. To see that (iii) $\Rightarrow$ (iv) note that to prove K_n projective is equivalent to showing that the functor $\text{Hom}_A(K_n, \text{—})$ is exact. So suppose that

$$0 \longrightarrow N' \longrightarrow N \longrightarrow N'' \longrightarrow 0$$

is an exact sequence of A-modules. The exact sequence (iv) may be despliced to a braid of short exact sequences

Since P_i is projective for $0 \leq i < n$, the iterations of the connecting homomorphisms, which are isomorphisms, yield a commutative diagram

$$\begin{array}{ccccccc}
\mathrm{Hom}_A(P_{n-1},\, N) & \longrightarrow & \mathrm{Hom}_A(K_n,\, N) & \longrightarrow & \mathrm{Ext}^n_A(M,\, N) & \longrightarrow & 0 \\
\downarrow & & \downarrow & & \downarrow & & \\
\mathrm{Hom}_A(P_{n-1},\, N'') & \longrightarrow & \mathrm{Hom}_A(K_n,\, N'') & \longrightarrow & \mathrm{Ext}^n_A(M,\, N'') & \longrightarrow & 0
\end{array}$$

Since P_{n-1} is projective the vertical arrow on the left is surjective. The vertical arrow on the right is surjective because $\mathrm{Ext}^n_A(M,\, \mbox{---})$ is right exact. Thus the middle vertical map is surjective by the 4-lemma [136], chapter 1, lemma 3.2. Finally, since (iv) clearly implies (i) we are done. $\square$

6.4   The Converse of Hilbert's Syzygy Theorem

Hilbert's Syzygy Theorem 6.3.1 says that polynomial algebras $\mathbb{F}[V]$ have finite global dimension. The converse of this result is due to J. -P. Serre. We intend to prove this by induction over the Krull dimension and we start with the case of Krull dimension zero.

LEMMA 6.4.1 : *Let A be a graded connected commutative algebra of Krull dimension zero over a field $\mathbb{F}$. If F', F'' are finitely generated free A-modules, $F' \neq 0$, and $\varphi : F' \longrightarrow F''$ an A-module homomorphism such that $Q(\varphi) = 0 : QF' \longrightarrow QF''$, then $\ker(\varphi) \neq 0$.*

PROOF :   Clearly it suffices to consider the case when QF' has rank 1. Let $x' \in F'$ be a generator, and $x''_1, \ldots, x''_n \in F''$ a basis. For degree reasons $\mathrm{Im}(\varphi)$ lies in the free submodule spanned by those x''_i with $\deg(x''_i) \leq \deg(x')$, and since $Q(\varphi) = 0$ in fact in the submodule spanned by those x''_i with $\deg(x''_i) < \deg(x')$. So without loss of generality we may suppose $\deg(x''_i) < \deg(x')$ for $i = 1, \ldots, n$. The Poincaré series of F'' is of the form

$$P(F'',\, t) = t^{\deg(x''_1)} P(A,\, t) + \cdots + t^{\deg(x''_n)} P(A,\, t)$$

while that of F' has the form

$$P(F',\, t) = t^{\deg(x')} P(A,\, t).$$

Since $\dim(A) = 0$, $P(A,\, t)$ is a polynomial. Thus

$$\begin{aligned}
\deg(P(F',\, t)) &= \deg(x') + \deg(P(A,\, t)) \\
&> \max\Big\{\deg(x''_1), \ldots, \deg(x''_n)\Big\} + \deg(P(A,\, t)) \\
&= \deg(P(F'',\, t)).
\end{aligned}$$

Therefore F' is nonzero in a degree where F'' is zero, so φ must certainly vanish in this degree. $\square$

THEOREM 6.4.2 (S. Eilenberg): *Let A be a graded connected commutative algebra of Krull dimension zero over a field $\mathbb{F}$. If $A \not\cong \mathbb{F}$ then A has infinite global dimension.*

PROOF : We suppose that $A \not\cong \mathbb{F}$ and we regard $\mathbb{F}$ as an A-module via the augmentation homomorphism $\varepsilon : A \longrightarrow \mathbb{F}$. We proceed to construct a minimal resolution for $\mathbb{F}$ as an A-module. To this end let $x_1, \ldots, x_n$ be a minimal ideal basis for the augmentation ideal $\bar{A} = \ker(\varepsilon)$, i.e. the images of $x_1, \ldots, x_n$ in $\mathbb{F} \otimes_A \bar{A}$ are an $\mathbb{F}$-basis, and set

$$\bigoplus_{i=1}^{n} A \cdot \bar{x}_i = F_1 \xrightarrow{d_1} F_0 \xrightarrow{\varepsilon} \mathbb{F} \longrightarrow 0$$

where ε is the augmentation, $A \cdot \bar{x}_i$ is a free A-module on the single generator $\bar{x}_i$ of degree $\deg(\bar{x}_i) = \deg(x_i)$ and d_1 is defined by $d_1(\bar{x}_i) = x_i$ for $i = 1, \ldots, n$. Note that $Q(d_1) = 0$ and $\ker(d_1) \subset \bar{A} \cdot F_1$. Suppose inductively that we have constructed a partial free resolution

$$F_s \xrightarrow{d_s} F_{s-1} \xrightarrow{d_{s-1}} \cdots \xrightarrow{d_1} F_0 \xrightarrow{\varepsilon} \mathbb{F} \longrightarrow 0$$

with $Q(d_i) = 0$ and $\ker(d_i) \subset \bar{A} \cdot F_i$, $i = 1, \ldots, s$. Since A is Noetherian $\ker(d_s)$ has a finite set of generators $x_{s,1}, \ldots, x_{s,r}$ projecting in $\mathbb{F} \otimes_A \ker(d_s)$ to an $\mathbb{F}$-basis. Since $\ker(d_s) \subset \bar{A} \cdot F_s$ it follows that the $x_{s,i}$ project to zero in $Q(F_s) = \mathbb{F} \otimes_A F_s$ for $i = 1, \ldots, r$. Thus we may define

$$\bigoplus_{j=1}^{r} A \cdot \bar{x}_{s,j} = F_{s+1} \xrightarrow{d_{s+1}} F_s$$

by $d_{s+1}(\bar{x}_{s,j}) = x_{s,j}$. Then $Q(d_{s+1}) = 0$ and since the elements $\bar{x}_{s,j}$ project to an $\mathbb{F}$-basis of $\mathbb{F} \otimes_A \ker(d_s)$ we have $\ker(d_{s+1}) \subset \bar{A} \cdot F_{s+1}$. Inductively we obtain a free resolution

$$(\mathcal{F}) \qquad \cdots \longrightarrow F_s \xrightarrow{d_s} \cdots \xrightarrow{d_1} F_0 \xrightarrow{\varepsilon} \mathbb{F} \longrightarrow 0$$

where

$$Q(F_s) \neq 0$$
$$0 = Q(d_s) : QF_s \longrightarrow QF_{s-1}$$

for all s. We may compute $\mathrm{Tor}_*^A(\mathbb{F}, \mathbb{F})$ from $\mathbb{F} \otimes_A \mathcal{F}$ by taking homology. This is the complex

$$\cdots \longrightarrow Q(F_s) \xrightarrow{Q(d_s)=0} \cdots \xrightarrow{Q(d_1)=0} Q(F_0) \longrightarrow 0$$

and hence

$$\mathrm{Tor}_*^A(\mathbb{F}, \mathbb{F}) = Q(F_s) \neq 0 \qquad \forall\, s \geq 0.$$

So A has infinite global dimension. $\square$

THEOREM 6.4.3 (J. -P. Serre): *Let A be a finitely generated graded connected commutative algebra over a field $\mathbb{F}$. If A has finite global dimension d then $A \cong \mathbb{F}[z_1, \ldots, z_d]$.*

PROOF : We proceed by induction on the Krull dimension. The case of dimension zero is Eilenberg's theorem 6.4.2, so we may suppose that A has positive Krull dimension. Choose a set of elements $z_1, \ldots, z_n$ that project to a basis for QA. Since $\dim(A) > 0$ it follows from 5.3.3 that at least one of these elements, say z_1, must be a nonzero divisor in A. Then the Koszul complex $\mathcal{K}_A(z_1)$ is acyclic so

$$\mathrm{Tor}_*^{\mathbb{F}[z_1]}(A, \ \mathbb{F}) = 0 \qquad * > 0$$

and A is a free $\mathbb{F}[z_1]$-module. Let $B = \mathbb{F} \otimes_{\mathbb{F}[z_1]} A \cong A/(z_1)$. Then B is generated by $z_2, \ldots, z_n$, so $\dim_{\mathbb{F}}(Q(B)) < n$ and by 5.2.6 $\dim(B) \leq n - 1$.

Assume A has finite global dimension. We claim B does also. For if M is a B-module, the quotient map $q : A \longrightarrow B$ allows us to view M as an A-module. As an A-module M has a finite free resolution, say,

$$(\mathcal{F}) \qquad 0 \longrightarrow F_d \longrightarrow \cdots \longrightarrow F_1 \longrightarrow F_0 \longrightarrow M \longrightarrow 0$$

of length at most $d = \mathrm{gl\text{-}dim}(A)$. Each F_i being free over A is by transitivity free over $\mathbb{F}[z_1]$ and hence

$$0 \longrightarrow \mathbb{F} \otimes_{\mathbb{F}[z_1]} F_d \longrightarrow \cdots \longrightarrow \mathbb{F} \otimes_{\mathbb{F}[z_1]} F_0 \longrightarrow \mathbb{F} \otimes_{\mathbb{F}[z_1]} M \longrightarrow 0$$

is exact. Since M is a trivial $\mathbb{F}[z_1]$-module, $\mathbb{F} \otimes_{\mathbb{F}[z_1]} M \cong M/z_1 \cdot M \cong M$. The modules $\mathbb{F} \otimes_{\mathbb{F}[z_1]} F_i$, $i = 0, \ldots, d$, are free over $\mathbb{F} \otimes_{\mathbb{F}[z_1]} A = B$. Thus we have constructed a free resolution of M as a B-module of length at most $d = \mathrm{gl\text{-}dim}(A)$ and hence B has finite global dimension as claimed. By the inductive hypothesis we therefore have $B \cong \mathbb{F}[\bar{z}_2, \ldots, \bar{z}_n]$, where $\bar{z}_i = 1 \otimes_{\mathbb{F}[z_1]} z_i \in B$, $i = 2, \ldots, n$. Define

$$\varphi : \mathbb{F}[z_1, \ \bar{z}_2, \ldots, \ \bar{z}_n] \longrightarrow A \text{ by } \begin{cases} \varphi(z_1) = z_1 \\ \varphi(\bar{z}_i) = z_i & \text{for } i = 2, \ldots, n \end{cases}$$

We claim that φ is an isomorphism. It is certainly epic by 5.2.4 because $Q(\varphi)$ is an isomorphism. To see it is also monic suppose that $f \neq 0 \in \mathbb{F}[z_1, \ \bar{z}_2, \ldots, \ \bar{z}_n]$ is of minimal degree in $\ker(\varphi)$. We may write $f = z_1 q + r$ with $q \in \mathbb{F}[z_1, \ \bar{z}_2, \ldots, \ \bar{z}_n]$ and $r \in \mathbb{F}[\bar{z}_2, \ldots, \bar{z}_n]$. Since the composite

$$\mathbb{F}[\bar{z}_2, \ldots, \bar{z}_n] \hookrightarrow \mathbb{F}[z_1, \ \bar{z}_2, \ldots, \ \bar{z}_n] \longrightarrow A \longrightarrow B$$

is the identity we see $r = 0$. But $z_1 \in A$ was chosen not to be a zero divisor so $q = 0$ and hence $f = 0$, a contradiction. Thus φ is an isomorphism. $\square$

We can reap a future benefit for invariant theory at this point (see chapter 7, theorem 7.4.1 and [214]).

COROLLARY 6.4.4 : *Suppose $A \subset \mathbb{F}[z_1, \ldots, z_n]$ is a graded subalgebra of the graded polynomial algebra $\mathbb{F}[z_1, \ldots, z_n]$. (We do not require that $z_1, \ldots, z_n$ have degree 1, only positive degree.) If $\mathbb{F}[z_1, \ldots, z_n]$ is free as an A-module then $A = \mathbb{F}[f_1, \ldots, f_k]$ with $f_1, \ldots, f_k$ a regular sequence in $\mathbb{F}[z_1, \ldots, z_n]$.*

PROOF : By 6.4.3 and 6.2.3 it suffices to show that A has finite global dimension. Set $S = \mathbb{F}[z_1, \ldots, z_n]$. Suppose that M is a graded A-module. Let

$$0 \longrightarrow K_n \longrightarrow P_{n-1} \longrightarrow \cdots \longrightarrow P_0 \longrightarrow M \longrightarrow 0$$

be a partial projective resolution of M, i.e. all but K_n are projective and the sequence is exact. Since S is free as an A-module the functor $- \otimes_A S$ is exact and sends projective A-modules to projective S-modules. Thus

$$0 \longrightarrow K_n \otimes_A S \longrightarrow P_{n-1} \otimes_A S \longrightarrow \cdots \longrightarrow M \otimes_A S \longrightarrow 0$$

is an exact sequence of S-modules, where all but perhaps the last term is S-projective. It follows from 6.3.3 that $K_n \otimes_A S$ is a projective S-module, because the global dimension of S is n. Since S is free as an A-module $K_n \otimes_A S$ is also projective over A. Moreover S being a free A module with $1 \in S$ as one of the free generators, there is an A-module splitting $\sigma : S \longrightarrow A$ to the inclusion $i : A \hookrightarrow S$. Thus

$$K_n = K_n \otimes_A A \hookrightarrow K_n \otimes_A S \underset{1 \otimes \sigma}{\overset{1 \otimes i}{\rightleftarrows}} K_n \otimes_A A$$

represents K_n as an A-module direct summand in the projective A-module $K_n \otimes_A S$ and hence K_n is projective as an A-module and the global dimension of A is at most n. $\square$

6.5   Poincaré Duality Algebras

The Koszul complex can be used to provide us with our first general result concerning rings of coinvariants, and again it is best to place this result in a setting more general than just that of invariant theory.

DEFINITION : *If A is a graded connected commutative algebra over a field $\mathbb{F}$ we say that A is a **Poincaré duality algebra** of dimension n if:*
   — *$A_i = 0$ for $i > n$,*
   — *$\dim_\mathbb{F}(A_n) = 1$,*
   — *the pairing $A_i \otimes_\mathbb{F} A_{n-i} \longrightarrow A_n$ given by multiplication is non-singular, i.e. a class $a \in A_i$ is zero if and only if $a \cdot b = 0$ for all $b \in A_{n-i}$.*

*If A is a Poincaré duality algebra of dimension n an element $[A] \neq 0 \in A_n$ is referred to as a **fundamental class** for A.*

If A is a graded connected commutative algebra over a field $\mathbb{F}$ and $[A] \in A$ an element of degree n then it is easy to see that A is a Poincaré duality algebra of dimension n with fundamental class $[A]$ if and only if

$$\text{Ann}_A([A]) = \bar{A}$$
$$\text{Ann}_A(\bar{A}) = \mathbb{F} \cdot [A]$$

where $\bar{A}$ is the augmentation ideal of A.

THEOREM 6.5.1 : *Let* $\mathbb{F}$ *be a field,* $f_1, \ldots, f_n \in \mathbb{F}[z_1, \ldots, z_n]$ *a regular sequence of maximal length and* $A = \mathbb{F}[z_1, \ldots, z_n]/(f_1, \ldots, f_n)$. *Write* $f_i = \sum a_{i,j} z_j$ *where* $a_{i,j} \in \mathbb{F}[z_1, \ldots, z_n]$ *and set* $[A] = \det(a_{i,j})$. *Then* $[A] \in A$ *is well defined and* A *is a Poincaré duality algebra with fundamental class* $[A]$.

PROOF : A simple calculation with determinants shows that the element $[A] \in A$ is well defined.

Introduce the Koszul complexes

$$\mathcal{E} = \mathbb{F}[z_1, \ldots, z_n] \otimes E[sf_1, \ldots, sf_n] \quad \text{where} \quad \partial(sf_i) = f_i, \ 1 \le i \le n$$
$$\mathcal{K} = \mathbb{F}[z_1, \ldots, z_n] \otimes E[sz_1, \ldots, sz_n] \quad \text{where} \quad \partial(sz_i) = z_i, \ 1 \le i \le n.$$

By theorem 6.2.3 $\mathcal{E}$ is a free acyclic resolution of A as an $\mathbb{F}[z_1, \ldots, z_n]$-module, and by corollary 6.2.5 $\mathcal{K}$ is a free acyclic resolution of $\mathbb{F}$ as an $\mathbb{F}[z_1, \ldots, z_n]$-module. Therefore $\text{Tor}_*^{\mathbb{F}[z_1, \ldots, z_n]}(A, \mathbb{F})$ is the homology of any of the three complexes

$$A \otimes_{\mathbb{F}[z_1, \ldots, z_n]} \mathcal{K}, \quad \mathcal{E} \otimes_{\mathbb{F}[z_1, \ldots, z_n]} \mathcal{K}, \quad \mathcal{E} \otimes_{\mathbb{F}[z_1, \ldots, z_n]} \mathbb{F}$$

and the augmentations provide chain homotopy equivalences between them:

$$A \otimes_{\mathbb{F}[z_1, \ldots, z_n]} \mathcal{K} \xleftarrow{\eta} \mathcal{E} \otimes_{\mathbb{F}[z_1, \ldots, z_n]} \mathcal{K} \xrightarrow{\varepsilon} \mathcal{E} \otimes_{\mathbb{F}[z_1, \ldots, z_n]} \mathbb{F}.$$

From the complex $\mathcal{E} \otimes_{\mathbb{F}[z_1, \ldots, z_n]} \mathbb{F}$ one readily sees

(A) $$\text{Tor}_*^{\mathbb{F}[z_1, \ldots, z_n]}(A, \mathbb{F}) = E[sf_1, \ldots, sf_n].$$

We next compute with the complex $A \otimes_{\mathbb{F}[z_1, \ldots, z_n]} \mathcal{K}$. Let

$$\alpha_i = \sum_{j=1}^n a_{i,j} sz_j \in A \otimes_{\mathbb{F}[z_1, \ldots, z_n]} \mathcal{K}.$$

and note that

$$\partial(\alpha_i) = \sum_{i=j}^n a_{i,j} z_j = 0 \in A$$

so α_i is a cycle. Examining the chain homotopy equivalences η and ε we see no linear combination of $\alpha_1, \ldots, \alpha_n$ is a boundary, and therefore $\alpha_1, \ldots, \alpha_n$ is a basis for $\mathrm{Tor}_1^{\mathbb{F}[z_1, \ldots, z_n]}(A, \mathbb{F})$. From (A) it then follows that

(B) $\qquad\qquad \mathrm{Tor}_*^{\mathbb{F}[z_1, \ldots, z_n]}(A, \mathbb{F}) = E[\alpha_1, \ldots, \alpha_n].$

Continuing to compute with this complex we examine $\mathrm{Tor}_n^{\mathbb{F}[z_1, \ldots, z_n]}(A, \mathbb{F})$. A typical chain of homological degree n is $\zeta = a \otimes sz_1 \cdots sz_n$ whose boundary is

$$\partial(\zeta) = \sum \pm a\bar{z}_i \otimes sz_1 \cdots \widehat{sz_i} \cdots sz_n$$

where $\bar{z}_i$ denotes the residue class of $z_i \in A$ and $\frown$ means the term under the $\frown$ is omitted. Since the classes $sz_1 \cdots \widehat{sz_i} \cdots sz_n$ are linearly independent, ζ is a cycle if and only if $a\bar{z}_i = 0$ for $i = 1, \ldots, n$, i.e. if and only if $a \in \mathrm{Ann}_A(\bar{A})$ so

(C) $\qquad\qquad \mathrm{Tor}_n^{\mathbb{F}[z_1, \ldots, z_n]}(A, \mathbb{F}) = \mathrm{Ann}_A(\bar{A}).$

But (B) provides an alternate computation of $\mathrm{Tor}_n^{\mathbb{F}[z_1, \ldots, z_n]}(A, \mathbb{F})$, namely

(D) $\qquad\qquad \mathrm{Tor}_n^{\mathbb{F}[z_1, \ldots, z_n]}(A, \mathbb{F}) = \mathrm{Span}_{\mathbb{F}}\{(\det(a_{i,j}))\},$

since

$$\left(\sum a_{1,j} sz_j\right) \cdots \left(\sum a_{n,j} sz_j\right) = \det(a_{i,j}) \otimes sz_1 \otimes \cdots \otimes sz_n$$

by the definition of the determinant. By combining (C) and (D) we get

(E) $\qquad\qquad 0 \neq \mathrm{Ann}_A(\bar{A}) = \mathrm{Span}_{\mathbb{F}}\{\det(a_{i,j})\}.$

Next apply Cramer's rule[3] to the linear system in A

$$\sum_{j=1}^{n} a_{i,j} x_j = 0 \qquad i = 1, \ldots, n$$

---

[3] One way to think about this is as follows. If

$$A \begin{bmatrix} y_1 \\ \vdots \\ y_n \end{bmatrix} = \begin{bmatrix} 0 \\ \vdots \\ 0 \end{bmatrix}$$

then multiplying both sides by the cofactor matrix ([210] page 204) of A we obtain

$$\begin{bmatrix} 0 \\ \vdots \\ 0 \end{bmatrix} = A^{\mathrm{cof}} A \begin{bmatrix} y_1 \\ \vdots \\ y_n \end{bmatrix} = \det(A) \begin{bmatrix} y_1 \\ \vdots \\ y_n \end{bmatrix}.$$

Therefore $\det(A) \in \mathrm{Ann}(y_1, \ldots, y_n)$.

to obtain
$$\mathrm{Ann}_A(\det(a_{i,j})) \supset (\bar{z}_1, \ldots, \bar{z}_n)$$
whence $\mathrm{Ann}_A(\det(a_{i,j})) \supset \bar{A}$, and since by (E) $\det(a_{i,j}) \neq 0$, the annihilator ideal cannot be any larger.

To complete the proof note that any element $a \in A$ of maximal degree annihilates $\bar{A}$ for degree reasons. By (E) it follows $A^i = 0$ for $i > d = \deg([A])$ and that A^d is a one dimensional vector space spanned by $[A]$. Moreover if $a \in A_i$ for $0 < i < d$ then there must exist an element $b \in A$ with $a \cdot b \neq 0$. If b has maximal degree then it follows that $\deg(b) = d - i$. $\square$

COROLLARY 6.5.2 : Let $\mathbb{F}$ be a field, $f_1, \ldots, f_n \in \mathbb{F}[z_1, \ldots, z_n]$ a regular sequence of maximal length and $A = \mathbb{F}[z_1, \ldots, z_n]/(f_1, \ldots, f_n)$. If $\deg(f_1) \cdots \deg(f_n)$ is relatively prime to the characteristic of $\mathbb{F}$ then A is a Poincaré duality algebra with fundamental class $\det[\frac{\partial f_i}{\partial z_j}]$.

PROOF : By Euler's formula
$$\deg(f_i)f_i = \sum \frac{\partial f_i}{\partial z_j} z_j$$

so the Jacobian determinant is a nonzero multiple of the fundamental class given by 6.5.1. $\square$

REMARK : If $\mathbb{F}$ is a field and $f_1, \ldots, f_m \in \mathbb{F}[z_1, \ldots, z_n]$ is a regular sequence then the quotient algebra $\mathbb{F}[z_1, \ldots, z_n]/(f_1, \ldots, f_m)$ is a graded analog of a complete intersection (see e.g. [61]).

Applied to representations $G \hookrightarrow \mathrm{GL}(n, \mathbb{F})$ whose rings of invariants are polynomial algebras (see chapters 7 and 8) such as the symmetric group in its tautological representation, these results yield our first concrete facts concerning $\mathbb{F}[V]_G$.

EXAMPLE 1 : Consider the representation $D_8 \hookrightarrow \mathrm{GL}(2, \mathbb{R})$ of the dihedral group D_8 of order 8 as the group of isometries of a square. (See example 3 in section 3.2.) The ring of invariants $\mathbb{R}[x, y]^{D_8}$ is $\mathbb{R}[x^2+y^2, x^2y^2]$. Therefore the ring of covariants is a Poincaré duality algebra of dimension 4 with fundamental class
$$[\mathbb{R}[x, y]_{D_8}] = \det \begin{bmatrix} \frac{\partial(x^2+y^2)}{\partial x} & \frac{\partial(x^2+y^2)}{\partial y} \\ \frac{\partial(x^2y^2)}{\partial x} & \frac{\partial(x^2y^2)}{\partial y} \end{bmatrix} = 4(x^3y - xy^3).$$

The ring $\mathbb{R}[x, y]$ is a free $\mathbb{R}[x^2+y^2, x^2y^2] = \mathbb{R}[x, y]^{D_8}$-module and therefore the Poincaré series of $\mathbb{R}[x, y]_{D_8}$ is given by
$$P(\mathbb{R}[x, y]_{D_8}, t) = \frac{(1-t^2)(1-t^4)}{(1-t)^2} = 1 + 2t + 2t^2 + 2t^3 + t^4.$$

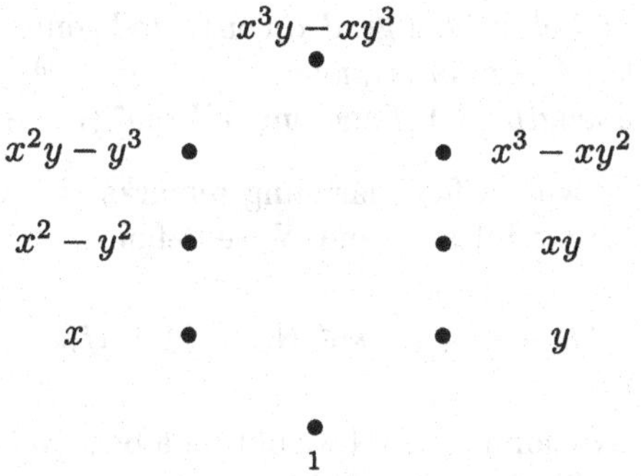

$$x^3y - xy^3$$
•

$$x^2y - y^3 \quad \bullet \qquad\qquad \bullet \quad x^3 - xy^2$$

$$x^2 - y^2 \quad \bullet \qquad\qquad \bullet \quad xy$$

$$x \quad \bullet \qquad\qquad \bullet \quad y$$

•
1

Diagram 6.5.1 : $\mathbb{F}[x,\ y]_{D_8}$

The diagram 6.5.1 is an aid to visualizing this ring of coinvariants.

As in chapter 1 section 1.2 example 2 the nodes on a horizontal level indicate basis vectors for the elements of degree equal to the height of the node above the node labeled 1, which has degree 0. The elements of $\mathbb{R}[x,\ y]_{D_8}$ satisfy numerous relations, e.g.

$$x^2 = -y^2$$
$$x^4 = x^2y^2 = -y^4\,.$$

6.6   Homological Codimension

The least upper bound of the lengths of regular sequences in a graded connected algebra A over a field $\mathbb{F}$ is another measure of finiteness called the **homological codimension** of A and denoted by codim(A). By 6.2.2 codim$(A) \leq$ dim(A). (The codimension is sometimes referred to in the literature as the **depth** of the algebra.)

If A is a graded connected commutative algebra over a field $\mathbb{F}$ we say a regular sequence $a_1, \ldots, a_r \in A$ is **maximal** if it cannot be extended to a regular sequence $a_1, \ldots, a_r,\ a_{r+1}$ in A with one more element. It is by no means clear that two maximal regular sequences contain the same number of elements, i.e. have the same length. A proof that this is the case is our next goal.

Suppose $a_1, \ldots, a_r \in A$ is a regular sequence. Then it can be extended to a regular sequence of length $r+1$ if and only if $A/(a_1, \ldots, a_r)$ contains a nonzero divisor, in which case we choose a_{r+1} to be one such. This means that for a maximal regular sequence $a_1, \ldots, a_r$ every element of positive degree in $A/(a_1, \ldots, a_r)$ is a zero divisor. We may reformulate this as follows.

LEMMA 6.6.1 : *Let A be a graded connected commutative noetherean algebra over a field $\mathbb{F}$. A regular sequence $a_1, \ldots, a_r \in A$ may be extended to a regular sequence of length $r+1$ if and only if* $\mathrm{Hom}_A^*(\mathbb{F},\ A/(a_1, \ldots, a_r)) = 0$.

We precede the proof with a few clarifying remarks about the graded Hom-functor. For graded A-modules M and N we define

$$\mathrm{Hom}_A^d(M,\ N) = \left\{ f : M_k \longrightarrow N_{k+d} \mid k \in \mathbb{N},\ f(ax) = af(x) \quad \forall\, a \in A,\ x \in M \right\}.$$

By restricting our attention to $d \in \mathbb{N}$ we obtain a positively graded A-module $\mathrm{Hom}_A^*(M,\ N)$ where the A-module action is given by

$$(a \cdot f)(x) = a \cdot f(x) \quad \forall\, a \in A,\ x \in M.$$

If it is clear from context that M, A, N are graded we may drop the $*$ from the notation.

In the lemma we regard $\mathbb{F}$ as a graded A-module concentrated in degree zero, i.e. with $\mathbb{F}$ the homogeneous component of degree zero and all other homogeneous components equal to 0, the action of A being given by the augmentation map $\varepsilon : A \longrightarrow \mathbb{F}$.

PROOF : The evaluation map at $1 \in \mathbb{F}$ provides a bijection

$$\mathrm{Hom}_A^*(\mathbb{F},\ A) \xrightarrow{\ e\ } \{a \in A \mid \bar{A} \cdot a = 0\}.$$

If $\mathrm{Hom}_A^*(\mathbb{F},\ A) \neq 0$ then there is an element $a \in A$ that annihilates every element of positive degree so there can be no regular sequence of length 1 in A. Conversely, if A contains no regular sequence of length 1 then every element of positive degree is a zero divisor. The set of zero divisors in A is the union of the associated primes of $(0) \subset A$ ([266] volume I § IV.6 corollary 3) so by 5.3.1 $\bar{A}$ itself is an associated prime of 0. Hence there is an element $a \in \bar{A}$ annihilated by $\bar{A}$ so $\mathrm{Hom}_A^*(\mathbb{F},\ A) \neq 0$. □

The following remarkable theorem of J. -P. Serre ([200] proposition IV.5) shows that $\mathrm{Hom}_A^*(\mathbb{F},\ A/(a_1, \ldots, a_r))$ is *independent* of $a_1, \ldots, a_r \in A$ so long as they form a regular sequence.

THEOREM 6.6.2 (J. -P. Serre): *Suppose $a_1, \ldots, a_r \in A$ is a regular sequence. Then* $\mathrm{Ext}_A^i(\mathbb{F},\ A) = 0$ *for $i < r$ and*

$$\mathrm{Ext}_A^r(\mathbb{F},\ A) \cong \mathrm{Ext}_A^{r-1}(\mathbb{F},\ A/(a_1)) \cong \cdots \cong \mathrm{Ext}_A^0(\mathbb{F}, A/(a_1, \ldots, a_r)).$$

(We use $\mathrm{Ext}_A^0(-,\ -)$ as an alternate notation for $\mathrm{Hom}_A(-,\ -)$.)

PROOF : Consider the case $r = 1$. There is the exact sequence

$$0 \longrightarrow A \xrightarrow{\alpha_1} A \longrightarrow A/(a_1) \longrightarrow 0$$

where α_1 denotes multiplication by $a_1 \in A$. Since a_1 has positive degree it annihilates $\mathbb{F}$ and hence $\mathrm{Ext}_A^*(\mathbb{F}, \alpha_1) = 0$. The long exact sequence for $\mathrm{Ext}_A^*(\mathbb{F}, -)$ applied to the preceding exact sequence therefore decomposes into short exact sequences

$$0 \longrightarrow \mathrm{Ext}_A^n(\mathbb{F}, A) \longrightarrow \mathrm{Ext}_A^n(\mathbb{F}, A/(a_1)) \longrightarrow \mathrm{Ext}_A^{n+1}(\mathbb{F}, A) \longrightarrow 0.$$

Put $n = 0$ and apply 6.6.1 to obtain

$$\mathrm{Hom}_A(\mathbb{F}, A/(a_1)) \cong \mathrm{Ext}_A^1(\mathbb{F}, A)$$

which when combined with 6.6.1 for $r = 1$ yields the desired conclusion.

We may therefore proceed by induction on r and assume that the result has been established for $0 < s < r$. For $n < r$ we then have

$$\mathrm{Ext}_A^n(\mathbb{F}, A) \cong \mathrm{Ext}_A^{n-1}(\mathbb{F}, A/(a_1)) \cong \cdots \cong \mathrm{Ext}_A^0(\mathbb{F}, A/(a_1, \ldots, a_n))$$

by 6.6.1, as the regular sequence $a_1, \ldots, a_n \in A$ extends to the regular sequence $a_1, \ldots, a_n, a_{n+1} \in A$.

Since $a_1, \ldots, a_r \in A$ is a regular sequence we have the exact sequences

$$0 \longrightarrow A \xrightarrow{\alpha_1} A \longrightarrow A/(a_1) \longrightarrow 0$$
$$0 \longrightarrow A/(a_1) \xrightarrow{\alpha_2} A/(a_1) \longrightarrow A/(a_1, a_2) \longrightarrow 0$$
$$\vdots$$
$$0 \longrightarrow A/(a_1, \ldots, a_{r-1}) \xrightarrow{\alpha_r} A/(a_1, \ldots, a_{r-1}) \longrightarrow A/(a_1, \ldots, a_r) \longrightarrow 0$$

where α_i denotes multiplication by a_i. The maps $\mathrm{Ext}_A^n(\mathbb{F}, \alpha_i)$ are all zero since they are multiplication by a_i which is of positive degree and $\bar{A}$ annihilates $\mathrm{Ext}_A^*(\mathbb{F}, -)$. Therefore the long exact sequence for $\mathrm{Ext}_A^*(\mathbb{F}, -)$ applied to the preceding exact sequences decompose into short exact sequences

$$0 \to \mathrm{Ext}_A^n(\mathbb{F}, A/(a_1,\ldots, a_i)) \to \mathrm{Ext}_A^n(\mathbb{F}, A/(a_1,\ldots, a_{i+1})) \to \mathrm{Ext}_A^{n+1}(\mathbb{F}, A/(a_1,\ldots, a_i)) \to 0.$$

By the induction hypothesis

$$\mathrm{Ext}_A^n(\mathbb{F}, A/(a_1, \ldots, a_i)) = 0$$

for $n < i$. If we put $n + 1 = r$, $i = 0$ in the preceding short exact sequence we obtain

$$0 \longrightarrow \mathrm{Ext}_A^{r-1}(\mathbb{F}, A) \longrightarrow \mathrm{Ext}_A^{r-1}(\mathbb{F}, A/(a_1)) \longrightarrow \mathrm{Ext}_A^r(\mathbb{F}, A) \longrightarrow 0$$
$$\parallel$$
$$0$$

and therefore we have an isomorphism

$$\text{Ext}_A^{r-1}(\mathbb{F}, \ A/(a_1)) \cong \text{Ext}_A^r(\mathbb{F}, \ A).$$

Successively putting $(n+1, \ i) = (r, \ 0), \ (r-1, \ 1), \ldots, \ (1, \ r-1)$ we obtain in this way the isomorphisms

$$\text{Ext}_A^{r-1}(\mathbb{F}, \ A/(a_1)) \cong \text{Ext}_A^r(\mathbb{F}, \ A)$$

$$\text{Ext}_A^{r-2}(\mathbb{F}, \ A/(a_1, \ a_2)) \cong \text{Ext}_A^{r-1}(\mathbb{F}, \ A/(a_1))$$

$$\vdots$$

$$\text{Hom}_A(\mathbb{F}, \ A/(a_1, \ldots, \ a_r)) \cong \text{Ext}_A^1(\mathbb{F}, \ A/(a_1, \ldots, \ a_{r-1}))$$

and the result follows. $\square$

COROLLARY 6.6.3 : *Let A be a graded commutative connected algebra over a field. Then A contains a regular sequence of length r if and only if $\text{Ext}_A^n(\mathbb{F}, \ A) = 0$ for $n < r$. (As before we use $\text{Ext}_A^0(—, \ —)$ as an alternate notation for $\text{Hom}_A(—, \ —)$.)*

PROOF : If A contains a regular sequence of length r, then $\text{Ext}_A^n(\mathbb{F}, \ A) = 0$ for $n < r$ by 6.6.2. Conversely, suppose that $\text{Ext}_A^n(\mathbb{F}, \ A) = 0$ for $n < r$. In the case $r = 1$ this means $\text{Hom}_A(\mathbb{F}, \ A) = 0$ so A contains a nonzero divisor by the argument used to prove 6.6.1. A nonzero divisor is a regular sequence of length 1 so we may proceed inductively, assuming the corollary established for $0 < s < r$. Therefore A contains a regular sequence of length $r - 1$. By 6.6.2

$$0 = \text{Ext}_A^{r-1}(\mathbb{F}, \ A) = \text{Hom}_A(\mathbb{F}, \ A/(a_1, \ldots, \ a_{r-1}))$$

and hence by 6.6.1 $a_1, \ldots, \ a_{r-1}$ extends to a regular sequence of length r. $\square$

COROLLARY 6.6.4 : *Let A be a graded connected commutative algebra over a field $\mathbb{F}$. Then*

$$\text{codim}(A) = \max_n \left\{ n \mid \text{Ext}_A^{n-1}(\mathbb{F}, \ A) = 0 \right\} = \min_n \left\{ n \mid \text{Ext}_A^n(\mathbb{F}, \ A) \neq 0 \right\}.$$

PROOF : This follows from the preceding and the definitions. $\square$

Theorem 6.6.2 has a number of consequences whose classical proofs are considerably more intricate (see for example [266] II appendix 6).

COROLLARY 6.6.5 : *In a graded connected commutative algebra over a field $\mathbb{F}$ any two maximal regular sequences have the same length. The length of a maximal regular sequence is the smallest integer r such that $\text{Ext}_A^r(\mathbb{F}, \ A) \neq 0$.* $\square$

COROLLARY 6.6.6 : *Let A be a graded connected commutative algebra over a field $\mathbb{F}$. A regular sequence of length r in A extends to a regular sequence of length $r + 1$ in A if and only if $\mathrm{Ext}_A^r(\mathbb{F}, A) = 0$.* □

6.7 Cohen-Macaulay Rings

By 6.2.2 the length of the longest possible regular sequence in a graded connected commutative algebra A over a field $\mathbb{F}$ is at most equal to its Krull dimension. As is often true, the extreme case is worthy of additional study.

DEFINITION : *A graded commutative algebra A over a field $\mathbb{F}$ is called* **Cohen-Macaulay** *if it contains a regular sequence of length $\dim(A)$, i.e. if and only if $\mathrm{codim}(A) = \dim(A)$.*

Not every algebra is Cohen-Macaulay. In the algebra $A = \mathbb{F}[x, y]/(x^2, xy)$, the ideal (x, y) in A is maximal, hence prime. It has height 1 but contains no regular sequence. Therefore A has Krull dimension 1 but codimension 0, hence is not Cohen-Macaulay.

If A is Cohen-Macaulay of Krull dimension d, then (see proposition 6.7.1) there is a regular sequence $a_1, \ldots, a_d \in A$ which is a system of parameters. By 6.2.1 these are algebraically independent and A is free (by 6.2.8) and finitely generated as a module over the subalgebra generated by $a_1, \ldots, a_d$. Therefore a Cohen-Macaulay algebra A is as a free finitely generated module over a polynomial subalgebra. But how to find a system of parameters with this property? In fact *any* system of parameters in a Cohen-Macaulay algebra has this property: we make the proof of this assertion our next major goal.

In 5.3.10 we saw that dividing a graded connected commutative algebra A over a field $\mathbb{F}$ by an ideal generated by k elements reduces the Krull dimension by at most k and that equality holds for a subset of a system of parameters. Here is the analog for regular sequences.

PROPOSITION 6.7.1 : *If A is a graded connected commutative algebra over a field $\mathbb{F}$ and $a_1, \ldots, a_t \in A$ a regular sequence, then $\dim(A/(a_1, \ldots, a_t))$ $= \dim(A) - t$. In particular, if $a_1, \ldots, a_d \in A$ is a regular sequence with $d = \dim(A)$, then $a_1, \ldots, a_d$ is a system of parameters and A is Cohen-Macaulay.*

PROOF : Let $B \subseteq A$ be the subalgebra generated by $a_1, \ldots, a_t$. Then $B \cong \mathbb{F}[a_1, \ldots, a_t]$ is a polynomial algebra and A is a free B-module. Thus as B-modules

$$A \cong \mathbb{F}[a_1, \ldots, a_t] \otimes_{\mathbb{F}} A/(a_1, \ldots, a_t),$$

so both sides have the same Poincaré series, and hence by Noether normalization the same Krull dimension. By 5.3.9 and 5.2.2 the right hand side has

Krull dimension $t + \dim(A/(a_1, \ldots, a_t))$ and the result follows. $\square$

COROLLARY 6.7.2 : *If A is a graded connected commutative algebra over a field $\mathbb{F}$ of Krull dimension d and $a_1, \ldots, a_t \in A$ a regular sequence, then they extend to a system of parameters $a_1, \ldots, a_t, a_{t+1}, \ldots, a_d$ for A and $\dim(A/(a_1, \ldots, a_i)) = \dim(A) - i$ for $i = 1, \ldots, d$.*

PROOF : Apply 5.3.10. $\square$

LEMMA 6.7.3 : *Let A be a graded connected commutative algebra over a field $\mathbb{F}$, $I \subset A$ a proper ideal and $\mathfrak{p}$ an associated prime ideal of I. If $a \in \bar{A}$ does not belong to any associated prime ideal of I, then there exists an associated prime ideal $\tilde{\mathfrak{p}}$ of $I + (a)$ such that $\tilde{\mathfrak{p}} \supset \mathfrak{p}$.*

PROOF : Suppose the contrary, namely that $\mathfrak{p} \not\subset \tilde{\mathfrak{p}}$ for any associated prime $\tilde{\mathfrak{p}}$ of $I + (a)$. Then by lemma 5.3.1 $\mathfrak{p} \not\subset \bigcup_{\tilde{\mathfrak{p}}_s \in \mathrm{Ass}(I)} \tilde{\mathfrak{p}}_s$, where $\mathrm{Ass}(I)$ denotes the set of associated primes of I. Let $x \in \mathfrak{p}$ be such that $x \notin \tilde{\mathfrak{p}}$ for any associated prime ideal $\tilde{\mathfrak{p}}$ of $I + (a)$. Since $x \in \mathfrak{p}$, an associated prime of I, x is a zero divisor in A/I. Choose $y \in A$ of minimal degree such that

$$y \neq 0 \in A/I \qquad xy = 0 \in A/I.$$

Then since x is not a zero divisor in $A/(I + (a))$ it follows that $y \in I + (a)$, say $y = b + ca$, $b \in I$, $c \in A$. Since $b = 0 \in A/I$ it follows that $(ca)x = 0 \in A/I$. However, x is not a zero divisor in A/I and therefore $ca = 0 \in A/I$. Since $a \neq 0 \in A/I$, c has positive degree smaller than that of y which is a contradiction. $\square$

LEMMA 6.7.4 : *Let A be a Cohen-Macaulay algebra and $\mathfrak{p}$ an associated prime ideal of (0). Then $\dim(A/\mathfrak{p}) = \dim(A)$. Hence the associated prime ideals of (0) are all isolated, i.e. minimal.*

PROOF : Choose a regular sequence $a_1, \ldots, a_d \in A$, $d = \dim(A)$. By repeated application of 6.7.3 we receive a proper chain

$$\mathfrak{p} = \mathfrak{p}_0 \subset \mathfrak{p}_1 \subset \cdots \subset \mathfrak{p}_d$$

of prime ideals in A. Then

$$0 \subset \mathfrak{p}_1/\mathfrak{p} \subset \cdots \subset \mathfrak{p}_d/\mathfrak{p}$$

is a proper chain of prime ideals in $A/\mathfrak{p}$ so $\dim(A/\mathfrak{p}) \geq d$. On the other hand the Poincaré series of A dominates the Poincaré series of $A/\mathfrak{p}$ term by term so $d(A) \geq d(A/\mathfrak{p})$ and therefore by the Noether normalization theorem we have

$$\dim(A) = d(A) \geq d(A/\mathfrak{p}) = \dim(A/\mathfrak{p}) \geq \dim(A)$$

and the result follows. $\square$

If A is a commutative Noetherian ring and $I \subset A$ is an ideal, then there is a bijective correspondence between associated prime ideals $\mathfrak{p}$ of I in A and associated primes $\mathfrak{p}/I$ of (0) in A/I. Hence 6.7.4 implies the following:

PROPOSITION 6.7.5 (Macaulay's Unmixedness Theorem): *If A is a Cohen-Macaulay algebra and $I \subset A$ an ideal that is generated by $\mathrm{ht}(I)$ elements [4], then all the associated prime ideals of I are isolated.* □

THEOREM 6.7.6 (F. S. Macaulay): *Let A be a Cohen-Macaulay algebra and $a_1, \ldots, a_r \in A$ such that $\dim(A/(a_1, \ldots, a_r)) = \dim(A) - r$. Then $a_1, \ldots, a_r$ is a regular sequence, and if $\mathfrak{p}$ is any associated prime of the ideal $(a_1, \ldots, a_r)$ we have*

$$\mathrm{ht}(\mathfrak{p}) = r$$
$$\dim(A/\mathfrak{p}) = \dim(A) - \mathrm{ht}(\mathfrak{p}) = \mathrm{dp}(\mathfrak{p})$$

and $A/(a_1, \ldots, a_r)$ is Cohen-Macaulay.

PROOF : We proceed by induction on r. The case $r = 0$ follows from 6.7.4, namely, in a Cohen-Macaulay algebra associated primes of zero satisfy $\dim(A/\mathfrak{p}) = \dim(A)$ while in any graded connected commutative algebra over a field $\mathbb{F}$

$$\dim(A) \geq \dim(A/\mathfrak{p}) + \mathrm{ht}(\mathfrak{p})$$

so if A is Cohen-Macaulay $\mathrm{ht}(\mathfrak{p}) = 0$. By repeated application of 6.7.3 we obtain a regular sequence $a_1, \ldots, a_d \in A$ whose residue classes are a regular sequence in $A/\mathfrak{p}$.

Suppose inductively the theorem is true for all $s < r$ and let $a_1, \ldots, a_r \in A$ satisfy $\dim(A/(a_1, \ldots, a_r)) = \dim(A) - r$. Then by 5.3.10 $a_1, \ldots, a_r$ is a subset of a system of parameters so by the same result $\dim(A/(a_1, \ldots, a_{r-1})) = \dim(A) - (r - 1)$. By the induction hypothesis $a_1, \ldots, a_{r-1} \in A$ is a regular sequence and for any associated prime ideal of $(a_1, \ldots, a_{r-1})$ we have $\mathrm{ht}(\mathfrak{p}) = r - 1$, and $\dim(A/\mathfrak{p}) = \dim(A) - \mathrm{ht}(\mathfrak{p})$.

If a_r is a zero divisor in $A/(a_1, \ldots, a_{r-1})$ then a_r belongs to some associated prime ideal $\mathfrak{p}$ of $(a_1, \ldots, a_{r-1})$. But then $(a_1, \ldots, a_r) \subset \mathfrak{p}$ whence

$$d - r = \dim(A/(a_1, \ldots, a_r)) \geq \dim(A/\mathfrak{p}) = d - r + 1$$

which is impossible. Therefore $a_1, \ldots, a_r$ is a regular sequence.

Since $a_1, \ldots, a_r$ is a regular sequence in a Cohen-Macaulay algebra they may be extended to a regular sequence $a_1, \ldots, a_r, a_{r+1}, \ldots, a_d$ of length $d = \dim(A)$ by 6.6.5. The quotient map $A \longrightarrow A/(a_1, \ldots, a_r)$ sets up a bijective correspondence between associated primes of $(a_1, \ldots, a_r)$ in A and associated primes of (0) in $A/(a_1, \ldots, a_r)$, so applying 6.7.4 yields $\mathrm{ht}(\mathfrak{p}) \geq r$, $\dim(A/\mathfrak{p}) = \dim(A) - \mathrm{ht}(\mathfrak{p})$ for any associated prime of $(a_1, \ldots, a_r)$. □

---

[4] The **height** of an ideal I is the minimal height of an associated prime ideal of I.

COROLLARY 6.7.7 (F. S. Macaulay): *If A is Cohen-Macaulay then every system of parameters is a regular sequence.*

PROOF : Note that $\dim(A/(a_1,\ldots,a_d)) = 0$ if $a_1,\ldots,a_d \in A$ is a system of parameters and apply 6.7.6. □

It follows from 6.7.7 that the algebra[5] of example 2 in section 6.2 is not a Cohen-Macaulay algebra.

Before drawing out further consequences of Macaulay's theorem we indicate the relevance of Cohen-Macaulay algebras for invariant theory.

THEOREM 6.7.8 (J.A. Eagon and M. Hochster): *Let $\varrho : G \hookrightarrow \mathrm{GL}(n,\,\mathbb{F})$ be a representation of a finite group G over a field $\mathbb{F}$. If $|G|$ is prime to the characteristic of $\mathbb{F}$ then $\mathbb{F}[V]^G$ is a Cohen-Macaulay algebra.*

PROOF : Let $f_1,\ldots,f_n \in \mathbb{F}[V]^G$ be a system of parameters. Then

$$\mathbb{F}[f_1,\ldots,f_n] \subset \mathbb{F}[V]^G \subset \mathbb{F}[V]$$

are finite extensions so $f_1,\ldots,f_n \in \mathbb{F}[V]$ is a system of parameters. $\mathbb{F}[V]$ is Cohen-Macaulay so by Macaulay's theorem 6.7.7 $f_1,\ldots,f_n \in \mathbb{F}[V]$ is a regular sequence. Hence by 6.2.3 $\mathbb{F}[V]$ is a free $\mathbb{F}[f_1,\ldots,f_n]$-module. The projection

$$\pi^G : \mathbb{F}[V] \longrightarrow \mathbb{F}[V]^G$$

derived from the transfer (see section 2.4) represents $\mathbb{F}[V]$ as an $\mathbb{F}[V]^G$-module, and *a fortiori* as an $\mathbb{F}[f_1,\ldots,f_n]$-module, direct summand in $\mathbb{F}[V]$. Hence $\mathbb{F}[V]^G$ is a free $\mathbb{F}[f_1,\ldots,f_n]$-module, so by 6.2.8 $f_1,\ldots,f_n \in \mathbb{F}[V]^G$ is a regular sequence. □

EXAMPLE 1 : The **generalized quaternion group** Q_{4k} of order $4k$ ([54], page 253) (also called the **dicyclic group** in [63]) is defined by two generators $S,\ T \in Q_{4k}$ and two relations

$$S^k = T^2, \qquad STS = T.$$

By iterating the second relation we obtain $S^k T S^k = T$ from which it follows that $T^4 = 1 = S^{2k}$, so a complete list without repetition of the elements of G consists of

$$\{S^\sigma T^\delta \mid \alpha = 0,\ldots,2k-1,\ \delta = 0,\ 1\}.$$

There is a faithful two dimensional representation $\varrho : Q_{4k} \hookrightarrow \mathrm{GL}(2,\,\mathbb{C})$ given by

$$\varrho(S) = \begin{bmatrix} \lambda & 0 \\ 0 & \lambda^{-1} \end{bmatrix} \qquad \varrho(T) = \begin{bmatrix} 0 & 1 \\ -1 & 0 \end{bmatrix}$$

---

[5] It also cannot occur as a ring of invariants, since it is not integrally closed in its field of fractions.

where $\lambda = \exp(\pi i/k)$. N.b. $\lambda^k = -1$ and $\lambda^{2k} = 1$. In fact, if we identify $\mathbb{C}^2$ with $\mathbb{H}$ (the quaternions) then Q_{4k} may be identified with the group of unit quaternions generated by

$$\exp(\frac{\pi i}{k}), \quad j$$

and ϱ with the representation induced by right multiplication. For $k = 1$ we have $Q_4 \cong \mathbb{Z}/4$ and the invariants were examined in section 5.3. For $k = 2$ the group Q_8 is the usual quaternion group, and we examined its invariants in section 4.3 example 1.

We begin our study of $\mathbb{C}[x, y]^{Q_{4k}}$ by applying Molien's theorem to compute the Poincaré series. To this end note

$$\det(1 - S^m t) = \det \begin{bmatrix} 1 - \lambda^m t & 0 \\ 0 & 1 - \lambda^{-m} t \end{bmatrix} = \frac{1}{(1 - \lambda^m t)(1 - \lambda^{-m} t)}$$

$$\det(1 - S^m T t) = \det \begin{bmatrix} 1 & -\lambda^m t \\ \lambda^{-m} t & 1 \end{bmatrix} = 1 + t^2.$$

Applying Molien's theorem then gives

$$P(\mathbb{C}[x, y]^{Q_{4k}}, t) = \frac{1}{4k} \left[\frac{2k}{1 + t^2} + \sum_{m=0}^{2k-1} \frac{1}{(1 - \lambda^m t)(1 - \lambda^{-m} t)} \right].$$

The second term in the formula is one we evaluated in section 4.3. Substituting from equation ($\spadesuit$) section 4.3 and simplifying further yields

$$
\begin{aligned}
P(\mathbb{C}[x, y]^{Q_{4k}}, t) &= \frac{1}{4k} \left[\frac{2k}{1 + t^2} + 2k \frac{1 + t^2 + \cdots + t^{4k-2}}{(1 - t^{2k})^2} \right] \\
&= \frac{1}{2} \left[\frac{1}{1 + t^2} + \frac{1 + t^2 + \cdots + t^{4k-2}}{(1 - t^{2k})^2} \right] \\
&= \frac{1}{2} \left[\frac{(1 - t^{2k})^2 + (1 + t^2)(1 + t^2 + \cdots + t^{4k-2})}{(1 + t^2)(1 - t^{2k})^2} \right] \\
&= \frac{1}{2} \left[\frac{1 - 2t^{2k} + t^{4k} + 1 + 2t^2 + \cdots + 2t^{4k-2} + t^{4k}}{(1 + t^2)(1 - t^{2k})^2} \right] \\
&= \frac{1}{2} \left[\frac{2 + 2t^2 + \cdots + \widehat{2t^{2k}} + \cdots + 2t^{4k}}{(1 + t^2)(1 - t^{2k})^2} \right] \\
&= \frac{1 + t^2 + \cdots + \widehat{t^{2k}} + \cdots + t^{4k}}{(1 + t^2)(1 - t^{2k})^2},
\end{aligned}
$$

where $\widehat{}$ means the term under the $\widehat{}$ is omitted.

We next write down some invariants. Since $-1 \in Q_{4k}$ the similarity subgroup has even order so the invariants have even degrees. Using the formulae

$$S(x) = \lambda x \qquad T(x) = -y$$
$$S(y) = \lambda^{-1} y \qquad T(y) = x$$

one readily verifies

$$S(x^2 y^2) = \lambda^2 x^2 \lambda^{-2} y^2 = x^2 y^2$$
$$T(x^2 y^2) = (-y)^2 (x)^2 = x^2 y^2$$

so $f_1 = x^2 y^2$ is an invariant of degree 4. Experimenting with small values of k leads to the discovery of further invariants:

$$S(x^{2k} + y^{2k}) = \lambda^{2k} x^{2k} + \lambda^{-2k} y^{2k} = x^{2k} + y^{2k}$$
$$T(x^{2k} + y^{2k}) = (-y)^{2k} + x^{2k} = x^{2k} + y^{2k}$$

and

$$S(x^{2k+1} y - x y^{2k+1}) = \lambda^{2k+1} x^{2k+1} \lambda^{-1} y - \lambda x \lambda^{-(2k+1)} y^{2k+1} = x^{2k+1} y - x y^{2k+1}$$
$$T(x^{2k+1} y - x y^{2k+1}) = (-y)^{2k+1} x + y x^{2k+1} = x^{2k+1} y - x y^{2k+1}.$$

Set $f_2 = x^{2k} + y^{2k}$ and $h = x^{2k+1} y - x y^{2k+1}$. These computations show

$$\mathbb{C}[f_1, \ f_2] \oplus \mathbb{C}[f_1, \ f_2] \cdot h \subset \mathbb{C}[x, \ y]^{Q_{4k}}$$

and suggests further simplifications of the Poincaré series

$$P(\mathbb{C}[x, \ y]^{Q_{4k}}, \ t) = \frac{(1 + t^2 + \cdots + t^{2k-2}) + (t^{2k+2} + \cdots + t^{4k})}{(1 + t^2)(1 - t^{2k})^2}$$

$$= \frac{(1 - t^2)(1 + \cdots + t^{2k-2}) + (1 - t^2)(t^{2k+2} + \cdots + t^{4k})}{(1 - t^2)(1 + t^2)(1 - t^{2k})^2}$$

$$= \frac{(1 - t^{2k}) + t^{2k+2}(1 - t^{2k})}{(1 - t^4)(1 - t^{2k})^2}$$

$$= \frac{1 + t^{2k+2}}{(1 - t^4)(1 - t^{2k})}.$$

Since

$$P(\mathbb{C}[f_1, \ f_2] \oplus \mathbb{C}[f_1, \ f_2] \cdot h, \ t) = \frac{1 + t^{2k+2}}{(1 - t^4)(1 - t^{2k})}$$

as well, we conclude

$$\mathbb{C}[x, \ y]^{Q_{4k}} = \mathbb{C}[f_1, \ f_2] \oplus \mathbb{C}[f_1, \ f_2] \cdot h.$$

The elements f_1, f_2 are a system of parameters and $\{1,\ h\}$ is a basis for $\mathbb{C}[x,\ y]^{Q_{4k}}$ over $\mathbb{C}[f_1,\ f_2]$. There is one syzygy provided by the relation

$$h^2 = (x^{2k+1}y - xy^{2k+1})^2 = x^{4k+2}y^2 - 2x^{2k+2}y^{2k+2} + x^2y^{4k+2}$$
$$= x^2y^2(x^{4k} - 2x^{2k}y^{2k} + y^{4k}) = x^2y^2((x^{2k} + y^{2k})^2 - 4x^{2k}y^{2k})$$
$$= f_1(f_2{}^2 - 4f_1{}^k)\ .$$

In the modular case the ring of invariants $\mathbb{F}[V]^G$ of a finite group G may fail to be Cohen-Macaulay. If $\mathbb{F}[V]^G$ is Cohen-Macaulay then combining 6.6.1 and 6.6.2 it follows that $\mathrm{Ext}^i_{\mathbb{F}[V]^G}(\mathbb{F},\ \mathbb{F}[V]^G) = 0$ for $i < \dim_{\mathbb{F}}(V)$ and conversely. Hence one way to show that $\mathbb{F}[V]^G$ is *not* Cohen-Macaulay is to show that some low dimensional Ext-module is nonzero. Examples constructed in this way may be found in [153], [11] and [30]. The simplest to describe is the ring of invariants of the regular representation of the cyclic group of order 5. The proof that this ring is not Cohen-Macaulay is on the other hand not simple. The following example is a bit more complicated to describe, but the verifications that it is not Cohen-Macaulay are easier. For an alternative analysis of this example due to M.D. Neusel see [220] § 4 example 3.

EXAMPLE 2 (H. E. A. Campbell and I. P. Hughes [43]): Consider the representation $\sigma : \mathbb{Z}/2 \hookrightarrow \mathrm{GL}(n,\ \mathbb{F}_2)$ afforded by the matrix

$$S = \begin{bmatrix} 1 & 1 \\ 0 & 1 \end{bmatrix} \in \mathrm{GL}(n,\ \mathbb{F}_2)\ .$$

This is the permutation representation on the two element set where the fixed point, which is the sum of the elements, has been chosen as one of the basis elements. In section 2.4 example 2 we examined the invariants of $\sigma \oplus \sigma \oplus \sigma : \mathbb{Z}/2 \hookrightarrow \mathrm{GL}(6,\ \mathbb{F}_2)$ in connection with Noether's bound on the degrees of generators for nonmodular rings of invariants. These invariants also provide an example of a ring of invariants that is not Cohen-Macaulay.

To see this recall that in the notations of section 2.4 example 2 the polynomials

$$\eta_1 = x_2y_3 + x_3y_2$$
$$\eta_2 = x_1y_3 + x_3y_1$$
$$\eta_3 = x_1y_2 + x_2y_1$$

are invariant. The invariants of $\tau : \mathbb{Z}/2 \oplus \mathbb{Z}/2 \oplus \mathbb{Z}/2 \hookrightarrow \mathrm{GL}(6,\ \mathbb{F}_2)$ afforded by

$$\begin{bmatrix} S & 0 & 0 \\ 0 & 1 & 0 \\ 0 & 0 & 1 \end{bmatrix},\ \begin{bmatrix} 1 & 0 & 0 \\ 0 & S & 0 \\ 0 & 0 & 1 \end{bmatrix},\ \begin{bmatrix} 1 & 0 & 0 \\ 0 & 1 & 0 \\ 0 & 0 & S \end{bmatrix} \in \mathrm{GL}(6,\ \mathbb{F}_2)$$

are

$$\mathbb{F}_2[x_1, \ y_1, \ x_2, \ y_2, \ x_3, \ y_3]^{\mathbb{Z}/2 \oplus \mathbb{Z}/2 \oplus \mathbb{Z}/2} = \mathbb{F}_2[y_1, \ \xi_1, \ y_2, \ \xi_2, \ y_3, \ \xi_3]$$

where $\xi_i = x_i(x_i + y_i)$ for $i = 1, \ 2, \ 3$. The representation σ is the restriction of τ to the diagonal subgroup, hence

$$\mathbb{F}_2[x_1, \ y_1, \ x_2, \ y_2, \ x_3, \ y_3]^{\mathbb{Z}/2 \oplus \mathbb{Z}/2 \oplus \mathbb{Z}/2} \subset \mathbb{F}_2[x_1, \ y_1, \ x_2, \ y_2, \ x_3, \ y_3]^{\mathbb{Z}/2}$$

is a finite extension. Therefore the polynomials

$$y_1, \ \xi_1, \ y_2, \ \xi_2, \ y_3, \ \xi_3 \in \mathbb{F}_2[x_1, \ y_1, \ x_2, \ y_2, \ x_3, \ y_3]^{\mathbb{Z}/2}$$

are a system of parameters.

If $\mathbb{F}_2[x_1, \ y_1, \ x_2, \ y_2, \ x_3, \ y_3]^{\mathbb{Z}/2}$ were Cohen-Macaulay then there would be a direct sum decomposition

$$\mathbb{F}_2[x_1, \ y_1, \ x_2, \ y_2, \ x_3, \ y_3]^{\mathbb{Z}/2} = \bigoplus_{i=0}^{k} \mathbb{F}_2[y_1, \ \xi_1, \ y_2, \ \xi_2, \ y_3, \ \xi_3] \cdot h_i$$

for certain classes $h_0, \ h_1, \dots, \ h_k \in \mathbb{F}_2[x_1, \ y_1, \ x_2, \ y_2, \ x_3, \ y_3]^{\mathbb{Z}/2}$, where without loss of generality we may suppose that $h_0 = 1$. Since $y_1, \ y_2, \ y_3$ is a basis for the invariants of degree 1 it follows that $\deg(h_i) \geq 2$ for $i = 1, \dots, k$. We have unique expressions

$$\eta_i = \sum_{j=0}^{k} \varphi_{i, \ j} \cdot h_j \qquad i = 1, \ 2, \ 3 \, .$$

Since $\deg(\eta_i) = 2$ for $i = 1, \ 2, \ 3$ and $\deg(h_j) \geq 2$ for $j = 1, \dots, k$ we have $\varphi_{i,j} \in \mathbb{F}_2$ for $j = 1, \dots, k$. The polynomials $\eta_1, \ \eta_2, \ \eta_3$ satisfy the relation

$$y_1 \eta_1 + y_2 \eta_2 + y_3 \eta_3 = 0 \in \mathbb{F}_2[x_1, \ y_1, \ x_2, \ y_2, \ x_3, \ y_3]^{\mathbb{Z}/2} \, .$$

Hence
$$0 = y_1 \eta_1 + y_2 \eta_2 + y_3 \eta_3$$

$$= \sum_{j=0}^{k} (y_1 \varphi_{1, \ j} + y_2 \varphi_{2, \ j} + y_3 \varphi_{3, \ j}) \cdot h_j \, .$$

Since $1 = h_0, \ h_1, \dots, \ h_k$ are linearly independent over $\mathbb{F}_2[y_1, \ \xi_1, y_2, \ \xi_2, y_3, \ \xi_3]$ it follows that all the coefficients must vanish. In particular

$$y_1 \varphi_{1, \ j} + y_2 \varphi_{2, \ j} + y_3 \varphi_{3, \ j} = 0 \qquad j = 1, \dots, k \, .$$

Since $\varphi_{1, \ j}, \ \varphi_{2, \ j}, \ \varphi_{3, \ j} \in \mathbb{F}_2$ for $j = 1, \dots, k$ this represents a linear relation between $y_1, \ y_2, \ y_3$. But $y_1, \ y_2, \ y_3$ are linearly independent. Hence

$$\varphi_{1, \ j} = \varphi_{2, \ j} = \varphi_{3, \ j} = 0 \qquad j = 1, \dots, k \, .$$

In particular

$$\eta_1 = \varphi_{1,\, 0} \in \mathbb{F}_2[y_1,\ \xi_1,\ y_2,\ \xi_2,\ y_3,\ \xi_3]$$

which is a contradiction. Therefore $\mathbb{F}_2[x_1,\ y_1,\ x_2,\ y_2,\ x_3,\ y_3]^{\mathbb{Z}/2}$ is not Cohen-Macaulay.

The following result provides a lower bound for the codimension.

PROPOSITION 6.7.9 : *Let G be a finite group acting on the n-dimensional vector space V over the field $\mathbb{F}$. If $n \geq 2$ then $\operatorname{codim}(\mathbb{F}[V]^G) \geq 2$.*

PROOF : Let $f_1, \ldots, f_n$ be a system of parameters for $\mathbb{F}[V]^G$. Since $\mathbb{F}[V] \supset \mathbb{F}[V]^G$ is a finite extension $f_1, \ldots, f_n$ are a system of parameters for $\mathbb{F}[V]$ and hence by Macaulay's theorem 6.7.7 are a regular sequence in $\mathbb{F}[V]$. Choose two distinct elements, say x and y, from this sequence. We suppose that $x,\ y \in \mathbb{F}[V]^G$ is not a regular sequence. Then, by interchanging the role of x and y if need be, we may suppose x is a zero divisor in $\mathbb{F}[V]^G/(y)$. So we may write $ux = vy$ for some nonzero polynomials $u,\ v \in \mathbb{F}[V]^G$. Since $\mathbb{F}[V]$ is a unique factorization ring this implies $x = vw$ for some $w \in \mathbb{F}[V]$ so $v(uw - y) = 0$. The ring $\mathbb{F}[V]$ has no zero divisors and $v \neq 0$, so $y = uw$. If $g \in G$ then since y and u are G-invariant we get $uw = y = gy = g(uw) = ug(w)$ which implies w is G-invariant and $x \in (v) \subset \mathbb{F}[V]^G$. Thus u cannot be a zero divisor in $\mathbb{F}[V]^G/(v)$ and hence $u,\ v \in \mathbb{F}[V]^G$ is a regular sequence. □

COROLLARY 6.7.10 : *Let G be a finite group and $\varrho : G \hookrightarrow \mathrm{GL}(2,\ \mathbb{F})$ a representation. Then $\mathbb{F}[V]^G$ is Cohen-Macaulay.* □

Macaulay's theorems 6.7.6 and 6.7.7 have a number of other very interesting consequences. We indicate a few of these.

COROLLARY 6.7.11 : *Suppose $f_1, \ldots, f_n \in \mathbb{F}[x_1, \ldots, x_n]$ and introduce the polynomial algebra $\mathbb{F}[z_1, \ldots, z_n]$ where $\deg(z_i) = \deg(f_i)$, $i = 1, \ldots, n$. Regard $\mathbb{F}[x_1, \ldots, x_n]$ as a module over the polynomial algebra $\mathbb{F}[z_1, \ldots, z_n]$ via the algebra homomorphism induced by $z_i \mapsto f_i, i = 1, \ldots, n$. Then the following are equivalent:*

(i) *$f_1, \ldots, f_n \in \mathbb{F}[x_1, \ldots, x_n]$ is a regular sequence.*
(ii) *$f_1, \ldots, f_n$ are algebraically independent and $\mathbb{F}[x_1, \ldots, x_n]$ is a finitely generated $\mathbb{F}[f_1, \ldots, f_n]$-module.*
(iii) *$\mathbb{F}[x_1, \ldots, x_n]$ is a finitely generated $\mathbb{F}[z_1, \ldots, z_n]$-module.*
(iv) *$\mathbb{F}[x_1, \ldots, x_n]$ is a free finitely generated $\mathbb{F}[z_1, \ldots, z_n]$-module.*

PROOF : (i) $\Rightarrow$ (ii) follows from 6.2.1 and 6.7.1. (ii) $\Rightarrow$ (iii) is clear. From (iii) it follows that $f_1, \ldots, f_n$ are a system of parameters so by Macaulay's theorem form a regular sequence and thus (iii) $\Rightarrow$ (iv). (iv) $\Rightarrow$ (i) follows from Koszul's theorem 6.2.8. □

We derive further benefits for invariant theory from Macaulay's theorem and the preceding corollary.

COROLLARY 6.7.12 : *Let* $G \hookrightarrow \mathrm{GL}(n, \, \mathbb{F})$ *be a representation of a finite group* G *over the field* $\mathbb{F}$. *If* $\mathbb{F}[V]^G$ *is generated as an algebra by* n *elements* $f_1, \ldots, f_n$, *then* $f_1, \ldots, f_n$ *is a regular sequence and* $\mathbb{F}[V]^G \cong \mathbb{F}[f_1, \ldots, f_n]$.

PROOF : This follows from the preceding corollary since $\mathbb{F}[V]$ is finitely generated over $\mathbb{F}[V]^G$. $\square$

By making use of 6.4.4 we obtain yet another stepping stone towards understanding when the ring of invariants is a polynomial algebra.

COROLLARY 6.7.13 : *Let* $G \hookrightarrow \mathrm{GL}(n, \, \mathbb{F})$ *be a representation of a finite group* G *over the field* $\mathbb{F}$. *Then* $\mathbb{F}[V]^G$ *is a polynomial algebra if and only if* $\mathbb{F}[V]$ *is a free* $\mathbb{F}[V]^G$-*module*.

PROOF : If $\mathbb{F}[V]$ is a free $\mathbb{F}[V]^G$-module then by 6.4.4 $\mathbb{F}[V]^G$ is a polynomial algebra. On the other hand if $\mathbb{F}[V]^G$ is a polynomial algebra, then a set of polynomial generators for $\mathbb{F}[V]^G$ are a system of parameters for $\mathbb{F}[V]$ and hence a regular sequence by Macaulay's theorem 6.7.6 and so by 6.2.8 $\mathbb{F}[V]$ is a free $\mathbb{F}[V]^G$-module. $\square$

6.8 Applications to Poincaré Series

We collect a number of applications of the preceding discussions to the Poincaré series of rings of invariants. For a detailed discussion of how these results may be used in a concrete case (to correct an 80 year old error) see [139].

The ring of invariants of a finite group is Cohen-Macaulay when the order of the group is relatively prime to the characteristic of the ground field by 6.7.8. It is reasonable to ask for estimates of the degrees of a system of parameters. If $f_1, \ldots, f_n$ is a system of parameters for $\mathbb{F}[V]^G$ then so is $f_1^{k_1}, \ldots, f_n^{k_n}$ for any positive integers $k_1, \ldots, k_n$, so it is only meaningful to ask for lower bounds on the degrees of the $f_1, \ldots, f_n$.

THEOREM 6.8.1 (L. Solomon [228]): *Let* $\varrho : G \hookrightarrow \mathrm{GL}(n, \, \mathbb{F})$ *be a representation of a finite group* G *whose order is prime to the characteristic of* $\mathbb{F}$. *If* $f_1, \ldots, f_n$ *is a system of parameters for* $\mathbb{F}[V]^G$ *with* $\deg(f_i) = d_i$, *then*

$$d_1 \cdots d_n = |G| \cdot \dim_{\mathbb{F}}(\mathrm{Tot}(\mathbb{F} \otimes_{\mathbb{F}[f_1, \ldots, f_n]} \mathbb{F}[V]^G))$$

so $d_1 \cdots d_n \equiv 0 \bmod |G|$. *The equality* $d_1 d_2 \cdots d_n = |G|$ *holds if and only if* $\mathbb{F}[V]^G = \mathbb{F}[f_1, \ldots, f_n]$.

PROOF :   Since $f_1, \ldots, f_n$ is a system of parameters for $\mathbb{F}[V]^G$, then $\mathbb{F}[V]^G$ is a free finitely generated module over $\mathbb{F}[f_1, \ldots, f_n]$. Therefore the Poincaré series of $\mathbb{F}[V]^G$ is

$$P(\mathbb{F}[V]^G, \, t) = \frac{p(t)}{\prod_{i=1}^{n}(1 - t^{d_i})}$$

where $p(t)$ is the Poincaré series of $\mathbb{F} \otimes_{\mathbb{F}[f_1, \ldots, f_n]} \mathbb{F}[V]^G$, which is a polynomial with non-negative integral coefficients. Since $\mathbb{F}[V]$ is a free $\mathbb{F}[f_1, \ldots, f_n]$-module $p(1) = \dim_{\mathbb{F}}(\text{Tot}(\mathbb{F} \otimes_{\mathbb{F}[f_1, \ldots, f_n]} \mathbb{F}[V]^G))$. By 5.5.3

$$(1 - t)^n P(\mathbb{F}[V]^G, \, t) \Big|_{t=1} = \frac{1}{|G|}$$

whereas

$$(1 - t)^n \frac{p(t)}{\prod_{i=1}^{n}(1 - t^{d_i})} \bigg|_{t=1} = \frac{p(1)}{d_1 \cdots d_n}$$

and hence cross multiplication gives $d_1 \cdots d_n \cdot p(1) = |G|$. If there is a system of parameters $f_1, \ldots, f_n$ for $\mathbb{F}[V]^G$ with $d_1 \cdots d_n = |G|$ then by Noether normalization 5.3.3 $f_1, \ldots, f_n$ are algebraically independent and $\mathbb{F}[V]^G = \mathbb{F}[f_1, \ldots, f_n]$ by 5.5.5.   □

This result may be interpreted as follows. Suppose that $\mathbb{F}[V]^G$ is Cohen-Macaulay and $f_1, \ldots, f_n \in \mathbb{F}[V]^G$ is a system of parameters with $d_i = \deg(f_i)$ for $i = 1, \ldots, n$. Then as $\mathbb{F}[f_1, \ldots, f_n]$-module $\mathbb{F}[V]^G$ is freely generated by $\frac{d_1 \cdots d_n}{|G|}$ classes. The degrees of these polynomials are often computable using the following result.

PROPOSITION 6.8.2 : Let $\varrho : G \hookrightarrow \mathrm{GL}(n, \, \mathbb{F})$ be a representation of a finite group G whose order is prime to the characteristic of the field $\mathbb{F}$. Let $f_1, \ldots, f_n \in \mathbb{F}[V]^G$ be a system of parameters and $h_1, \ldots, h_d \in \mathbb{F}[V]^G$ elements such that

$$\mathbb{F}[V]^G = \bigoplus_{i=1}^{d} \mathbb{F}[f_1, \ldots, f_n] \cdot h_i$$

as $\mathbb{F}[f_1, \ldots, f_n]$-module. Let $\deg(f_i) = d_i$ for $i = 1, \ldots, n$ and $\deg(h_i) = e_i$ for $i = 1, \ldots, d$. Then

$$P(\mathbb{F}[V]^G, \, t) \prod_{i=1}^{n}(1 - t^{d_i}) = t^{e_1} + t^{e_2} + \cdots + t^{e_d}$$

and $d = \dim_{\mathbb{F}} (\text{Tot}(\mathbb{F}[V]/(f_1, \ldots, f_n)))$.

PROOF : The second conclusion follows from the previous result and the first by simply computing the Poincaré series of $\bigoplus\limits_{i=1}^{d} \mathbb{F}[f_1, \ldots, f_d] \cdot h_i.$ $\quad\square$

If $G \hookrightarrow \mathrm{GL}(n, \mathbb{F})$ is a representation in coprime characteristic, then 4.4.3 provides information about the second coefficient in the Laurent expansion of the Poincaré series $P(\mathbb{F}[V]^G, t)$ about $t = 1$ in terms of the number of pseudoreflections contained in G. By 6.7.8 there is a system of parameters in the coprime case, and with a bit of calculus we obtain a relation between the degrees of the system of parameters and the number of pseudoreflections.

THEOREM 6.8.3 : *Suppose* $\varrho : G \hookrightarrow \mathrm{GL}(n, \mathbb{F})$ *is a representation of a finite group* G *over a field* $\mathbb{F}$ *whose characteristic is prime to the order of* G. *Let* $f_1, \ldots, f_n$ *be a system of parameters for* $\mathbb{F}[V]^G$ *with* $\deg(f_i) = d_i$ *and choose elements* $h_1, \ldots, h_d \in \mathbb{F}[V]^G$, $d = \dim_{\mathbb{F}}(\mathrm{Tot}(\mathbb{F} \otimes_{\mathbb{F}[f_1,\ldots,f_n]} \mathbb{F}[V]^G))$, *such that*

$$\mathbb{F}[V]^G \cong \bigoplus_{j=1}^{d} \mathbb{F}[f_1, \ldots, f_n] \cdot h_j.$$

Then

$$|s(G)| = \sum_{i=1}^{n}(d_i - 1) - \frac{2}{d}\sum_{j=1}^{d} e_j$$

where $e_j = \deg(h_j)$ *for* $j = 1, \ldots, d$.

PROOF : We have

$$P(\mathbb{F}[V]^G, t) = \frac{p(t)}{\prod_{i=1}^{n}(1 - t^{d_i})}.$$

where $p(t)$ is the Poincaré series of $\mathbb{F} \otimes_{\mathbb{F}[f_1,\ldots,f_n]} \mathbb{F}[V]^G$. Set

$$f(t) = \prod_{i=1}^{n} \frac{1}{(1 - t^{d_i})} = \sum_{j=-n}^{\infty} a_j(1 - t)^j$$

$$p(t) = \sum_{j=0}^{d} b_j(1 - t)^j$$

and take their Cauchy product to get

$$P(\mathbb{F}[V]^G, t) = \frac{a_{-n}b_0}{(1 - t)^n} + \frac{a_{-n}b_1 + a_{-n+1}b_0}{(1 - t)^{n-1}} + \cdots$$

The values of a_{-n} and a_{-n+1} are given by the binomial theorem and the values of b_0 and b_1 by Taylor's theorem. Substituting into the preceding formula and using 4.4.1 yields the result by equating coefficients. $\quad\square$

In the preceding theorem the degrees of the generators $h_1, \ldots, h_d$ of $\mathbb{F}[V]^G$ as a module over a system of a parameters made an appearance. Naturally one should ask for estimates of these degrees also.

THEOREM 6.8.4 (R. P. Stanley [235]): *Suppose* $\varrho : G \hookrightarrow GL(n, \mathbb{F})$ *is a representation of a finite group G over a field $\mathbb{F}$ whose characteristic is prime to the order of G. Let $f_1, \ldots, f_n$ be a system of parameters for $\mathbb{F}[V]^G$ with $\deg(f_i) = d_i$ and choose elements $h_1, \ldots, h_d \in \mathbb{F}[V]^G$, $d = \dim_{\mathbb{F}}(\mathrm{Tot}(\mathbb{F} \otimes_{\mathbb{F}[f_1, \ldots, f_n]} \mathbb{F}[V]^G))$, such that*

$$\mathbb{F}[V]^G \cong \bigoplus_{j=1}^{d} \mathbb{F}[f_1, \ldots, f_n] \cdot h_j.$$

Let $e_i = \deg(h_i)$ for $i = 1, \ldots, d$ and suppose $e_1 \leq \cdots \leq e_d$. Then

$$e_d = \sum_{i=1}^{n} (d_i - 1) - \ell$$

where ℓ is the minimal degree of a $\det^{-1}$ relative invariant of G.

PROOF : Molien's theorem says

$$P(\mathbb{F}[V]^G, t) = \frac{1}{|G|} \sum_{g \in G} \frac{1}{\det(1 - gt)},$$

so

$$P(\mathbb{F}[V]^G, 1/t) = \frac{1}{|G|} \sum_{g \in G} \frac{1}{\det(1 - gt^{-1})}$$

$$= \frac{(-1)^n t^n}{|G|} \sum_{g \in G} \frac{\det(g)^{-1}}{\det(1 - g^{-1}t)}$$

$$= (-1)^n t^n P(\mathbb{F}[V]_{\det^{-1}}^G, t).$$

We have for the Poincaré series of $\mathbb{F}[V]^G$

$$P(\mathbb{F}[V]^G, t) = \frac{\sum_{j=1}^{d} t^{e_j}}{\prod_{i=1}^{n} 1 - t^{d_i}},$$

so

$$P(\mathbb{F}[V]^G, 1/t) = \frac{(-1)^n \sum_{j=1}^{d} t^{d_1 + \cdots + d_n - e_j}}{\prod_{i=1}^{n} 1 - t^{d_i}}.$$

Equating these two expressions and looking at terms of least degree gives $d_1 + \cdots + d_n - e_d = n + \ell$ as claimed. □

COROLLARY 6.8.5 (R. P. Stanley): *With the hypotheses and notations of the preceding theorem*

$$e_d \leq \sum_{i=1}^{n} (d_i - 1)$$

with equality if and only if $\varrho(G) \subset SL(n, \mathbb{F})$. □

Chapter 7
Groups Generated by Reflections

Many of the examples of rings of invariants that we have computed have turned out to be polynomial algebras, and we have seen several necessary conditions for the ring of invariants to be polynomial, such as 5.5.5 and 4.4.3. At present no necessary and sufficient conditions are known which assure that the ring of invariants of a finite group is a polynomial algebra. In the non-modular case, however, i.e. when the group order is prime to the characteristic of the field, e.g. if the field has characteristic zero, the groups with polynomial algebras of invariants are precisely those generated by pseudoreflections. This was originally observed by Shephard-Todd based on computing all examples with the aid of their classification theorem. See sections 7.3 for a discussion of the classification and section 7.4 for a proof of the theorem of Shephard-Todd without the use of the classification. This chapter is devoted to the nonmodular invariant theory of finite pseudoreflection groups. It represents only a small part of the known, unknown and interesting phenomena connected with these groups. See for example the many books, survey articles and research papers listed in the reference list.

7.1 Pseudoreflections

Let V be a finite dimensional vector space over the field $\mathbb{F}$. Recall from section 4.4 that a linear automorphism $s : V \longrightarrow V$ is called a **pseudoreflection** if it is not the identity, has finite order, and leaves a codimension 1 subspace, denoted by H_s, or V^s, and called the **hyperplane of** s, pointwise fixed. The subspace $\text{Im}(1 - s)$ of V is therefore of dimension 1. It is called the **direction** of s. The characteristic polynomial of a pseudoreflection s has 1 as a root of multiplicity $\dim_{\mathbb{F}}(V) - 1$, and one other root, say λ_s, which is a root of unity in $\mathbb{F}$. The subspace $H_s = \ker(1 - s)$ is the $+1$-eigenspace of s and if $\lambda_s \neq 1$ then s is diagonalizable and we call λ_s the **root of** s. If s has order prime to p, the characteristic of $\mathbb{F}$, then s is diagonalizable. If s is diagonalizable and $\lambda_s = -1$ the pseudoreflection has order 2 and is called simply a **reflection** (sometimes a **real reflection**). Note that s is a reflection if and only if

$s|_{(1-s)(V)} = -1$. If s is not diagonalizable then $\lambda_s = 1$ and s has Jordan form

$$
\begin{bmatrix}
1 & 0 & 0 & \cdots & 0 \\
0 & 1 & 0 & \cdots & 0 \\
\vdots & \vdots & & \ddots & \vdots \\
0 & 0 & \cdots & 1 & 1 \\
0 & 0 & 0 & \cdots & 1
\end{bmatrix}.
$$

In this case s is also called a **transvection**. This happens if and only if the characteristic of $\mathbb{F}$ is $p \neq 0$ and s is of order p. When s is a transvection $s|_{(1-s)(V)} = 1$. Therefore for $v \in V$ we may write $s(v) = v + \tau(v)$ where $\tau(v) \in H_s$ (just set $\tau = s - 1$) and $\tau : V \longrightarrow H_s$ is a linear transformation with $\dim_{\mathbb{F}}(\mathrm{Im}(\tau)) = 1$. Transvections play an important role in the study of finite geometries and the classification of their groups of motions. See for example [12], [154] and the references there. We will look at their role in invariant theory in more detail in chapter 8.

If $s : V \longrightarrow V$ is a diagonalizable pseudoreflection with root $\lambda_s \neq 1$ we call an eigenvector of s corresponding to λ_s a **root vector of** s. A root vector of s is well defined only up to nonzero scalar multiple. The one dimensional subspace spanned by a root vector of s is the λ_s-eigenspace and is referred to as the **root space of** s and will be denoted by R_s.

In characteristic zero there are a number of elementary properties of roots and reflecting hyperplanes that are easily verified using an inner product on V (see e.g. [97]). In characteristic $p \neq 0$, where an inner product is not available, these require different proofs. We collect what we will need in the sequel here.

LEMMA 7.1.1 : *If $s : V \longrightarrow V$ is a pseudoreflection and $T \in \mathrm{GL}(V)$ then $TH_s = H_{TsT^{-1}}$.*

PROOF : Since TsT^{-1} is similar to s it is a pseudoreflection. H_s and $H_{TsT^{-1}}$ are both codimension 1 subspaces so it will suffice to show $TH_s \subseteq H_{TsT^{-1}}$. Let $x \in TH_s$ and write $x = T(y)$. Then

$$
TsT^{-1}(x) = TsT^{-1}(T(y)) = T(s(y)) = T(y) = x
$$

so $x \in H_{TsT^{-1}}$. $\square$

If $\varrho : G \hookrightarrow \mathrm{GL}(n, \mathbb{F})$ is a representation we denote by $s(G)$ **the set of all pseudoreflections** in $\varrho(G)$. This depends of course on ϱ and when necessary we write $s(\varrho)$ to indicate this dependence. Similarly we denote by $\mathcal{H}(G)$ the set of all reflecting hyperplanes for elements of $s(G)$, i.e. $\mathcal{H}(G) = \{H_s \mid s \in s(G)\}$. Lemma 7.1.1 says that G acts as a permutation group on $\mathcal{H}(G)$. The set of diagonalizable pseudoreflections of G will be denoted by $s_\Delta(G)$ and the set of transvections by $s_{\underline{\Delta}}(G)$.

Associated to a diagonalizable pseudoreflection $s \in s_\Delta(G)$ is the reflecting hyperplane H_s, the root space R_s and the root λ_s. For a fixed reflecting hyperplane H_s and root space R_s the set

$$C_{(H_s, R_s)} = \{t \in s_\Delta \mid H_s = H_t, \ R_s = R_t\} \cup \{1\}$$

consists of all diagonalizable pseudoreflections with reflecting hyperplane H_s and root space R_s together with the identity element. $C_{(H_s, R_s)}$ is a cyclic group isomorphic to a subgroup of $\mathbb{F}^\times$: an isomorphism is given by $t \mapsto \lambda_t$ and $1 \mapsto 1$. More generally we have:

LEMMA 7.1.2 : *Let $\varrho : G \hookrightarrow GL(n, \mathbb{F})$ be a representation of a finite group G and $H \in \mathcal{H}(G)$ a reflecting hyperplane of G. Set*

$$C_H = \{s \in G \mid s = 1 \text{ or } V^s = H\}.$$

If the order of G is relatively prime to the characteristic of $\mathbb{F}$ then C_H is a cyclic group and the map

$$\lambda : C_H \longrightarrow \mathbb{F}^\times \qquad s \mapsto \begin{cases} \lambda_s & s \neq 1 \\ 1 & s = 1 \end{cases}$$

is an injective homomorphism.

PROOF : The first step is to show that C_H is a group. Let $s, \ t \in C_H$. Then $H_s = H = H_t$ so

$$st(x) = s(x) = x \qquad \forall \, x \in H.$$

Therefore $st = 1$ or st is a pseudoreflection with reflecting hyperplane H so C_H is closed under multiplication. If $1 \neq s \in C_H$ then $V^{s^{-1}} = H$ so s^{-1} is also a pseudoreflection with reflecting hyperplane H, i.e. $s^{-1} \in C_H$. Since $1 \in C_H$ by definition, it therefore follows that C_H is a group.

For $s \in C_H$ the order of s is a divisor of $|G|$ and hence the order of s is relatively prime to the characteristic of $\mathbb{F}$, so s is diagonalizable. The map λ is a homomorphism since $\lambda(s) = \det(s)$. If $s \neq 1$ then $\lambda_s \neq 1$ and hence the homomorphism λ is an injection. Since a finite subgroup of $\mathbb{F}^\times$ is cyclic the result follows. $\square$

REMARK : If $H < V$ is a codimension one subspace we set

$$G_H = \{s \in GL(n, \mathbb{F}) \mid s = 1 \text{ or } V^s = H\}.$$

Thus G_H consists of all pseudoreflections with fixed hyperplane H together with the identity element. If the characteristic of $\mathbb{F}$ is $p \neq 0$ then G_H is **not** a cyclic group for $n > 1$. Instead, if we choose a basis $x_1, \ldots, x_{n-1}$ for H and extend it by x_n to a basis for $\mathbb{F}^n$,

$$G_H = (\underset{n-1}{\oplus} \mathbb{F}) \rtimes \mathbb{F}^\times,$$

where $\underset{n-1}{\oplus}$ $\mathbb{F}$ consists of the matrices

$$S_i(\alpha) = \begin{bmatrix} 1 & 0 & \cdots & & 0 \\ 0 & 1 & 0\cdots & & 0 \\ \vdots & & \ddots & & \cdots \\ 0 & \cdots & & 1 \cdots 0 & \alpha \\ \vdots & & \cdots & \cdots & \vdots \\ 0 & \cdots & & 0 & 1 \end{bmatrix} \in GL(n, \mathbb{F}) \quad i = 1, \ldots, n-1,$$

were the off diagonal $\alpha \in \mathbb{F}$ is in the ith row and the last column. In other words, if $E_{i,j}$ denotes the $n \times n$ matrix with a 1 in the ith row and j^{th} column, and zeros elsewhere for $i, j \in \{1, \ldots, n\}$, then $S_i = I + \alpha E_{i,n}$ for $i = 1, \ldots, n-1$. These matrices are transvections with common hyperplane H. The remaining generators of G_H are the matrices

$$T_\zeta = \begin{bmatrix} 1 & 0 & \cdots & 0 \\ \vdots & & & \vdots \\ 0 & \cdots & 1 & 0 \\ 0 & \cdots & 0 & \zeta \end{bmatrix} \in GL(n, \mathbb{F}) \quad \zeta \in \mathbb{F}^\times.$$

See chapter 8 section 8.2 for a more detailed discussion. The following example clarifies the essential point.

EXAMPLE 1 : Let $\mathbb{F}$ be a Galois field of characteristic not equal to 2. Denote by $\{x, y\}$ the standard basis for $V = \mathbb{F}^2$. Define s, $t \in GL(n, \mathbb{F})$ by

$$s(x) = x \qquad t(x) = x$$
$$s(y) = -y \qquad t(x+y) = -x - y.$$

Then s and t are diagonalizable pseudoreflections, with nonidentity eigenvalue -1 and reflecting hyperplane $H = \text{Span}_{\mathbb{F}}(x)$. A root vector for s is y and a root vector for t is $x + y$. Since

$$t(y) = t(x + y - x) = -2x - y,$$

if follows that

$$ts(y) = 2x + y$$
$$st(y) = -2x + y.$$

Therefore $ts \neq st$ and hence the group G_H is not abelian, and certainly not cyclic. The subgroup of $GL(n, \mathbb{F})$ generated by s and t is a dihedral group of order $2p$, and st, ts are transvections, i.e. non-diagonalizable pseudo-reflections, with reflecting hyperplane H. Therefore the set

$$\{s \in GL(n, \mathbb{F}) \mid s = 1 \text{ or } V^s = H \text{ and } s \text{ is diagonalizable}\}$$

is **not** a subgroup of $GL(n, \mathbb{F})$ because the product of two diagonalizable reflections with the same reflecting hyperplane need not be diagonalizable.

LEMMA 7.1.3 : Let $\varrho : G \hookrightarrow \mathrm{GL}(n, \; \mathbb{F})$ be a representation of a finite group. Assume that the order of G is relatively prime to the characteristic of $\mathbb{F}$. If s, $t \in s(G)$ and $H_s = H_t$ then $R_s = R_t$.

PROOF : Set $H_s = H = H_t$. By lemma 7.1.2 the group C_H is cyclic. If $s_H \in C_H$ is a generator then s and t are powers of s_H and therefore $R_s = R_{s_H} = R_t$. $\square$

Let $s \, : \, V \longrightarrow V$ be a pseudoreflection with reflecting hyperplane $H_s = \ker(1 - s)$. The subspace $\mathrm{Im}(1 - s) \subset V$ is 1-dimensional so we may choose a nonzero vector $x_s \in \mathrm{Im}(1 - s)$. Then

$$s(v) = v + \ell_s(v)x_s \qquad \forall \, v \in V$$

where $\ell_s \, : \, V \longrightarrow \mathbb{F}$ is a linear functional with $\ker(\ell_s) = H_s$. The linear functional ℓ_s depends[1] only on the choice of x_s so is also well defined up to a nonzero scalar. If s is a diagonalizable pseudoreflection then $\mathrm{Im}(1 - s)$ is the root space R_s of s and x_s is a root vector, so $\ell_s(x_s) = \lambda_s - 1 \neq 0$ (since $\lambda_s \neq 1$), whereas if s is a transvection $\mathrm{Im}(1 - s) \subset \ker(1 - s)$ so $x_s \in H_s$ and $\ell_s(x_s) = 0$. For $f \in \mathbb{F}[V]$ and $u \in H_s$ we have

$$(sf - f)(u) = f(s^{-1}(u)) - f(u) = f(u) - f(u) = 0$$

and therefore ℓ_s divides $sf - f \in \mathbb{F}[V]$. We define $\Delta_s(f)$ to be the quotient, i.e.

$$sf - f = \Delta_s(f)\ell_s$$

in $\mathbb{F}[V]$. So $\Delta_s(f) \in \mathbb{F}[V]$ has degree $\deg(f) - 1$. Note that Δ_s depends on the choice of x_s and ℓ_s, so is well defined up to a nonzero scalar, and called the **Δ-operator** or **twisted derivation** associated to s. The following lemma is clear.

LEMMA 7.1.4 : If $s \in \mathrm{GL}(n, \; \mathbb{F})$ is a pseudoreflection and $f \in \mathbb{F}[V]$, then $f \in \mathbb{F}[V]^S$, where S is the subgroup of $\mathrm{GL}(n, \; \mathbb{F})$ generated by s, if and only if $\Delta_s(f) = 0$. $\square$

Thus if $\varrho \, : \, G \hookrightarrow \mathrm{GL}(n, \; \mathbb{F})$ is a representation and $\varrho(G)$ is generated by pseudoreflections, then $f \in \mathbb{F}[V]^G$ if and only if $\Delta_s(f) = 0$ for all $s \in s_\varrho(G)$. The operators Δ_s have played an important role in the study of groups generated by pseudoreflections and their invariants. See for example [65], [116], [117] and [164].

---

[1] If $\mathbb{F} = \mathbb{R}$ with inner product $\langle - \mid - \rangle$ preserved by s, then $\ell_s(x) = -2 \dfrac{\langle x \mid x_s \rangle}{\langle x_s \mid x_s \rangle}$ is a canonical choice for ℓ_s.

LEMMA 7.1.5 : *Let $s \in \mathrm{GL}(n, \mathbb{F})$ be a pseudoreflection and ℓ_s a linear form defining H_s. Then*

$$s(\ell_s) = \lambda_s \cdot \ell_s$$
$$\Delta_s(\ell_s) = \lambda_s - 1.$$

PROOF : Choose a basis $x_1, \ldots, x_{n-1}$ for H_s and extend this by x_n to a basis for V with respect to which s is in Jordan normal form. The space of linear forms vanishing on H_s is 1-dimensional, so without loss of generality we may assume that

$$\ell_s(x_i) = \begin{cases} 0 & \text{for } i = 1, \ldots, n-1 \\ 1 & \text{for } i = n. \end{cases}$$

Then if s is diagonalizable

$$s(\ell_s)(x_i) = \ell_s(s^{-1}x_i) = \begin{cases} \ell_s(x_i) = 0 & \text{for } i = 1, \ldots, n-1 \\ \ell_s(\lambda_s^{-1}x_n) = \lambda_s^{-1} & \text{for } i = n \end{cases}$$
$$= \lambda_s^{-1}\ell_s(x_i) = (\lambda_s \cdot \ell_s)(x_n) \qquad \text{for } i = 1, \ldots, n.$$

If s is a transvection we have

$$s(\ell_s)(x_i) = \ell_s(s^{-1}x_i)$$
$$= \begin{cases} \ell_s(x_i) = 0 & \text{for } i = 1, \ldots, n-1 \\ \ell_s(x_n + x_{n-1}) = \ell_s(x_n) + \ell_s(x_{n-1}) = 1 & \text{for } i = n \end{cases}$$
$$= \ell_s(x_i) \qquad \text{for } i = 1, \ldots, n.$$

Hence $s\ell_s = \lambda_s \cdot \ell_s$, where in the case of a transvection we abuse notations to write $\lambda_s = 1$. By definition of Δ_s we have

$$\ell_s \Delta_s(\ell_s) = s\ell_s - \ell_s = \lambda_s \cdot \ell_s - \ell_s$$

and the result follows. $\square$

LEMMA 7.1.6 : *If s, $t \in \mathrm{GL}(V)$ are diagonalizable pseudoreflections, then $\Delta_s(\ell_t) = 0$ if and only if $H_s \neq H_t$ and $H_t = H_{sts^{-1}}$.*

PROOF : Suppose $\Delta_s(\ell_t) = 0$. Then $s\ell_t = \ell_t$. Let $x \in H_t$. Then

$$0 = \ell_t(x) = s\ell_t(x) = \ell_t(s^{-1}x)$$

so $s^{-1}(x) \in \ker(\ell_t) = H_t$. Applying s to this inclusion yields $x \in sH_t = H_{sts^{-1}}$, so $H_t \subseteq H_{sts^{-1}}$. Since H_t and $H_{sts^{-1}}$ are both hyperplanes it follows that $H_t = H_{sts^{-1}}$.

If $H_s = H_t$ then $\ell_s = \alpha\ell_t$ for some $\alpha \in \mathbb{F}^\times$ and by the preceding lemma we have (n.b. $\lambda_s \neq 1$ since s is diagonalizable)

$$\Delta_s(\ell_t) = \Delta_s(\alpha\ell_s) = \alpha(\lambda_s - 1) \neq 0.$$

Hence $H_s \neq H_t$.

Conversely if $H_s \neq H_t$ and $H_t = H_{sts^{-1}}$ we may choose a vector x in H_t that is not in H_s. Since $H_t = H_{sts^{-1}} = sH_t$ it follows $s^{-1}(x) \in H_t$. Hence

$$\Delta_s(\ell_t)\ell_s(x) = s\ell_t(x) - \ell_t(x) = \ell_t(s^{-1}x) - \ell_t(x)$$
$$= 0 - 0 = 0.$$

However $\ell_s(x) \neq 0$ since $x \notin H_s = \ker(\ell_s)$. Therefore $\Delta_s(\ell_t) = 0$ as required.
□

The following result provides a key connection between pseudoreflections and invariant theory.

PROPOSITION 7.1.7 : *Suppose $\varrho : G \hookrightarrow \mathrm{GL}(n, \mathbb{F})$ is a representation of a finite group G for which $\varrho(G)$ is generated by pseudoreflections and G has order prime to the characteristic of $\mathbb{F}$. Then all pseudoreflections of G are diagonalizable and $\mathcal{L} = \prod_{s \in s(G)} \ell_s$ is a $\det^{-1} : G \longrightarrow \mathbb{F}$ relative invariant.*

PROOF : The elements of $s(G)$ have order prime to the characteristic of $\mathbb{F}$ so are diagonalizable. Since $|G|$ is relatively prime to the characteristic of $\mathbb{F}$ it follows from 7.1.2 that C_H is a cyclic group for all $H \in \mathcal{H}(G)$.

Choose generators $s_H \in C_H$ for each $H \in \mathcal{H}(G)$. Since every $s \in s(G)$ belongs to a unique C_H for $H \in \mathcal{H}(G)$ it follows that we may choose $\ell_s = \ell_{s_H}$ when $s \in C_H$. Hence

$$\mathcal{L} = \prod_{s \in s(G)} \ell_s = \prod_{H \in \mathcal{H}(G)} \ell_{s_H}^{c_H - 1}$$

where $c_H = |C_H|$.

If $H \in \mathcal{H}(G)$ then s_H acts on $\mathcal{H}(G)$ by $H' \mapsto s_H(H')$ and permutes the elements of $\mathcal{H}(G) \setminus \{H\}$. Therefore for $H' \in \mathcal{H}(G) \setminus \{H\}$ there is an $H'' \in \mathcal{H} \setminus \{H\}$ and an $\alpha' \in \mathbb{F}^\times$ such that $s_H(\ell_{H'}) = \alpha'\ell_{H''}$. Set

$$\mathcal{L}' = \prod_{H' \neq H} \ell_{s_{H'}}^{c_{H'} - 1},$$

so (the orders of $s_{H'}$ and $s_{H''}$ are the same since the subgroups that they generate are conjugate)

$$s_H(\mathcal{L}') = \alpha \prod_{H'' \neq H} \ell_{s_{H''}}^{c_{H''} - 1} = \alpha\mathcal{L}'.$$

Therefore

$$(\alpha - 1)\mathcal{L}' = s_H\mathcal{L}' - \mathcal{L}' = \Delta_{s_H}(\mathcal{L}') \cdot \ell_{s_H}.$$

If $\alpha \neq 1$ this equation implies that ℓ_{s_H} divides $\mathcal{L}'$. However, the linear forms $\ell_{s_{H'}}$ for $H' \in \mathcal{H}(G)$ are all relatively prime since they have distinct kernels. Therefore ℓ_{s_H} cannot divide $\mathcal{L}'$ and hence $\alpha = 1$.

By lemma 7.1.6 we have

$$s_H(\mathcal{L}) = s_H(\ell_{s_H}^{c_H-1}\mathcal{L}') = s_H(\ell_{s_H}^{c_H-1})s_H(\mathcal{L}')$$
$$= (\lambda_{s_H}^{c_H-1}\ell_{s_H}^{c_H-1})\mathcal{L}' = \det(s_H)^{-1}\mathcal{L}$$

as required. $\square$

7.2   Real and Crystallographic Reflection Groups

Let V be a finite dimensional vector space over the field $\mathbb{F}$. If $\mathbb{F} = \mathbb{Q}$, the field of rational numbers, then ± 1 are the only roots of unity and pseudoreflections are reflections. The product of two distinct reflections is a rotation in the plane determined by their root vectors and since its trace is rational, this product has order two, three, four or six [97]. The finite groups generated by reflections over $\mathbb{Q}$ are called the **crystallographic reflection groups** and they are in one-one correspondence with disjoint unions of Dynkin diagrams (see below) $\mathbf{A}_n$, $\mathbf{B}_n$, $\mathbf{D}_n$, $\mathbf{G}_2$, $\mathbf{F}_4$, $\mathbf{E}_6$, $\mathbf{E}_7$, and $\mathbf{E}_8$ (the reflection groups corresponding to $\mathbf{C}_n$ are the same as those for $\mathbf{B}_n$; only the root lengths are different). The crystallographic groups have been well studied and there are many books such as Benson and Grove [97], Conway and Sloane [60], Hiller [105], Chapter 1, Humphreys [114], Chapter III, or Bourbaki [39] that deal with them.

Over the real numbers $\mathbb{R}$, once again all pseudoreflections are reflections. The classification of the finite real reflection groups, often called **Coxeter groups**, was obtained by Coxeter[62] (see also Benson and Grove [97]).

Suppose $\varrho : G \hookrightarrow \mathrm{GL}(n, \mathbb{C})$ is a representation of a finite group G for which $\varrho(G)$ is generated by pseudoreflections. The ring of invariants is a polynomial algebra (see theorem 7.4.1). Inspection of the table of complex pseudoreflection groups (see section 7.3) shows that those ϱ which may be realized as the complexification of a real representation are precisely the pseudoreflection groups with a nonzero quadratic invariant. This is not hard to prove without the table and has nothing to do with pseudoreflection groups per se.

THEOREM 7.2.1 (G. Frobenius, I. Schur): *Let $\varrho : G \hookrightarrow \mathrm{GL}(n, \mathbb{C})$ be an irreducible representation of a finite group G. Then ϱ is the complexification of a real representation if and only if $\mathbb{C}[V]^G$ contains a nonzero quadratic polynomial.*

PROOF :   If the representation ϱ is the complexification of a real representation $\varrho_{\mathbb{R}} : G \hookrightarrow \mathrm{GL}(n, \mathbb{R})$ then averaging an inner product $\langle - \mid - \rangle$

on $\mathbb{R}^n$ over G yields a G-invariant inner product, i.e. we define a new inner product by

$$\langle x \mid y \rangle_G = \sum_{g \in G} \langle gx \mid gy \rangle \qquad \forall\, x,\, y \in \mathbb{R}^n,$$

which remains invariant under the G-action. Thus the square of the norm

$$q : \mathbb{R}^n \longrightarrow \mathbb{R} \qquad q(v) = \|v\|^2 = \langle v \mid v \rangle_G \qquad v \in \mathbb{R}^n$$

is a nonzero G-invariant quadratic polynomial, and remains so after complexification.

Conversely, if $q \in \mathbb{C}[V]^G$ is a nonzero quadratic invariant, we may define a symmetric bilinear form $B(-, \; -)$ on V by

$$2B(x,\; y) = q(x+y) - q(x) - q(y) \qquad x,\; y \in V.$$

From the G-invariance of q it follows that

$$B(gx,\; gy) = B(x,\; y) \qquad \forall\, g \in G \qquad x,\; y \in V.$$

Let $\langle - \mid - \rangle$ be a G-invariant Hermitian form on V. By the principal axis theorem [210] $B(x,\; y) = \langle x \mid T(y) \rangle$ for some *semi-linear* $T : V \longrightarrow V$, i.e. $T(x+y) = T(x) + T(y)$ and $T(\lambda x) = \bar{\lambda} T(x)$ for $x,\; y \in V$ and $\lambda \in \mathbb{C}$. The G-invariance of B implies that T commutes with the G-action. The transformation T^2 is complex linear and commutes with the G-action. By Schur's lemma $T^2 = \alpha I$ since V is irreducible. Since T^2 is Hermitian and positive definite it follows that $\alpha > 0 \in \mathbb{R}$ so by rescaling q we may suppose $\alpha = 1$, i.e. $T^2 = I$. Set $V_{\mathbb{R}} = \{v \in V \mid T(v) = v\}$. Then $V_{\mathbb{R}}$ is a real subspace of V, stable under the G-action with $V_{\mathbb{R}} + iV_{\mathbb{R}} = V$ and $V_{\mathbb{R}} \cap iV_{\mathbb{R}} = \{0\}$. Thus ϱ is the complexification of the real representation of G afforded by $V_{\mathbb{R}}$. $\square$

A representation $\varrho : G \hookrightarrow GL(n,\; \mathbb{R})$ is called a **real reflection representation** if $\varrho(G)$ is generated by reflections. One consequence of Coxeter's classification is that the representation ϱ is intrinsic to G, i.e., is unique, and may be reconstructed from the abstract group G starting only with an appropriate presentation of G in terms of generators and relations. For this reason we allow ourselves to say G **is a real reflection group** when there is a real reflection representation for G. We describe enough of the proof of the classification theorem to obtain one result (see theorem 7.2.11) of significance for invariant theory.

If $\varrho : G \hookrightarrow GL(n,\; \mathbb{R})$ is a finite real reflection group and $\langle - \mid - \rangle$ is an inner product on $\mathbb{R}^n$, then averaging $\langle - \mid - \rangle$ over the group yields a G-invariant inner product:

$$\langle x \mid y \rangle_G = \sum_{g \in G} \langle gx \mid gy \rangle \qquad \forall\, x,\; y \in \mathbb{R}^n.$$

So we may suppose that G acts orthogonally, i.e. $\varrho : G \hookrightarrow \mathbb{O}(n)$. If $s \in s(G)$ has reflecting hyperplane H_s and root vector r_s then $r_s \perp H_s$ and

$$s(v) = v - 2\frac{\langle v \mid r_s \rangle}{\langle r_s \mid r_s \rangle} \cdot r_s.$$

LEMMA 7.2.2 : *If s', $s'' \in \mathbb{O}(n)$ are orthogonal reflections then the composition $s's''$ is a rotation through an angle equal to twice the angle between the lines through a pair of corresponding root vectors.*

PROOF : The determinant of $s's''$ is $+1$ so it is a rotation. Let $r_{s'}$, $r_{s''}$ be unit length root vectors for s', s'' and let θ be the angle between the lines that they span in $\mathbb{R}^n$. Then $\cos(\theta) = \langle r_{s'} \mid r_{s''} \rangle$. To compute the angle of the rotation $s's''$ we compute the angle φ between $r_{s''}$ and $s's''(r_{s''})$. We find

$$\begin{aligned}
\cos(\varphi) &= \langle r_{s''} \mid s's''(r_{s''}) \rangle = \langle r_{s''} \mid s'(-r_{s''}) \rangle \\
&= -\langle r_{s''} \mid s'(r_{s''}) \rangle = -\langle r_{s''} \mid r_{s''} - 2\langle r_{s''} \mid r_{s'} \rangle \cdot r_{s'} \rangle \\
&= -\left(\langle r_{s''} \mid r_{s''} \rangle - 2\langle r_{s''} \mid r_{s'} \rangle^2 \right) = 2\cos^2(\theta) - 1 \\
&= \cos(2\theta)
\end{aligned}$$

by the double angle formula. $\square$

If $s \in \mathbb{O}(n)$ is an orthogonal reflection then the linear form $\ell_s : \mathbb{R}^n \longrightarrow \mathbb{R}$ defined by

$$\ell_s(v) = \frac{-2\langle v \mid r_s \rangle}{\langle r_s \mid r_s \rangle} \qquad v \in \mathbb{R}^n$$

represents a canonical choice (well defined up to sign) for a linear form defining H_s. The analog of 7.1.5 - 7.1.7 may be summarized as follows. (The proof involves manipulating inner products. See e.g. [105], [97].)

LEMMA 7.2.3 : *Suppose s', s, $s'' \in \mathbb{O}(n)$ are orthogonal reflections. Then*
    (i) $s\ell_s = -\ell_s$,
    (ii) $s'\ell_{s''} = \ell_{s''}$ if $s' \neq s''$. $\square$

PROPOSITION 7.2.4 : *If $G \hookrightarrow \mathbb{O}(n)$ is a finite real reflection group, then*

$$\mathcal{L} = \prod_{s \in s(G)} \ell_s \in \mathbb{R}[V]^G_{\det}.$$

If $f \in \mathbb{F}[V]^G_{\det}$ then $\mathcal{L}$ divides f.

PROOF : The first assertion follows from 7.2.3. To prove the second let $s \in s(G)$ with reflecting hyperplane $H_s \subset \mathbb{R}^n = V$. For $v \in H_s$ we have

$$f(v) = f(sv) = \det(s) \cdot f(v) = -f(v)$$

so $f\big|_{H_s} = 0$ and hence ℓ_s divides f. Since distinct reflections have distinct reflecting hyperplanes the linear forms $\{\ell_s \mid s \in s(G)\}$ are relatively prime, so $\mathcal{L}$, which is their product, divides f $\square$

We agree to choose all root vectors to have unit length (n.b. this is not always the best choice, particularly for the crystallographic groups, but it will serve us), so each reflection $s \in s(G)$ has two root vectors that differ only in sign and are both orthogonal to the reflecting hyperplane H_s. Since G permutes $\mathcal{H}(G)$ it also permutes the unit normals to the hyperplanes of G and hence permutes the set $\mathcal{R}(G)$ of unit root vectors.

Conversely, suppose given a finite set of vectors $\mathcal{R} \subset \mathbb{R}^n$ not containing 0, and satisfying

$$r \in \mathcal{R} \Longrightarrow s_r(\mathcal{R}) \subset \mathcal{R},$$

where s_r is the reflection with root vector r. Such a collection $\mathcal{R}$ is referred to as a **root system**. Associated to a root system is a reflection group $W(\mathcal{R}) \subset GL(n, \mathbb{R})$ defined as the subgroup generated by the reflections $\{s_r \mid r \in \mathcal{R}\}$. Clearly $W(\mathcal{R}(G)) = G$ so the root system of a reflection group G may be used to describe G completely.

LEMMA 7.2.5 : Let $\varrho : G \hookrightarrow \mathbb{O}(n)$ be a reflection representation of the finite group G. Then the set of fixed points $V^G = \{0\}$ if and only if $\mathcal{R}(G)$ spans $V = \mathbb{R}^n$.

PROOF : Suppose $\mathcal{R}(G)$ spans V. If $v \in V^G$ then for $s \in s(G)$ we have $s(v) = v$ so $v \perp r_s$. Since $s(G)$ generates G, $v \perp \mathcal{R}(G)$ so $v = 0$. On the other hand, if $\mathrm{Span}_{\mathbb{R}}(\mathcal{R}(G))$ is a proper subspace of V, then for $0 \neq v \in \mathrm{Span}_{\mathbb{R}}(\mathcal{R}(G))^{\perp}$ we have $s(v) = v$ and since $s(G)$ generates G, $g(v) = v$ for all $g \in G$ so $0 \neq v \in V^G$. $\square$

We need a way to distinguish the two root vectors of a reflecting hyperplane. To do so we choose an **arbitrary vector** (in this day and age a non-functorial arbitrary choice is indeed a rarity!) $z \in V$ such that $\langle z \mid r_s \rangle \neq 0$ for all $r_s \in \mathcal{R}(G)$. Notice that $\langle z \mid r_s \rangle \neq 0$ for all $r_s \in \mathcal{R}(G)$ is equivalent to $z \notin H$ for any $H \in \mathcal{H}(G)$. Since $\mathcal{H}(G)$ is finite, $V \neq \cup_{H \in \mathcal{H}(G)} H$, so such a choice is always possible. With respect to z we partition $\mathcal{R}(G)$ into two disjoint subsets

$$\mathcal{R}_+(G) = \{r \in \mathcal{R}(G) \mid \langle z \mid r \rangle > 0\}$$
$$\mathcal{R}_-(G) = \{r \in \mathcal{R}(G) \mid \langle z \mid r \rangle < 0\}$$

called the **positive roots** and **negative roots** respectively.

DEFINITION : *Let $\varrho : G \hookrightarrow \mathbb{O}(n)$ be a fixed point free reflection representation of a finite group G. Let $0 \neq z \in \mathbb{R}^n$ satisfy $\langle z \mid r \rangle \neq 0$ for all $r \in \mathcal{R}(G)$. A subset $\mathcal{F}(G) \subset \mathcal{R}_+(G)$ is called a* **fundamental system of positive roots for G with respect to z,** *or* **simple roots with respect to z,** *if every element of $\mathcal{R}_+(G)$ may be written as a linear combination of elements of $\mathcal{F}(G)$ with non-negative coefficients, and $\mathcal{F}(G)$ is minimal with respect to this property.*

The existence of a fundamental system of positive roots with respect to z is clear: one removes elements from $\mathcal{R}_+(G)$ so long as every element of $\mathcal{R}_+(G)$ can be written as a linear combination with non-negative coefficients of the elements that remain. The uniqueness of such a fundamental system of roots is by no means clear. We should indicate the dependence of $\mathcal{F}(G)$ on the choice of the vector $z \in \mathbb{R}^n$ by writing $\mathcal{F}_z(G)$. We prefer not to clutter up our notation, and so we agree that for the rest of this section $G \hookrightarrow \mathbb{O}(n)$ is a fixed point free reflection group and $z \in V = \mathbb{R}^n$ a fixed vector with nontrivial inner product with respect to all the root vectors of G. $\mathcal{R}_+(G)$ and $\mathcal{R}_-(G)$ are the positive and negative roots, with $\mathcal{F}(G)$ a fundamental system of positive roots, all with respect to z. We could try to use the characteristic property of the fundamental roots to separate out the vectors of V into a union of two disjoint sets, one of positive and one of negative vectors: however this will not work as the following lemma shows.

LEMMA 7.2.6 : *If $r' \neq r'' \in \mathcal{F}(G)$ and λ', λ'' are positive real numbers then the vector $v = \lambda' r' - \lambda'' r''$ is neither positive nor negative.*

PROOF : Suppose that v were positive in the sense that

$$\lambda' r' - \lambda'' r'' = \mu' r' + \mu'' r'' + \sum_{i=1}^{k} \mu_i r_i$$

where $r_1, \ldots, r_k$ are the elements of $\mathcal{F}(G)$ different from r', r'' and $\mu', \mu'', \mu_1, \ldots, \mu_k \geq 0$. Then $\lambda' \leq \mu'$ for otherwise we could rewrite the above equation in the form

$$(\lambda' - \mu')r' = (\mu'' + \lambda'')r'' + \sum_{i=1}^{k} \mu_i r_i$$

which says that $\mathcal{F}(G) \setminus \{r'\}$ has the characteristic property for a fundamental system of positive roots contrary to the minimality of $\mathcal{F}(G)$ with respect to this property. Therefore $\lambda' \leq \mu'$ and if rewrite our relation in the form

$$0 = (\mu' - \lambda')r' + (\mu'' + \lambda'')r'' + \sum_{i=1}^{k} \mu_i r_i$$

we get upon applying $\langle z \mid - \rangle$

$$0 = \langle z \mid (\mu' - \lambda')r' + (\mu'' + \lambda'')r'' + \sum_{i=1}^{k} \mu_i r_i \rangle$$

$$= (\mu' - \lambda')\langle z \mid r' \rangle + (\mu'' + \lambda'')\langle z \mid r'' \rangle + \sum_{i=1}^{k} \mu_i \langle z \mid r_i \rangle,$$

which since $\langle z \mid r' \rangle$, $\langle z \mid r'' \rangle$, $\langle z \mid r_1 \rangle, \ldots, \langle z \mid r_k \rangle$ are strictly positive implies that all the coefficients vanish, so $\lambda'' = 0$ contrary to hypothesis. $\square$

A system of fundamental roots $\mathcal{F}(G)$ has a remarkable geometric property which is the key to the classification of Coxeter groups.

PROPOSITION 7.2.7 : Let $\varrho : G \hookrightarrow \mathbf{O}(n)$ be a fixed point free reflection representation of the finite group G. Let $0 \neq z \in \mathbb{R}^n$ satisfy $\langle z \mid r \rangle \neq 0$ for all $r \in \mathcal{R}(G)$. Let $\mathcal{F}(G)$ be a system of fundamental roots with respect to z. Then $\langle r' \mid r'' \rangle \leq 0$ for all $r' \neq r'' \in \mathcal{F}(G)$.

PROOF : Let s' be the reflection corresponding to r'. Since G permutes the roots, $s'r''$ lies in $\mathcal{R}(G)$ and is thus either positive or negative (with respect to z). We have

$$s'r'' = r'' - 2\langle r' \mid r'' \rangle r'$$

which has one positive coefficient, 1, and hence by 7.2.6 the other coefficient must be non-negative. $\square$

LEMMA 7.2.8 : Let V be a vector space over $\mathbb{R}$ with scalar product $\langle - \mid - \rangle$ and $H \subset V$ a hyperplane. If $v_1, \ldots, v_m \in V$ all lie on one side of H and $\langle v_i \mid v_j \rangle \leq 0$ for $i \neq j$ then $v_1, \ldots, v_m$ are linearly independent.

PROOF : To say that $v_1, \ldots, v_m$ all lie on one side of H means that for a suitable choice of unit normal u to H we have $\langle u \mid v_i \rangle > 0$ for $i = 1, \ldots, m$. Suppose that we have a linear relation between the $v_1, \ldots, v_m$. By reordering we may suppose

$$\sum_{i=1}^{k} a_i v_i = \sum_{j=k+1}^{m} a_j v_j$$

where all the a_i are non-negative. Then

$$\left\| \sum_{i=1}^{k} a_i v_i \right\|^2 = \left\langle \sum_{i=1}^{k} a_i v_i \mid \sum_{j=1}^{k} a_j v_j \right\rangle$$

$$= \left\langle \sum_{i=1}^{k} a_i v_i \mid \sum_{j=k+1}^{m} a_j v_j \right\rangle = \sum_{i=1, j=k+1}^{k,m} a_i a_j \langle v_i \mid v_j \rangle \leq 0$$

so $\left\| \sum_{i=1}^{k} a_i v_i \right\|^2 = 0 = \left\| \sum_{j=k+1}^{m} a_j v_j \right\|^2$ and $a_1 = \cdots = a_k = 0 = a_{k+1} = \cdots = a_m$. $\square$

THEOREM 7.2.9 : *Let* $\varrho : G \hookrightarrow \mathbb{O}(n)$ *be a fixed point free reflection representation of a finite group* G. *Let* $0 \neq z \in \mathbb{R}^n$ *satisfy* $\langle z \mid r \rangle \neq 0$ *for all* $r \in \mathcal{R}(G)$. *Then there is a unique system of fundamental roots* $\mathcal{F}(G)$ *with respect to* z *and it is a basis for* $\mathbb{R}^n$.

PROOF : By 7.2.5 and the definitions $\mathcal{F}(G)$ spans V and all the fundamental roots lie on one side of the hyperplane orthogonal to z so by 7.2.7 and 7.2.8 are linearly independent.

If $\mathcal{F}'(G)$ and $\mathcal{F}''(G)$ are two fundamental systems of positive roots for G with respect to z then they are both bases for $V = \mathbb{R}^n$. Let $T : V \longrightarrow V$ be a linear extension of a bijection between $\mathcal{F}'(G)$ and $\mathcal{F}''(G)$. Since $\mathcal{F}'(G)$ and $\mathcal{F}''(G)$ consist of positive roots the elements of one of these sets can be written as non-negative integral linear combinations of the elements of the other set. Hence the matrix A of this linear transformation with respect to either of these basis of $\mathbb{R}^n$ has non-negative integral entries. The same applies to A^{-1}. A matrix with non-negative integral entries whose inverse also has non-negative integral entries is a permutation matrix and the result follows. $\square$

LEMMA 7.2.10 : *Let* $\varrho : G \hookrightarrow \mathbb{O}(n)$ *be a fixed point free reflection representation of the finite group* G. *Let* $0 \neq z \in \mathbb{R}^n = V$ *satisfy* $\langle z \mid r \rangle \neq 0$ *for all* $r \in \mathcal{R}(G)$. *If* r *is a positive root with respect to* z *and* s *is a reflection corresponding to a simple root with respect to* z *different from* r, *then* $s(r)$ *is a positive root with respect to* z.

PROOF : Let the simple roots be $\{r_1, \ldots, r_n\}$ and the corresponding reflections $\{s_1, \ldots, s_n\}$ be chosen so that $s_1 = s$. Since $r \in \mathcal{R}_+(G)$ we may write

$$r = \lambda_1 r_1 + \lambda_2 r_2 + \cdots + \lambda_n r_n$$

with $\lambda_1, \ldots, \lambda_n \geq 0$ and as $r \neq r_1$ at least one of the coefficients $\lambda_2, \ldots, \lambda_n$ is nonzero. Without loss of generality we may suppose $\lambda_2 \neq 0$. Then we find

$$s(r) = s_1(r) = -\lambda_1 r_1 + (\lambda_2 r_2 - 2\langle r_2 \mid r_1 \rangle r_1) + \cdots + (\lambda_n r_n - 2\langle r_n \mid r_1 \rangle r_1)$$

$$= \mu_1 r_1 + \lambda_2 r_2 + \lambda_3 r_3 + \cdots + \lambda_n r_n.$$

Since $s(r)$ is again a root it must be either positive or negative. But one of the coefficients in the preceding expression is positive, namely λ_2, and hence $s(r) \in \mathcal{R}_+(G)$. $\square$

THEOREM 7.2.11 : *Let $\varrho : G \hookrightarrow \mathbb{O}(n)$ be a fixed point free reflection representation of a finite group G. Let $0 \neq z \in \mathbb{R}^n = V$ satisfy $\langle z \mid r \rangle \neq 0$ for all $r \in \mathcal{R}(G)$. Then G is generated by the n reflections corresponding to the simple roots with respect to z.*

PROOF : Let W be the subgroup of G generated by the reflections corresponding to the simple roots. If $r \in \mathcal{R}_+(G)$ then by 7.2.5 there is a simple root r_1 with $\langle r \mid r_1 \rangle \neq 0$ and hence by the definitions involved $\langle r \mid r_1 \rangle > 0$. Consider

$$x_1 = s_1(r) = r - 2\langle r \mid r_1 \rangle r_1$$

where $s_1 \in W$ is the reflection corresponding to the simple root r_1. If $r \notin \mathcal{F}(G)$ then by 7.2.10 $x_1 \in \mathcal{R}_+(G)$. Note

$$\langle z \mid x_1 \rangle = \langle z \mid r - 2\langle r \mid r_1 \rangle r_1 \rangle = \langle z \mid r \rangle - 2\langle r \mid r_1 \rangle \langle z \mid r_1 \rangle \; < \; \langle z \mid r \rangle.$$

By repeating the above process we obtain a sequence of elements $r = x_0, x_1, \ldots, x_k \in \mathcal{R}_+(G)$ with

$$x_i = s_i s_{i-1} \cdots s_1(r) \qquad i = 1, \ldots, k$$

and

$$\langle z \mid x_0 \rangle \; > \; \langle z \mid x_1 \rangle \; > \; \cdots \; > \; \langle z \mid x_k \rangle.$$

The group W is finite so there are only a finite number of compositions $s_k \cdots s_1$ and hence the above process must stop with an element $x_k = s_k \cdots s_1(r) = t \in \mathcal{F}(G)$. By lemma 7.1.1 we then have

$$s_r = s_k \cdots s_1 t s_1^{-1} \cdots s_k^{-1} \in W$$

where s_r is the reflection corresponding to r. Since $s(G)$ generates G we have $G = W$ as required. $\square$

The reflections corresponding to the fundamental roots are called **fundamental reflections**. By 7.2.9 the number of fundamental reflections is the dimension of the reflection representation ϱ of G. Thus a Coxeter group in $\mathbb{O}(n)$ has a presentation in terms of the n fundamental reflections $\{s_1, \ldots, s_n\}$ as generators, and certain relations. In fact it is enough to specify the orders $m_{i,j}$ of the products $s_i s_j$ for $i, j = 1, \ldots, n$. As we will see later this result has consequences for the invariant theory of real reflection groups.

DEFINITION : *Let $\varrho : G \hookrightarrow \mathbb{O}(n)$ be a fixed point free reflection representation of a finite group G. Let $0 \neq z \in \mathbb{R}^n = V$ satisfy $\langle z \mid r \rangle \neq 0$ for all $r \in \mathcal{R}(G)$ and let $\{s_1, \ldots, s_n\}$ be the fundamental reflections of G with respect to z and $\{r_1, \ldots, r_n\}$ be the corresponding roots of unit length. Let $\langle r_i \mid r_j \rangle = -\cos(\pi/m_{i,j})$ where $m_{i,j} \in \mathbb{N}$ for $i, j = 1, \ldots, n$. (This is the cosine of half the angle of rotation of $s_i s_j$ by 7.2.2 and hence $m_{i,j}$ is the order of this rotation.) The matrix $(m_{i,j})$ is called the* **Coxeter matrix** *of*

G. The **Coxeter graph** *of G is the labeled graph with vertices* $\{1,\ldots,n\}$ *where two vertices, i and j, are connected by a labeled edge if and only if* $\langle r_i \mid r_j \rangle = -\cos(\pi/m_{i,j})$ *where* $m_{i,j} > 2$, *i.e. if and only if* $r_i \not\perp r_j$; *the edge is labeled with* $m_{i,j}$ *if* $m_{i,j} > 3$, *otherwise we suppress the label.*

The definition of the Coxeter graph has been chosen so that the irreducible (reflection) representations correspond to the connected graphs. The classification theorem of Coxeter [62] [97] [105] says that G and its reflection representation are classified up to conjugation by their Coxeter graph. The table 7.2.1 lists the connected Coxeter graphs. To the left of each graph is the label by which the graph and the corresponding Coxeter groups are referred to. (N.b. There are several different conventions for drawing Coxeter graphs: Bourbaki [39] uses one of the alternate conventions. The labels however are standard.) If in addition the nodes of the Coxeter graph are labeled with the corresponding root vector, the resulting graph is called a **Dynkin diagram**.

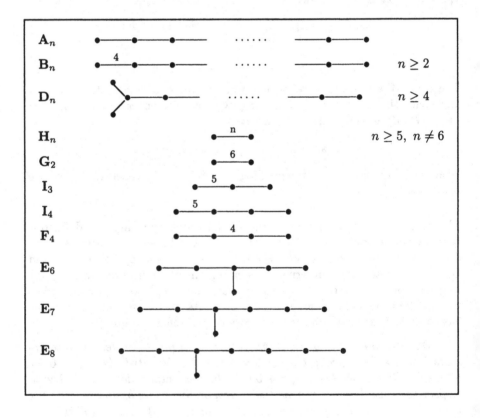

Table 7.2.1: The connected Coxeter graphs

To construct the diagram from the root system requires the angles $\theta_{i,\,j}$ between pairs of roots r_i, $r_j \in \mathcal{R}$. The angle $\theta_{i,\,j}$ is half the angle of rotation $\Theta_{i,\,j}$ of $s_i s_j$. The angle $\Theta_{i,\,j}$ is a fraction $\frac{2\pi}{m_{i,\,j}}$ where $m_{i,\,j}$ is the order of $s_i s_j$. For example, for the root system $\mathbf{A}_n$ we may choose as fundamental roots the vectors

$$\mathcal{F}(\mathbf{A}_n) = \{E_i - E_j \in \mathbb{R}^{n+1} \mid i \neq j,\ 1 \leq i,\ j \leq n+1\}.$$

For the roots $r_1 = E_1 - E_2$, $r_2 = E_2 - E_3$ we have

$$\cos(\theta_{r_1,\,r_2}) = \frac{\langle E_1 - E_2 \mid E_2 - E_3 \rangle}{\|E_1 - E_2\|^2 \|E_2 - E_3\|^2} = \frac{-1}{\sqrt{2}\sqrt{2}} = -\frac{1}{2}.$$

Therefore $\theta_{1,\,2} = 120°$ so r_1 and r_2 are joined by a single edge in the diagram.

The groups of type $\mathbf{A}_n$ are the symmetric groups Σ_{n+1} acting on $\mathbb{R}^n \subset \mathbb{R}^{n+1}$ by restricting the tautological representation on $\mathbb{R}^{n+1}$ to the hyperplane of vectors with coordinate sum zero. Those of type $\mathbf{B}_n$ the signed permutation groups acting on $\mathbb{R}^n$ and $\mathbf{D}_n$ the subgroup of $\mathbf{B}_n$ consisting of elements changing only an even number of signs. The groups of type $\mathbf{H}_n$ are the dihedral groups of order $2n$ acting as the group of symmetries of a regular n-gon in the plane. The remaining groups in the list are somewhat more complicated to describe. Those of type $\mathbf{G}_2$, $\mathbf{F}_4$, $\mathbf{E}_i$, $i = 6,\ 7,\ 8$, are the Weyl groups (in their canonical representation) of the compact simply connected Lie groups of the same name. The crystallographic groups are the Weyl groups of the compact Lie groups and correspond to the types $\mathbf{A}_n$, $\mathbf{B}_n$, $\mathbf{D}_n$, $\mathbf{G}_2$, $\mathbf{F}_4$, $\mathbf{E}_6$, $\mathbf{E}_7$, and $\mathbf{E}_8$. The groups $\mathbf{A}_2$, $\mathbf{B}_2$ and $\mathbf{G}_2$ are isomorphic to the dihedral groups of orders six, eight and twelve.

Let us examine some of the crystallographic groups in more detail. To this end we introduce the notations $W(\mathbf{\Gamma})$ for the Coxeter group associated with the Coxeter graph $\mathbf{\Gamma}$. The corresponding root system is denoted by $\mathcal{R}(\mathbf{\Gamma})$ and a fundamental system of roots by $\mathcal{F}(\mathbf{\Gamma})$.

EXAMPLE 1 : The Coxeter group $W(\mathbf{D}_n)$ is the Weyl group of the compact simply connected Lie group $\mathrm{Spin}(2n)$. The root system $\mathcal{R}(\mathbf{D}_n)$ of type $\mathbf{D}_n$ consists of the $2n(n-1)$ vectors $\pm E_i \pm E_j \in \mathbb{R}^n$ where $1 \leq i < j \leq n$ and $E_1, \ldots, E_n$ is an orthonormal basis. A basis $\mathcal{F}(\mathbf{D}_n)$ is given by the vectors

$$\mathcal{F}(\mathbf{D}_n) = \{E_1 - E_2,\ E_2 - E_3, \ldots, E_{n-1} - E_n,\ E_{n-1} + E_n\}.$$

The group $W(\mathbf{D}_n)$ is generated by the reflections

$$s_{E_i \pm E_j}(v) = v - \langle v \mid E_i \pm E_j \rangle (E_i \pm E_j) \qquad 0 \leq i < j \leq n.$$

From these formulae it easily follows that $s_{E_i - E_j}$ interchanges E_i and E_j leaving E_k fixed for $k \neq i,\ j$. Likewise, $s_{E_i + E_j}$ sends E_i to $-E_j$ and E_j to

$-E_i$ and leaves E_k fixed for $k \neq i,\ j$. Therefore the composite $s_{E_i-E_j}s_{E_i+E_j}$ changes the sign of E_i and E_j leaving E_k fixed for $k \neq i,\ j$. Since any even number of sign changes can be effected by changing pairs of signs we see that $W(\mathbf{D}_n)$ consists of all permutations of $\{E_1,\dots,E_n\}$ and all the even numbers of sign changes of $\{E_1,\dots,E_n\}$. Thus $W(\mathbf{D}_n) \cong (\mathbb{Z}/2)^{n-1} \rtimes \Sigma_n$ represented in the group of signed permutation matrices as the subgroup of index 2 changing an even number of signs. It has order $2^{n-1} \cdot n!$.

EXAMPLE 2 : The Coxeter group $W(\mathbf{B}_n)$ is the Weyl group of the compact simply connected Lie group $\mathrm{Spin}(2n+1)$. The root system $\mathcal{R}(\mathbf{B}_n)$ of type $\mathbf{B}_n$ consists of the $2n^2$ vectors

$$\pm E_i \pm E_j \quad 1 \leq i < j \leq n$$
$$\pm E_i \qquad i = 1,\dots,n$$

in $\mathbb{R}^n$, where $E_1,\dots,E_n$ is an orthonormal basis. A fundamental system of roots of type $\mathbf{B}_n$ is given by the vectors

$$\mathcal{F}(\mathbf{B}_n) = \{E_1 - E_2,\ E_2 - E_3,\dots,E_{n-1} - E_n,\ E_n\}.$$

The group $W(\mathbf{B}_n)$ is generated by the reflections $\{s_{E_i \pm E_j}, s_{E_k} \mid 1 \leq i < j \leq n,\ k = 1,\dots,n\}$. Thus $W(\mathbf{B}_n)$ consists of all signed permutations of $\{E_1,\dots,E_n\}$, has order $2^n \cdot n!$, and $W(\mathbf{B}_n) = (\mathbb{Z}/2)^n \rtimes \Sigma_n = \mathbb{Z}/2 \int \Sigma_n$. Thus $W(\mathbf{D}_n) \lhd W(\mathbf{B}_n)$ and $W(\mathbf{B}_n)$ is generated by $W(\mathbf{D}_n)$ and any one of the reflections s_{E_i} for $0 \leq i \leq n$.

EXAMPLE 3 : To describe the root system of type $\mathbf{G}_2$ we identify $\mathbb{R}^2$ with the complex numbers $\mathbb{C}$. Let $\zeta = \exp(2\pi i/6)$ be a primitive 6-th root of unity. The points $\{\zeta^j \mid j = 0,\dots,5\}$ form the vertices of a regular hexagon. For the root system $\mathcal{R}(\mathbf{G}_2)$ we take the 6 vertices and the 6 midpoints of the 6 sides of this regular hexagon.

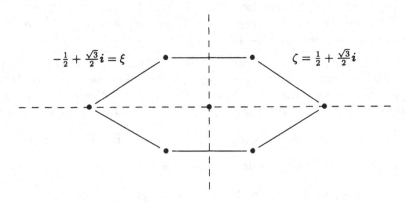

The $\mathbf{G}_2$ Hexagon

The group $W(\mathbf{G}_2)$ is then generated by the reflections in the lines joining opposite pairs of vertices and midpoints of opposite sides. This is the full symmetry group of the hexagon and is isomorphic to the dihedral group of order 12, D_{12}, generated by a rotation D and the reflection S defined by

$$D(1) = \zeta = -\xi^2 = 1 + \xi \qquad S(1) = 1$$
$$D(\xi) = -1 \qquad\qquad S(\xi) = \xi^2 = -\xi - 1$$

where $\xi = \zeta^2$ and we have employed 1, ξ as a basis for $\mathbb{C}$ over $\mathbb{R}$. The matrices of D and S with respect to the basis 1, ξ are

$$D = \begin{bmatrix} 1 & -1 \\ 1 & 0 \end{bmatrix}, \quad S = \begin{bmatrix} 1 & -1 \\ 0 & -1 \end{bmatrix} \in \mathrm{GL}(2, \mathbb{Z})$$

and provide an integral representation of $D_{12} \cong W(\mathbf{G}_2)$ as a Coxeter group.

EXAMPLE 4 : For our last detailed example we consider the Coxeter group $W(\mathbf{F}_4)$ corresponding to the Dynkin diagram $\mathbf{F}_4$. The root system $\mathcal{R}(\mathbf{F}_4)$ consists of the 48 vectors

$$\begin{array}{ll} \pm E_i \pm E_j & 1 \leq i < j \leq 4 \\ \pm E_i & i = 1,\ 2,\ 3,\ 4 \\ \tfrac{1}{2}(\pm E_1 \pm E_2 \pm E_3 \pm E_4) & \end{array}$$

in $\mathbb{R}^4$, where $E_1, \ldots, E_4$ is the standard orthonormal basis for $\mathbb{R}^4$. Thus $\mathcal{R}(\mathbf{D}_4) \subset \mathcal{R}(\mathbf{B}_4) \subset \mathcal{R}(\mathbf{F}_4)$ and there are corresponding inclusions of groups $W(\mathbf{D}_4) < W(\mathbf{B}_4) < W(\mathbf{F}_4)$. A basis for $\mathcal{R}(\mathbf{F}_4)$ consists of

$$\mathcal{F}(\mathbf{F}_4) = \{E_1 - E_2,\ E_2 - E_3,\ E_3,\ \tfrac{1}{2}(E_1 - E_2 - E_3 - E_4)\}$$

The group $W(\mathbf{F}_4)$ acts on $\mathbb{R}^4$ orthogonally and hence maps the long roots (i.e. $\pm E_i \pm E_j$, $1 \leq i < j \leq 4$ which have length $\sqrt{2}$) amongst themselves. The formula of lemma 7.1.1 may be rewritten

$$s_{s_{r'}(r'')} = s_{r'} s_{r''} s_{r'}^{-1} .$$

If $r'' \in \mathcal{R}(\mathbf{F}_4)$ is a long root then so is $s_{r'}(r'')$ for any root $r' \in \mathcal{R}(\mathbf{F}_4)$. Hence

$$s_r W(\mathbf{D}_4) s_r^{-1} \subset W(\mathbf{D}_4) \qquad \forall\, r \in \mathcal{R}(\mathbf{F}_4)$$

and hence $W(\mathbf{D}_4)$ is a normal subgroup of $W(\mathbf{F}_4)$.

The group $W(\mathbf{F}_4)$ acts on the Dynkin diagram

$$\Delta_1 = E_1 - E_2$$
$$\Delta_2 = E_2 - E_3$$
$$\Delta_3 = E_3 - E_4$$
$$\Delta_4 = E_3 + E_4$$

of $\mathbf{D}_4$. The diagram has a rotational symmetry, the **triality**, given by the cyclic permutation of the end nodes. The action of $W(\mathbf{F}_4)$ on the Dynkin diagram $\mathbf{D}_4$ induces this automorphism as the next discussion shows.

The group $W(\mathbf{F}_4)$ is generated by $W(\mathbf{B}_4)$ and any one of the reflections with root vector

$$\tfrac{1}{2}(\pm E_1 \pm E_2 \pm E_3 \pm E_4)$$

for some fixed choice of the signs. Thus $W(\mathbf{F}_4)$ is represented as the subgroup of $\mathrm{GL}(4, \mathbb{Q})$ generated by the signed permutation matrices (this is $W(\mathbf{B}_4)$) and the Hadamard matrix

$$\frac{1}{2}\begin{bmatrix} 1 & 1 & 1 & 1 \\ 1 & -1 & 1 & -1 \\ 1 & 1 & -1 & -1 \\ 1 & -1 & -1 & 1 \end{bmatrix} \in \mathrm{GL}(4,\ \mathbb{Z}[\tfrac{1}{2}]) \,.$$

(In the notations of [63] this is the reflection group [3, 4, 3].)

Let $T = s_{(E_1 - E_2 + E_3 - E_4)}$ be the reflection in the hyperplane

$$x_1 - x_2 + x_3 - x_4 = 0$$

which has root vector

$$r = \tfrac{1}{2}(E_1 - E_2 + E_3 - E_4) \,.$$

As basis for $\mathbb{R}^4$ choose the fundamental roots

$$\Delta_1 = E_1 - E_2, \ \Delta_2 = E_2 - E_3, \ \Delta_3 = E_3 - E_4, \ \Delta_4 = E_3 + E_4$$

of $\mathbf{D}_4$. The reflection T is given by the formula

$$\begin{aligned} T(v) &= v - \frac{2\langle v \mid E_1 - E_2 + E_3 - E_4 \rangle}{\|E_1 - E_2 + E_3 - E_4\|^2}(E_1 - E_2 + E_3 - E_4) \\ &= v - \tfrac{1}{2}\langle v \mid E_1 - E_2 + E_3 - E_4 \rangle(E_1 - E_2 + E_3 - E_4) \,. \end{aligned}$$

Thus we find:

$$T(\Delta_1) = \Delta_3, \quad T(\Delta_2) = \Delta_2, \quad T(\Delta_3) = \Delta_1, \quad T(\Delta_4) = \Delta_4$$

so T exchanges Δ_1 and Δ_3 leaving Δ_2 and Δ_4 fixed. Let $S = s_{E_4}$. Then

$$S(E_i) = \begin{cases} E_i & i = 1,\, 2,\, 3 \\ -E_4 & i = 4 \,. \end{cases}$$

Therefore

$$S(\Delta_1) = \Delta_1, \quad S(\Delta_2) = \Delta_2, \quad S(\Delta_3) = \Delta_4, \quad S(\Delta_4) = \Delta_3 \,.$$

Let $D = TS$. Then

$$D(\Delta_1) = \Delta_3, \quad D(\Delta_2) = \Delta_2, \quad D(\Delta_3) = \Delta_4, \quad D(\Delta_4) = \Delta_1$$

so D cyclically permutes Δ_1, Δ_3, Δ_4 and leaves Δ_2 fixed.

We next examine the subgroup of $GL(4, \mathbb{R})$ generated by S and D. From the preceding formulas we find:

$$SDS(\Delta_1) = SD(\Delta_1) = S(\Delta_3) = \Delta_4 = D^{-1}(\Delta_1)$$

$$SDS(\Delta_2) = SD(\Delta_2) = S(\Delta_2) = \Delta_2 = D^{-1}(\Delta_2)$$

$$SDS(\Delta_3) = SD(\Delta_4) = S(\Delta_1) = \Delta_1 = D^{-1}(\Delta_3)$$

$$SDS(\Delta_4) = SD(\Delta_3) = S(\Delta_4) = \Delta_3 = D^{-1}(\Delta_4).$$

Hence $SDS = D^2$ and the subgroup of $GL(4, \mathbb{R})$ generated by S and D is therefore isomorphic to $D_6 = \Sigma_3$.

Since D cyclically permutes Δ_1, Δ_3, Δ_4 leaving Δ_2 fixed it maps the root system $\mathcal{R}(\mathbf{D}_4)$ of $\mathbf{D}_4$ into itself. The involution S exchanges Δ_3 and Δ_4 leaving Δ_1 and Δ_2 fixed, so it too maps $\mathcal{R}(\mathbf{D}_4)$ into itself. From 7.1.1 it follows that the subgroup generated by S and D belongs to the normalizer of $W(\mathbf{D}_4)$ in $GL(4, \mathbb{R})$ and $W(\mathbf{F}_4)$ is generated by $W(\mathbf{D}_4)$, S and D. Therefore we have shown:

$$W(\mathbf{D}_4) \lhd W(\mathbf{F}_4)$$
$$W(\mathbf{F}_4)/W(\mathbf{D}_4) \cong \Sigma_3$$
$$W(\mathbf{F}_4) \cong \left((\mathbb{Z}/2)^3 \rtimes \Sigma_4\right) \rtimes \Sigma_3.$$

The element $D \in W(\mathbf{F}_4)$ of order 3 induces the triality of the Dynkin diagram of $\mathbf{D}_4$. The group $W(\mathbf{F}_4)$ has order $1152 = 2^7 \cdot 3^2$.

For a many-sided discussion of the Coxeter groups of type $\mathbf{E}_6$, $\mathbf{E}_7$ and particularly $\mathbf{E}_8$ the reader is referred to [60].

The list of irreducible Coxeter groups contains in addition to the crystallographic groups and the dihedral groups the group $\mathbf{I}_3$ which is the group of symmetries of a regular icosahedron embedded in $\mathbb{R}^3$. Abstractly this group is isomorphic to the direct product of the alternating group A_5 with $\mathbb{Z}/2$ and has order 120. The alternating group is of course the simple group of order 60 and in turn is isomorphic to $PGL(2, \mathbb{F}_5)$. (See for example [71].) The last group in the list, $\mathbf{I}_4$, is a group containing the double cover of $A_5 \times A_5$ as a subgroup of index two, acting on $\mathbb{R}^4$. It may be described as follows. The inverse image of A_5 under the double cover map $SU(2) \longrightarrow SO(3)$ is a proper double cover of A_5, written $2A_5$, and called the binary icosahedral group. It sits in a diagram:

$$
\begin{array}{ccccccccc}
1 \longrightarrow & \mathbb{Z}/2 & \longrightarrow & SU(2) & \longrightarrow & SO(3) & \longrightarrow 1 \\
& \| & & \cup & & \cup & \\
1 \longrightarrow & \mathbb{Z}/2 & \longrightarrow & 2A_5 & \longrightarrow & A_5 & \longrightarrow 1.
\end{array}
$$

There is the double covering $SU(2) \times SU(2) \longrightarrow SO(4)$, and the image under this map of $2A_5 \times 2A_5$ is the quotient by the diagonal of the two central elements, and hence is a diagonal double cover of $A_5 \times A_5$. So we have the diagram

$$
\begin{array}{ccccccc}
1 \longrightarrow & \mathbb{Z}/2 & \longrightarrow & SU(2) \times SU(2) & \longrightarrow & SO(4) & \longrightarrow 1 \\
 & \| & & \cup & & \cup & \\
1 \longrightarrow & \mathbb{Z}/2 & \longrightarrow & 2A_5 \times 2A_5 & \longrightarrow & 2(A_5 \times A_5) & \longrightarrow 1.
\end{array}
$$

The group $O(4)$ is a split extension of $SO(4)$ by an element of order two exchanging the two copies of $SU(2)$, and adjoining this automorphism to $2(A_5 \times A_5)$ gives the desired reflection group I_4. It has order 14,400.

7.3   Groups Generated by Pseudoreflections

The groups generated by complex pseudoreflections were classified by Shephard and Todd [206]. The situation is considerably more complicated than in the real or rational case. A list of the indecomposable complex pseudoreflection groups is given in the table on the following page, which is taken from Clark and Ewing [56], who computed the character fields and partially (see also section 10.1) extended the classification to pseudoreflection groups over fields whose characteristic does not divide the group order.

The table lists equivalence classes of *subgroups* of $GL(n, \mathbb{C})$, two subgroups being equivalent if some automorphism of $GL(n, \mathbb{C})$, not necessarily a conjugation, sends one isomorphically onto the other. Two subgroups that differ by complex conjugation (an outer automorphism of $\mathbb{C}$) are therefore considered as equivalent, and appear only once in the table. An abstract group may occur more than once in the table. For example the dihedral group D_6 occurs twice in the table: once as a real reflection group of type 1, and once as a complex reflection group of type 2a. See the discussion of the RIGIDITY THEOREM in section 10.1 for an elaboration of this point.

Using their classification theorem and case by case checking, Shephard and Todd were able to verify that in all cases the rings of invariants of the complex pseudoreflection groups are polynomial algebras. The first column in the table corresponds to the numbering of the groups used by Shephard-Todd, and the second column gives the dimension of the pseudoreflection representation. The column headed *degrees* lists the degrees of the polynomial generators for the ring of invariants. These integers are called the **degrees of G**. In the table of Clark and Ewing the degrees have been doubled because they are interested in topological applications where the vector space V naturally occurs in degree two.

In reading the table, the last column lists those primes that do not divide the order of G and for which the character field lies in $\mathbb{Q}_p^\wedge$. The next to

	DIM	$|G|$	DEGREES	χ – FIELD	GOOD PRIMES
1	n	$(n+1)!$	$2,3, \quad ..., \quad n+1$	$\mathbb{Q}$	$> n+1$
2a	n	$qm^{n-1}n!$ $(q|m$ and $m>1)$	$m,2m,...,(n-1)m,qn$	$\mathbb{Q}(\zeta_m)$	$1 \pmod m$, $>n$
2b	2	$2m \quad (m>2)$	$2,m$	$\mathbb{Q}(\zeta_m+\zeta_m^{-1})$	$\pm 1 \pmod m$
3	1	m	m	$\mathbb{Q}(\zeta_m)$	$1 \pmod m$
4	2	$24=2^3 3$	$4,6$	$\mathbb{Q}(\zeta_3)$	$1 \pmod 3$
5	2	$72=2^3 3^2$	$6,12$	$\mathbb{Q}(\zeta_3)$	$1 \pmod 3$
6	2	$48=2^4 3$	$4,12$	$\mathbb{Q}(\zeta_{12})$	$1 \pmod{12}$
7	2	$144=2^4 3^2$	$12,12$	$\mathbb{Q}(\zeta_{12})$	$1 \pmod{12}$
8	2	$96=2^5 3$	$8,12$	$\mathbb{Q}(i)$	$1 \pmod 4$
9	2	$192=2^6 3$	$8,24$	$\mathbb{Q}(\zeta_8)$	$1 \pmod 8$
10	2	$288=2^5 3^2$	$12,24$	$\mathbb{Q}(\zeta_{12})$	$1 \pmod{12}$
11	2	$576=2^6 3^2$	$24,24$	$\mathbb{Q}(\zeta_{24})$	$1 \pmod{24}$
12	2	$48=2^4 3$	$6,8$	$\mathbb{Q}(\sqrt{-2})$	$1,3 \pmod 8$, $\neq 3$
13	2	$96=2^5 3$	$8,12$	$\mathbb{Q}(\zeta_8)$	$1 \pmod 8$
14	2	$144=2^4 3^2$	$6,24$	$\mathbb{Q}(\zeta_3,\sqrt{-2})$	$1,19 \pmod{24}$
15	2	$288=2^5 3^2$	$12,24$	$\mathbb{Q}(\zeta_{24})$	$1 \pmod{24}$
16	2	$600=2^3 3 \cdot 5^2$	$20,30$	$\mathbb{Q}(\zeta_5)$	$1 \pmod 5$
17	2	$1200=2^4 3 \cdot 5^2$	$20,60$	$\mathbb{Q}(\zeta_{20})$	$1 \pmod{20}$
18	2	$1800=2^3 3^2 5^2$	$30,60$	$\mathbb{Q}(\zeta_{15})$	$1 \pmod{15}$
19	2	$3600=2^4 3^2 5^2$	$60,60$	$\mathbb{Q}(\zeta_{60})$	$1 \pmod{60}$
20	2	$360=2^3 3^2 5$	$12,30$	$\mathbb{Q}(\zeta_3,\sqrt{5})$	$1,4 \pmod{15}$
21	2	$720=2^4 3^2 5$	$12,60$	$\mathbb{Q}(\zeta_{12},\sqrt{5})$	$1,49 \pmod{60}$
22	2	$240=2^4 3 \cdot 5$	$12,20$	$\mathbb{Q}(i,\sqrt{5})$	$1,9 \pmod{20}$
23	3	$120=2^3 3 \cdot 5$	$2,6,10$	$\mathbb{Q}(\sqrt{5})$	$1,4 \pmod 5$
24	3	$336=2^4 3 \cdot 7$	$4,6,14$	$\mathbb{Q}(\sqrt{-7})$	$1,2,4 \pmod 7$
25	3	$648=2^3 3^4$	$6,9,12$	$\mathbb{Q}(\zeta_3)$	$1 \pmod 3$
26	3	$1296=2^4 3^4$	$6,12,18$	$\mathbb{Q}(\zeta_3)$	$1 \pmod 3$
27	3	$2160=2^4 3^3 \cdot 5$	$6,12,30$	$\mathbb{Q}(\zeta_3,\sqrt{5})$	$1,4 \pmod{15}$
28	4	$1152=2^7 3^2$	$2,6,8,12$	$\mathbb{Q}$	$\neq 2$ or 3
29	4	$7680=2^9 3 \cdot 5$	$4,8,12,20$	$\mathbb{Q}(i)$	$1 \pmod 4$, $\neq 5$
30	4	$2^6 3^2 5^2$	$2,12,20,30$	$\mathbb{Q}(\sqrt{5})$	$1,4 \pmod 5$
31	4	$2^{10} 3^2 5$	$8,12,20,24$	$\mathbb{Q}(i)$	$1 \pmod 4$, $\neq 5$
32	4	$2^7 3^5 5$	$12,18,24,30$	$\mathbb{Q}(\zeta_3)$	$1 \pmod 3$
33	5	$2^7 3^4 5$	$4,6,10,12,18$	$\mathbb{Q}(\zeta_3)$	$1 \pmod 3$
34	6	$2^9 3^7 5 \cdot 7$	$6,12,18,24,30,42$	$\mathbb{Q}(\zeta_3)$	$1 \pmod 3$, $\neq 7$
35	6	$2^7 3^4 5$	$2,5,6,8,9,12$	$\mathbb{Q}$	$\neq 2,3$ or 5
36	7	$2^{10} 3^4 5 \cdot 7$	$2,6,8,10,12,14,18$	$\mathbb{Q}$	$\neq 2,3,5$ or 7
37	8	$2^{14} 3^5 5^2 7$	$2,8,12,14,18,20,24,30$	$\mathbb{Q}$	$\neq 2,3,5$ or 7

Table 7.3.1: The finite irreducible complex pseudoreflection groups

last column gives the character field of the pseudoreflection representation where ζ_m denotes a primitive m-th root of unity. In general, a complex representation of a finite group is not necessarily equivalent to a representation whose matrix entries are in the field generated by the character values. The Schur index measures the extent to which this fails. However, for a group generated by pseudoreflections, the Schur index is always 1 as we show in the next theorem. Hence for those primes listed as *good primes* in the table there is a mod p reduction of the representation, which is again generated by pseudoreflections. See chapter 10 for a further discussion of this point.

THEOREM 7.3.1   (A. Clark and J. Ewing): *If $\varrho : G \hookrightarrow \mathrm{GL}(n, \mathbb{C})$ represents a finite group G as a group generated by pseudoreflections then ϱ is equivalent to a representation defined over the subfield of $\mathbb{C}$ generated by the characters of G on $\mathbb{C}^n$.*

PROOF :   Let $\mathbb{K}$ be the subfield of $\mathbb{C}$ generated by the character values of G on $\mathbb{C}^n$. Then the Wedderburn component ([64] § 26) of the group algebra $\mathbb{K}(G)$ corresponding to $\mathbb{C}^n$ is of the form $\Gamma \cong \mathrm{Mat}_m(\Delta)$, where Δ is a division algebra whose center is $\mathbb{K}$ and $m \in \mathbb{N}$. We have $|\Delta : \mathbb{K}| = d^2$ (in fact d is just the Schur index of $\mathbb{C}^n$), and since $\mathbb{C} \otimes_{\mathbb{K}} \Gamma \cong \mathrm{Mat}_n(\mathbb{C})$, we have $n^2 = \dim_{\mathbb{K}} \Gamma = m^2 d^2$ and so $n = md$. If V is the irreducible $\mathbb{K}(G)$-module corresponding to Γ, then $\dim_{\Delta}(V) = m$, $\dim_{\mathbb{K}}(V) = md^2$ and $\mathbb{C} \otimes_{\mathbb{K}} V \cong \oplus_d \mathbb{C}^n$. The fixed points of an element $g \in G$ on V form a Δ-linear subspace, and hence have dimension divisible by d^2. So the fixed points of g on $\mathbb{C}^n$ have dimension divisible by d. In particular, if g is a pseudoreflection then $n - 1$ is divisible by d and so, since d also divides n, $d = 1$. Thus $m = n$, $\Gamma \cong \mathrm{Mat}_n(\mathbb{K})$, and $\mathbb{C} \otimes_{\mathbb{K}} V \cong \mathbb{C}^n$. $\square$

From this result and the preceding table we may read off over which fields $\mathbb{Q}_p^{\wedge}$ the pseudoreflection representation may be defined. It is not apparent how many such p-adic representations there are, nor whether these yield pseudoreflection groups. By clearing denominators (there are only finitely many) we also get representations defined over $\mathbb{Z}_p^{\wedge}$ and by reduction mod p representations in $\mathrm{GL}(n, \mathbb{F}_p)$. Again the same questions occur. In chapter 10, theorem 10.2.1, we will settle the uniqueness problems involved here.

The complex pseudoreflection groups are all remarkable in one way or another but it is more difficult to describe them than the real reflection groups. The symmetric groups occur as the family of type 1. The groups of type 2a contain the Coxeter groups of type $\mathbf{B}_n$ and their invariant theory is discussed in section 7.4. The dihedral groups build the family 2b, and the cyclic groups the family 3. The crystallographic and Coxeter groups may be located in the table by looking for groups with $\mathbb{Q}$, respectively $\mathbb{Q}(\alpha)$ with α real, as character field.

If $G < \mathbf{S}\mathbf{O}(3)$ is the group of rotations of one of the Platonic solids (tetrahedron, octahedron, cube, dodecahedron or icosahedron) then lifting G across the double covering $\mathbf{S}\mathbf{U}(2) \longrightarrow \mathbf{S}\mathbf{O}(3)$ defines one of the three (the octahedron and the cube, and the dodecahedron and the icosahedron have the same group of symmetries) **binary polyhedral groups** $\mathcal{T}^*$, $\mathcal{W}^*$, $\mathcal{I}^* \subset \mathbf{S}\mathbf{U}(2)$. These groups are used to construct several of the 2-dimensional examples by adjoining matrices of the form $\zeta \cdot I$ where ζ is a root of unity (see [206], section 4) to obtain a finite subgroup of $\mathbf{U}(2)$ generated by pseudoreflections. For ex-

ample, the group of type 12, which is isomorphic to $GL(2, \mathbb{F}_3)$, is obtained
from the octahedral group in this way and is generated by the matrices

$$\begin{bmatrix} -1 & 0 \\ 0 & -1 \end{bmatrix} \qquad \frac{1}{\sqrt{2}}\begin{bmatrix} -1 & i \\ -i & 1 \end{bmatrix} \qquad \frac{1}{\sqrt{2}}\begin{bmatrix} \varepsilon & \varepsilon \\ \varepsilon^3 & \varepsilon^7 \end{bmatrix}$$

where $\varepsilon = \exp(2\pi i/8)$. These matrices however do not lie in the character
field of the pseudoreflection representation. Alternatively:

EXAMPLE 1 : The group $GL(2, \mathbb{F}_3)$ contains a normal chain

$$Q_8 \ \triangleleft \ SL(2, \mathbb{F}_3) \ \triangleleft \ GL(2, \mathbb{F}_3)$$
$$\downarrow \qquad\qquad \downarrow$$
$$\mathbb{Z}/3 \qquad\quad \mathbb{Z}/2\,.$$

where Q_8 is the quaternion group of order 8. (See [71] § 107 page 86. The
inclusion $Q_8 \hookrightarrow GL(2, \mathbb{F}_3)$ is also described in section 3.1 example 1.) Let

$$\tilde{A} = \begin{bmatrix} 0 & 1 \\ -1 & 0 \end{bmatrix}, \qquad \tilde{B} = \frac{1}{\sqrt{-2}}\begin{bmatrix} -1 & 1 \\ 1 & 1 \end{bmatrix} \in GL(2, \mathbb{Q}(\sqrt{-2}))$$

and set

$$\tilde{C} = \tilde{A}\tilde{B} = \frac{1}{\sqrt{-2}}\begin{bmatrix} 1 & 1 \\ 1 & -1 \end{bmatrix}.$$

One readily verifies that

$$\tilde{A}^2 = \tilde{B}^2 = \tilde{C}^2 = -I$$

so that the subgroup of $GL(2, \mathbb{Q}(\sqrt{-2}))$ generated by $\tilde{A}$ and $\tilde{B}$ is isomorphic
to Q_8. Denote by $\mathbb{Z}_{(3)}$ the integers localized at the prime ideal (3) and by
$\mathbb{Z}_{(3)}(\sqrt{-2})$ the subring of $\mathbb{Q}(\sqrt{-2})$ consisting of elements of the form $a+b\sqrt{-2}$
where $a, b \in \mathbb{Z}_{(3)}$. The principal ideal $(-1 + \sqrt{-2}) \subset \mathbb{Z}_{(3)}(\sqrt{-2})$ is prime
and the natural map

$$\mathbb{Z}_{(3)}(\sqrt{-2}) \overset{q}{\longrightarrow} \mathbb{F}_3 \qquad a + b\sqrt{-2} \ \longmapsto \ a + b$$

identifies the quotient $\mathbb{Z}_{(3)}(\sqrt{-2})/(-1 + \sqrt{-2})$ with $\mathbb{F}_3$. Under the map
q the matrices $\tilde{A}$ and $\tilde{B}$ map to the matrices A and B of section 3.1 example 1 respectively. The matrices $A, B \in GL(2, \mathbb{F}_3)$ generate $Q_8 =$
$Syl_2(SL(2, \mathbb{F}_3))$, so q induces an isomorphism from $Q_8 < GL(2, \mathbb{Z}_{(3)}(\sqrt{-2}))$
to $Q_8 < SL(2, \mathbb{F}_3)$.

Let

$$\tilde{S} = \begin{bmatrix} \omega & \frac{1}{2} \\ -\frac{1}{2} & \bar{\omega} \end{bmatrix} \in GL(2, \mathbb{Q}(\sqrt{-2})) \qquad \omega = \frac{-1+\sqrt{-2}}{2}, \ \bar{\omega} = \frac{-1-\sqrt{-2}}{2}.$$

The matrix $\tilde{S}$ satisfies

$$\tilde{S}\tilde{A}\tilde{S}^{-1} = -\tilde{B}$$

$$\tilde{S}\tilde{B}\tilde{S}^{-1} = -\tilde{C}$$

and has order[2] 3, and so $\tilde{S}$ belongs to the normalizer of Q_8 in $\mathrm{GL}(2, \mathbb{Q}(\sqrt{-2}))$. Hence the subgroup of $\mathrm{GL}(2, \mathbb{Q}(\sqrt{-2}))$ generated by $\tilde{A}$, $\tilde{B}$ and $\tilde{S}$ has order 24. Under the reduction map $\mathbb{Z}_{(3)}(\sqrt{-2}) \xrightarrow{q} \mathbb{F}_3$, the matrix $\tilde{S}$ maps to

$$S = \begin{bmatrix} 0 & -1 \\ 1 & -1 \end{bmatrix} \in \mathrm{SL}(2, \mathbb{F}_3),$$

which generates the Sylow 3-subgroup of $\mathrm{SL}(2, \mathbb{F}_3)$. Therefore the reduction map q induces an isomorphism of the subgroup of $\mathrm{GL}(2, \mathbb{Q}(\sqrt{-2}))$ generated by $\tilde{A}$, $\tilde{B}$ and $\tilde{S}$ with $\mathrm{SL}(2, \mathbb{F}_3)$.

Finally, let

$$\tilde{T} = \begin{bmatrix} 0 & 1 \\ 1 & 0 \end{bmatrix} \in \mathrm{GL}(2, \mathbb{Q}(\sqrt{-2})).$$

The matrix $\tilde{T}$ is of order 2 and determinant -1. One has

$$\tilde{T}\begin{bmatrix} a & b \\ c & d \end{bmatrix}\tilde{T}^{-1} = \begin{bmatrix} d & c \\ b & a \end{bmatrix}$$

from which it follows that

$$\tilde{T}\tilde{S}\tilde{T}^{-1} = \begin{bmatrix} \bar{\omega} & -\frac{1}{2} \\ \frac{1}{2} & \omega \end{bmatrix} = \tilde{S}^{-1}$$

$$\tilde{T}\tilde{B}\tilde{T}^{-1} = \frac{1}{\sqrt{-2}}\begin{bmatrix} 1 & 1 \\ 1 & -1 \end{bmatrix} = \tilde{C}$$

$$\tilde{T}\tilde{A}\tilde{T}^{-1} = \begin{bmatrix} 0 & -1 \\ 1 & 0 \end{bmatrix} = \tilde{A}^{-1},$$

---

[2] This may be verified by multiplication, or by making use of:

    LEMMA : *Let $\mathbb{F}$ be a field containing no cube root of 1 other than 1, and let $I \neq T \in \mathrm{GL}(2, \mathbb{F})$. Then T has order 3 if and only if $\mathrm{Tr}(T) = -1$ and $\det(T) = 1$.*

    PROOF : Since $T^3 = I$ the minimal polynomial $m_T(t)$ of T is a divisor of $t^3 - 1$. Since $\mathbb{F}$ contains no third root of unity apart from 1,

$$t^3 - 1 = (t - 1)(t^2 + t + 1)$$

is the decomposition of $t^3 - 1$ into irreducible factors. Therefore the minimal polynomial of T is $t^2 + t + 1$. By the theorem of Cayley-Hamilton [210] the characteristic polynomial

$$\Delta_T(t) = \det(T - tI) = t^2 - \mathrm{Tr}(T)t + \det(T)$$

is divisible by the minimal polynomial. Both $m_T(t)$ and $\Delta_T(t)$ are monic quadratic polynomials, so must be equal. $\square$

hence $\tilde{T}$ belongs to the normalizer of $\mathrm{SL}(2,\ \mathbb{F}_3)$ in $\mathrm{GL}(2,\ \mathbb{Q}(\sqrt{-2}))$. Therefore the subgroup of $\mathrm{GL}(2,\ \mathbb{Q}(\sqrt{-2}))$ generated by $\tilde{A}$, $\tilde{B}$, $\tilde{S}$ and $\tilde{T}$ has order 48 and is mapped by the reduction map q isomorphically onto $\mathrm{GL}(2,\ \mathbb{F}_3)$. Thus the matrices $\tilde{A}$, $\tilde{B}$, $\tilde{S}$ and $\tilde{T}$ afford a lift, $\varrho : \mathrm{GL}(2,\ \mathbb{F}_3) \hookrightarrow \mathrm{GL}(2,\ \mathbb{Q}(\sqrt{-2}))$, of the tautological representation of $\mathrm{GL}(2,\ \mathbb{F}_3)$ on $\mathbb{F}_3^2$ to characteristic zero. The twelve matrices

$$\pm\tilde{T} = \begin{bmatrix} 0 & \pm 1 \\ \pm 1 & 0 \end{bmatrix}, \ \pm\tilde{T}\tilde{S} = \pm\begin{bmatrix} -\frac{1}{2} & \bar{\omega} \\ \omega & \frac{1}{2} \end{bmatrix}, \ \pm\tilde{T}\tilde{S}^{-1} = \pm\begin{bmatrix} \frac{1}{2} & \omega \\ \bar{\omega} & -\frac{1}{2} \end{bmatrix},$$

$$\pm\tilde{A}\tilde{T} = \begin{bmatrix} \pm 1 & 0 \\ 0 & \mp 1 \end{bmatrix}, \ \pm\tilde{S}\tilde{T} = \pm\begin{bmatrix} \frac{1}{2} & \omega \\ \bar{\omega} & -\frac{1}{2} \end{bmatrix}, \ \pm\tilde{S}^{-1}\tilde{T} = \pm\begin{bmatrix} -\frac{1}{2} & \bar{\omega} \\ \omega & \frac{1}{2} \end{bmatrix}$$

are pseudoreflections, belong to $\varrho(\mathrm{GL}(2,\ \mathbb{F}_3)) < \mathrm{GL}(2,\ \mathbb{Q}(\sqrt{-2}))$ and generate $\varrho(\mathrm{GL}(2,\ \mathbb{F}_3))$ as a subgroup of $\mathrm{GL}(2,\ \mathbb{Q}(\sqrt{-2}))$.

EXAMPLE 2 : The group number 24 in the list is also of this type. Abstractly it is isomorphic to $\mathbb{Z}/2 \times \mathrm{GL}(3,\ \mathbb{F}_2)$ which has order 336. The finite[3] linear group $\mathrm{GL}(3,\ \mathbb{F}_2)$ has a representation over the field $\mathbb{Q}(\sqrt{-7})$ (see [257], Bd. II, Abschnitt 131) given by the matrices

$$\tau = \begin{bmatrix} \varepsilon & 0 & 0 \\ 0 & \varepsilon^2 & 0 \\ 0 & 0 & \varepsilon^4 \end{bmatrix} \qquad \chi = \begin{bmatrix} 0 & 0 & 1 \\ 1 & 0 & 0 \\ 0 & 1 & 0 \end{bmatrix}$$

$$\omega = \begin{bmatrix} \alpha & \beta & \gamma \\ \beta & \gamma & \alpha \\ \gamma & \alpha & \beta \end{bmatrix} \qquad \Theta = \begin{bmatrix} \gamma\varepsilon^2 & \alpha\varepsilon^6 & \beta \\ \alpha & \beta\varepsilon^4 & \gamma\varepsilon^5 \\ \beta\varepsilon^3 & \gamma & \alpha\varepsilon \end{bmatrix}$$

where $\varepsilon = \exp(2\pi i/7)$, $\alpha = \frac{-2\sin(8\pi/7)}{\sqrt{7}}$, $\beta = \frac{-2\sin(4\pi/7)}{\sqrt{7}}$, $\gamma = \frac{-2\sin(2\pi/7)}{\sqrt{7}}$ and to quote Weber *"Hiernach ist es nun sehr einfach, die charakterische Relationen des Theorems I, § 88 für unsere Gruppe durch wirkliche Ausrechnung zu bestätigen, und damit ist also eine ternäre Gruppe 168*[sten]* Grades hergestellt."* The characteristic property he refers to is a characterization of $\mathrm{GL}(3,\ \mathbb{F}_2)$ in terms of four generators τ, χ, ω and Θ and nine relations. Using this theorem it is easy to see that the mod 2 reduction of this representation is the tautological representation of $\mathrm{GL}(3,\ \mathbb{F}_2)$ on $\mathbb{F}_2^3$. Shephard and Todd write down a simpler form of the representation for $\mathbb{Z}/2 \times \mathrm{GL}(3,\ \mathbb{F}_2)$ over $\mathbb{Q}(\sqrt{-7})$: it is generated by the three pseudoreflection matrices

$$\begin{bmatrix} 1 & 0 & 0 \\ 0 & 0 & 1 \\ 0 & 1 & 0 \end{bmatrix} \qquad \begin{bmatrix} 1 & 0 & 0 \\ 0 & 1 & 0 \\ 0 & 0 & -1 \end{bmatrix} \qquad \frac{1}{2}\begin{bmatrix} 1 & -1 & -\zeta \\ -1 & 1 & -\zeta \\ -\bar{\zeta} & -\bar{\zeta} & 0 \end{bmatrix}$$

---

[3] The group $\mathrm{GL}(3,\ \mathbb{F}_2)$ is isomorphic to $PSL(2,\ \mathbb{F}_7)$ and was studied by F. Klein in connection with his investigations on solvability of polynomial equations. The isomorphism is attributed by Weber to Kronecker. There is an extensive literature on this group, see for example [20] and the references there.

where $\zeta = \frac{1}{2}(1 + \sqrt{-7})$. The reflections in this representation are all of order 2 but the representation is not the complexification of a real representation because there is no quadratic invariant. The fundamental invariants for this group were computed by F. Klein, and may taken to be [60] page 203:

$$f_1 = z_1^3 z_2 + z_2^3 z_3 + z_3^3 z_1$$

$$f_2 = 5z_1^2 z_2^2 z_3^2 - \sum_{i=1}^{3} z_i^5 z_{i+2}$$

$$f_3 = \sum_{i=1}^{3} [z_i^{14} - 34z_i^{11} z_{i+1}^2 z_{i+2} + 18z_i^7 z_{i+1}^7 - 250z_i^4 z_{i+1}^9 z_{i+2}$$
$$+ 375z_i^8 z_{i+1}^4 z_{i+2}^2 - 126z_i^5 z_{i+1}^6 z_{i+2}^3],$$

where the indices in the summations are taken modulo 3.

Many of the remaining groups in the table seem to be similar low dimensional accidents. We will look at some of them in the coming sections.

A pseudoreflection group $\varrho : G \hookrightarrow GL(n, \mathbb{C})$ need not be generated by n elements as a real reflection group $\varrho : G \hookrightarrow GL(n, \mathbb{R})$ is (compare 7.2.11). Therefore it is not immediately obvious how to generalize the Coxeter matrix (or graph) to the complex case. In addition, the order of the pseudoreflections need not be 2 in the complex case so, at a minimum, the vertices of the graph (resp. the diagonal entries in the matrix) need nontrivial labels. Despite these obstacles a proof of the classification theorem of Shephard-Todd using a generalization of the Coxeter graph is to be found in [57].

The completeness of the tables of Coxeter and of Shephard-Todd and the additional information they supply in their paper [206] allows one to observe a number of interesting phenomena. Here is a short list of such observations:

OBSERVATION 7.3.2 : If $G \hookrightarrow O(n)$ is a real reflection group with degrees $d_1 \leq d_2 \leq \cdots \leq d_n$ then $d_i + d_{n+1-i}$ is a constant independent of i. This was observed by Coxeter [62] and first proved by Coleman [59].

OBSERVATION 7.3.3 (Coxeter [62]): If $G \hookrightarrow O(n)$ is a real reflection group and $s_1, \ldots, s_n \in G$ a system of fundamental reflections then the conjugacy class of $C = s_1 \cdots s_n$ is independent of the order and choice of the fundamental reflections. If h is its order then $|s(G)| = \frac{1}{2}nh$. This was first explained by R. Steinberg [237]. If $\zeta = \exp(2\pi i/h)$ then the eigenvalues of C are $\zeta^{d_1-1}, \ldots, \zeta^{d_n-1}$, where $d_1, \ldots, d_n$ are the degrees of G. This is the origin of the term **exponents** for the numbers $m_i = d_i - 1$ for $i = 1, \ldots, n$. This was explained by Coleman [59].

OBSERVATION 7.3.4 : *Shephard and Todd observed that if a complex pseudoreflection group $G \hookrightarrow \mathrm{GL}(n, \mathbb{C})$ happens to be generated by n elements (always true in the real case by 7.2.11) then there is a choice for the generating reflections $s_1, \ldots, s_n \in G$ such that $C = s_1 \cdot s_2 \cdots s_n$ has order $h = m_n + 1 = d_n$ and the eigenvalues of C are $\zeta^{m_1}, \ldots, \zeta^{m_n}$, where $d_1 \leq d_2 \leq \cdots \leq d_n$ are the degrees of G.*

OBSERVATION 7.3.5 : *Let $G \hookrightarrow \mathrm{GL}(n, \mathbb{C})$ be a complex pseudoreflection group. For $g \in G$ denote by $L(g)$ the codimension in V of the fixed point set V^g. Then, if $m_1 \leq m_2 \leq \cdots \leq m_n$ are the exponents of G*

$$\sum_{g \in G} t^{L(g)} = \prod_{i=1}^{n} (1 + m_i t)$$

or, put in other words, m_i is the count of the number of elements of G whose fixed point set has codimension i. This was first explained by L. Solomon [226] by studying the invariant theory of $\mathbb{F}[V] \otimes E[V]$ (see chapter 9).

There are many other such observations one can make on the basis of the tables. Such observations can often be germinal for an entire theory. A good example of this comes from algebraic topology and Lie theory. By circa 1930, based on extensive case by case construction of cell structures for the compact simply connected Lie groups, it had been observed that their rational cohomology is always an exterior algebra. It was E. Cartan who pointed out that this observation was not a *proof*. It was H. Hopf who first explained and proved this result: to do so he invented Hopf algebras and H-spaces.

7.4   The Shephard-Todd Theorem

A uniform proof that the rings of invariants of the complex pseudoreflection groups are polynomial, without the classification (but heavily computational and unmotivated), was first given by Chevalley [55]. We follow [214] in proving this result without classification for all nonmodular pseudoreflection representations.

THEOREM 7.4.1 (G. C. Shephard and J. A.Todd, C. Chevalley): *Let V be a finite dimensional vector space over the field $\mathbb{F}$ and $\varrho : G \hookrightarrow \mathrm{GL}(V)$ a representation of a finite group G. Assume that $|G|$ is relatively prime to the characteristic of $\mathbb{F}$. Then the following are equivalent:*

(i) *G is generated by pseudoreflections,*

(ii) *$\mathbb{F}[V]^G$ is a polynomial algebra.*

PROOF : Suppose that G is generated by pseudoreflections in the representation ϱ. By 6.4.4 it is sufficient to show that $\mathbb{F}[V]$ is a free $\mathbb{F}[V]^G$-module to conclude $\mathbb{F}[V]^G$ is a polynomial algebra, so by 6.1.1 we need only show that

$$\mathrm{Tor}_1^{\mathbb{F}[V]^G}(\mathbb{F},\ \mathbb{F}[V]) = 0\,.$$

(The following argument is lifted from [39] V.5.2 Thm 1/(b).) The group G acts on $\mathrm{Tor}_1^{\mathbb{F}[V]^G}(\mathbb{F},\ \mathbb{F}[V])$ through its action on $\mathbb{F}[V]$. From the definition of Δ_s in section 7.1 we obtain

$$s(f) - f = \ell_s \Delta_s(f) \qquad \forall\, f \in \mathbb{F}[V]\,.$$

The endomorphism Δ_s of $\mathbb{F}[V]$ is $\mathbb{F}[V]^G$-linear and therefore

$$s(\zeta) - \zeta = \ell_s \cdot \mathrm{Tor}_1^{\mathbb{F}[V]^G}(\mathbb{F}, \Delta_s)(\zeta) \qquad \forall\, \zeta \in \mathrm{Tor}_1^{\mathbb{F}[V]^G}(\mathbb{F},\ \mathbb{F}[V])\,.$$

If $\mathrm{Tor}_1^{\mathbb{F}[V]^G}(\mathbb{F},\mathbb{F}[V])$ were nonzero and ζ a nonzero element of minimal degree in $\mathrm{Tor}_1^{\mathbb{F}[V]^G}(\mathbb{F},\ \mathbb{F}[V])$, then $\mathrm{Tor}_1^{\mathbb{F}[V]^G}(\mathbb{F},\ \Delta_s)(\zeta) = 0$ for degree reasons, since Δ_s lowers degrees by 1. So the preceding equation says $s(\zeta) = \zeta$ for all pseudoreflections $s \in G$. Since G is generated by pseudoreflections this means ζ is G-invariant.

If M is a G-module, then the group G acts on $\mathrm{Tor}_1^{\mathbb{F}[V]^G}(\mathbb{F},\ M)$ via the action of G on M. The inclusion $i : \mathbb{F}[V]^G \hookrightarrow \mathbb{F}[V]$ is a morphism of G-modules, where G acts trivially on $\mathbb{F}[V]^G$. Since $|G|$ is relatively prime to the characteristic of $\mathbb{F}$ we may form the averaging operator

$$\pi^G = \frac{1}{|G|} \sum_{g \in G} g : \mathbb{F}[V] \longrightarrow \mathbb{F}[V]^G\,,$$

which is an $\mathbb{F}[V]^G$-module homomorphism. Hence we have induced maps

$$\mathrm{Tor}_1^{\mathbb{F}[V]^G}(\mathbb{F},\ \mathbb{F}[V])$$
$$\searrow \quad {\scriptstyle \mathrm{Tor}_1^{\mathbb{F}[V]^G}(\mathbb{F},\ \pi^G)}$$
$$\mathrm{Tor}_1^{\mathbb{F}[V]^G}(\mathbb{F},\ \mathbb{F}[V]^G)$$
$$\searrow \quad {\scriptstyle \mathrm{Tor}_1^{\mathbb{F}[G]}(\mathbb{F},\ i)}$$
$$\mathrm{Tor}_1^{\mathbb{F}[V]^G}(\mathbb{F},\ \mathbb{F}[V])\,.$$

The composition $i \cdot \pi^G \cdot$ is the identity when restricted to invariant elements. Hence

$$\zeta = \mathrm{Tor}_1^{\mathbb{F}[V]^G}(\mathbb{F},\ i) \cdot \mathrm{Tor}_1^{\mathbb{F}[V]^G}(\mathbb{F},\ \pi^G)(\zeta) = 0$$

because

$$\mathrm{Tor}_1^{\mathbb{F}[V]^G}(\mathbb{F},\ \mathbb{F}[V]^G) = 0$$

since $\mathbb{F}[V]^G$ is a free module over itself. Hence no nonzero ζ could exist and therefore $\mathrm{Tor}_1^{\mathbb{F}[V]^G}(\mathbb{F}, \mathbb{F}[V]) = 0$ as required.

Suppose on the other hand that $\mathbb{F}[V]^G$ is a polynomial algebra, with polynomial generators $f_1, \ldots, f_n$ of degrees $d_1 \leq \cdots \leq d_n$. Let $H \leq G$ be the subgroup of G generated by $s(G)$, the set of pseudoreflections in G. By what has already been shown $\mathbb{F}[V]^H$ is also a polynomial algebra. Let $\mathbb{F}[V]^H$ have generators $h_1, \ldots, h_n$ of degrees $e_1 \leq \cdots \leq e_n$. (There are the same number of generators by Noether normalization since $\dim(\mathbb{F}[V]^G) = n = \dim(\mathbb{F}[V]^H$.) Of course $f_i \in \mathbb{F}[h_1, \ldots, h_n]$ for $i = 1, \ldots, n$. By 4.4.3 we have

$$|G| = d_1 \cdots d_n \qquad s(G) = \sum_{i=1}^n (d_i - 1)$$
$$|H| = e_1 \cdots e_n \qquad s(H) = \sum_{i=1}^n (e_i - 1).$$

We claim that $d_i \geq e_i$ for $i = 1, \ldots, n$. For $i = 1$ this is clear. Assume that $e_i \leq d_i$ for $i = 1, \ldots, m$ and consider f_{m+1}. If $d_{m+1} < e_{m+1}$ then f_{m+1} must be a polynomial in $h_1, \ldots, h_m$. Since $d_i \leq d_{m+1} < e_{m+1}$ for $i = 1, \ldots, m$, $f_1, \ldots, f_m$ are also polynomials in $h_1, \ldots, h_m$ and hence $f_1, \ldots, f_{m+1} \in \mathbb{F}[h_1, \ldots, h_m]$ would be algebraically independent which is impossible. Therefore $d_i \geq e_i$ for $i = 1, \ldots, n$ as claimed. Since G and H have the same pseudoreflections we obtain

$$\sum_{i=1}^n (d_i - 1) = \sum_{i=1}^n (e_i - 1)$$

from which it follows that $d_i = e_i$ for $i = 1, \ldots, n$. But then $|G| = d_1 \cdots d_n = e_1 \cdots e_n = |H|$ so $H = G$ and G is generated by pseudoreflections. $\square$

If $\varrho : G \hookrightarrow \mathrm{GL}(n, \mathbb{F})$ is a representation of a finite group whose order is prime to the characteristic of the field $\mathbb{F}$ and $\varrho(G)$ is generated by pseudoreflections, then we may write $\mathbb{F}[V]^G = \mathbb{F}[f_1, \ldots, f_n]$. The polynomials $f_1, \ldots, f_n$ are called **fundamental invariants** of G (or ϱ to be more precise), the integers $d_1 = \deg(f_1), \ldots, d_n = \deg(f_n)$ are called the **degrees** of G and the integers $m_1 = d_1 - 1, \ldots, m_n = d_n - 1$ the **exponents** of G. The fundamental invariants are not unique, but their degrees are completely determined by the Poincaré series of $\mathbb{F}[V]^G$. By 4.4.3 the product of the degrees is $|G|$ and the sum of the exponents is $|s(G)|$. A number of interesting properties of the degrees and exponents of complex pseudoreflection groups may be found in the papers of L. Solomon [226], [227], and the references there. (For the relation between complex reflection groups and the topology of the complement of arrangements of hyperplanes see [176], [177], and [178].)

A portion of the proof of 7.4.1 may be used to improve 5.5.5 to obtain a criterion readily verified in examples (e.g. section 5.5 example 1, section 5.6 example 6, etc.) for $\mathbb{F}[V]^G$ to be a polynomial algebra in the nonmodular case.

PROPOSITION 7.4.2 : Let $\varrho : G \hookrightarrow \mathrm{GL}(n, \mathbb{F})$ be a representation of a finite group G over a field $\mathbb{F}$ of characteristic prime to the order of G. Suppose $f_1, \ldots, f_n \in \mathbb{F}[V]^G$, $\deg(f_i) = d_i$ for $i = 1, \ldots, n$, satisfy

(i) $\displaystyle\prod_{i=1}^{n} d_i = |G|$,

(ii) $f_1, \ldots, f_n$ are algebraically independent.
Then $\mathbb{F}[V]^G = \mathbb{F}[f_1, \ldots, f_n]$.

PROOF : Let H be the subgroup of G generated by $s(G)$. By 7.4.1 we have $\mathbb{F}[V]^H = \mathbb{F}[h_1, \ldots, h_n]$. Set $e_i = \deg(h_i)$ for $i = 1, \ldots, n$ and arrange $d_1, \ldots, d_n$ and $e_1, \ldots, e_n$ in nondecreasing order. Then as in the proof of 7.4.1, the inclusions

$$\mathbb{F}[f_1, \ldots, f_n] \subseteq \mathbb{F}[V]^G \subseteq \mathbb{F}[V]^H = \mathbb{F}[h_1, \ldots, h_n]$$

imply $d_i \geq e_i$ for $i = 1, \ldots, n$. Both G and H have the same pseudoreflections so by 4.4.2 and 7.4.1 we have

$$\sum_{i=1}^{n}(d_i - 1) \leq |s(G)| = |s(H)| = \sum_{i=1}^{n}(e_i - 1).$$

Therefore $d_i = e_i$ for $i = 1, \ldots, n$ and the result follows. $\square$

REMARK : In the non-coprime situation, if $G \hookrightarrow \mathrm{GL}(n, \mathbb{F})$ and $\mathbb{F}[V]^G$ is a polynomial algebra then G is still generated by pseudoreflections. The proof depends on a result from commutative algebra called the *purity of the branch locus* [15]. (A breakdown of this result into a series of exercises may be found in [39] Exercises Chapitre 5 § 6 exercises 7 and 8 on pages 138 - 139.) In non-coprime characteristic the converse can fail, namely a group $G \hookrightarrow \mathrm{GL}(n, \mathbb{F})$ generated by pseudoreflections whose order is divisible by the characteristic of the field can fail to have polynomial invariants. (See example 4 below.)

THEOREM 7.4.3 : Let V be a finite dimensional vector space over the field $\mathbb{F}$ and $\varrho : G \hookrightarrow \mathrm{GL}(V)$ a representation of a finite group G. Assume that $\varrho(G)$ is generated by pseudoreflections and $|G|$ is relatively prime to the characteristic of $\mathbb{F}$. Then there is an isomorphism

$$\mathbb{F}[V] \cong \mathbb{F}[V]^G \otimes \mathbb{F}[V]_G$$

as both $\mathbb{F}[V]^G$- and $\mathbb{F}(G)$-modules.

PROOF : By 7.4.1 $\mathbb{F}[V]$ is a free $\mathbb{F}[V]^G$-module. (Alternatively note that from the corollary to Macaulay's theorem 6.7.11 a system of polynomial generators for $\mathbb{F}[V]^G$ forms a regular sequence in $\mathbb{F}[V]$ and hence by

6.2.8 $\mathbb{F}[V]$ is a free $\mathbb{F}[V]^G$-module.) Since the order of G is prime to the characteristic we may choose an $\mathbb{F}(G)$-module splitting

$$\alpha : \mathbb{F}[V]_G \longrightarrow \mathbb{F}[V]$$

to the projection

$$\mathbb{F}[V] \longrightarrow \mathbb{F}[V]_G$$

and the map

$$\mathbb{F}[V]^G \otimes \mathbb{F}[V]_G \longrightarrow \mathbb{F}[V], \qquad f \otimes h \mapsto f \cdot \alpha(h)$$

is the desired isomorphism. $\quad\square$

EXAMPLE 1 : By examination of the table of Shephard-Todd we see that there are four infinite families of groups. The groups of type 1 are the reduced tautological representations (i.e. the tautological representation divided by the fixed line) of the symmetric groups, those of 2b the dihedral groups and those of type 3 the cyclic groups. In this example we examine the groups of type 2a. These are the complex analogues of the familiar signed permutation groups and are defined as follows.

Let q, m, $n \in \mathbb{N}$ be positive integers with q dividing m, and set $k = m/q$. Let $\tau : \Sigma_n \hookrightarrow \mathrm{GL}(n,\ \mathbb{C})$ be the tautological representation of the symmetric group and $A(m,\ q,\ n) < \mathrm{GL}(n,\ \mathbb{C})$ be the subgroup consisting of diagonal matrices

$$D(\theta_1,\ \theta_2,\ldots,\ \theta_n) = \begin{bmatrix} \theta_1 & 0 & \ldots & 0 \\ 0 & \theta_2 & \ldots & 0 \\ \vdots & \vdots & & \vdots \\ 0 & 0 & \ldots & \theta_n \end{bmatrix}$$

where θ_i is an m-th root of unity and for which $(\theta_1 \theta_2 \cdots \theta_n)^k = \det[D(\theta_1,\ \theta_2,\ldots,\ \theta_n)]^k = 1$. The group $G(m,\ q,\ n) < \mathrm{GL}(n,\ \mathbb{C})$ is the subgroup generated by $\tau(\Sigma_n)$ and $A(m,\ q,\ n)$. The group Σ_n acts on the elements of $A(m,\ q,\ n)$ by permuting the diagonal entries and it is not hard to see that $G(m,\ q,\ n)$ is the semi-direct product. If $e_1,\ldots,\ e_n \in \mathbb{C}[V]$ are the elementary symmetric polynomials then

$$f_i = e_i(x_1^m,\ldots,\ x_n^m) \qquad 1 \le i < n$$
$$f_n = e_n(x_1,\ldots,\ x_n)^q$$

are algebraically independent invariants of $G(m,\ q,\ n)$ whose degrees are $m,\ 2m,\ldots,\ (n-1)m,\ qn$ and hence

$$\prod \deg(f_i) = qm^{n-1}n! = |G(m,\ q,\ n)|$$

so

$$\mathbb{C}[V]^{G(m,\ q,\ n)} = \mathbb{C}[f_1,\ldots,\ f_n]$$

by 7.4.1.

The preceding discussion carries through unchanged with $\mathbb{C}$ replaced by any field of characteristic p, where p is a good prime for $G(m, q, n)$, i.e. $p \equiv 1 \bmod m$, $p > n$. If p is an arbitrary prime satisfying $p \equiv 1 \bmod m$, then 7.4.1 need not apply, for if $p \leq n$ then p is not a good prime for $G(m, q, n)$. However, a field $\mathbb{F}$ of characteristic p contains a primitive m-th root of unity, since $m \mid p - 1$, and hence the pseudoreflection representation of $G(m, q, n) \hookrightarrow \mathrm{GL}(n, \mathbb{F})$ is defined exactly as above. The polynomials $f_i = e_i(x_1^m, \ldots, x_n^m)$ for $1 \leq i < n$ and $f_n = e_n(x_1, \ldots, x_n)^q$ belong to $\mathbb{F}[V]^{G(m, q, n)}$ just as before. Since $\prod_{i=1}^{n} \deg(f_i) = |G(m, q, n)|$ we may apply 7.4.2 to conclude $\mathbb{F}[V]^{G(m, q, n)} = \mathbb{F}[f_1, \ldots, f_n]$ even if p is not a good prime, but satisfies $p \equiv 1 \bmod m$.

EXAMPLE 2 : The group number 9 in the Shephard-Todd list occurs in coding theory (see e.g. [235] example 5.2 and the survey article [209]). This group has a complex 2-dimensional representation given by the two matrices

$$S = \frac{1}{\sqrt{2}} \begin{bmatrix} 1 & 1 \\ 1 & -1 \end{bmatrix} \qquad T = \begin{bmatrix} 1 & 0 \\ 0 & i \end{bmatrix}$$

which are pseudoreflections of order 2 and 4 respectively, so the group G generated by S and T is a complex pseudoreflection group. A bit of matrix multiplication yields a complete list of its elements, viz.

$$\varepsilon^i \begin{bmatrix} 1 & 0 \\ 0 & \alpha \end{bmatrix} \qquad \varepsilon^i \begin{bmatrix} 0 & 1 \\ \alpha & 0 \end{bmatrix} \qquad \frac{\varepsilon^i}{\sqrt{2}} \begin{bmatrix} 1 & \beta \\ \alpha & -\alpha\beta \end{bmatrix} ,$$

where $\varepsilon = \exp(2\pi i/8) = \frac{1+i}{\sqrt{2}}$ and $\alpha, \beta \in \{\pm 1, \pm i\}$, with $i = 0, \ldots, 7$, from which it follows that G has order 192. The product ST has order 24 and its cube is the matrix

$$\frac{1+i}{\sqrt{2}} \begin{bmatrix} 1 & 0 \\ 0 & 1 \end{bmatrix}$$

which is diagonal of order 8. Hence $A(G)$, the similarity subgroup of G, (see section 2.4) has order divisible by 8 and by 2.4.4 the degrees of the invariants of G are divisible by 8. So by 7.4.1

$$\mathbb{C}[x, y]^G = \mathbb{C}[f_1, f_2], \qquad \deg(f_1), \deg(f_2) \equiv 0 \bmod 8 .$$

If $d_1 = \deg(f_1) \leq \deg(f_2) = d_2$ then by 4.4.3 $d_1 d_2 = 192$ and the only solution with $d_1, d_2 \equiv 0 \bmod 8$ is $d_1 = 8$ and $d_2 = 24$.

The subgroup generated by S and T^2 is a dihedral group of order 16 acting on $\mathbb{R}^2$ as the symmetries of a regular octagon. In the example at the end of section 4.4 we saw that

$$\mathbb{R}[x, y]^{D_{16}} = \mathbb{R}[x^2 + y^2, \, x^2 y^2 (x^2 - y^2)^2] .$$

For future reference we set

$$h_1 = x^2 + y^2 \qquad h_2 = x^2 y^2 (x^2 - y^2)^2 \,.$$

Since $D_{16} < G$ we have

$$\mathbb{C}[f_1, \; f_2] = \mathbb{C}[x,y]^G \subset \mathbb{C}[x, \; y]^{D_{16}} = \mathbb{C}[x^2 + y^2, \; x^2 y^2 (x^2 - y^2)^2]$$

hence f_1 must be a linear combination of the two polynomials h_1^4 and h_2. Any invariant of D_{16} is invariant under S and T^2 so is a G-invariant if and only if it is invariant under T. Setting

$$f_1 = \lambda(x^2 + y^2)^4 + \mu x^2 y^2 (x^2 - y^2)^2$$
$$= \lambda x^8 + (4\lambda + \mu)x^2 y^6 + (6\lambda - 2\mu)x^4 y^4 + (4\lambda + \mu)x^6 y^2 + \lambda y^8$$

with $\lambda, \; \mu \in \mathbb{C}$ one computes

$$T(f_1) = \lambda x^8 - (4\lambda + \mu)x^2 y^6 + (6\lambda - 2\mu)x^4 y^4 - (4\lambda + \mu)x^6 y^2 + \lambda y^8$$

and equating coefficients yields $4\lambda + \mu = 0$ and hence setting $\lambda = 1$ we get

$$f_1 = x^8 + 14x^4 y^4 + y^8$$

as a fundamental invariant of G of degree 8. To find f_2, an invariant of degree 24, we could proceed in the same manner, however this is not necessary, as the formulae

$$T(x^2 y^2) = - x^2 y^2$$
$$T(x^2 + y^2) = x^2 - y^2$$
$$T(x^2 - y^2) = x^2 + y^2$$

which were part of the preceding calculation yield directly that

$$T((h_1^2 h_2)^2) = T(((x^2 + y^2)^2 x^2 y^2 (x^2 - y^2)^2)^2)$$
$$= ((x^2 - y^2)^2 (-x^2 y^2)(x^2 + y^2)^2)^2 = (h_1^2 h_2)^2$$

and since $(h_1^2 h_2)^2$ is algebraically independent from f_1 we may choose $f_2 = (h_1^2 h_2)^2$ and hence we have after a bit of computation

$$\mathbb{C}[x, \; y]^G = \mathbb{C}[x^8 + 14x^4 y^4 + y^8, \; x^{10} y^2 - 2x^6 y^6 + x^2 y^{10}] \,.$$

EXAMPLE 3 : Consider the Coxeter group $W(\mathbf{G}_2)$ defined by the root system of type $\mathbf{G}_2$ (see section 7.2 example 3). Recall that $W(\mathbf{G}_2)$ is isomorphic to the dihedral group D_{12} represented as the group of symmetries of a regular hexagon. The primes dividing $|W(\mathbf{G}_2)|$ are 2 and 3. A representation of $W(\mathbf{G}_2) = D_{12}$ as a Coxeter group is afforded by the matrices

$$D = \begin{bmatrix} 0 & -1 \\ 1 & 1 \end{bmatrix}, \quad S = \begin{bmatrix} -1 & 1 \\ 0 & 1 \end{bmatrix} \in \mathrm{GL}(2, \mathbb{Z}).$$

The group $W(\mathbf{G}_2)$ is the group of type 2b with $m = 6$ in the Shephard-Todd list, so 3 is not a good prime for $W(\mathbf{G}_2)$. However, the above matrices reduced mod p afford a faithful representation of $W(\mathbf{G}_2)$ in $\mathrm{GL}(2, \mathbb{F}_p)$ for any odd prime p including $p = 3$. (This is a special case of the more general result 10.7.1.) We will see that $p = 3$, while not good, is also not bad for $W(\mathbf{G}_2)$.

We begin our discussion of the invariants of $W(\mathbf{G}_2)$ by recalling from section 4.3 example 2 that

$$P(\mathbb{Q}[x, y]^{W(\mathbf{G}_2)}, t) = \frac{1}{(1 - t^2)(1 - t^6)},$$

so from that same example and the theorem of Shephard-Todd-Chevalley

$$\mathbb{F}[x, y]^{W(\mathbf{G}_2)} = \mathbb{F}[f, h] \qquad \deg(f) = 2, \ \deg(h) = 6$$

for any field $\mathbb{F}$ of characteristic not equal to 2 or 3.

To find suitable polynomials f and h, note that the 6 vertices of the hexagon are an orbit of the action of $W(\mathbf{G}_2)$. Choose as basis for $\mathbb{R}^2$ the elements $1, \zeta = \exp(2\pi i/6) \in \mathbb{C}$ as in example 3 of section 7.2. In terms of the dual basis x, y the orbit $\{\pm 1, \pm \zeta, \pm \zeta^2\}$ corresponds to the orbit $\{\pm x, \pm y, \pm(y - x)\} = B$. At this point it is important to note that the 6 points B in the dual of $\mathbb{F}^2$ are an orbit of $W(\mathbf{G}_2)$ regardless of the characteristic of $\mathbb{F}$. (For characteristic 2 there are only 3 points, but the mod 2 reduction of the pseudoreflection representation is not faithful, so we exclude this case from the discussion.) The total Chern class of B is

$$c(B) = (t + x)(t - x)(t + y)(t - y)(t + (y - x))(t - (y - x))$$
$$= (t^2 - x^2)(t^2 - y^2)(t^2 - (y - x)^2)$$

so

$$c_i = \begin{cases} 0 & \text{for } i \text{ odd} \\ e_{i/2}(x^2, y^2, (y - x)^2) & \text{for } i \text{ even} \end{cases}$$

where e_j denotes the j-th elementary symmetric polynomial. Expanding and simplifying gives:

$$c_2 = 2(x^2 - xy + y^2)$$
$$c_4 = 4(x^2 - yx + y^2)^2 = c_2^2$$
$$c_6 = (xy^2 - x^2 y)^2.$$

Note $\deg(c_2)\deg(c_6) = 2 \cdot 6 = 12 = |W(\mathbf{G}_2)|$ and c_2, c_6 are a system of parameters, even if the characteristic of $\mathbb{F}$ is 3, so by 5.5.5 we may conclude

$$\mathbb{F}[x, \ y]^{W(\mathbf{G}_2)} = \mathbb{F}[x^2 - xy + y^2, \ (xy^2 - xy^2)^2]$$

for any field of characteristic not equal to 2. Therefore, even though 3 is not a good prime for $W(\mathbf{G}_2)$ it is also not a bad prime.

If G is a group generated by pseudoreflections and p is a prime dividing the order of G then p is not a good prime for G. However, this does not mean that p is a bad prime for G, i.e. that, either no mod p reduction of the pseudoreflection representation exists, or such a representation exists, but the invariants are not polynomial. For example, the symmetric group (or more generally the groups of type 2a in the Shephard-Todd list) have mod p pseudoreflection representations for all primes p with polynomial invariants. Things can however go awry.

EXAMPLE 4 : The Coxeter group $W(\mathbf{F}_4)$ corresponding to the root system of type $\mathbf{F}_4$ was examined in section 7.2 example 4. The group $W(\mathbf{F}_4)$ contains $W(\mathbf{D}_4)$ as a normal subgroup and $W(\mathbf{F}_4)/W(\mathbf{D}_4) \cong \Sigma_3$. The invariants of $W(\mathbf{D}_4) \cong (\mathbb{Z}/2)^3 \rtimes \Sigma_4$ occur as a special case of example 1: the group $W(\mathbf{D}_4)$ is the group $G(2, 1, 2)$ in the notations of that example. We find

$$\mathbb{F}[z_1, \ z_2, \ z_3, \ z_4]^{W(\mathbf{D}_4)} = \mathbb{F}[\mathfrak{p}_1, \ \mathfrak{p}_2, \ \mathfrak{p}_3, \ X]$$

where[4]

$$\mathfrak{p}_i = e_i(z_1^2, \ z_2^2, \ z_3^2, \ z_4^2) \quad i = 1, \ 2, \ 3, \ 4$$

$$X = z_1 z_2 z_3 z_4$$

$$\mathfrak{p}_4 = X^2$$

and $\mathbb{F}$ is any field of characteristic not equal to 2. The order of $W(\mathbf{D}_4)$ is $192 = 64 \cdot 3$ so neither 2 nor 3 is a good prime for $W(\mathbf{D}_4)$. The prime 3 is however not bad, i.e. the mod 3 invariants of $W(\mathbf{D}_4)$ are still a polynomial algebra. This no longer holds true for $W(\mathbf{F}_4)$ as we proceed to show.

The representation of $W(\mathbf{F}_4)$ as a Coxeter group in $\mathrm{GL}(4, \ \mathbb{Q})$ is given by the signed permutation matrices (corresponding to $W(\mathbf{B}_4)$) and the Hadamard matrix

$$\frac{1}{2}\begin{bmatrix} 1 & 1 & 1 & 1 \\ 1 & -1 & 1 & -1 \\ 1 & 1 & -1 & -1 \\ 1 & -1 & -1 & 1 \end{bmatrix} \in \mathrm{GL}(4, \ \mathbb{Z}[\tfrac{1}{2}]),$$

and the reduction mod p of these matrices provide a faithful representation of $W(\mathbf{F}_4)$ for all odd primes p.

---

[4] We have chosen to use a notation suggested by algebraic topology. The $\mathfrak{p}_i$ should be thought of as Pontrjagin classes and X as the Euler class.

The order of $W(\mathbf{F}_4)$ is $1152 = 2^7 \cdot 3^2$ so by the theorem of Shephard-Todd-Chevalley

$$\mathbb{F}[z_1,\ z_2,\ z_3,\ z_4]^{W(\mathbf{F}_4)} = \mathbb{F}[f_1,\ f_2,\ f_3,\ f_4]$$

when $\mathbb{F}$ is a field of characteristic not equal to 2 or 3. The degrees $d_i = \deg(f_i)$ for $i = 1,\ 2,\ 3,\ 4$ satisfy

$$d_1 d_2 d_3 d_4 = |W(\mathbf{F}_4)| = 1152$$
$$(d_1 - 1) + (d_2 - 1) + (d_3 - 1) + (d_4 - 1) = |s(W(\mathbf{F}_4))| = 24.$$

(The number of reflections in a real reflection group is half the number of roots, and $|\mathcal{R}(\mathbf{F}_4)| = 48$.) The action of $W(\mathbf{F}_4)$ on $\mathbb{R}^4$ preserves the inner product so there is a quadratic invariant and we may suppose $2 = d_1 \leq d_2 \leq d_3 \leq d_4$. A short computation (using a suitable computer program) shows that

$$d_1 = 2, \quad d_2 = 6, \quad d_3 = 8, \quad d_4 = 12$$

is the unique solution in positive integers to the preceding equations. Thus for any field $\mathbb{F}$ of characteristic not equal to 2 or 3

$$P(\mathbb{F}[z_1,\ z_2,\ z_3,\ z_4]^{W(\mathbf{F}_4)},\ t) = \frac{1}{(1 - t^2)(1 - t^6)(1 - t^8)(1 - t^{12})}$$
$$= 1 + t^2 + t^4 + 2t^6 + 3t^8 + \cdots.$$

The group $W(\mathbf{D}_4)$ is a normal subgroup of $W(\mathbf{F}_4)$ with quotient Σ_3. To find explicit generators for $\mathbb{F}[z_1,\ z_2,\ z_3,\ z_4]^{W(\mathbf{F}_4)}$ we may exploit the isomorphism

$$\mathbb{F}[z_1,\ z_2,\ z_3,\ z_4]^{W(\mathbf{F}_4)} = \left(\mathbb{F}[z_1,\ z_2,\ z_3,\ z_4]^{W(\mathbf{D}_4)} \right)^{\Sigma_3}$$
$$= \mathbb{F}[\mathfrak{p}_1,\ \mathfrak{p}_2,\ \mathfrak{p}_3,\ X]^{\Sigma_3}.$$

The key to computing further is to understand the action of Σ_3 on the homogeneous component of degree 4 of $\mathbb{F}[\mathfrak{p}_1,\ \mathfrak{p}_2,\ \mathfrak{p}_3,\ X]$, which is spanned by the three polynomials $\mathfrak{p}_1{}^2$, $\mathfrak{p}_2$, X. We require a lemma.

LEMMA 7.4.4 : *Let $\varrho : G \hookrightarrow \mathrm{GL}(n,\ \mathbb{F})$ be a representation of a finite group G over a field $\mathbb{F}$ whose characteristic is prime to the order of G. Suppose $f_1, \ldots, f_r \in \mathbb{F}[V]^G$. Then G acts on the quotient ring $\mathbb{F}[V]/(f_1, \ldots, f_r)$ by automorphisms, and the natural map*

$$q : \mathbb{F}[V]^G \longrightarrow (\mathbb{F}[V]/(f_1, \ldots, f_r))^G$$

is an epimorphism.

PROOF :   Since $f_1, \ldots, f_r \in \mathbb{F}[V]^G$, the ideal $(f_1, \ldots, f_r)$ is invariant under the action of G and hence G acts on the quotient ring $\mathbb{F}[V]/(f_1, \ldots, f_r)$. Suppose $h \in (\mathbb{F}[V]/(f_1, \ldots, f_r))^G$. Choose a lift $\bar{h} \in \mathbb{F}[V]$ of h, i.e. $\bar{q}(\bar{h}) = h$ where

$$\bar{q} : \mathbb{F}[V] \longrightarrow \mathbb{F}[V]/(f_1, \ldots, f_r)$$

is the quotient map. Since $\bar{q}$ is G-equivariant $g\bar{h}$ is also a lift of h for any $g \in G$. Hence

$$\bar{q}\left(\frac{1}{|G|}\sum_{g \in G} g\bar{h}\right) = h,$$

and since $\frac{1}{|G|}\sum_{g \in G} g\bar{h} \in \mathbb{F}[V]^G$ the lemma is established. $\square$

REMARK :   The preceding lemma is false if the order of G is not relatively prime to the characteristic of $\mathbb{F}$. A generic counterexample may be constructed as follows. Let G be a nontrivial finite p-group, p a prime, and $\mathbb{F}$ a finite field of characteristic p. Suppose that $\varrho : G \hookrightarrow \mathrm{GL}(n, \mathbb{F})$ is a representation. Decompose $V = (\mathbb{F}^n)^*$ into a disjoint union

$$V = V^G \amalg \coprod_{i=1}^{r} B_i$$

where V^G is the fixed point set and B_i, $i = 1, \ldots, r$, the nontrivial orbits (i.e. containing more than one element) of G. Since G is a p-group $|B_i| \equiv 0 \bmod p$ for $i = 1, \ldots, r$. The field $\mathbb{F}$ has characteristic p and is finite so $|V| \equiv 0 \bmod p$ also. Hence $|V^G| \equiv 0 \bmod p$. Clearly $0 \in V^G$ so V^G is not empty. Since the number of elements in V^G is divisible by p, V^G must be a subspace of positive dimension. However, ϱ is not the trivial representation, so $V^G \neq V$. Let $f_1, \ldots, f_s \in V^G$ be a basis. Then $\mathbb{F}[V]/(f_1, \ldots, f_s) \cong \mathbb{F}[V/V^G]$ and $V/V^G \neq \{0\}$. By the preceding argument $(V/V^G)^G$ contains nonzero elements, and hence $\mathbb{F}[V/V^G]^G$ contains nonzero linear invariants. But

$$q : \mathbb{F}[V]^G \longrightarrow \mathbb{F}[V/V^G]^G$$

maps the linear invariants of $\mathbb{F}[V]^G$ to zero in $\mathbb{F}[V/V^G]^G$. Hence none of the linear invariants in $(\mathbb{F}[V]/(f_1, \ldots, f_s))^G$ lift to invariants of G on $F[V]$.

To apply the preceding lemma to our analysis of $\mathbb{F}[z_1, z_2, z_3, z_4]^{W(\mathbf{D_4})} = \mathbb{F}[\mathfrak{p}_1, \mathfrak{p}_2, \mathfrak{p}_3, X]$ as Σ_3-module we suppose that the field $\mathbb{F}$ has characteristic $p \neq 2, 3$. A direct computation using the formulas developed in section 7.2 example 4 shows that the elements

$$\mathfrak{p}_1, \ \bar{\mathfrak{p}}_3 = -6\mathfrak{p}_3 + \mathfrak{p}_1\mathfrak{p}_2 \in \mathbb{F}[\mathfrak{p}_1, \mathfrak{p}_2, \mathfrak{p}_3, X]$$

are invariant under the action of Σ_3. By the preceding lemma

$$q : \mathbb{F}[\mathfrak{p}_1, \ \mathfrak{p}_2, \ \mathfrak{p}_3, \ X]^{\Sigma_3} \longrightarrow \left(\frac{\mathbb{F}[\mathfrak{p}_1, \ \mathfrak{p}_2, \ \mathfrak{p}_3, \ X]}{(\mathfrak{p}_1, \ \bar{\mathfrak{p}}_3)} \right)^{\Sigma_3}$$

is an epimorphism. Moreover

$$\frac{\mathbb{F}[\mathfrak{p}_1, \ \mathfrak{p}_2, \ \mathfrak{p}_3, \ X]}{(\mathfrak{p}_1, \ \bar{\mathfrak{p}}_3)} \cong \mathbb{F}[\bar{\mathfrak{p}}_2, \ \bar{X}]$$

where $\bar{\mathfrak{p}}_2 = q(\mathfrak{p}_2)$, $\bar{X} = q(X)$. The group Σ_3 acts by automorphisms on $\mathbb{F}[\bar{\mathfrak{p}}_2, \ \bar{X}]$ and hence is completely determined by the representation

$$\bar{\varrho} : \Sigma_3 \longrightarrow GL(2, \ \mathbb{F})$$

given by restricting the Σ_3 action to the $\mathbb{F}$-span of $\bar{\mathfrak{p}}_2$ and $\bar{X}$, which is $\mathbb{F}[\bar{\mathfrak{p}}_2, \ \bar{X}]_4$.

We intend to use character theory to determine the action of Σ_3 on $\mathbb{F}[\bar{\mathfrak{p}}_2, \ \bar{X}]$, so we assume that the field $\mathbb{F}$ is algebraically closed of characteristic not equal to 2 or 3. To more readily recognize the results of computations we set $\bar{Y} = q(z_1{}^4 + z_2{}^4 + z_3{}^4 + z_4{}^4)$ and employ $\bar{X}$, $\bar{Y}$ as a basis for $\mathbb{F}[\bar{\mathfrak{p}}_2, \ \bar{X}]_4$. The matrix $D \in \Sigma_3$ of the triality in the notations of section 7.2 example 4 is given by

$$D = TS = \frac{1}{2} \begin{bmatrix} 1 & 1 & -1 & 1 \\ 1 & 1 & 1 & -1 \\ -1 & 1 & 1 & 1 \\ 1 & -1 & 1 & 1 \end{bmatrix} \begin{bmatrix} 1 & 0 & 0 & 0 \\ 0 & 1 & 0 & 0 \\ 0 & 0 & 1 & 0 \\ 0 & 0 & 0 & -1 \end{bmatrix} = \frac{1}{2} \begin{bmatrix} 1 & 1 & -1 & -1 \\ 1 & 1 & 1 & 1 \\ -1 & 1 & 1 & -1 \\ 1 & -1 & 1 & -1 \end{bmatrix}.$$

The action of the group Σ_3 on $\mathbb{F}[\bar{\mathfrak{p}}_2, \ \bar{X}]$ is given by the matrices D^{tr} and S^{tr}. We have

$$S^{\mathrm{tr}}(\bar{X}) = -\bar{X}, \qquad D^{\mathrm{tr}}(\bar{X}) = -\tfrac{1}{16}(8\bar{X} + 2\bar{Y})$$
$$S^{\mathrm{tr}}(\bar{Y}) = \bar{Y}, \qquad D^{\mathrm{tr}}(\bar{Y}) = \tfrac{1}{16}(96\bar{X} - 8\bar{Y}).$$

The computation for S^{tr} is elementary, and the one for D^{tr} is best done with a computer algebra program. The matrices of S^{tr} and D^{tr} with respect to the basis $\{\bar{X}, \ \bar{Y}\}$ for $\mathbb{F}[\bar{\mathfrak{p}}_2, \ \bar{Y}]_4$ are

$$S^{\mathrm{tr}} = \begin{bmatrix} -1 & 0 \\ 0 & 1 \end{bmatrix}, \qquad D^{\mathrm{tr}} = \begin{bmatrix} \frac{-1}{2} & 6 \\ \frac{-1}{8} & \frac{-1}{2} \end{bmatrix}.$$

The group Σ_3 has three conjugation classes of elements for which $\{1, \ D, \ S\}$ are a complete set of representatives. Therefore the character $\chi_{\bar{\varrho}}$ of $\bar{\varrho}$ is given by

$$\chi_{\bar{\varrho}}(1) = 2, \quad \chi_{\bar{\varrho}}(D) = -1, \quad \chi_{\bar{\varrho}}(S) = 0.$$

The group Σ_3 has three irreducible representations, and the character table [202] of Σ_3 is given in Table 7.4.1. In the table ϵ denotes the trivial one-dimensional representation, det the determinant representation, and $\overline{\mathrm{reg}}$ the irreducible 2-dimensional representation which is the quotient of the regular representation by the fixed point set. Inspection[5] of the table shows that $\chi_{\bar\varrho} = \chi_{\overline{reg}}$.

Σ_3	1	D	S
χ_ϵ	1	1	1
χ_{det}	1	1	-1
$\chi_{\overline{reg}}$	2	-1	0

Table 7.4.1: The character table of Σ_3

Therefore the invariants $\mathbb{F}[\bar{p}_2, \ \bar{X}]$ are, apart from a change of basis and a shift of grading, as given in section 3.2 example 1 and we find

$$\mathbb{F}[\bar{p}_2, \ \bar{X}]^{\Sigma_3} = \mathbb{F}[h_8, \ h_{12}]$$

where h_8, h_{12} have degrees 8 and 12 respectively. By lemma 7.4.4 these polynomials may be lifted to invariant polynomials $\bar{h}_8, \bar{h}_{12} \in \mathbb{F}[z_1, \ z_2, \ z_3, \ z_4]^{W(\mathbf{F}_4)}$. Explicit polynomials may be found in [60] page 203 or [137]. For example

$$h_8 = z_1^8 + z_2^8 + z_3^8 + z_4^8 + 14\mathrm{sym}(z_1^4 z_2^4) + 168X^2$$
$$h_{12} = z_1^{12} + z_2^{12} + z_3^{12} + z_4^{12} + 22\mathrm{sym}(z_1^6 z_2^6) + 330\mathrm{sym}(z_1^6 z_2^2 z_3^2 z_4^2)$$
$$+ 165\mathrm{sym}(z_1^4 z_2^4 z_3^4) \ + 330\mathrm{sym}(z_1^4 z_2^4 z_2^2 z_4^2),$$

where sym(—) denotes the symmetrization of the indicated monomial, i.e. the smallest symmetric polynomial in the lexicographic order that contains the indicated monomial with coefficient 1. (Or, put another way the orbit sum of the orbit containing the indicated monomial.)

To summarize, for a field $\mathbb{F}$ not of characteristic 2 or 3,

$$\mathbb{F}[z_1, \ z_2, \ z_3, \ z_4]^{W(\mathbf{F}_4)} = \mathbb{F}[h_2, \ h_6, \ h_8, \ h_{12}],$$

---

[5] An alternative method to show that $\bar\varrho$ is the irreducible representation of Σ_3 of dimension 2 uses the Poincaré series

$$1 + t^2 + t^4 + 2t^6 + 3t^8 + \cdots$$

of $\mathbb{F}[z_1, \ z_2, \ z_3, \ z_4]^{W(\mathbf{F}_4)}$ and case by case analysis of the possible 2-dimensional representations of Σ_3. One finds that the only choice of $\bar\varrho$ compatible with the Poincaré series is $\overline{reg}$.

where h_i has degree i and

$$h_2 = \mathfrak{p}_1 = z_1{}^2 + z_2{}^2 + z_3{}^2 + z_4{}^2$$

$$h_6 = \bar{\mathfrak{p}}_3 = -6\mathrm{sym}(z_1{}^2 z_2{}^2 z_3{}^2) + (z_1{}^2 + z_2{}^2 + z_3{}^2 + z_4{}^2)\mathrm{sym}(z_1{}^2 z_2{}^2)$$

$$h_8 = z_1{}^8 + z_2{}^8 + z_3{}^8 + z_4{}^8 + 14\mathrm{sym}(z_1{}^4 z_2{}^4) + 168X^2$$

$$h_{12} = z_1{}^{12} + z_2{}^{12} + z_3{}^{12} + z_4{}^{12} + 22\mathrm{sym}(z_1{}^6 z_2{}^6) + 330\mathrm{sym}(z_1{}^6 z_2{}^3 z_2{}^2 z_4{}^2)$$

$$+ 165\mathrm{sym}(z_1{}^4 z_2{}^4 z_3{}^4) + 330\mathrm{sym}(z_1{}^4 z_2{}^4 z_2{}^2 z_4{}^2).$$

If the characteristic of $\mathbb{F}$ is 3 then the preceding analysis does not apply because the representation theory of Σ_3 over $\mathbb{F}$ is more complicated. In fact a direct computation (see section 10.3 for an explanation of why one should think of this computation) shows that (remember $\mathbb{F}$ has characteristic 3)

$$f = z_1{}^4 + z_2{}^4 + z_3{}^4 + z_4{}^4 \in \mathbb{F}[z_1,\ z_2,\ z_3,\ z_4]^{W(\mathbf{F}_4)}.$$

Since f does not belong to the subalgebra generated by $\mathfrak{p}_1$ we see that $\mathbb{F}[z_1,\ z_2,\ z_3,\ z_4]^{W(\mathbf{F}_4)}$ contains at least two linearly independent invariants of degree 4. Hence the Poincaré series of the invariants of $W(\mathbf{F}_4)$ in characteristic 3 starts off

$$1 + t^2 + at^4 + \cdots$$

where a is at least 2. This is different from the Poincaré series for good primes or in characteristic zero. From this it is not hard to see that $p = 3$ is a bad prime for $W(\mathbf{F}_4)$, i.e. the invariants in characteristic 3 are no longer a polynomial algebra. The group $W(\mathbf{F}_4) < \mathrm{GL}(4,\ \mathbb{F})$ is however generated by reflections, and hence we see that theorem 7.4.1 does not remain true when the characteristic of $\mathbb{F}$ divides the order of G.

The invariants of $W(\mathbf{F}_4)$ in characteristic 3 were computed by H. Toda [249]. The result is

$$\mathbb{F}[z_1,\ z_2,\ z_3,\ z_4]^{W(\mathbf{F}_4)} \cong \frac{\mathbb{F}[\mathfrak{p}_1,\ \bar{\mathfrak{p}}_2,\ \bar{\mathfrak{p}}_5,\ \bar{\mathfrak{p}}_9,\ \bar{\mathfrak{p}}_{12}]}{(\mathfrak{r}_{15})}$$

where

$$\bar{\mathfrak{p}}_2 = \mathfrak{p}_2 - \mathfrak{p}_1{}^2$$

$$\bar{\mathfrak{p}}_5 = \mathfrak{p}_4 \mathfrak{p}_1 + \mathfrak{p}_3 \bar{\mathfrak{p}}_2$$

$$\bar{\mathfrak{p}}_9 = \mathfrak{p}_3{}^3 - \mathfrak{p}_4 \mathfrak{p}_3 \mathfrak{p}_1{}^2 + \mathfrak{p}_3 \bar{\mathfrak{p}}_2 \mathfrak{p}_1 - \mathfrak{p}_4 \bar{\mathfrak{p}}_2 \mathfrak{p}_1{}^3$$

$$\bar{\mathfrak{p}}_{12} = \mathfrak{p}_4{}^3 + \mathfrak{p}_4{}^2 \mathfrak{p}_2{}^2 + \mathfrak{p}_4 \bar{\mathfrak{p}}_2{}^4$$

$$\mathfrak{r}_{15} = \bar{\mathfrak{p}}_5{}^3 + \bar{\mathfrak{p}}_5{}^2 \bar{\mathfrak{p}}_2{}^2 \mathfrak{p}_1 - \bar{\mathfrak{p}}_{12} \mathfrak{p}_1{}^3 - \bar{\mathfrak{p}}_9 \bar{\mathfrak{p}}_2{}^3$$

with the classes $\mathfrak{p}_i$ having degree $2i$ and the relation $\mathfrak{r}_{15}$ having degree 30. This ring is not polynomial but it is Cohen-Macaulay.

EXAMPLE 5 : The group of type 12 in the Shephard-Todd list is the finite linear group $\mathrm{GL}(2, \mathbb{F}_3)$. This group has a complex representation of degree 2 lifting the tautological representation on $\mathbb{F}_3^2$ to $\mathbb{Z}(\sqrt{-2})$ which we examined in section 7.3 example 1. The prime 3 is not a good prime, but the computations in chapter 5 section 5.6 example 4 show the invariants mod 3 are a polynomial algebra, namely

$$\mathbb{F}_3[x, y]^{\mathrm{GL}(2, \ \mathbb{F}_3)} = \mathbb{F}_3\left[(xy^3 - x^3y)^2, \ \frac{xy^9 - x^9y}{xy^3 - x^3y}\right].$$

Thus, while not a good prime, the prime 3 is also not a bad prime. For a study of how to detect such phenomena see [65], [117] and [164].

7.5 Coinvariants

If $G \hookrightarrow \mathrm{GL}(n, \mathbb{F})$ is a finite pseudoreflection group in coprime characteristic then the algebra of coinvariants $\mathbb{F}[V]_G$ has a number of special properties.

THEOREM 7.5.1 : Suppose that $G \hookrightarrow \mathrm{GL}(n, \mathbb{F})$ is a finite pseudo-reflection group and $|G| \in \mathbb{F}^\times$. Then $\mathbb{F}[V]_G$ is a Poincaré duality algebra and $\dim_\mathbb{F}(\mathrm{Tot}(\mathbb{F}[V]_G)) = |G|$. The degree of the fundamental class is $|s(G)| = \sum_{i=1}^n (d_i - 1)$ where $d_1, \ldots, d_n$ are the degrees of G.

PROOF : By the theorem of Shephard-Todd 7.4.1 $\mathbb{F}[V]^G$ is a polynomial algebra so by 6.5.1 $\mathbb{F}[V]_G$ is a Poincaré duality algebra. To compute its dimension note by 7.4.3 that

$$\mathbb{F}[V] \cong \mathbb{F}[V]^G \otimes \mathbb{F}[V]_G$$

as $\mathbb{F}[V]^G$-module. If the degrees of G are $d_1, \ldots, d_n$ then taking Poincaré series gives

$$P(\mathbb{F}[V]_G, t) = \frac{1}{(1-t)^n} / \prod_{i=1}^n \frac{1}{1 - t^{d_i}}$$

$$= \frac{\prod_{i=1}^n 1 - t^{d_i}}{(1-t)^n}$$

$$= \prod_{i=1}^n 1 + t + t^2 + \cdots + t^{d_i - 1}$$

and evaluating at $t = 1$ gives $d_1 \cdots d_n = \dim_\mathbb{F}(\mathrm{Tot}(\mathbb{F}[V]_G))$, so the result follows from 4.4.3. $\square$

REMARK : If $G \hookrightarrow \mathrm{GL}(n, \; \mathbb{F})$ is a finite group generated by pseudoreflections whose order is prime to the characteristic of $\mathbb{F}$, then by the theorem of Shephard-Todd 7.4.1 $\mathbb{F}[V]^G = \mathbb{F}[f_1, \ldots, f_n]$ and hence if $z_1, \ldots, z_n$ is a basis for V^* then by 6.5.2 the Jacobian determinant $\det[\frac{\partial f_i}{\partial z_j}]$ is a fundamental class for $\mathbb{F}[V]_G$. In characteristic zero the converse of theorem 7.5.1 holds. Namely, Steinberg [238] and Kane [118] have shown, $\mathbb{F}[V]_G$ is a Poincaré duality algebra if and only if $\varrho(G)$ is generated by pseudoreflections.

For a representation $\varrho : G \hookrightarrow \mathrm{GL}(n, \; \mathbb{F})$ the ideal $(\overline{\mathbb{F}[V]^G})$ is invariant under the operation of G on $\mathbb{F}[V]$ so the action of G passes to the quotient $\mathbb{F}[V]_G$. This gives us a *finite dimensional* representation of G which in the case of a pseudoreflection group in coprime characteristic was first computed by Chevalley [55]:

THEOREM 7.5.2 (C. Chevalley): *Let $G \hookrightarrow \mathrm{GL}(n, \; \mathbb{F})$ be a finite group generated by pseudoreflections whose order is prime to the characteristic of $\mathbb{F}$. Then $\mathrm{Tot}(\mathbb{F}[V]_G)$ is the regular representation of G.*

PROOF : By the theorem of Shephard-Todd 7.4.1 $\mathbb{F}[V]^G$ is a polynomial algebra on $n = \dim_{\mathbb{F}}(V)$ generators so by 6.7.7 $\mathbb{F}[V]$ is a free module over $\mathbb{F}[V]^G$ and there is an isomorphism

$$\mathbb{F}[V] \cong \mathbb{F}[V]^G \otimes \mathbb{F}[V]_G$$

as graded $\mathbb{F}(G)$-modules, i.e. as graded G-representations. Choose elements $h_1, \ldots, h_d \in \mathbb{F}[V]$ that generate $\mathbb{F}[V]$ freely as an $\mathbb{F}[V]^G$-module and note that they project to a vector space basis $\bar{h}_1, \ldots, \bar{h}_d$ in $\mathbb{F}[V]_G$. Write

$$g \cdot h_i = \sum a_{i,j}(g) h_i$$

where $(a_{i,j}(g))$ is a $d \times d$ matrix with entries in $\mathbb{F}[V]$. By 1.2.4 the extension of fields of fractions $\mathbb{F}(V) \mid \mathbb{F}(V)^G$ is Galois so $\mathbb{F}(V)$ regarded as a G-representation over $\mathbb{F}(V)^G$ is the regular representation, and hence taking traces gives

$$\sum_{i=1}^{n} a_{i,i}(g) \mathrm{tr}(a_{i,j}(g)) = \chi_{\mathbf{reg}}(g)$$

where **reg** denotes the regular representation of G and $\chi_{\mathbf{reg}}$ its character. The action of $g \in G$ on $\mathbb{F}[V]_G$ is represented by the matrix $(\varepsilon(a_{i,j}(g)))$ where $\varepsilon : \mathbb{F}[V] \longrightarrow \mathbb{F}$ is the augmentation. By homogeneity the entries $a_{i,i}(g)$ have degree zero and hence

$$\mathrm{tr}(\varepsilon(a_{i,j}(g))) = \mathrm{tr}(a_{i,j}(g)) = \chi_{\mathbf{reg}}$$

and the result follows. (For the preceding use of character theory see for example [201] § 2.3 corollary 3.) $\square$

REMARK : There are nice generalizations of this result due to R. Stanley [235], proposition 4.9, using Molien's theorem rather than Galois theory, and S. Mitchell, [147] theorem 1.4, in a modular setting using modular characters. If $\mathbb{F}[V]^G$ is a polynomial algebra, but the characteristic of $\mathbb{F}$ divides $|G|$, then $\mathbb{F}[V]_G$ cannot be the regular representation. To see this, note that the unit 1 in $\mathbb{F}[V]_G$ spans a trivial one-dimensional representation of G and is the degree 0 component of $\mathbb{F}[V]_G$. If $\mathbb{F}[V]_G$ were the regular representation, then it would be a free $\mathbb{F}(G)$-module. The action of G on $\mathbb{F}[V]_G$ preserves the grading, so the homogeneous components are direct summands, and hence projective. Therefore the trivial one-dimensional representation $1 \cdot \mathbb{F} \subset \mathbb{F}[V]_G$ would be projective, which is impossible.

EXAMPLE 1 : Consider the dihedral group D_8 of order 8 in its natural 2-dimensional representation as the group of symmetries of a square. In section 3.2 example 3 we found the invariants to be

$$\mathbb{R}[x,\ y]^{D_8} = \mathbb{R}[x^2 + y^2,\ x^2 y^2],$$

where the action on the dual vector space is given by the matrices

$$D = \begin{bmatrix} 0 & -1 \\ 1 & 0 \end{bmatrix} \qquad S = \begin{bmatrix} -1 & 0 \\ 0 & 1 \end{bmatrix}$$

with respect to the basis $\{x,\ y\}$. The fundamental class of $\mathbb{R}[x,y]_{D_8}$ is given by

$$J = \det \begin{bmatrix} \frac{\partial(x^2+y^2)}{\partial x} & \frac{\partial(x^2+y^2)}{\partial y} \\ \frac{\partial(x^2 y^2)}{\partial x} & \frac{\partial(x^2 y^2)}{\partial y} \end{bmatrix} = 4(x^3 y - x y^3)$$

which is of degree 4. The coinvariants of degree 1 have as a basis $\{x,\ y\}$ so Poincaré duality says that the coinvariants of degree 3 have a basis $\{(x^3 y - x y^3)/x,\ (x^3 y - x y^3)/y\}$. The dimension of the homogeneous component of $\mathbb{R}[x,\ y]$ of degree 2 is 3 and one of the basis elements is an invariant. Thus $\mathbb{R}[x,\ y]_{D_8}$ has a basis consisting of $\{x^2 - y^2,\ xy\}$ in degree 2. The algebra $\mathbb{R}[x,\ y]_{D_8}$ may be visualized as in example 2 of section 1.2 with the aid of the following diagram

where the nodes on a horizontal level are basis vectors for the elements of degree equal to their height above the node labeled 1, which has degree 0. The group D_8 has 4 irreducible representations of dimension 1 obtained by assigning ± 1 to D and S in all possible ways and the single irreducible 2-dimensional representation given by D and S themselves. Examining the structure of $\mathbb{R}[x,\ y]_{D_8}$ laid out in the preceding diagram, we see that the irreducible 2-dimensional representation occurs twice: once in degree 1 and once in degree 3. The 4 irreducible representations of degree 1 are given by

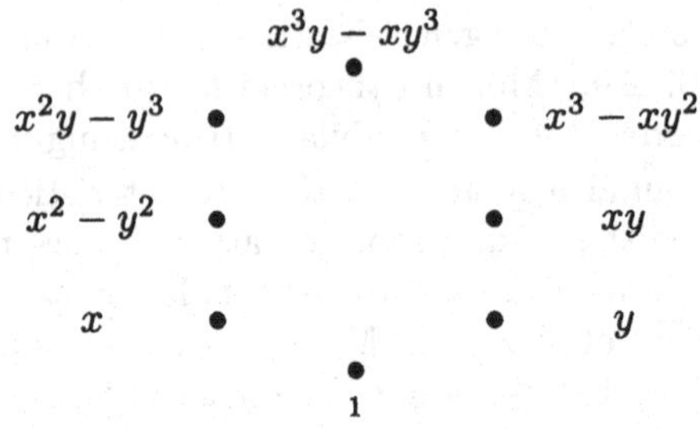

$$x^3y - xy^3$$
$$\bullet$$

$x^2y - y^3$ $\bullet$ $\bullet$ $x^3 - xy^2$

$x^2 - y^2$ $\bullet$ $\bullet$ xy

x $\bullet$ $\bullet$ y

$\bullet$
1

Diagram 7.5.1: $\mathbb{R}[x, \, y]_{D_8}$

the following table:

$$x^3y - xy^3 \quad \begin{cases} D \mapsto 1 \\ S \mapsto -1 \end{cases}$$

$$x^2 - y^2 \quad \begin{cases} D \mapsto -1 \\ S \mapsto 1 \end{cases}$$

$$xy \quad \begin{cases} D \mapsto -1 \\ S \mapsto -1 \end{cases}$$

$$1 \quad \begin{cases} D \mapsto 1 \\ S \mapsto 1. \end{cases}$$

The sum of the squares of the dimensions of irreducible components is $(2)^2 + 4 = 8 = |D_8|$ as it should be for the regular representation.

If $\varrho : G \hookrightarrow \mathrm{GL}(n, \, \mathbb{F})$ is a representation the grading of $\mathbb{F}[V]_G$ is preserved by the action of G. Hence the irreducible representations of G lie in homogeneous components of $\mathbb{F}[V]_G$. If $\varrho(G)$ is generated by pseudoreflections and $\mathbb{F}$ has characteristic prime to the order of G then $\mathbb{F}[V]$ is free as a $\mathbb{F}[V]^G$-module and one has $\mathbb{F}[V] \cong \mathbb{F}[V]^G \otimes \mathbb{F}[V]_G$ as $\mathbb{F}(G)$-modules. From Molien's theorem we thus obtain

$$P((\mathbb{F}[V]_G)\Big|_S, t) = \left(\frac{\dim_{\mathbb{F}}(S)}{|G|} \prod_{i=1}^{n}(1 - t^{d_i})\right) \sum_{g \in G} \frac{\mathrm{tr}(g, \, S)}{\det(1 - gt)}.$$

Here $(\mathbb{F}[V]_G)\Big|_S$ denotes the S-isotypical subspace of $\mathbb{F}[V]_G$, which in theory allows one to read off where in $\mathbb{F}[V]_G$ summands isomorphic to S occur. In the case of the Weyl groups this has been studied in detail. See e.g. [31], [227] and section 9.3. The one-dimensional representations of G occur exactly once, and for the $\det^{-1}$ representation we have:

PROPOSITION 7.5.3 : *Let $G \hookrightarrow \mathrm{GL}(n, \ \mathbb{F})$ be a finite group generated by pseudoreflections, whose order is prime to the characteristic of $\mathbb{F}$. Then the fundamental class $[\mathbb{F}[V]_G]$ of $\mathbb{F}[V]_G$ satisfies*

$$g \cdot [\mathbb{F}[V]_G] = \det g^{-1} \cdot [\mathbb{F}[V]_G]$$

and hence $(\mathbb{F}[V]_G)_{\Sigma_{i=1}^n (d_i - 1)} \cong \det^{-1}$, where $d_1, \ldots, d_n$ are the degrees of G.

PROOF : Choose a basis $x_1, \ldots, \ x_n$ for V^* and let $\mathbb{F}[V]^G = \mathbb{F}[f_1, \ldots, \ f_n]$. By 6.5.2

$$[\mathbb{F}[V]_G] = \det \left[\frac{\partial f_i}{\partial x_j} \right]$$

is a fundamental class for $\mathbb{F}[V]_G$ so what we need to do is show that this class is a $\det^{-1}$-relative invariant for the action of G on $\mathbb{F}[V]_G$. We can actually prove a bit more and we formulate this as a lemma.

LEMMA 7.5.4 : *Let $G \hookrightarrow \mathrm{GL}(n, \ \mathbb{F})$ be a finite group. Choose a basis $\{x_1, \ldots, x_n\}$ for V (n.b. V, not V^*) and let $f_1, \ldots, f_n \in \mathbb{F}[V]^G$. Then $\det \left[\frac{\partial f_i}{\partial x_j} \right] \in \mathbb{F}[V]$ is a relative invariant for the determinant representation* $\det : G \longrightarrow \mathrm{GL}(1, \ \mathbb{F}) = \mathbb{F}^\times$.

PROOF OF LEMMA : The polynomials $f_1, \ldots, f_n$ are invariant so for $v \in V$

(♣) $f_i(g \cdot v) = f_i(v)$ $i = 1, \ldots, n$.

Consider the composition

$$V \xrightarrow{\ g\ } V \xrightarrow{\ f\ } \mathbb{F}$$

and apply the chain rule to compute partial derivatives to obtain

$$\frac{\partial f_i}{\partial x_j} = \sum_{k=1}^n \frac{\partial f_i}{\partial (g \cdot x_k)} \frac{\partial (g \cdot x_k)}{\partial x_j}$$

which we may rewrite as the matrix equation

(♠) $\left[\dfrac{\partial f_i}{\partial x_j} \right] = \left[\dfrac{\partial f_i}{\partial (g \cdot x_k)} \right] \cdot g$

since $\left[\frac{\partial (g \cdot x_k)}{\partial x_j} \right] = g$ as matrices. From (♣) it follows that

$$\left[\frac{\partial f_i}{\partial (g \cdot x_k)} \right] = \left[\frac{\partial f_i}{\partial x_k} \right]$$

so taking determinants in (♠) yields the result. □

Returning to the proof of the proposition we note that as G acts on $\mathbb{F}[V]^G$ via the contragredient representation (i.e. the dual of the inclusion $G \hookrightarrow \mathrm{GL}(V)$) we have that

$$g \cdot \det \left[\frac{\partial f_i}{\partial z_j} \right] = \det(g)^{-1} \det \left[\frac{\partial f_i}{\partial z_j} \right] \in \mathbb{F}[V]$$

By passing to the quotient $\mathbb{F}[V]_G$, the result follows. $\square$

The dual of 7.5.4 will come in handy so we record it explicitly for future reference.

LEMMA 7.5.5 : *Let $G \hookrightarrow \mathrm{GL}(n, \mathbb{F})$ be a finite group generated by pseudoreflections, whose order is prime to the characteristic of $\mathbb{F}$. Choose a basis $\{z_1, \ldots, z_n\}$ for V^* and $f_1, \ldots, f_n \in \mathbb{F}[V]^G$ with $\mathbb{F}[V]^G = \mathbb{F}[f_1, \ldots, f_n]$. Then $\det \left[\frac{\partial f_i}{\partial z_j} \right] \in \mathbb{F}[V]$ is a relative invariant for the inverse determinant representation $\det^{-1} : G \longrightarrow \mathrm{GL}(1, \mathbb{F}) = \mathbb{F}^\times$.* $\square$

By reversing viewpoints we may obtain a result of Steinberg [237] yielding another formula for the fundamental class of $\mathbb{F}[V]_G$ in the non-modular pseudoreflection case. (See also section 7.7.)

PROPOSITION 7.5.6 (R. Steinberg): *Let $G \hookrightarrow \mathrm{GL}(n, \mathbb{F})$ be a finite pseudoreflection group whose order is prime to the characteristic of $\mathbb{F}$. Choose polynomials $f_1, \ldots, f_n \in \mathbb{F}[V]^G$ with $\mathbb{F}[V]^G = \mathbb{F}[f_1, \ldots, f_n]$ and let $z_1, \ldots, z_n$ be a basis for V^*. Then there is a nonzero $c \in \mathbb{F}$ such that*

$$\det \left[\frac{\partial f_i}{\partial z_j} \right] = c \prod_{s \in s(G)} \ell_s .$$

PROOF : For convenience set $J = \det \left[\frac{\partial f_i}{\partial z_j} \right]$. Then for $s \in s(G)$ we have

$$(\lambda_s - 1)J = sJ - J = \Delta_s(J)\ell_s$$

and therefore ℓ_s divides J since $\lambda_s - 1 \neq 0$. Write $J = \ell_s^a K$ where K is relatively prime to ℓ_s and $a \in \mathbb{N}$. Since s maps both J and ℓ_s^a into multiples of themselves it does the same to K, so we may write $sK = \alpha K$ for some $\alpha \in \mathbb{F}$. Then

$$\ell_s \Delta_s(K) = sK - K = (\alpha - 1)K .$$

If $\alpha \neq 1$ then this says that ℓ_s divides K contrary to the choices of $a \in \mathbb{N}$ and K. Therefore $sK = K$.

The set of pseudoreflections $t \in G$ with $H_t = H_s$ form the non-identity elements of a cyclic subgroup C_{H_s} of G by 7.1.2. Let r be a generator of this group, and choose $\ell_r = \ell_s$. By 7.1.5 we have

$$\lambda_r^{-1}J = r(J) = \lambda_r^a \ell_r^a K = \lambda_r^a J$$

and hence $a \equiv -1 \bmod |C_{H_s}|$. (N.b. λ_r has order $|C_{H_s}|$ in $\mathbb{C}$.) Therefore

$$J = \beta \cdot \left(\ell_r^{|C_{H_s}|-1} \right)^b K = \beta \cdot \left(\prod_{H_t = H_s} \ell_t \right)^b K$$

where $b = a/|C_{H_s}|$ and $\beta \neq 0 \in \mathbb{F}$. Since the linear forms $\ell_{t'}$ and $\ell_{t''}$ are relatively prime when $H_{t'} \neq H_{t''}$ we conclude that $\mathcal{L} = \prod_{s \in s(G)} \ell_s = \prod_{H \in \mathcal{H}(G)} \ell_t^{|C_H|-1}$ divides J, where for each $H \in \mathcal{H}(G)$ we choose one pseudo-reflection $\ell_t \in C_H$. Since both $\mathcal{L}$ and J are nonzero elements of the same degree the result follows. $\square$

Since the set of zeros of $\prod_{s \in s(G)} \ell_s$ is precisely the set of hyperplanes $\mathcal{H}(G)$, this says that $\det \left[\frac{\partial f_i}{\partial z_j} \right]$ is nonzero on the complement of the set $\bigcup_{H \in \mathcal{H}(G)} H \subset V$. The topological space $V \smallsetminus \bigcup_{H \in \mathcal{H}(G)} H$ has been extensively studied. See [178] and the references there.

7.6   Relative Invariants

For a finite group G, a representation $\varrho : G \hookrightarrow \mathrm{GL}(n, \mathbb{F})$ and a linear representation $\chi : G \longrightarrow \mathbb{F}^\times$ we have the $\mathbb{F}[V]^G$-module $\mathbb{F}[V]_\chi^G$ of χ-relative invariants. For a pseudoreflection group in the non-modular case these take a particularly simple form and were described by Stanley [233] and Springer [231] in characteristic zero (see also [248]) and in the nonmodular case in [41].

If $\varrho : G \hookrightarrow \mathrm{GL}(n, \mathbb{F})$ is a pseudoreflection representation and $|G| \in \mathbb{F}^\times$ then $\mathbb{F}[V]$ is a free $\mathbb{F}[V]^G$-module and

$$(*) \qquad\qquad \mathbb{F}[V] \cong \mathbb{F}[V]^G \otimes \mathbb{F}[V]_G$$

as $\mathbb{F}(G)$-modules by 7.4.3. By Chevalley's theorem 7.5.2 $\mathbb{F}[V]_G$ is the regular representation of G. If $\chi : G \longrightarrow \mathbb{F}^\times$ is a linear character of G then $(\mathbb{F}[V]_G)_\chi^G$ is 1-dimensional since every irreducible representation of G occurs in the regular representation as often as its dimension. From $(*)$ it therefore follows:

PROPOSITION 7.6.1 : Let $\varrho : G \hookrightarrow \mathrm{GL}(n, \mathbb{F})$ be a representation of a finite group and $\chi : G \longrightarrow \mathbb{F}^\times$ a one-dimensional representation. Assume that $\varrho(G)$ is generated by pseudoreflections and the order of G is prime to the characteristic of $\mathbb{F}$. Then $\mathbb{F}[V]_\chi^G \cong \mathbb{F}[V]^G \cdot f_\chi$, where f_χ lifts a basis element of $(\mathbb{F}[V]_G)_\chi^G$ to $\mathbb{F}[V]_\chi^G$. $\square$

The isomorphism given in the preceding proposition can be made more explicit. The first step is to construct a χ-relative invariant. To this end recall from section 7.1 that

$$s_\Delta(G) = \coprod_{H \in \mathcal{H}(G)} C_H \smallsetminus \{1\}$$

Let $s_H \in C_H$ be a generator and $\chi(s_H) = \det^{-k_s}(s_H)$ and set

$$(\star) \qquad\qquad f_\chi = \prod_{H \in \mathcal{H}(G)} \ell_{s_H}^{k_s}.$$

LEMMA 7.6.2 : *With the preceding notations*

$$s f_\chi = \chi(s) f_\chi$$

for all $s \in s_\Delta(G)$.

PROOF : This follows as in 7.1.7 and the definition of f_χ. $\square$

LEMMA 7.6.3 : *With the preceding notations, any* $f \in \mathbb{F}[V]_\chi^G$ *is divisible by* f_χ.

PROOF : For $s \in s_\Delta(G)$ with $k_s \neq 0$ we have

$$(\chi(s_H) - 1)f = s_H f - f = \Delta_{s_H}(f) \ell_{s_H}$$

and the result follows since $\chi(s_H) \neq 1$. $\square$

THEOREM 7.6.4 (R. P. Stanley, T. A. Springer): *Let* $\varrho : G \hookrightarrow \mathrm{GL}(n, \mathbb{F})$ *be a representation of a finite group* G *and* $\chi : G \longrightarrow \mathbb{F}^\times$ *a one-dimensional representation. Assume that* $\varrho(G)$ *is generated by pseudoreflections and that* G *has order prime to the characteristic of* $\mathbb{F}$. *Then* $\mathbb{F}[V]_\chi^G \cong \mathbb{F}[V]^G \cdot f_\chi$, *i.e.* $\mathbb{F}[V]_\chi^G$ *is a free module of rank 1 over* $\mathbb{F}[V]^G$ *with generator* f_χ *given by the formula* $(\star)$ *preceding lemma 7.6.2.*

PROOF : Since $|G| \in \mathbb{F}^\times$, $s_\Delta(G) = s(G)$, so by 7.6.2 f_χ is a χ-relative invariant. By 7.6.3 if $f \in \mathbb{F}[V]_\chi^G$ then f_χ divides f so $f/f_\chi \in \mathbb{F}[V]^G$. Therefore the map of degree $\deg(f_\chi)$

$$\varphi_\chi : \mathbb{F}[V]^G \longrightarrow \mathbb{F}[V]_\chi^G$$

given by $\varphi_\chi(h) = h \cdot f_\chi$ is surjective. Since $\mathbb{F}[V]$ is an integral domain, φ_χ is also monic and hence an isomorphism. $\square$

There are two special cases worthy of consideration: namely $\chi = \det$ and $\chi = \det^{-1}$. We consider these in more detail.

PROPOSITION 7.6.5 : *Let* $\varrho : G \hookrightarrow \mathrm{GL}(n, \mathbb{F})$ *be a representation of a finite group* G *and* $\det^{-1} : G \longrightarrow \mathbb{F}^{\times}$ *the inverse determinant representation. Assume that* $\varrho(G)$ *is generated by pseudoreflections and has order prime to the characteristic of* $\mathbb{F}$*. Then*

$$\mathbb{F}[V]^G_{\det^{-1}} \cong \mathbb{F}[V]^G \cdot \det \left[\frac{\partial f_i}{\partial z_j} \right]$$

where $\mathbb{F}[V]^G = \mathbb{F}[f_1, \ldots, f_n]$ *and* $z_1, \ldots, z_n$ *is a basis for* V^**. In addition there is a* $c \neq 0 \in \mathbb{F}$ *such that*

$$\det \left[\frac{\partial f_i}{\partial z_j} \right] = c \prod_{s \in s(G)} \ell_s$$

and $\det \left[\frac{\partial f_i}{\partial z_j} \right]$ *has degree* $|s(G)|$*.*

PROOF : This follows from 7.6.4 and 7.5.6. □

PROPOSITION 7.6.6 : *Let* $\varrho : G \hookrightarrow \mathrm{GL}(n, \mathbb{F})$ *be a representation of a finite group and* $\det : G \longrightarrow \mathbb{F}^{\times}$ *the determinant representation. Assume that* $\varrho(G)$ *is generated by pseudoreflections and has order prime to the characteristic of* $\mathbb{F}$*. Then*

$$\mathbb{F}[V]^G_{\det} \cong \mathbb{F}[V]^G \cdot \prod_{H_s \in \mathcal{H}(G)} \ell_s$$

where the product is taken over one pseudoreflection $s \in C(H_s, R_s) \smallsetminus \{1\}$ *for each reflecting hyperplane* $H_s \in \mathcal{H}(G)$*. The degree of* $\prod_{H_s \in \mathcal{H}(G)} \ell_s$ *is* $|\mathcal{H}(G)|$*, the number of reflecting hyperplanes in* G*.*

PROOF : This follows from 7.6.4 since in the notation preceding $(\star)$ each $k_s = 1$. □

EXAMPLE 1 : Consider the dihedral group D_{2k} of order $2k$ represented in $\mathrm{GL}(2, \mathbb{R})$ as the group of symmetries of a regular k-gon centered at the origin. The group D_{2k} in this representation is generated by the matrices

$$D = \begin{bmatrix} \cos \frac{2\pi}{k} & -\sin \frac{2\pi}{k} \\ \sin \frac{2\pi}{k} & \cos \frac{2\pi}{k} \end{bmatrix} \qquad S = \begin{bmatrix} 1 & 0 \\ 0 & -1 \end{bmatrix}$$

where D is a rotation through $2\pi/k$ radians and S a reflection in an axis. In section 5.5 we found the invariants of this group. Recall that $\mathbb{R}[x, y]^{D_{2k}} = \mathbb{R}[f_1, f_2]$ where $f_1 = x^2 + y^2$ is an invariant quadratic polynomial, and if we introduce the complex variable $z = x + iy$ $f_2 = \mathfrak{Re}(z^k)$ is a polynomial[6] in x and y that is an invariant of degree k.

---

[6] As usual $\mathfrak{Re}(-)$, resp. $\mathfrak{Im}(-)$, denote the real, resp. imaginary, part of $-$.

The determinant $\det : D_{2k} \longrightarrow \{\pm 1\}$ provides a one-dimensional representation whose kernel is the rotation subgroup generated by the matrix D. There are k reflections in G, namely $\{D^i S \mid i = 0, \ldots, k-1\}$. Thus from 7.6.5 we obtain

$$f_{\det} = \det \begin{bmatrix} \frac{\partial f_1}{\partial x} & \frac{\partial f_2}{\partial x} \\ \frac{\partial f_1}{\partial y} & \frac{\partial f_2}{\partial y} \end{bmatrix} = -2k\Im\mathrm{m}(z^k) \neq 0 \,.$$

and $\mathbb{R}[x, \, y]_{\det}^{D_{2k}} = \mathbb{R}[f_1, \, f_2] \cdot f_{\det}$. For example when $k = 4$ we saw in section 3.2 that $\mathbb{R}[x, \, y]^{D_8} = \mathbb{R}[x^2 + y^2, \, x^2 y^2]$ so

$$f_{\det} = \det \begin{bmatrix} 2x & 2xy^2 \\ 2y & 2x^2 y \end{bmatrix} = 4x^3 y - 4xy^3$$

and $\mathbb{R}[x, \, y]_{\det}^{D_8} = \mathbb{R}[x^2 + y^2, \, x^2 y^2] \cdot (4x^3 y - 4xy^3)$.

If $\varrho : G \hookrightarrow \mathrm{GL}(n, \, \mathbb{F})$ is a representation and $\chi : G \longrightarrow \mathbb{F}^\times$ a one-dimensional representation we set $G_\chi = \ker(\chi) < G$. We then have $\mathbb{F}[V]^G \subset \mathbb{F}[V]^{G_\chi}$. In the case when $\varrho(G)$ is generated by pseudoreflections and the order of G is prime to the characteristic of $\mathbb{F}$ this extension of rings takes a very simple form.

THEOREM 7.6.7 : *Let $\varrho : G \hookrightarrow \mathrm{GL}(n, \, \mathbb{F})$ be a representation of a finite group G and $\chi : G \longrightarrow \mathbb{F}^\times$ a one-dimensional representation. Assume that $\varrho(G)$ is generated by pseudoreflections and that G has order prime to the characteristic of $\mathbb{F}$. Then*

$$\mathbb{F}[V]^{G_\chi} \cong \bigoplus_{i=0}^{m-1} \mathbb{F}[V]^G_{\chi^i} \cong \bigoplus_{i=0}^{m-1} \mathbb{F}[V]^G \cdot f_{\chi^i}$$

where the f_{χ^i} are given by

$$f_\chi = \prod_{C(H_s, \, R_s)} \prod_{t \in C(H_s, \, R_s) \smallsetminus \{1\}} \ell_t^{k_s(i)} \,,$$

m is the order of χ, and $k_s(i)$ is the residue of ik_s modulo m.

PROOF : The subgroup G_χ of G is normal and the quotient is isomorphic to $\mathbb{Z}/m$ where m is the order of χ and is prime to the characteristic of $\mathbb{F}$. The order of G_χ is also prime to the characteristic of $\mathbb{F}$ so $\mathbb{F}[V]^{G_\chi}$ is an $\mathbb{F}[V]^G$-module direct summand in $\mathbb{F}[V]$, a direct summand splitting being given by the transfer. Thus by 6.7.13 $\mathbb{F}[V]^{G_\chi}$ is a free finitely generated $\mathbb{F}[V]^G$-module. Since the corresponding extension of quotient fields $FF(\mathbb{F}[V]^G) \subset FF(\mathbb{F}[V]^{G_\chi})$ is Galois of degree m, $\mathbb{F}[V]^{G_\chi}$ has rank m as $\mathbb{F}[V]^G$-module. The modules $\mathbb{F}[V]^G_{\chi^i}$ for $i = 0, \ldots, m-1$ are all contained in $\mathbb{F}[V]^{G_\chi}$ and their sum is direct since one-dimensional representations are irreducible. Thus the inclusion $\overset{m-1}{\underset{i=0}{\oplus}} \mathbb{F}[V]^G_{\chi^i} \subset \mathbb{F}[V]^{G_\chi}$ is an isomorphism. $\square$

REMARK : Let $\varrho : G \hookrightarrow \mathrm{GL}(n, \; \mathbb{F})$ be a representation of a finite group G and $\chi : G \longrightarrow \mathbb{F}^\times$ a one-dimensional representation. Assume that $\varrho(G)$ is generated by pseudoreflections and that G has order prime to the characteristic of $\mathbb{F}$. Then the theorem of Eagon-Hochster 6.7.8 says that $\mathbb{F}[V]^{G_\chi}$ is Cohen-Macaulay, i.e. is free and finitely generated over a polynomial subalgebra. The theorem 7.6.7 gives an explicit system of parameters: the theorem of Shephard-Todd 7.4.1 says $\mathbb{F}[V]^G = \mathbb{F}[f_1, \ldots, f_n]$ and by 7.6.7 $f_1, \ldots, f_n$ are a system of parameters for $\mathbb{F}[V]^{G_\chi}$ and $1, f_\chi, \ldots, f_{\chi^{m-1}}$ generate $\mathbb{F}[V]^{G_\chi}$ freely as an $\mathbb{F}[f_1, \ldots, f_n]$-module.

Here is a special case to close out this section.

COROLLARY 7.6.8 : Let $\varrho : G \hookrightarrow \mathrm{GL}(n, \; \mathbb{F})$ be a representation of a finite group G and $\chi : G \longrightarrow \mathbb{F}^\times$ a linear representation. Assume that $\varrho(G)$ is generated by pseudoreflections and that G has order prime to the characteristic of $\mathbb{F}$. If there is a pseudoreflection s of order $m = |\chi|$, where $|\chi|$ denotes the order of χ, such that $\chi(s) \in \mathbb{F}^\times$ also has order m then $\mathbb{F}[V]^G \subset \mathbb{F}[V]^{G_\chi}$ is a radical extension, i.e., there is an element $h_\chi \in \mathbb{F}[V]^{G_\chi}$ with $h_\chi^m = f_\chi \in \mathbb{F}[V]^G$ and $\mathbb{F}[V]^{G_\chi}$ is generated as an algebra by $\mathbb{F}[V]^G$ and h_χ, so we may write

$$\mathbb{F}[V]^{G_\chi} \cong \mathbb{F}[V]^G(\sqrt[m]{f_\chi}) \, .$$

PROOF : This follows from 7.6.7 after noting that under the stated hypotheses $f_{\chi^i} = (f_\chi)^i$. $\quad \Box$

EXAMPLE 2 : Consider the group of rotations $\mathcal{I}$ and symmetries $\mathbf{I}_3$ of a regular icosohedron. The group $\mathbf{I}_3 \subset \mathrm{GL}(3, \; \mathbb{R})$ is generated by the 15 reflections in the hyperplanes through the origin orthogonal to the axes of rotation joining the midpoints of opposite edges of the icosohedron. The group $\mathcal{I}$ is the kernel of $\det : \mathbf{I}_3 \longrightarrow \mathbb{Z}/2$. The invariants of $\mathbf{I}_3$ are a polynomial algebra by theorem 7.4.1 on forms of degree $d_1 \leq d_2 \leq d_3$ which satisfy

$$d_1 d_2 d_3 = |\mathbf{I}_3| = 120$$
$$(d_1 - 1) + (d_2 - 1) + (d_3 - 1) = |s(\mathbf{I}_3)| = 15 \, .$$

Since $\mathbf{I}_3$ acts orthogonally the square of the norm is invariant so $d_1 = 2$ and we obtain

$$d_2 d_3 = 60$$

$$d_2 + d_3 = 16 \, .$$

The only integral solution to these equations with $d_2 \leq d_3$ is $d_2 = 6$, $d_3 = 10$. Therefore

$$\mathbb{R}[x, \; y, \; z]^{\mathbf{I}_3} = \mathbb{R}[f, \; h, \; k]$$

with $\deg(f) = 2$, $\deg(h) = 6$, $\deg(k) = 10$. By theorem 7.6.4

$$\mathbb{R}[x, \, y, \, z]_{\det}^{\mathbf{I}_3} = \mathbb{R}[f, \, h, \, k] \cdot r$$

where $r = \prod\limits_{s \in s(\mathbf{I}_3)} \ell_s$, so $\deg(r) = 15$. Since $\mathcal{I} = \ker\{\det : \mathbf{I}_3 \longrightarrow \mathbb{Z}/2\}$ the polynomial r belongs to $\mathbb{R}[x, \, y, \, z]^{\mathcal{I}}$, the square of r belongs to $\mathbb{R}[x, \, y, \, z]^{\mathbf{I}_3}$ and f, h, k and r generate $\mathbb{R}[x, \, y, \, z]^{\mathcal{I}}$. Finally by 7.6.8

$$\mathbb{R}[x, \, y, \, z]^{\mathcal{I}} = \mathbb{R}[x, \, y, \, z]^{\mathbf{I}_3} \oplus \mathbb{R}[x, \, y, \, z]_{\det}^{\mathbf{I}_3} \cdot r.$$

Therefore the Poincaré series of $\mathbb{R}[x, \, y, \, z]^{\mathcal{I}}$ is

$$P(\mathbb{R}[x, \, y, \, z]^{\mathcal{I}}, \, t) = \frac{1 + t^{15}}{(1 - t^2)(1 - t^6)(1 - t^{10})}$$

verifying the computation of Molien (see section 4.3 example 5).

For a further study of examples of this type see [106].

7.7   Automorphisms of Rings of Coinvariants

If $\varrho : G \hookrightarrow \mathrm{GL}(n, \, \mathbb{F})$ is a representation of a finite group G we denote by $N_{\mathrm{GL}(n, \, \mathbb{F})}(\varrho)$ the normalizer of $\varrho(G)$ in $\mathrm{GL}(n, \, \mathbb{F})$. For $T \in \mathrm{GL}(n, \, \mathbb{F})$ there is an induced automorphism (also denoted by T) of $\mathbb{F}[V]$ given by

$$(Tf)(v) = f(T^{-1}(v)).$$

If $T \in N_{\mathrm{GL}(n, \, \mathbb{F})}(\varrho)$ one readily verifies that $T(\mathbb{F}[V]^G) \subset \mathbb{F}[V]^G$ so that T induces an automorphism of $\mathbb{F}[V]_G$. On the other hand, an automorphism α of the algebra $\mathbb{F}[V]_G$ lifts to an algebra automorphism of $\mathbb{F}[V]$ and if G has no linear invariants the lift is unique (since in this case the quotient map $\mathbb{F}[V] \longrightarrow \mathbb{F}[V]_G$ is an isomorphism on the homogeneous components of degree 1). By taking the dual of the restriction of the lift to the linear invariants we receive an element of $\mathrm{GL}(V)$. Thus there are homomorphisms

$$N_{\mathrm{GL}(n, \, \mathbb{F})}(\varrho) \overset{\mu}{\longrightarrow} \mathrm{Aut}(\mathbb{F}[V]_G) \overset{\lambda}{\longrightarrow} \mathrm{GL}(V)$$

where $\mathrm{Aut}(\mathbb{F}[V]_G)$ denotes the group of algebra automorphisms of $\mathbb{F}[V]_G$.

For real reflection groups Papadima [179] and Sperlich [230] have shown that $\mathrm{Im}\lambda \subset N_{\mathrm{GL}(n, \, \mathbb{F})}(\varrho)$ and hence μ is an isomorphism. By reorganizing the proof we are able to achieve the same result for all pseudoreflection groups provided only that the field is large enough (compare e.g. 3.1.10).

We suppose that $\varrho : G \hookrightarrow \mathrm{GL}(n, \, \mathbb{F})$ is a finite pseudoreflection group whose order is prime to the characteristic of $\mathbb{F}$. Then $\mathbb{F}[V]^G = \mathbb{F}[f_1, \ldots, f_n]$ and

$\mathbb{F}[V]_G$ is a Poincaré duality algebra with fundamental class $J = \det[\frac{\partial f_i}{\partial z_j}]$, where $z_1, \ldots, z_n \in V^*$ is a basis. Let the degree of J be $d = |s(G)|$. We begin with an alternate description of a fundamental class for $\mathbb{F}[V]_G$. To this end choose an isomorphism $\xi : V \longrightarrow V^*$ such that $\xi(gv) = g^{-1}\xi(v)$. For $\ell \in V^*$, regarded as a linear polynomial on V, we have

$$\ell^d = \bar{f}(\ell) \cdot J \in \mathbb{F}[V]_G$$

where $\bar{f}(\ell) \in \mathbb{F}$. Composing with ξ we get

$$f : V \xrightarrow{\xi} V^* \xrightarrow{\bar{f}} \mathbb{F}$$

which is a polynomial function.

LEMMA 7.7.1 : *With the preceding notations the function f is a polynomial function of degree d on V, hence an element of $\mathbb{F}[V]$ of degree d.*

PROOF : Since ξ is a liner transformation it will suffice to show that $\bar{f}$ is a homogeneous polynomial function on V^* of degree d. By definition $\bar{f}(\ell) \cdot J = \ell^d$ for any $\ell \in V^*$, so

$$\bar{f}(\lambda\ell) \cdot J = (\lambda\ell)^d = \lambda^d \ell^d = \lambda^d \bar{f}(\ell) J.$$

Divide by J to get $\bar{f}(\lambda\ell) = \lambda^d \bar{f}(\ell)$. $\square$

LEMMA 7.7.2 : *With the preceding notations $gf = (\det g)^{-1}f$, i.e. f is a $\det^{-1}$-relative invariant.*

PROOF : We have

$$\begin{aligned}
gf(v) \cdot J &= f(g^{-1}v) \cdot J = \xi(g^{-1}v)^d \\
&= (g\xi)(v)^d = (g\xi)^d(v) = (g \cdot (\xi^d))(v) \\
&= g \cdot f(v)J = f(v)gJ = \det(g^{-1})f(v)J
\end{aligned}$$

and the result follows. $\square$

From 7.6.5 we see that J divides f and, since they are of the same degree, $f = bJ$ for some $b \in \mathbb{F}$. To see that $b \neq 0$ it suffices to recall that by the proof of 2.4.1 J may be written as a sum of d-th powers provided $d! \neq 0 \in \mathbb{F}$. Thus we have shown:

LEMMA 7.7.3 : *Let $\varrho : G \hookrightarrow \mathrm{GL}(n, \mathbb{F})$ be a finite group generated by pseudoreflections. Assume $|G|$ and $|s(G)|!$ are relatively prime to the characteristic of $\mathbb{F}$. Then with the preceding notations $f = bJ$ for some constant $b \neq 0 \in \mathbb{F}$.* $\square$

If the field $\mathbb{F}$ is large enough we thus have three ways to construct a fundamental class of $\mathbb{F}[V]_G$ and a generator for the $\det^{-1}$-relative invariants $\mathbb{F}[V]_{\det^{-1}}^G$, namely f, J and $\prod_{s \in s(G)} \ell_s$. They are scalar multiples of each other so $f = b \cdot J = bc \prod_{s \in s(G)} \ell_s$ where $bc \neq 0 \in \mathbb{F}$.

LEMMA 7.7.4 : *Suppose that $\alpha \in \mathrm{Aut}(\mathbb{F}[V]_G)$ where the action of G on V^* has no fixed points. Then α is induced by a unique automorphism of V and hence lifts uniquely to an algebra automorphism of $\mathbb{F}[V]$.*

PROOF : Take the dual of α restricted to the linear elements, which since $V^{* \, G} = \{0\}$ defines the required automorphism of V. $\square$

LEMMA 7.7.5 : *If $\alpha \in \mathrm{Aut}(\mathbb{F}[V]_G)$ then $\alpha f = af$ for some $a \neq 0 \in \mathbb{F}$.*

PROOF : In $\mathbb{F}[V]_G$ we have

$$
\begin{aligned}
(\alpha f)(v) \cdot J &= f(\alpha^{-1}v) \cdot J = (\xi(\alpha^{-1}v))^d \\
&= (\alpha \xi(v))^d = \alpha \xi^d(v) \\
&= \alpha(f(v)J) = f(v)a \cdot J
\end{aligned}
$$

where $\alpha(J) = a \cdot J$, $a \neq 0 \in \mathbb{F}$. $\square$

THEOREM 7.7.6 : *Let $\varrho : G \hookrightarrow \mathrm{GL}(n, \mathbb{F})$ be a finite pseudoreflection group with $V^{* \, G} = \{0\}$ and both $|G|$ and $|s(G)|!$ relatively prime to the characteristic of $\mathbb{F}$. Then*

$$
\mathrm{Aut}(\mathbb{F}[V]_G) \cong N_{\mathrm{GL}(n, \, \mathbb{F})}(\varrho) \, .
$$

PROOF : We have already seen that there are maps

$$
N_{\mathrm{GL}(n, \, \mathbb{F})}(\varrho) \xrightarrow{\mu} \mathrm{Aut}(\mathbb{F}[V]_G) \xrightarrow{\lambda} \mathrm{GL}(V)
$$

with $\lambda \circ \mu = 1$ and λ injective. It therefore suffices to show that $\lambda(\alpha) \in N_{\mathrm{GL}(n, \, \mathbb{F})}(\varrho)$ for all $\alpha \in \mathrm{Aut}(\mathbb{F}[V]_G)$. To this end recall that we have shown:

$$
f \neq 0
$$
$$
f \in \mathbb{F}[V]_{\det^{-1}}^G
$$
$$
f = b \cdot J \qquad b \neq 0 \in \mathbb{F}.
$$

So by 7.5.6 we have

$$
f = b \cdot J = bc \prod_{s \in s(G)} \ell_s
$$

where $c \neq 0 \in \mathbb{F}$. The roots of ℓ_s are the hyperplanes $\{H_s\} = \mathcal{H}_\varrho(G)$. Therefore α permutes the hyperplanes of G. By 7.1.1 we have

$$\alpha(H_s) = H_{\alpha s \alpha^{-1}} \in \mathcal{H}_\varrho(G) \qquad s \in s(G)$$

so $\alpha s \alpha^{-1} \in s(G) \subset G$ for all $s \in s(G)$. Since $s(G)$ generates G this says $\alpha G \alpha^{-1} = G$ and so $\alpha \in N_{\mathrm{GL}(n, \ \mathbb{F})}(\varrho)$. $\square$

For $\varrho : G \hookrightarrow \mathrm{GL}(n, \ \mathbb{F})$, the field $\mathbb{F}$ is sufficiently large in the sense needed in 7.7.6 provided for example the characteristic of $\mathbb{F}$ is zero or larger than $|G|$. Notice that 7.7.6 implies that an automorphism $\alpha \in \mathrm{Aut}(\mathbb{F}[V]_G)$ lifts (uniquely) to an automorphism of $\mathbb{F}[V]$ *that maps the subalgebra* $\mathbb{F}[V]^G$ *into itself*. In other words α induces an automorphism of the coexact sequence

$$1 \longrightarrow \mathbb{F}[V]^G \longrightarrow \mathbb{F}[V] \longrightarrow \mathbb{F}[V]_G \longrightarrow 1$$

of graded commutative algebras over $\mathbb{F}$.

Chapter 8
Modular Invariants

By modular invariant theory[1] we understand the study of invariants of finite groups over fields of nonzero characteristic, where the characteristic may divide the order of the group. If the group is a finite p-group and the field has characteristic p we speak of purely modular invariant theory. The invariant theory of finite groups over finite fields presents a number of special features and problems, and this chapter examines some of these. (See also chapter 10 and 11 where we pursue others.)

If $\varrho : G \hookrightarrow \mathrm{GL}(n,\ \mathrm{I\!F})$ is a representation of a finite group whose order is divisible by the characteristic of $\mathrm{I\!F}$ then the averaging operator derived from the transfer (see section 2.4) is no longer defined. This loss is partially compensated by the fact that for a Galois field $\mathrm{I\!F}$ the general linear group $\mathrm{GL}(n,\ \mathrm{I\!F})$ is a finite group, and the ring of invariants $\mathrm{I\!F}[V]^{\mathrm{GL}(V)}$ is nontrivial and contained in $\mathrm{I\!F}[V]^G$ for any $\varrho : G \hookrightarrow \mathrm{GL}(n,\ \mathrm{I\!F})$, where as usual $V = \mathrm{I\!F}^n$. Thus there exist *universal invariants*, i.e. invariants present in the ring of invariants of all finite group over a finite field.

Let $\mathrm{I\!F}_q$ denote the Galois field with $q = p^s$ elements. The general linear group $\mathrm{GL}(n,\ \mathrm{I\!F}_q)$ is a finite group of order $(q^n - 1)(q^n - q) \cdots (q^n - q^{n-1})$ which acts on $V = \mathrm{I\!F}_q^n$. We denote by $\mathbf{D}^*(n)$ the ring of invariants $\mathrm{I\!F}_q[V]^{\mathrm{GL}(n,\ \mathrm{I\!F}_q)}$ and call it the **Dickson algebra** after L.E. Dickson who originally computed it in 1911 [68]. For any finite group G acting on V the ring of invariants $\mathrm{I\!F}_q[V]^G$ is a finite extension of $\mathrm{I\!F}_q[V]^{\mathrm{GL}(n,\ \mathrm{I\!F}_q)}$. The Dickson algebra therefore plays a key role in the invariant theory of finite groups in characteristic p. In section 5.6 we worked out the invariants of $\mathrm{GL}(2,\ \mathrm{I\!F}_p)$ and our first goal will be to describe $\mathrm{I\!F}_q[V]^{\mathrm{GL}(n,\ \mathrm{I\!F}_q)}$ in the general case. We do so following Tamagawa [246]. As in section 5.6 we then consider the special linear groups and the family of subgroups $\mathrm{SL}_m(n,\ \mathrm{I\!F}_q)$ lying between $\mathrm{SL}(n,\ \mathrm{I\!F}_q)$ and $\mathrm{GL}(n,\ \mathrm{I\!F}_q)$.

---

[1] The term modular invariant is used in the classical literature in a different sense. A polynomial $f \in \mathrm{I\!F}[V]$ is a modular invariant for the group G acting on V in the classical sense if $f(gv) = f(v)$ for all $g \in G$ and $v \in V$, *where, however f is restricted to be a function from V to $\mathrm{I\!F}$ and not $\mathrm{I\!F}$, so may not be an invariant in our sense.*

For other proofs of Dickson's theorem see [239], [224] and chapter 10 section 10.6.

The groups generated by pseudoreflections have played a significant role in the development of invariant theory. For a field $\mathbb{F}$ of finite characteristic a pseudoreflection $s \in \mathrm{GL}(n, \mathbb{F})$ need not be diagonalizable, and the non-diagonalizable pseudoreflections play an important role. We will examine them in[2] section 8.2 in connection with certain families of subgroups of $\mathrm{GL}(n, \mathbb{F})$ whose invariant theory was studied by Nakajima and, independently, by Landweber and Stong.

In the purely modular case, i.e. for p-groups P in characteristic p, the fixed point set V^P cannot consist of the zero vector alone. (For Galois fields this is an easy consequence of the class equation.) The dimension of the fixed point set is a measure of how complicated the action of P on V is: the higher the dimension the simpler the action and this is nicely illustrated in section 8.2, [78] and [166].

If the characteristic of the field divides the order of the group, then the ring of invariants $\mathbb{F}[V]^G$ need not be Cohen-Macaulay. The Cohen-Macaulay property for $\mathbb{F}[V]^G$ in characteristic p is controlled by the p-Sylow subgroup of G. We examine this and several related facts in section 8.3

If $\varrho : G \hookrightarrow \mathrm{GL}(n, \mathbb{F})$ is generated by pseudoreflections, but the characteristic of $\mathbb{F}$ divides the order of G, then the ring of invariants $\mathbb{F}[V]^G$ need not be a polynomial algebra. In section 8.4 we examine a bit of what can be said about the rings of invariants of groups generated by pseudoreflections in the modular case. This is a current area of active research, see e.g. [43] – [49], [117], [153] – [161] and [164] – [168].

8.1   The Dickson Algebra

Fix a prime p and a power $q = p^s$ of it. As usual denote by $\mathbb{F}_q$ the Galois field with q elements and let $\mathbf{D}^*(n) = \mathbb{F}_q[V]^{\mathrm{GL}(n,\ \mathbb{F}_q)}$ denote the **Dickson algebra** of degree n. We will follow T. Tamagawa [246] in this section in delineating the structure of $\mathbf{D}^*(n)$ aided by a suggestion of N. Jacobson who told us of Ore's paper [175].

Denote by $\mathbb{F}$ an algebraically closed field of characteristic p of sufficiently large transcendence degree[3] over $\mathbb{F}_q$ where $q = p^s$.

---

[2] Portions of this and section 8.2 are based on [131] and [222]. I am indebted to P. S. Landweber and R. E. Stong for permission to include this material here.
[3] It will be apparent from the context how large the transcendence degree must be to make the various choices we must make.

DEFINITION : *A polynomial $p(x) \in \mathbb{F}[x]$ is called a q-**polynomial** if*
$$p(x) = a_0 x + a_1 x^q + \cdots + a_n x^{q^n}$$
for some $n \in \mathbb{N}$.

If $\lambda \in \mathbb{F}$ and $p(x) \in \mathbb{F}[x]$ is a q-polynomial then
$$p(\lambda) = a_0 \lambda + a_1 \lambda^q + \cdots + a_n \lambda^{q^n}$$
and since $\lambda \mapsto \lambda^q$ (the Frobenius) is a linear map we get
$$p(\lambda' + \lambda'') = p(\lambda') + p(\lambda'') \qquad \lambda', \ \lambda'' \in \mathbb{F}.$$
If $\mathbb{F}_q < \mathbb{F}$ is the field of q elements, then $\lambda^q = \lambda$ for all $\lambda \in \mathbb{F}_q$ and therefore
$$p(\lambda \cdot \mu) = \lambda p(\mu) \qquad \lambda \in \mathbb{F}_q, \quad \mu \in \mathbb{F}$$
so $p : \mathbb{F} \longrightarrow \mathbb{F}$ defines an $\mathbb{F}_q$-linear transformation, and we get:

LEMMA 8.1.1 : *If $p(x) \in \mathbb{F}[x]$ is a q-polynomial, then the set of zeros $V_{p(x)}$ of $p(x)$ in $\mathbb{F}$, i.e.*
$$V_{p(x)} = \left\{ \lambda \in \mathbb{F} \mid p(\lambda) = 0 \right\},$$
is a finite dimensional vector space over the Galois field $\mathbb{F}_q$. □

Here is a simple example of a p-polynomial. Based on this example it is easy to construct q-polynomials.

LEMMA 8.1.2 : *Let p be a prime and $q = p^n$. The polynomial*
$$\Delta_n(x) = \prod_{\lambda \in \mathbb{F}_q} (x + \lambda) \in \mathbb{F}_q[x]$$
is a p-polynomial. Thus we may write
$$\Delta_n(x) = x^{p^n} + d_1 x^{p^{n-1}} + \cdots + d_n x.$$

PROOF : We prove this by induction on n. For $n = 1$ we have $\Delta_n(x) = x^p - x$ which is a p-polynomial. Suppose that the result is true for $m < n$. Choose $z \in \mathbb{F}_q \smallsetminus \mathbb{F}_{p^{n-1}}$ and note
$$\Delta_n(x) = \prod_{v \in \mathbb{F}_q} (x + v) = \prod_{\substack{v \in \mathbb{F}_{p^{n-1}} \\ \mu \in \mathbb{F}_p}} (x + v + \mu z)$$
$$= \prod_{\mu \in \mathbb{F}_p} \prod_{v \in \mathbb{F}_{p^{n-1}}} (x + v + \mu z)$$
$$= \prod_{\mu \in \mathbb{F}_p} \Delta_{n-1}(x + \mu z) = \prod_{\mu \in \mathbb{F}_p} (\Delta_{n-1}(x) + \mu \Delta_{n-1}(z))$$
$$= \Delta_{n-1}(x)^p - \Delta_{n-1}(z)^{p-1} \Delta_{n-1}(x)$$

and the result follows. $\square$

Since $\mathbb{F}$ has sufficiently large transcendence degree over $\mathbb{F}_q$ we may choose elements $y_1, \ldots, y_n \in \mathbb{F}$ which are algebraically independent over $\mathbb{F}_q$. We let

$$\varphi(x) = x^{q^n} + y_1 x^{q^{n-1}} + \cdots + y_n x \in \mathbb{F}[x].$$

This is a q-polynomial and hence its set of zeros, denoted by V, is a finite dimensional $\mathbb{F}_q$ vector space by 8.1.1.

LEMMA 8.1.3 : *With the preceding notations V is an n-dimensional vector space over $\mathbb{F}_q$.*

PROOF : We have

$$\frac{d}{dx}\varphi(x) = y_n \neq 0$$

so there are no multiple roots. Therefore there are q^n roots. By 8.1.1 V is a vector space over $\mathbb{F}_q$ so its dimension is determined by the number of elements it contains. $\square$

Continuing with the previous discussion let $x_1, \ldots, x_n \in V$ be a basis for V as an $\mathbb{F}_q$ vector space and consider the field extension $\mathbb{F}_q(x_1, \ldots, x_n) \mid \mathbb{F}_q$. Note that $y_1, \ldots, y_n$ satisfy

$$(\star) \qquad \varphi(x) = \prod_{v \in V}(x + v) = x^{q^n} + y_1 x^{q^{n-1}} + \cdots + y_n x,$$

which implies that $y_1, \ldots, y_n \in \mathbb{F}_q(x_1, \ldots, x_n)$. Since $y_1, \ldots, y_n$ are algebraically independent over $\mathbb{F}_q$ it follows that $x_1, \ldots, x_n$ must be algebraically independent over $\mathbb{F}_q$. Finally note that $(\star)$ says that the field extension $\mathbb{F}_q(x_1, \ldots, x_n) \mid \mathbb{F}_q(y_1, \ldots, y_n)$ is in fact the splitting field of the polynomial $\varphi(x) \in \mathbb{F}_q(y_1, \ldots, y_n)[x]$, and therefore

$$\mathbb{F}_q(x_1, \ldots, x_n) \mid \mathbb{F}_q(y_1, \ldots, y_n)$$

is a Galois extension.

PROPOSITION 8.1.4 : *With the preceding notations the Galois group of the extension $\mathbb{F}_q(x_1, \ldots, x_n) \mid \mathbb{F}_q(y_1, \ldots, y_n)$ is $\mathrm{GL}(n, \mathbb{F}_q)$.*

PROOF : Recall we defined V as the set of zeros of the polynomial

$$\varphi(x) = x^{q^n} + y_1 x^{q^{n-1}} + \cdots + y_n x = \prod_{v \in V}(x + v)$$

and by 8.1.3 $V = \underset{n}{\oplus} \mathbb{F}_q$, so evidently the Galois group G acts as a set of linear transformations of V, which implies G is a subgroup of $\mathrm{GL}(n, \mathbb{F}_q)$.

On the other hand, if $T \in \mathrm{GL}(n, \; \mathbb{F}_q)$, then it extends uniquely to an automorphism of $\mathbb{F}_q(x_1, \ldots, x_n)$ because $x_1, \ldots, x_n$ are algebraically independent over $\mathbb{F}_q$. Since y_i is the $(q^n - q^{n-i})$-th elementary symmetric function in the elements of $V \smallsetminus \{0\}$, it follows that the extended automorphism fixes $y_1, \ldots, y_n$ and T defines a Galois automorphism of $\mathbb{F}_q(x_1, \ldots, x_n)$ over $\mathbb{F}_q(y_1, \ldots, y_n)$, so $\mathrm{GL}(n, \; \mathbb{F}_q)$ is a subgroup of G. $\square$

The proof of Dickson's theorem that follows is due to T. Tamagawa [246].

THEOREM 8.1.5 (L. E. Dickson): *Suppose* $n \in \mathbb{N}$, *p a prime,* $q = p^s$ *and* $V = \mathbb{F}_q^n$. *Then*

$$\mathbf{D}^*(n) = \mathbb{F}_q[V]^{\mathrm{GL}(n, \; \mathbb{F}_q)} \cong \mathbb{F}_q[y_1, , \ldots, y_n]$$

where $\deg(y_i) = q^n - q^{n-i}$ *for* $i = 1, \ldots, n$.

PROOF : Using the preceding proposition and notations

$$\mathbb{F}_q(y_1, \ldots, y_n) = \mathbb{F}_q(x_1, \ldots, x_n)^{\mathrm{GL}(V)} .$$

Suppose $f(x) \in \mathbb{F}_q[x_1, \ldots, x_n]^{\mathrm{GL}(V)} = \mathbf{D}^*(n)$. If we pass to fields of fractions then $f(x_1, \ldots, x_n) \in \mathbb{F}_q(x_1, \ldots, x_n)^{\mathrm{GL}(V)} = \mathbb{F}_q(y_1, \ldots, y_n)$. Since $x_1, \ldots, x_n$ are integral over $\mathbb{F}_q(y_1, \ldots, y_n)$ (they are the roots of the polynomial $\varphi(x)$ in the notations of the preceding discussions) so is $f(x_1, \ldots, x_n)$. Summarizing we have:

(i) $f(x_1, \ldots, x_n) \in \mathbb{F}_q(y_1, \ldots, y_n)$,

(ii) $f(x_1, \ldots, x_n)$ is integral over $\mathbb{F}_q[y_1, \ldots, y_n]$.

Since a polynomial ring is integrally closed in its field of fractions (see for example [266] page 261 example 1) we conclude that $f(x_1, \ldots, x_n) \in \mathbb{F}_q[y_1, \ldots, y_n]$. Thus

$$\mathbb{F}_q[x_1, \ldots, x_n]^{\mathrm{GL}(V)} \subseteq \mathbb{F}_q[y_1, \ldots, y_n]$$

and since the reverse inclusion is trivial the result follows. $\square$

If $\{x_1, \ldots, x_n\}$ is a basis of an n-dimensional vector space V over the Galois field $\mathbb{F}_q$, Dickson gave rather complicated formulae for the polynomial generators of $\mathbf{D}^*(n) = \mathbb{F}_q[x_1, \ldots, x_n]^{\mathrm{GL}(n, \; \mathbb{F}_q)}$ involving determinants of certain matrices. The entries of the matrices are powers of the x_i that are hard to remember. Therefore the following elegant formulae communicated to the author by R. E. Stong[4] and T. Tamagawa[5] are a vast improvement.

---

[4] In a letter to the author dated 22 September 1983.
[5] In a lecture at Yale in the spring of 1991.

THEOREM 8.1.6 (R. E. Stong - T. Tamagawa): *Let $n \in \mathbb{N}$, p be a prime and q a power of p. Then*

$$\mathbf{D}^*(n) = \mathbb{F}_q[\mathbf{d}_{n,\,0}, \dots, \mathbf{d}_{n,\,n-1}]$$

where

$$\mathbf{d}_{n,\,i} = \sum_{\substack{W \leq V \\ \dim(W)=i}} \prod_{v \notin W} v.$$

The polynomial $\mathbf{d}_{n,\,i} \in \mathbf{D}^*(n)$ has degree $q^n - q^i$.

PROOF : The Poincaré series of $\mathbf{D}^*(n)$ is

$$P(\mathbf{D}^*(n),\ t) = \prod_{i=0}^{n-1} \frac{1}{1 - t^{q^n - q^i}}$$

and since $(q^n - q^i) + (q^n - q^j) > q^n - 1$ it follows that

$$P(\mathbf{D}^*(n),\ t) = t^{q^n - q^{n-1}} + t^{q^n - q^{n-2}} + \cdots + t^{q^n - 1} + \cdots.$$

Hence the homogeneous component of $\mathbf{D}^*(n)$ of degree $q^n - q^{n-i}$ is one-dimensional. It therefore suffices to show that $\mathbf{d}_{n,\,i} \neq 0$ since it clearly belongs to $\mathbf{D}^*(n)$. We do this by induction on n. The class $\mathbf{d}_{n,\,0}$ is the product of all the nonzero elements of V, so is always nonzero, which establishes the result for $n = 1$. For $n > 1$ it only remains to show that $\mathbf{d}_{n,\,1}, \dots, \mathbf{d}_{n,\,n-1}$ are nonzero. To this end let $W < V$ be a codimension one subspace. The inclusion induces a restriction map $\varrho : \mathbf{D}^*(n) \longrightarrow \mathbf{D}^*(n-1)$ and one readily checks

$$\varrho(\mathbf{d}_{n,\,i}) = \begin{cases} \mathbf{d}_{n-1,\,i-1}^q & 1 \leq i \leq n-1 \\ 0 & i = 0, \end{cases}$$

and so by induction the result follows. $\square$

If we regard the symmetric group on n elements as the general linear group of a vector space of dimension n over the field with <u>one</u> element then 8.1.6 gives the usual formula for the elementary symmetric polynomials.

The classes $\mathbf{d}_{n,\,i}$ are called the **Dickson polynomials** and the formulae for them in 8.1.6 are referred to as **the formulae of Stong and Tamagawa**.

For a finite group G acting on an n-dimensional vector space V over $\mathbb{F}_q$ the Dickson algebra $\mathbf{D}^*(n)$ is a subalgebra of $\mathbb{F}[V]^G$ and $\mathbf{D}^*(n) \subset \mathbb{F}[V]^G$ is a finite integral extension. Hence by Noether normalization 5.3.3 the Dickson polynomials $\mathbf{d}_{n,\,0}, \dots, \mathbf{d}_{n,\,n-1} \in \mathbb{F}[V]^G$ are a system of parameters and therefore by Macaulay's theorem 6.7.6 we have:

PROPOSITION 8.1.7 : *Let G be a finite group acting on an n-dimensional vector space V over the Galois field $\mathbb{F}_q$. Then the following conditions are equivalent*

(i) $\mathbb{F}[V]^G$ *is Cohen-Macaulay,*

(ii) *the Dickson polynomials $\mathbf{d}_{n,\,0}, \ldots, \mathbf{d}_{n,\,n-1} \in \mathbb{F}_q[V]^G$ are a regular sequence,*

(iii) $\mathbb{F}[V]^G$ *is a free finitely generated module over $\mathbf{D}^*(n)$.*

PROOF : This follows from the preceding proposition, Macaulay's theorem 6.7.7 and 6.2.8. □

One attractive feature of these results is the close connection between the Dickson polynomials and one of the other special features of characteristic p invariant theory, namely the Steenrod operations (see chapter 10). The connection between the homological codimension of rings of invariants and the Dickson polynomials has been stimulated by a conjecture Landweber and Stong [130] that has been recently solved by Dora Bourguiba and Said Zarati [37] (see also [190] and [220]).

It is a rather easy matter to deduce the invariants of the special linear group $\mathrm{SL}(n, \mathbb{F}_q)$ from Dickson's theorem 8.1.5. To this end, for each codimension 1 subspace $W < V$ choose a linear functional $\ell_W : V \longrightarrow \mathbb{F}_q$ with kernel W and define the polynomial (this is the **Euler class** of the orbit V^* [221])

$$L_n = \prod_{\{W < V \mid \dim(W)=1\}} \ell_W \quad \in \mathbb{F}[V].$$

In other words L_n is the product of one nonzero element chosen from each line in V^*. Since $\prod_{\lambda \in \mathbb{F}_q^\times} \lambda = 1$ it follows that

$$L_n^{q-1} = \mathbf{d}_{n,\,0} = \prod_{v \in V^* \setminus \{0\}} v,$$

the Dickson polynomial of degree $q^n - 1$.

THEOREM 8.1.8 (L. E. Dickson): *Suppose $n \in \mathbb{N}$, p a prime $q = p^s$ and $V = \mathbb{F}_q^n$. Then*

$$\mathbb{F}_q[V]^{\mathrm{SL}(n,\,\mathbb{F}_q)} = \mathbb{F}_q[L_n, \mathbf{d}_{n,\,1}, \ldots, \mathbf{d}_{n,\,n-1}].$$

PROOF : First of all notice for $g \in \mathrm{GL}(n, \mathbb{F}_q)$, that

$$g(L_n)^{q-1} = g\mathbf{d}_{n,\,0} = \mathbf{d}_{n,\,0} = L_n^{q-1}.$$

and therefore $gL_n = \chi(g)L_n$ for some one-dimensional representation $\chi : \mathrm{GL}(n, \mathbb{F}_q) \longrightarrow \mathbb{F}_q$. Since χ is a power of the det representation it follows that L_n is an $\mathrm{SL}(n, \mathbb{F}_q)$ invariant polynomial.

The elements $\mathbf{d}_{n,0}, \mathbf{d}_{n,1}, \ldots, \mathbf{d}_{n,n-1} \in \mathbb{F}_q[V]$ are a system of parameters and hence so are $L_n, \mathbf{d}_{n,1}, \ldots, \mathbf{d}_{n,n-1}$. Since they also belong to $\mathbb{F}_q[V]^{\mathrm{SL}(n, \mathbb{F}_q)}$ they likewise are a system of parameters for $\mathbb{F}_q[V]^{\mathrm{SL}(n, \mathbb{F}_q)}$. Moreover

$$|\mathrm{SL}(n, \mathbb{F}_q)| = \frac{1}{q-1} \prod_{i=0}^{n-1} (q^n - q^i)$$

$$= \deg(L_n) \cdot \deg(\mathbf{d}_{n,1}) \cdots \deg(\mathbf{d}_{n,n-1})$$

and the result follows from 5.5.5. $\square$

The computations of the invariants of the subgroups

$$\mathrm{SL}_m(n, \mathbb{F}_q) = \{T \in \mathrm{GL}(n, \mathbb{F}_q) \mid (\det(T))^m = 1\}$$

where $m \mid q - 1$, is completely analogous. See for example [39] page 137–138, exercise 6.

8.2   Transvections

In the representation theory of finite groups over fields of characteristic p the finite p-groups play an especially important role. Let $\varrho : P \hookrightarrow \mathrm{GL}(n, \mathbb{F})$ be a representation of a finite p-group over a field of characteristic $p \neq 0$. The class equation (see section 4.2) implies the center $Z(P)$ of P is nontrivial. Choose an element ζ of order p in the center of P. Since the prime field $\mathbb{F}_p$ is a splitting field for $\mathbb{Z}/p$ the element $\varrho(\zeta)$ may be put in Jordan normal form,

$$\varrho(\zeta) = \begin{bmatrix} 1 & \epsilon_1 & 0 & \cdots & & 0 \\ 0 & 1 & \epsilon_2 & 0\cdots & & 0 \\ \vdots & \ddots & \ddots & \cdots & & \vdots \\ 0 & \cdots & 0 & 1 & & \epsilon_{n-1} \\ 0 & \cdots & \cdots & & \cdots & 1 \end{bmatrix} \qquad \epsilon_1, \ldots, \epsilon_{n-1} = 0 \text{ or } 1$$

over $\mathbb{F}$. In this basis the first basis vector is a fixed point of ζ, so $V^\zeta \neq \{0\}$. Since ζ is central in P it follows that V^ζ is a P-invariant subspace. Moreover the action of P on V^ζ factors through the quotient map $P \longrightarrow P/<\zeta>$, and hence inducting on the order of P we conclude that $(V^\zeta)^{P/<\zeta>} \neq \{0\}$. Since central subgroups are normal $(V^\zeta)^{P/<\zeta>} = V^P$ and therefore we have shown:

LEMMA 8.2.1 : *If* $\varrho : P \hookrightarrow \mathrm{GL}(n, \ \mathbb{F})$ *is a representation of a finite p group over a field of characteristic p then* $V^P \neq \{0\}$. □

It is natural to suppose that the larger the dimension of the fixed point set is, the simpler the invariants $\mathbb{F}[V]^P$ will be. The case where the codimension of the fixed point set is 1 was studied by Landweber and Stong [130], who showed that the ring of invariants is a polynomial algebra. (See [166] for a study of the codimension 2 case.)

LEMMA 8.2.2 : *Let* $\varrho : G \hookrightarrow \mathrm{GL}(n, \ \mathbb{F})$ *be a representation of a finite group. The fixed point set* $V^G \hookrightarrow V$ *has codimension 1 if and only if every nonidentity element of* $\varrho(G)$ *is a pseudoreflection with hyperplane* V^G.

PROOF : Let $g \neq 1 \in G$, then $V^g \neq V$ and since $V^g \supset V^G$ it follows that $V^g = V^G$. Hence every nonidentity element of $\varrho(G)$ is a pseudoreflection with V^G as reflecting hyperplane. The converse is equally clear. □

The dual situation, namely where $(V^*)^G \hookrightarrow V^*$ has codimension 1, was studied in the abelian p-group case by Nakajima [158]. In both cases transvections play an important role. For ease in formulating results it will be convenient to introduce some definitions and notations.

DEFINITION : *Let the finite group G act on the vector space V over the field* $\mathbb{F}$. *We make no assumption about the characteristic of* $\mathbb{F}$. *Set*

$$V_G = V/\mathrm{Span}_{\mathbb{F}} \{gv - v \mid g \in G, \ v \in V\}.$$

The vector space V_G is called the **vector space of covariants**. *(Not to be confused with the ring* $\mathbb{F}[V]_G$ *of coinvariants.)*

LEMMA 8.2.3 : *Let the finite group G act on the finite dimensional vector space V over the field* $\mathbb{F}$. *(We make no assumption about the characteristic of* $\mathbb{F}$.*) Then*

$$\mathrm{Hom}_{\mathbb{F}}(V^G, \ \mathbb{F}) \cong \mathrm{Hom}_{\mathbb{F}}(V, \ \mathbb{F})_G = (V^*)_G.$$

(In other words: the covariants are dual to the fixed points.)

PROOF : Let us write U^* as usual for the dual of a vector space U. The inclusion $V^G \hookrightarrow V$ induces a surjection $q : V^* \longrightarrow (V^G)^*$. Clearly what we must show is that

$$\ker(q) = \mathrm{Span}_{\mathbb{F}} \{gv^* - v^* \mid g \in G, \ v^* \in V^*\}.$$

Note that

$$\ker(q) = \left\{v^* \in V^* \mid v^*|_{V^G} = 0\right\}.$$

First we show: $\mathrm{Span}_{\mathbb{F}} \{gv^* - v^* \mid g \in G, \ v^* \in V^*\} \subseteq \ker(q)$.

To verify this, suppose $v \in V^G$ and $v^* \in V^*$. Then

$$(gv^* - v^*)(v) = v^*(g^{-1}v) - v^*(v)$$
$$= v^*(v) - v^*(v) = 0.$$

Next we show: $\ker(q) \subseteq \mathrm{Span}_{\mathbb{F}} \{gv^* - v^* \mid g \in G, \ v^* \in V^*\}$.

To verify this let $w^* \in \ker(q)$. If $w^* \neq 0$ there exists $v_1 \in V$ such that $w^*(v_1) \neq 0$. Since $w^*\big|_{V^G} = 0$ it follows that $v_1 \notin V^G$, so there exists a $g_1 \in G$ such that $g_1^{-1} \cdot v \neq v$. Let $\ell_1^* \in V^*$ be a linear functional such that

$$\ell_1^*(g_1^{-1} \cdot v_1 - v_1) = w^*(v_1).$$

Consider $\bar{w}^* = w^* - (g_1 \ell_1^* - \ell_1^*)$. By (i) $\bar{w}^*\big|_{V^G} = 0$. In addition

$$\bar{w}^*(v_1) = w^*(v_1) - (g_1\ell_1^* - \ell_1^*)(v_1)$$
$$= w^*(v_1) - (\ell_1^*(g_1^{-1} \cdot v_1) - \ell_1^*(v_1))$$
$$= w^*(v_1) - \ell_1^*(g_1^{-1} \cdot v_1 - v_1)$$
$$= w^*(v_1) - w^*(v_1) = 0$$

so $\bar{w}^*$ is zero on $\mathrm{Span}\{V^G, \ v_1\}$. We may repeat the above construction with w^* replaced by $\bar{w}^*$ finding $v_2 \notin \mathrm{Span}\{V^G, v_1\}$, $g_2 \in G$ and $\ell_2^* \in V^*$ such that $\bar{\bar{w}}^* = \bar{w}^* - (g_2\ell_2^* - \ell_2^*)$ is zero on $\mathrm{Span}\{V^G, v_2, v_2\}$. Since V is finite dimensional sufficient repetition of the above construction leads to an equation

$$0 = w^* - \sum_{i=1}^{m}(g_i\ell_i^* - \ell_i^*)$$

in V^*. Hence

$$w^* = \sum_{i=1}^{m}(g_i\ell_i^* - \ell_i^*) \in \mathrm{Span}_{\mathbb{F}} \{gv^* - v^* \mid g \in G, \ v^* \in V^*\},$$

and the lemma follows from (i) and (ii). $\quad\square$

If $\mathbb{F}$ is a field of characteristic different from zero, then a pseudoreflection $s \in \mathrm{GL}(n, \mathbb{F})$ need not be diagonalizable. If s is a non-diagonalizable pseudo-reflection then (see section 7.1)

$$s(v) = v + \tau(v)$$

where $\tau : V \longrightarrow \ker(1 - s) = H_s$ is a linear transformation. Since s is a pseudoreflection $\dim_{\mathbb{F}}(\mathrm{Im}(1 - s)) = 1$ so $\mathrm{Im}(\tau) \subseteq H_s$ is a 1-dimensional subspace. If $0 \neq x_s \in \mathrm{Im}(\tau)$, then we may choose a linear functional $\ell_s :$ $V \longrightarrow \mathbb{F}$, with $\ker(\ell_s) = H_s$, and we may write

$$s(v) = v + \ell_s(v) \cdot x_s \qquad \forall\, v \in V.$$

Note that as $x_s \in H_s$ one has $\ell_s(x_s) = 0$. Classically, independent of the characteristic of $\mathbb{F}$, a linear transformation $T \in \mathrm{GL}(n, \mathbb{F})$ with a representation of the form

$$T(v) = v + \varphi(v) \cdot x \qquad \forall\, v \in V,$$

for some nonzero linear functional $\varphi : V \longrightarrow \mathbb{F}$ and nonzero vector $x \in \ker(\varphi)$ is called a transvection (in the German literature an *Überschiebung*).

DEFINITION : *An element $T \in \mathrm{GL}(n, \mathbb{F})$ is called a* **transvection** *with* **hyperplane** H_T, **transvector** $0 \neq x \in H_T$, *and* **direction** $\mathrm{Span}_{\mathbb{F}}\{x\}$, *if there is a linear functional $\varphi_T : V \longrightarrow \mathbb{F}$ such that $H_T = \ker(\varphi_T)$ and $T(v) = v + \varphi_T(v) \cdot x$ for all $v \in V$.*

The hyperplane H_T of a transvection T is the fixed point set of T, i.e. the $+1$-eigenspace of T, and has codimension 1. The direction of T is $\mathrm{Im}(1 - T)$ and is a one-dimensional subspace of H_T. Thus *if a transvection has finite order* it is a non-diagonalizable pseudoreflection. If $\mathbb{F}$ has characteristic zero then transvections have infinite order, such as for example,

$$\begin{bmatrix} 1 & 1 \\ 0 & 1 \end{bmatrix} \in \mathrm{GL}(2, \mathbb{C}).$$

If the field $\mathbb{F}$ has nonzero characteristic p then every transvection has order p, so is a pseudoreflection. This is easily seen by passing to the algebraic closure and examining the Jordan canonical form of T. Hence in nonzero characteristic the transvections are exactly the non-diagonalizable pseudoreflections and in characteristic zero they are never pseudoreflections.

If $T \in \mathrm{GL}(n, \mathbb{F})$ is a transvection, then so is T^{Tr} when regarded as a linear transformation on the dual vector space V^* of $V = \mathbb{F}^n$. There is a canonical isomorphism of V with V^{**}. With the aid of this isomorphism we may interpret a transvector x_T of T as a linear form $x_T : V^* \longrightarrow \mathbb{F}$. The kernel of this linear form in V^* is the hyperplane $H_{T^{\mathrm{Tr}}}$ of T^{Tr}. The subspace $H_T \subset V$ defines a linear form $V \longrightarrow \mathbb{F}$ by identifying V/H_T with $\mathbb{F}$. This linear form, interpreted as an element of V^*, is a transvector for T^{Tr}.

Let $\mathbb{F}$ be a field of characteristic $p \neq 0$. For $n \in \mathbb{N}$ and $\alpha \in \mathbb{F}$ consider the matrices

$$S_i(\alpha) = \begin{bmatrix} 1 & 0 & \dots & \dots & 0 \\ 0 & 1 & 0 & \dots & 0 \\ 0 & \dots & \ddots & \dots & \alpha \\ 0 & \dots & \dots & \ddots & 0 \\ 0 & \dots & \dots & 0 & 1 \end{bmatrix} \in \mathrm{GL}(n, \mathbb{F})$$

where the off diagonal α is in the i-th row and the last column. In other words, if $E_{i,j}$ denotes the $n \times n$ matrix with a 1 in the ith row and j^{th} column, and zeros elsewhere for i, $j \in \{1, \ldots, n\}$, then $S_i(\alpha) = I + \alpha E_{i,n}$ for $i = 1, \ldots, n - 1$. For $\alpha \neq 0$ these matrices are transvections with a common hyperplane $H = \mathbb{F}^{n-1}$, the codimension 1 subspace spanned by the first $n - 1$ standard basis vectors.

LEMMA 8.2.4 : *Let $\mathbb{F}$ be a field of characteristic $p \neq 0$. For each integer $i = 1, \ldots, n-1$ the set $\{S_i(\alpha) \mid \alpha \in \mathbb{F}\}$ is a subgroup of $\mathrm{GL}(n, \mathbb{F})$. It consists of all the transvections with hyperplane $H = \mathbb{F}^{n-1} = \mathrm{Span}_{\mathbb{F}}\{E_1, \ldots, E_{n-1}\}$ and direction E_i.*

PROOF : This follows from the definitions. $\square$

LEMMA 8.2.5 : *Let $\mathbb{F}$ be a field of characteristic $p \neq 0$. For $r = 1, \ldots, n - 1$ let $E_{\mathbb{F}}(r)$ denote the subgroup generated by the set $\{S_i(\alpha) \in \mathrm{GL}(n, \mathbb{F}) \mid \alpha \in \mathbb{F}, \ i = 1, \ldots, r\}$. Then $E_{\mathbb{F}}(r) \subset \mathrm{GL}(n, \mathbb{F})$ is a subgroup isomorphic to $\mathbb{F}^r$.*

PROOF : Matrix multiplication shows

$$E_{i,j}E_{k,l} = \begin{cases} 0 & \text{for } j \neq k \\ E_{i,l} & \text{for } j = k. \end{cases}$$

Therefore for $1 \leq i, \ j \leq n$,

$$S_i(\alpha)S_j(\beta) = (I + \alpha E_{i,n})(I + \beta E_{j,n}) = I + \alpha E_{i,n} + \beta E_{j,n}$$
$$= (I + \beta E_{j,n})(I + \alpha E_{i,n}) = S_j(\beta)S_i(\alpha).$$

Hence the matrices $S_i(\alpha)$, $S_j(\beta)$ where i, $j = 1, \ldots, n - 1$ and α, $\beta \in \mathbb{F}$ commute pairwise so $E_{\mathbb{F}}(r)$ is a commutative group. For $i = j$ we obtain from the preceding formulae

$$S_i(\alpha)S_i(\beta) = S_i(\alpha + \beta)$$

and hence the matrices $\{S_i(\alpha) \mid \alpha \in \mathbb{F}\}$ for fixed i form a subgroup of $\mathrm{GL}(n, \mathbb{F})$ isomorphic to $\mathbb{F}$. The subgroup generated by $S_i(\alpha)$ and $S_j(\beta)$ for $i \neq j$ and α, $\beta \in \mathbb{F}$ have only the identity element in common. Therefore the map

$$\sigma_r : \mathbb{F}^r \longrightarrow E_{\mathbb{F}}(r)$$

defined by

$$\sigma_r(\alpha_1, \ldots, \alpha_r) = S_1(\alpha_1) \cdots S_r(\alpha_r)$$

is an isomorphism. $\square$

Denote by $\sigma_r : E_{\mathbb{F}}(r) \hookrightarrow \mathrm{GL}(n, \mathbb{F})$ the representation of $E_{\mathbb{F}}(r)$ afforded by the matrices $\{S_1(\alpha_1), \ldots, S_r(\alpha_r) \mid \alpha_1, \ldots, \alpha_r \in \mathbb{F}\}$. Note that the fixed point set $V^{E_{\mathbb{F}}(r)}$ has codimension 1 in V and that every nonidentity element of $E_{\mathbb{F}}(r)$ is a pseudoreflection with reflecting hyperplane H. For $\mathbb{F} = \mathbb{F}_p$ we denote $E_{\mathbb{F}_p}(r)$ simply by $E(r)$. It is an elementary abelian p-group of order p^r.

PROPOSITION 8.2.6 : *Let $\mathbb{F}$ be a finite field of nonzero characteristic p and p^ℓ elements, and $\sigma_r : E_{\mathbb{F}}(r) \hookrightarrow \mathrm{GL}(n, \mathbb{F})$ the representation afforded by the matrices $S_1(\alpha_1), \ldots, S_r(\alpha_r)$ where $\alpha_1, \ldots, \alpha_r \in \mathbb{F}$. Then $V^{E_{\mathbb{F}}(r)} \hookrightarrow V$ has codimension 1 and*

$$\mathbb{F}[z_1, \ldots, z_n]^{E_{\mathbb{F}}(r)} = \mathbb{F}[f_1, \ldots, f_r, z_{r+1}, \ldots, z_n]$$

where

$$f_i = \prod_{\lambda \in \mathbb{F}} (z_i + \lambda z_n) = c_{p^\ell}([z_i])$$

is the p^ℓ-th Chern class of the orbit of z_i for $i = 1, \ldots, r$.

PROOF : With respect to the standard basis $\{z_1, \ldots, z_n\}$ for V^* the action of $E(r)$ is given by

$$S_i(\alpha)(z_j) = \begin{cases} z_j & j \neq i \\ z_i + \alpha z_n & j = i. \end{cases}$$

Hence f_i is the p^ℓ-th Chern class of the orbit of z_i so $f_i \in \mathbb{F}[z_1, \ldots, z_n]^{E_{\mathbb{F}}(r)}$. An application of 5.3.7 shows that the polynomials $f_1, \ldots, f_r, z_{r+1}, \ldots, z_n$ are a system of parameters. In addition

$$\deg(f_1) \cdots \deg(f_r) \deg(z_{r+1}) \cdots \deg(z_n) = p^{\ell r} = |E_{\mathbb{F}}(r)|$$

and the result follows from 5.5.5. $\square$

The transposed matrices $\{S_i(\alpha)^{\mathrm{tr}} \mid \alpha \in \mathbb{F}, i = 1, \ldots, r\}$ afford the dual representations

$$\sigma_r^* : E_{\mathbb{F}}(r) \hookrightarrow \mathrm{GL}(n, \mathbb{F}) \qquad r = 1, \ldots, n - 1.$$

By lemma 8.2.6 and proposition 8.2.3 all the elements of $E_{\mathbb{F}}^*(r) := \sigma_r^*(E_{\mathbb{F}}(r))$ are transvections with a common transvector $E_n = (0, \ldots, 0, 1) \in V = \mathbb{F}^n$, and $\dim_{\mathbb{F}}(V_{E_{\mathbb{F}}^*(r)}^*) = n - 1$ (i.e. has codimension 1). The dual of the preceding result is:

PROPOSITION 8.2.7 : *Let* $\mathbb{F}$ *be a finite field of nonzero characteristic* p *and* p^ℓ *elements, and* $\sigma_r^* : E_\mathbb{F}(r) \hookrightarrow \mathrm{GL}(n, \mathbb{F})$ *the representation afforded by the matrices* $\{S_i(\alpha)^{\mathrm{tr}} \mid \alpha \in \mathbb{F}, \ i = 1, \ldots, r\}$ *Then* $V_{E_\mathbb{F}^*(r)}$ *has dimension* $n - 1$ *and*

$$\mathbb{F}[z_1, \ldots, z_n]^{E_\mathbb{F}(r)} = \mathbb{F}[z_1, \ldots, z_{n-1}, \ f]$$

where

$$f = \prod_{a_1, \ldots, a_r \in \mathbb{F}} (z_n + a_r z_r + \cdots + a_1 z_1) = c_{p^{r\ell}}([z_n])$$

is the top Chern class of the orbit of z_n.

PROOF : By 8.2.3 $(V^*)^{E_\mathbb{F}^*(r)}$ has codimension 1 spanned by $z_1, \ldots, z_{n-1}$. The orbit of z_n is

$$[z_n] = \{z_n + a_r z_r + \cdots + a_1 z_1 \mid a_1, \ldots, a_r \in \mathbb{F}\}$$

and has $p^{r\ell} = |E_\mathbb{F}^*(r)|$ elements, so the result follows from 5.5.6. $\square$

It is perhaps worth noting that when f is written as a polynomial in z_n then the coefficients of the powers of z_n are the Dickson polynomials. Specifically

$$f = \sum_{i=0}^{r-1} \mathbf{d}_{r, i} z_n^{q^{r-i}} .$$

This is a p-polynomial and a Galois resolvent for $\mathbb{F}(V)^{E_\mathbb{F}(r)}$ over $FF(\mathbf{D}^*(n)) = \mathbb{F}(V)^{\mathrm{GL}(n, \mathbb{F})}$.

We next show that up to conjugation the representations

$$\sigma_r, \ \sigma_r^* : E_\mathbb{F}(r) \hookrightarrow \mathrm{GL}(n, \mathbb{F}) \qquad r = 1, \ldots, n-1$$

contain every representations where every nonidentity element is represented by a transvection.

LEMMA 8.2.8 : *Let* $S, T \in \mathrm{GL}(n, \mathbb{F})$ *be transvections with the same hyperplane. Then* S *and* T *commute and* ST *is the identity or a transvection with the same hyperplane and transvector* $x + y$, *where* x *is a transvector for* S *and* y *is a transvector for* T. *(ST = I when* $x + y = 0$.)

PROOF : Since S and T have the same hyperplane H we may choose a linear functional $\varphi : V \longrightarrow \mathbb{F}$ such that $H = \ker(\varphi)$ and

$$S(v) = v + \varphi(v) \cdot x$$
$$T(v) = v + \varphi(v) \cdot y$$

for all $v \in V$. Then

$$\begin{aligned} ST(v) &= S(v + \varphi(v) \cdot y) = S(v) + \varphi(v)S(y) \\ &= v + \varphi(v) \cdot x + \varphi(v) \cdot y (\text{because } y \in H) \\ &= v + \varphi(v)(x + y) = v + \varphi(v)(y + x) = TS(v) \end{aligned}$$

and the result follows. □

LEMMA 8.2.9 : *Let S, $T \in \mathrm{GL}(n, \mathbb{F})$ be transvections with the same direction. Then S and T commute and ST is a transvection with the same direction or the identity.*

PROOF : Choose linear functionals φ, ψ such that $H_S = \ker(\varphi)$, $H_T = \ker(\psi)$ and $x \in H_S \cap H_T$ is a transvector for both S and T. Then

$$S(v) = v + \varphi(v) \cdot x$$
$$T(v) = v + \psi(v) \cdot x \,,$$

so

$$ST(v) = S(v + \psi(v) \cdot x) = v + \varphi(v) \cdot x + \psi(v)S(x)$$
$$= v + \varphi(v) \cdot x + \psi(v) \cdot x \qquad (\text{because } x \in H_S)$$
$$= v + (\varphi(v) + \psi(v)) \cdot x = v + (\psi(v) + \varphi(v)) \cdot x = TS(v)$$

and the result follows. □

With the aid of these lemmas it is easy to construct examples of elementary abelian p-subgroups of $\mathrm{GL}(n, \mathbb{F})$, all of whose nonidentity elements are transvections. Consider a finite collection $\mathcal{H} = \{H_\alpha \mid \alpha \in \mathcal{A}\}$ of hyperplanes in V such that $\dim_{\mathbb{F}}(\bigcap_{\alpha \in \mathcal{A}} H_\alpha) > 0$. Let $\varphi_\alpha : V \longrightarrow \mathbb{F}$ be a collection of linear functionals with $\ker(\varphi_\alpha) = H_\alpha$ for $\alpha \in \mathcal{A}$. Choose $0 \neq x \in \bigcap_{\alpha \in \mathcal{A}} H_\alpha$, and define transvections

$$T_\alpha(v) = v + \varphi_\alpha(v)x \qquad \forall \, \alpha \in \mathcal{A}.$$

All these transvections have a common direction x and hence by 8.2.8 the subgroup of $\mathrm{GL}(n, \mathbb{F})$ generated by $\{T_\alpha \mid \alpha \in \mathcal{A}\}$ is an elementary abelian p-subgroup. For a discussion of the invariant theory of such groups when the hyperplanes are in general position see [168].

LEMMA 8.2.10 : *Let S, $T \in \mathrm{GL}(n, \mathbb{F})$ be transvections such that ST is again a transvection. Then S and T have either the same hyperplane or the same direction.*

PROOF : Choose linear functionals φ, ψ with $H_S = \ker(\varphi)$, $H_T = \ker(\psi)$ and transvectors x for T and y for S. Then for $v \in V$

$$S(v) = v + \varphi(v) \cdot x$$
$$T(v) = v + \psi(v) \cdot y \,.$$

Suppose $H_S \neq H_T$. Then H_S and H_T are distinct hyperplanes so neither is contained in the other. Hence we may choose vectors $A \in H_S$ and $B \in H_T$

such that $A \notin H_T$ and $B \notin H_S$. Then $\psi(A) \neq 0 \neq \varphi(B)$ so without loss of generality we may suppose

$$\psi(A) = 1 \quad \varphi(A) = 0$$
$$\psi(B) = 0 \quad \varphi(B) = 1.$$

Therefore

$$S(B) = B + x \quad S(A) = A$$
$$T(A) = A + y \quad T(B) = B.$$

Hence

$$ST(A) = A + \varphi(y) \cdot x + y$$
$$ST(B) = B + x$$

so $\text{Im}(1 - ST) \supset \text{Span}_{\mathbb{F}}\{x, y\}$. Since ST is a pseudoreflection $\text{Im}(1 - ST)$ is one-dimensional, so x and y must be linearly dependent and S and T have the same direction. $\square$

REMARK : Lemma 8.2.10 holds with *transvection* replaced by *pseudo-reflection*. The proof remains unchanged.

PROPOSITION 8.2.11 : *Let $\mathbb{F}$ be a field of characteristic $p \neq 0$. If $\varrho : G \hookrightarrow \text{GL}(n, \mathbb{F})$ is a representation of a finite group G and every nonidentity element of $\varrho(G)$ is a transvection then G is an elementary abelian p-group.*

PROOF : The elements of G are all of order p. If g, $h \in G$ then by 8.2.10 either g and h have the same hyperplane or the same direction. Hence by either lemma 8.2.8 or lemma 8.2.9 g and h commute. $\square$

PROPOSITION 8.2.12 : *Let $\mathbb{F}$ be a finite field of characteristic $p \neq 0$. If $\varrho : G \hookrightarrow \text{GL}(n, \mathbb{F})$ is a representation of a finite group G and every non-identity element of $\varrho(G)$ is a transvection, then the elements of $\varrho(G)$ all have either the same hyperplane or the same direction. Hence, up to conjugation in $\text{GL}(n, \mathbb{F})$, G is a subgroup of $E_{\mathbb{F}}(r)$ or $E_{\mathbb{F}}^*(r)$.*

PROOF : Suppose $\varrho(G)$ contains elements g', g'' with different directions. Then by lemma 8.2.10 g' and g'' have the same hyperplane. If $g \in G$ then the direction of g must be different from either the direction of g' or the direction of g'' and hence by lemma 8.2.10 the hyperplane of g must agree with the hyperplane of g' or g''. But g' and g'' have the same hyperplane H, so all the elements of G have H as hyperplane. $\square$

A slight reformulation of this result is possible.

PROPOSITION 8.2.13 : Let $\varrho : P \hookrightarrow \mathrm{GL}(n, \, \mathbb{F})$ be a representation of a finite p-group over a field of characteristic p. If V^P has dimension $n - 1$ then G is conjugate to a subgroup of $E_{\mathbb{F}}(r)$. If V_P has dimension $n - 1$ then P is conjugate to a subgroup of $E_{\mathbb{F}}^*(r)$. $\quad \square$

The following result is due to Nakajima [155] and Landweber-Stong [130] in the case $\mathbb{F} = \mathbb{F}_p$.

THEOREM 8.2.14 (H. Nakajima, P. S. Landweber and R. E. Stong): Let $\mathbb{F}$ be a finite field of characteristic $p \neq 0$. If $\varrho : G \hookrightarrow \mathrm{GL}(n, \, \mathbb{F})$ is a representation of a finite group G such that every nonidentity element of $\varrho(G)$ is a transvection, then G is an elementary abelian p-group which is conjugate to a subgroup $E_{\mathbb{F}}(r)$ or $E_{\mathbb{F}}^*(r)$ for some $r \in \{1, \ldots, n-1\}$ and either V^G or V_G has dimension $n - 1$. Moreover both $\mathbb{F}[V]^G$ and $\mathbb{F}[V^*]^G$ are polynomial algebras as described in 8.2.7 and 8.2.6.

PROOF : G is p-elementary by 8.2.12. By proposition 8.2.12 either all the nonidentity elements of G have the same direction L or the same hyperplane H. In the former case $L = \mathrm{Span}\{g(v) - v \mid g \in G, \, v \in V\}$ and V_G has dimension $n - 1$. In the latter case $V^G = H$ and has codimension 1. From either 8.2.6 or 8.2.7 we obtain that $\varrho(G)$ is conjugate to a subgroup of either $E_{\mathbb{F}}(r)$ or $E_{\mathbb{F}}^*(r)$.

It remains to show that $\mathbb{F}[V]^G$ and $\mathbb{F}[V^*]^G$ are polynomial algebras. We consider the case $V^G = H = \mathrm{Span}_{\mathbb{F}}\{E_1, \ldots, E_{n-1}\}$ as the other case is similar. We require a lemma.

LEMMA 8.2.15 : Let $\mathbb{F}$ be a finite field of characteristic p and E an r-dimensional vector space over $\mathbb{F}$. If $A \subseteq E$ is an additive subgroup then there exists a basis $E_1, \ldots, E_r$ for E and subgroups $A_1, \ldots, A_r$ of $\mathbb{F}$ such that

$$A \cong A_1 \oplus \cdots \oplus A_r \subseteq \mathbb{F}^r \cong E,$$

where the isomorphism $\mathbb{F}^r \cong E$ is afforded by the basis $E_1, \ldots, E_r$.

PROOF : For $r = 1$ the result is trivially true. Assume the result established for all $A \subseteq E$ with $\dim_{\mathbb{F}}(E) \leq r$. Let E have dimension $r + 1$ over $\mathbb{F}$ and assume that $A \subseteq E$ is an additive subgroup. Let $E' \subset E$ be an r-dimensional subspace and consider the diagram

$$
\begin{array}{ccccccccc}
0 & \longrightarrow & A \cap E' & \overset{\alpha'}{\longrightarrow} & A & \overset{\alpha''}{\longrightarrow} & A'' & \longrightarrow & 0 \\
 & & \downarrow \varphi' & & \downarrow \varphi & & \downarrow \varphi'' & & \\
0 & \longrightarrow & E' & \overset{\beta'}{\longrightarrow} & E & \overset{\beta''}{\longrightarrow} & E'' & \longrightarrow & 0
\end{array}
$$

in which the rows are exact and φ', φ are monomorphisms. We claim that φ'' is a monomorphism. To verify this suppose $a'' \in \ker(\varphi'')$. Choose $a \in A$ with $\alpha''(a) = a''$. Then

$$\beta'' \varphi(a) = \varphi'' \alpha''(a) = 0$$

so $\varphi(a) \in \ker(\beta'') = \text{Im}(\beta')$. This however says that $\varphi(a) \in E'$ and hence that $a \in A \cap E'$, whence $\alpha''(a) = 0$ as claimed.

The additive group of E is an elementary abelian p-group, and hence A as subgroup is also an elementary abelian p-group. Therefore the sequence of abelian groups

$$0 \longrightarrow A \cap E' \xrightarrow{\alpha'} A \xrightarrow{\alpha''} A'' \longrightarrow 0$$

is an exact sequence of $\mathbb{F}_p$-vector spaces, and hence split. By the induction hypothesis there is a basis $E_1, \ldots, E_r$ for E' and subgroups $A_1, \ldots, A_r \subseteq \mathbb{F}$ such that

$$\mathbb{F}^r \supseteq A_1 \oplus \cdots \oplus A_r \cong A \cap E' \subseteq E'$$

with respect to the isomorphism afforded by the basis $E_1, \ldots, E_r$. If we extend this basis to a basis $E_1, \ldots, E_{r+1}$ for E then

$$A \cong A_1 \oplus \cdots \oplus A_r \oplus A'' \subseteq \mathbb{F}^r \cong E$$

completing the induction step. $\quad\square$

Returning to the proof of 8.2.14 we apply 8.2.12 to conclude that up to conjugation G is a subgroup of $E_{\mathbb{F}}(r)$. If we identify $E_{\mathbb{F}}(r)$ with $\mathbb{F}^r$ via σ_r then as a consequence of 8.2.15 we may choose subgroups $A_1, \ldots, A_r \subseteq \mathbb{F}$ such that

$$\varrho(G) = \left\{ \begin{bmatrix} 1 & 0 & \cdots & \cdots & \alpha_1 \\ 0 & 1 & 0 & \cdots & \alpha_2 \\ 0 & \cdots & \ddots & \cdots & \alpha_r \\ 0 & \cdots & \cdots & \ddots & 0 \\ 0 & \cdots & \cdots & 0 & 1 \end{bmatrix} \;\middle|\; \alpha_i \in A_i \; i = 1, \ldots, r \right\}.$$

If $z_1, \ldots, z_n \in V^*$ is the dual basis then the orbits are

$$[z_i] = \begin{cases} \{z_i + \alpha_i z_n \mid \alpha_i \in A_i\} & \text{for } i = 1, \ldots, r \\ \{z_i\} & \text{for } i = r+1, \ldots, n. \end{cases}$$

Let

$$f_i = \prod_{\alpha_i \in A_i} (z_i + \alpha_i z_n).$$

be the top Chern class of the orbit of z_i for $i = 1, \ldots, r$. Then the polynomials $f_1, \ldots, f_r, z_{r+1}, \ldots, z_n \in \mathbb{F}[V]^G$ are a system of parameters. Since $\deg(f_i) = |A_i|$ for $i = 1, \ldots, r$ it follows that

$$\deg(f_1) \cdots \deg(f_r) \deg(z_{r+1}) \cdots \deg(z_n) = |A|$$

and hence by 5.5.6 it follows that

$$\mathbb{F}[V]^G = \mathbb{F}[f_1, \ldots, f_r, z_{r+1}, \ldots, z_n]$$

as required. $\quad\square$

Having considered representations $\varrho : G \hookrightarrow \mathrm{GL}(n, \ \mathbb{F})$ in which every non-identity element of $\varrho(G)$ is a transvection, it is natural to consider the larger family of representations for which every nonidentity element of $\varrho(G)$ is a pseudoreflection. These turn out to be precisely the representations studied by Landweber and Stong (see [130]) or their duals.

Let $\zeta \in \mathbb{F}^{\times}$ be a primitive s-th root of unity. For $1 \leq r \leq n$ denote by $G(\zeta, \ r)$ the subgroup of $\mathrm{GL}(n, \ \mathbb{F})$ generated by $\sigma_r(E_{\mathbb{F}}(r))$ and the matrix

$$
S_{\zeta} = \begin{bmatrix} 1 & 0 & \cdots & 0 \\ \vdots & & & \vdots \\ 0 & \cdots & 1 & 0 \\ 0 & \cdots & 0 & \zeta \end{bmatrix} \in \mathrm{GL}(n, \ \mathbb{F}).
$$

Note that S_{ζ} is a pseudoreflection of order s. The subgroup $E_{\mathbb{F}}(r) < G(\zeta, \ r)$ is seen to be normal and there is an evident split exact sequence

$$
1 \longrightarrow E_{\mathbb{F}}(r) \longrightarrow G(\zeta, \ r) \longrightarrow \mathbb{Z}/s \longrightarrow 1 .
$$

PROPOSITION 8.2.16 (P. S. Landweber and R. E. Stong): *Let $\mathbb{F}$ be a Galois field with p^{ℓ} elements, and $\varrho : G(\zeta, \ r) \hookrightarrow \mathrm{GL}(n, \ \mathbb{F})$ the inclusion. Then $V^{G(\zeta, \ r)}$ has codimension 1 and*

$$
\mathbb{F}[z_1, \ldots, z_n]^{G(\zeta, \ r)} = \mathbb{F}[f_1, \ldots, f_r, \ z_{r+1}, \ldots, z_{n-1}, \ z_n^s]
$$

where

$$
f_i = \prod_{\lambda \in \mathbb{F}} (z_i + \lambda z_n) = c_{p^{\ell}}([z_i])
$$

is the p^{ℓ}-th Chern class of the orbit of z_i for $i = 1, \ldots, r$ and z_n^s is the s-th Chern class of the orbit of z_n.

PROOF : Since $E_{\mathbb{F}}(r) \triangleleft G(\zeta, \ r)$ we have

$$
\mathbb{F}[z_1, \ldots, z_n]^{G(\zeta, \ r)} \cong (\mathbb{F}[z_1, \ldots, z_n]^{E(r)})^{\mathbb{Z}/s} .
$$

By 8.2.6 we obtain

$$
\mathbb{F}[z_1, \ldots, z_n]^{G(\zeta, \ r)} = \mathbb{F}[f_1, \ldots, f_r, \ z_{r+1}, \ldots, z_n]^{\mathbb{Z}/s}
$$

$$
= \mathbb{F}[f_1, \ldots, f_r, \ z_{r+1}, \ldots, z_{n-1}] \otimes \mathbb{F}[z_n]^{\mathbb{Z}/s}
$$

$$
= \mathbb{F}[f_1, \ldots, f_r, \ z_{r+1}, \ldots, z_{n-1}] \otimes \mathbb{F}[z_n^s]
$$

and the result follows. $\square$

In a completely analagous fashion we obtain:

PROPOSITION 8.2.17 : *Let $\mathbb{F}$ be a finite field of nonzero characteristic p and p^{ℓ} elements, and $\varrho^* : G(\zeta, r)^* \hookrightarrow \mathrm{GL}(n, \mathbb{F})$ the inclusion. Then $V_{G(\zeta, r)^*}$ has dimension $n - 1$ and*

$$\mathbb{F}[z_1, \ldots, z_n]^{G(\zeta, r)^*} = \mathbb{F}[z_1, \ldots, z_{n-1}, f^s]$$

where

$$f = \prod_{\alpha_1, \ldots, \alpha_r \in \mathbb{F}} (z_n + a_r z_r + \cdots a_1 z_1) = c_{p^{r\ell}}([z_n])$$

is the top Chern class of the orbit of z_n. $\square$

The groups $G(\zeta, r) \hookrightarrow \mathrm{GL}(n, \mathbb{F})$ satisfy $\dim_{\mathbb{F}}(V^{G(\zeta, r)}) = n - 1$, whereas the groups $G(\zeta, r)^* \hookrightarrow \mathrm{GL}(n, \mathbb{F})$ satisfy $\dim_{\mathbb{F}}(V_{G(\zeta, r)^*}) = n - 1$. These representations provide a complete list up to conjugation [222], [131] of groups with one or the other of these properties.

PROPOSITION 8.2.18 : *Let $\mathbb{F}$ be a finite field of characteristic $p \neq 0$. Suppose $\varrho : G \hookrightarrow \mathrm{GL}(n, \mathbb{F})$ is a representation of a finite group G and every nonidentity element of $\varrho(G)$ is a pseudoreflection. Then, either V^G has dimension $n - 1$ and $\varrho(G)$ is conjugate to a subgroup of $G(\zeta, r)$, or V_G has dimension $n - 1$ and $\varrho(G)$ is conjugate to a subgroup of $G(\zeta, r)^*$ for some $r \in \{1, \ldots, n - 1\}$ and primitive s^{th}-root of unity $\zeta \in \mathbb{F}^{\times}$.*

PROOF : By 8.2.10 and the remark following, either all the nonidentity elements of G have the same hyperplane or the same direction. We consider the case where they all have the same hyperplane V^G. Since all the diagonalizable pseudoreflections in G have the same reflecting hyperplane, it follows from 7.1.2 and 7.1.3 that they have a common root vector $z_n \notin V^G$. If G contains no diagonalizable pseudoreflections choose z_n to be any vector not in V^G. Choose a basis $z_1, \ldots, z_{n-1}$ for V^G. Then $z_1, \ldots, z_{n-1}, z_n$ is a basis for V and with respect to this basis every element of G has a matrix representative of the form:

$$\varrho(g) = \begin{bmatrix} 1 & 0 & \cdots & 0 & \alpha_1(g) \\ 0 & 1 & \cdots & 0 & \alpha_2(g) \\ \vdots & 0 & \cdots & \ddots & \vdots \\ 0 & \cdots & \cdots & \cdots & \alpha_n(g) \end{bmatrix}$$

where $\alpha_1, \ldots, \alpha_{n-1} \in \mathbb{F}$ and $\alpha_n \in \mathbb{F}^{\times}$, and therefore up to conjugation G is a subgroup of $G(\zeta, r)$ where ζ generates the subgroup $\{\alpha_n(g) \mid g \in G\}$. $\square$

THEOREM 8.2.19 (P. S. Landweber and R. E. Stong): *Let $\mathbb{F}$ be a finite field $\mathbb{F}$ of characteristic p and $\varrho : G \hookrightarrow GL(n, \mathbb{F})$ a representation of a finite group. Suppose that the fixed point set V^G has codimension 1 in V. Then there exist integers r, s and an s-th root of unity ζ such that $\varrho(G)$ is conjugate to a subgroup of $G(\zeta, r)$. Moreover $\mathbb{F}[V]^G = \mathbb{F}[f_1, \ldots, f_r, z_{r+1}, \ldots, z_{n-1}, z_n^s]$ is a polynomial algebra.*

PROOF : From lemma 8.2.3 it follows that $\dim_{\mathbb{F}}(V^*)_G = \dim_{\mathbb{F}}(V) - 1$. Therefore the subspace $\mathrm{Span}_{\mathbb{F}} \{gv^* - v^* \mid g \in G, \, v^* \in V^*\} \subset V^*$ has dimension 1. Choose a nonzero element $w^* \in \mathrm{Span}_{\mathbb{F}} \{gv^* - v^* \mid g \in G, \, v^* \in V^*\}$. Then for any vector $v^* \in V^*$ and any $g \in G$ we have

$$g \cdot v^* - v^* = \lambda_{(v^*, \, g)} w^*.$$

In particular

$$g \cdot w^* = w^* + \lambda_{(w^*, \, g)} w^* = (1 + \lambda_{(w^*, \, g)}) w^*,$$

so $\mathrm{Span}_{\mathbb{F}} \{gv^* - v^* \mid g \in G, \, v^* \in V^*\}$ is a G-invariant subspace of V^*. Note that $\lambda_{(v^*, \, g)} = 0$ if and only if $gv^* = v^*$, i.e. if and only if $v^* \in (V^*)^g$. Since the representation ϱ is faithful $(V^*)^g = V^*$ if and only if $g = 1 \in G$. Therefore for an element $g \neq 1 \in G$ there exists a $v^* \in V^*$ with $\lambda_{(v^*, \, g)} \neq 0$. Hence

$$\{0\} \neq \mathrm{Im}(g - 1) \subset \mathrm{Span}\{w^*\}$$

so $\dim_{\mathbb{F}}(\mathrm{Im}(g - 1)) = 1$ and $\dim_{\mathbb{F}}(\ker(g - 1)) = n - 1$. This says that the nonidentity elements of G are all pseudoreflections.

Define a one-dimensional representation $\zeta : G \longrightarrow \mathbb{F}^\times$ by the requirement

$$g \cdot w^* = \zeta(g) w^* \qquad \forall \, g \in G.$$

This is possible since $\mathrm{Span}\{w^*\}$ is an invariant subspace. Of course $\zeta(g) = 1 + \lambda_{(w^*, \, g)}$ for all $g \in G$. The kernel of ζ consists of those elements $g \in G$ for which

$$1 + \lambda_{(w^*, \, g)} = 1 \in \mathbb{F}^\times,$$

i.e., $\lambda_{(w^*, \, g)} = 0 \in \mathbb{F}$. In other words $g \in \ker(\zeta)$ if and only if $g \cdot w^* = w^*$ and therefore g is a non-diagonalizable pseudoreflection. Since non-diagonalizable pseudoreflections have order p it follows that every element $g \neq 1 \in \ker(\zeta)$ has order p. Hence $\ker(\zeta)$ is an elementary abelian p-group, and with respect to a suitably chosen basis of V^*

$$\ker(\zeta) = \left\{ \begin{bmatrix} 1 & 0 & \ldots & \ldots & \alpha_1 \\ 0 & 1 & 0 & \ldots & \alpha_2 \\ 0 & \ldots & \ddots & \ldots & \alpha_r \\ 0 & \ldots & \ldots & \ddots & 0 \\ 0 & \ldots & \ldots & 0 & 1 \end{bmatrix} \middle| \; \alpha_i \in A_i \; i = 1, \ldots, r \right\}.$$

If $z_1, \ldots, z_n \in V^*$ is the dual basis then the orbits are

$$[z_i] = \begin{cases} \{z_i + \alpha_i z_n \mid \alpha_i \in A_i\} & \text{for } i = 1, \ldots, r \\ \{z_i\} & \text{for } i = r+1, \ldots, n. \end{cases}$$

as in 8.2.15.

Next consider $\mathrm{Im}(\zeta) < \mathbb{F}^\times$. Since G is a finite group, $\mathrm{Im}(\zeta)$ is cyclic, generated by an s-th root of unity ξ. If $g \in G$ satisfies $\zeta(g) = \xi$, then $\zeta(g^p) = \xi^p$ is some other generator for $\mathrm{Im}(\zeta)$ and in addition g^p has order s. Thus we have constructed a split exact sequence

$$1 \longrightarrow \ker(\zeta) \longrightarrow G \longrightarrow \mathbb{Z}/s \longrightarrow 1$$

inside of $\mathrm{GL}(n, \mathbb{F})$. Hence $G = G(\zeta, r)$ and the representation ϱ is dual to the inclusion $G(\zeta, r) < \mathrm{GL}(n, \mathbb{F})$, so the result follows from 8.2.16. □

Completely analogous is the following:

THEOREM 8.2.20 : *Let $\varrho : G \hookrightarrow \mathrm{GL}(n, \mathbb{F})$ be a representation of a finite group over a finite field $\mathbb{F}$ of nonzero characteristic p. Suppose that the covariants V_G have dimension $n - 1$. Then there exist integers r, s and an s-th root of unity ζ such that $\varrho(G)$ is conjugate to a subgroup of $G(\zeta, r)^*$. Moreover $\mathbb{F}[V]^G = \mathbb{F}[z_1, \ldots, z_{n-1}, f^s]$ is a polynomial algebra.* □

By proposition 8.2.16 for $\mathbb{F} = \mathbb{F}_p$

$$\mathbb{F}[z_1, \ldots, z_n]^{G(\zeta, r)} = \mathbb{F}[f_1, \ldots, f_r, z_{r+1}, \ldots, z_{n-1}, z_n^s]$$

where $\deg(f_i) = p$ for $i = 1, \ldots, r$. Therefore the degrees of $G(\zeta, r)$ are

$$(p, \ldots, p, 1, \ldots, 1, s) = (d_1, \ldots, d_r, d_{r+1}, \ldots, d_{n-1}, d_n)$$

where p occurs r times, 1 occurs $n - r - 1$ times, and s once. By the proof of 8.2.19 every nonidentity element of $G(\zeta, r)$ is a pseudoreflection, so $|s(G(\zeta, r))| = sp^r - 1$, and therefore

$$\sum_{i=1}^{n}(d_i - 1) = r(p - 1) + s - 1 \neq |s(G(\zeta, r))|.$$

This shows that the formula in 4.4.3 for $|s(G)|$ need not hold for a group G whose order is divisible by the characteristic of the ground field. (We already saw this in section 5.6 example 1 for D_{2p}. Since $\varrho^*(D_{2p}) = G(-1, 1)$ that example was not an isolated instance, but rather part of a general phenomenon.)

REMARK : Notice that the groups $G(\zeta, \ r)$ come equipped with a representation $\zeta : G(\zeta, \ r) \longrightarrow \mathbb{F}^{\times}$ of dimension 1. It is not hard to see that

$$\mathbb{F}[z_1, \ldots, z_n]_{\zeta}^{G(\zeta, \ r)} = \mathbb{F}[f_1, \ldots, f_r, \ z_{r+1}, \ldots, z_{n-1}, \ z_n^s] \cdot z_n$$

in the notations of 8.2.16 and dually

$$\mathbb{F}[z_1, \ldots, z_n]_{\zeta}^{G(\zeta, \ r)^*} = \mathbb{F}[z_1, \ldots, z_{n-1}, \ f^s] \cdot z_n \ .$$

8.3 p-Sylow Subgroups and the Cohen-Macaulay Property

Let G be a finite group and $\varrho : G \hookrightarrow \mathrm{GL}(n, \ \mathbb{F})$ a representation over a field of nonzero characteristic p. Set $V = \mathbb{F}^n$ and as usual denote by V^G the fixed point set of the action of G on V. If V^G has codimension at least 2 in V Ellingsrud and Skjelbred [78] have shown that $\mathrm{codim}(\mathbb{F}[V]^G) \geq 2 + \dim_{\mathbb{F}}(V^G)$, where $\mathrm{codim}(\text{---})$ denotes homological codimension (see section 6.6), and if G is cyclic of prime power order then one has equality. (See also [88].) If $p > 3$ and V is the regular representation of $\mathbb{Z}/p$ over the Galois field of p elements then $\dim_{\mathbb{F}}(V^{\mathbb{Z}/p}) = 1$, so $\mathrm{codim}_{\mathbb{F}}(V^G) = p - 1 > 2$ and hence

$$\mathrm{codim}(\mathbb{F}[V]^{\mathbb{Z}/p}) = 2 + \dim_{\mathbb{F}}(V^{\mathbb{Z}/p}) = 3 < p = \dim(\mathbb{F}[V]^{\mathbb{Z}/p})$$

and $\mathbb{F}[V]^{\mathbb{Z}/p}$ is not Cohen-Macaulay. (See [220] for a discussion of recent results on the codimension of rings of invariants.)

The case where the codimension of the fixed point set is 1 was studied in section 8.2 where we saw that the ring of invariants is a polynomial algebra and hence Cohen-Macaulay. In general, the Cohen-Macaulay property in characteristic p is controlled by the p-Sylow subgroup. If G is a finite group and p is a prime denote by $\mathrm{Syl}_p(G)$ a p-Sylow subgroup of G.

PROPOSITION 8.3.1 : Let $\mathbb{F}$ be a field of characteristic $p \neq 0$ and $\varrho : G \hookrightarrow \mathrm{GL}(n, \ \mathbb{F})$ a representation of a finite group G. If $\mathbb{F}[V]^{\mathrm{Syl}_p(G)}$ is Cohen-Macaulay then so is $\mathbb{F}[V]^G$.

PROOF : The index $|G : \mathrm{Syl}_p(G)|$ is relatively prime to p. Therefore we have the projection operator (see section 2.4)

$$\pi_{\mathrm{Syl}_p(G)}^G : \mathbb{F}[V]^{\mathrm{Syl}_p(G)} \longrightarrow \mathbb{F}[V]^G$$

which is an $\mathbb{F}[V]^G$-module homomorphism such that the composite

$$(*) \qquad \mathbb{F}[V]^G \hookrightarrow \mathbb{F}[V]^{\mathrm{Syl}_p(G)} \xrightarrow{\pi_{\mathrm{Syl}_p(G)}^G} \mathbb{F}[V]^G$$

is the identity. Let $f_1, \ldots, f_n \in \mathbb{F}[V]^G$ be a system of parameters. $\mathbb{F}[V]^{\mathrm{Syl}_p(G)}$ is finitely generated as $\mathbb{F}[V]^G$-module. Hence $f_1, \ldots, f_n \in \mathbb{F}[V]^{\mathrm{Syl}_p(G)}$ is also a system of parameters for $\mathbb{F}[V]^{\mathrm{Syl}_p(G)}$. By hypothesis $\mathbb{F}[V]^{\mathrm{Syl}_p(G)}$ is Cohen-Macaulay, so by Macaulay's theorem (theorem 6.7.7) $f_1, \ldots, f_n \in \mathbb{F}[V]^{\mathrm{Syl}_p(G)}$ is a regular sequence. Therefore by corollary 6.2.8 $\mathbb{F}[V]^{\mathrm{Syl}_p(G)}$ is a free $\mathbb{F}[f_1, \ldots, f_n]$-module. It follows from (*) that $\mathbb{F}[V]^G$ is a direct summand of $\mathbb{F}[V]^{\mathrm{Syl}_p(G)}$ as $\mathbb{F}[V]^G$-module, and hence *a fortiori* as $\mathbb{F}[f_1, \ldots, f_n]$-module. Therefore $\mathbb{F}[V]^G$ is a projective $\mathbb{F}[f_1, \ldots, f_n]$-module, and hence by proposition 6.1.1 a free $\mathbb{F}[f_1, \ldots, f_n]$-module, which makes it Cohen-Macaulay. $\quad\square$

COROLLARY 8.3.2 : *Let $\mathbb{F}$ be a field of characteristic $p \neq 0$ and $\varrho : G \hookrightarrow \mathrm{GL}(n, \mathbb{F})$ a representation of a finite group G. If the fixed point set of a p-Sylow subgroup $\mathrm{Syl}_p(G)$ of G, $V^{\mathrm{Syl}_p(G)}$, has codimension 1 in V, or the covariants $V_{\mathrm{Syl}_p(G)}$ have dimension $n - 1$, then $\mathbb{F}[V]^G$ is Cohen-Macaulay.*

PROOF : This follows from 8.2.19, its dual 8.2.20, and the preceding proposition. $\quad\square$

This leads to an alternate proof that rings of invariants in two variables are Cohen-Macaulay. (See [216] for the case of three variables.)

COROLLARY 8.3.3 : *Let $\mathbb{F}$ be a field and $\varrho : G \hookrightarrow \mathrm{GL}(2, \mathbb{F})$ a representation of a finite group G. Then $\mathbb{F}[x, y]^G$ is Cohen-Macaulay.*

PROOF (M. D. Neusel) : By 8.2.1 $(V^*)^P \neq \{0\}$ and by 1.5.8 $\mathbb{F}[V]^P$ is a unique factorization domain. If $\ell \in \mathbb{F}[V]^P$ is a nonzero P-invariant linear form, it is indecomposable, and hence $(\ell) \subset \mathbb{F}[V]^P$ is a prime ideal of height 1, so $\mathbb{F}[V]^P/(\ell)$ is an integral domain of Krull dimension 1. Therefore every nonzero element is a nonzero divisor. If $f \in \mathbb{F}[V]^P$ projects nonzero in $\mathbb{F}[V]^P/(\ell)$ then $\ell, f \in \mathbb{F}[V]^P$ is a regular sequence. $\quad\square$

REMARK : It is not hard to show (see e.g. [220] theorem 4.6) that $\mathrm{codim}(\mathbb{F}[V]^{\mathrm{Syl}_p(G)}) \leq \mathrm{codim}(\mathbb{F}[V]^G)$ and the preceeding argument then shows that rings of invariants in at least two variables always have homological codimension at least 2.

PROPOSITION 8.3.4 : *Let $\mathbb{F}$ be a field of characteristic $p \neq 0$ and $\varrho : G \hookrightarrow \mathrm{GL}(n, \mathbb{F})$ a representation of a finite group G. Suppose that all the nonidentity elements of $\varrho(\mathrm{Syl}_p(G))$ are transvections. Then $\mathbb{F}[V]^G$ is Cohen-Macaulay.*

PROOF : This follows from 8.3.1 and 8.2.14. $\quad\square$

The p-Sylow subgroup of GL$(n,\ \mathbb{F}_p)$ is the group of unipotent matrices

$$\mathrm{Uni}(n,\ \mathbb{F}_p) = \left\{ \begin{bmatrix} 1 & & \\ 0 & \ddots & * \\ & & 1 \end{bmatrix} \in \mathrm{GL}(n,\ \mathbb{F}_p) \mid * \in \mathbb{F}_p \right\}.$$

It is a finite p-group of order $p^{n-1} \cdot p^{n-2} \cdots p = p^{\binom{n}{2}}$. The elements of Uni$(n,\ \mathbb{F}_p)$ all have determinant 1, so Uni$(n,\ \mathbb{F}_p)$ is also a p-Sylow subgroup of SL$(n,\ \mathbb{F}_p)$ and any subgroup between SL$(n,\ \mathbb{F}_p)$ and GL$(n,\ \mathbb{F}_p)$.

Let $E_{i,j}$ denote the $n \times n$ matrix with a 1 in the i-th row and the j-th column, and otherwise 0. The matrices $u_{i,j} = I + E_{i,j}$ belong to Uni$(n,\ \mathbb{F}_p)$ for $i < j$. If $\{x_1, \ldots, x_n\}$ is the standard basis for $V = \mathbb{F}_p^{\,n}$ then

$$E_{i,j}(x_k) = \begin{cases} 0 & j \neq k \\ x_i & j = k \end{cases}$$

$$u_{i,j}(x_k) = \begin{cases} x_k & j \neq k \\ x_k + x_i & j = k. \end{cases}$$

Passing to the dual vector space V^* and the dual basis $\{z_1, \ldots, z_n\}$ we find

$$E_{i,j}(z_k) = \begin{cases} 0 & i \neq k \\ z_j & i = k \end{cases}$$

$$u_{i,j}(z_k) = \begin{cases} x_k & i \neq k \\ z_k + z_j & i = k. \end{cases}$$

The elements of Uni$(n,\ \mathbb{F}_p)$ can be written uniquely in the form

$$u = I + \sum_{i<j} a_{i,j} E_{i,j} \qquad a_{i,j} \in \mathbb{F}_p$$

so we have

$$u(z_k) = \left(I + \sum_{i<j} a_{i,j} E_{i,j} \right)(z_k) = \begin{cases} z_k + \sum_{k<j} a_{k,j} z_j & k < n \\ z_n & k = n. \end{cases}$$

Therefore the orbit of z_k under the action of the group Uni$(n,\ \mathbb{F}_p)$ is

$$[z_k] = \begin{cases} \{z_k + b_{k+1} z_{k+1} + \cdots + b_n z_n \mid b_{k+1}, \ldots, b_n \in \mathbb{F}_p\} & k < n \\ \{z_n\} & k = n. \end{cases}$$

Hence $|\,[z_k]\,| = p^{n-k}$. If we set (take the orbit Chern class of top degree)

$$\mathsf{h}_{n,\,n-k} = \prod_{z \in [z_k]} z = \begin{cases} \prod_{b_{k+1}, \ldots, b_n \in \mathbb{F}_p} (z_k + b_{k+1} z_{k+1} +, \ldots, + b_n z_n) & k < n \\ z_n & k = n \end{cases}$$

for $k = 0, \ldots, n-1$ (note the indexing is analogous to that of the Dickson polynomials) then

$$\deg(\mathbf{h}_{n,i}) = p^{n-i-1} \qquad i = 0, \ldots, n-1.$$

The classes $\mathbf{h}_{n,0}, \ldots, \mathbf{h}_{n,n-1}$ are a regular sequence and

$$\prod_{i=0}^{n-1} \deg(\mathbf{h}_{n,i}) = p^{\binom{n}{2}} = |\mathrm{Uni}(n, \ \mathbb{F}_p)|.$$

So by the theorem of Nakajima-Stong 5.5.6 we obtain:

THEOREM 8.3.5 : *With the preceding notations*

$$\mathbb{F}_p[z_1, \ldots, z_n]^{\mathrm{Uni}(n, \ \mathbb{F}_p)} = \mathbb{F}_p[\mathbf{h}_{n,0}, \ldots, \mathbf{h}_{n,n-1}]$$

for any $n \in \mathbb{N}$ and any prime p. $\square$

COROLLARY 8.3.6 : *Let $\varrho : G \hookrightarrow \mathrm{GL}(n, \ \mathbb{F}_p)$ be a representation of a finite group. If $\varrho(\mathrm{Syl}_p(G)) = \mathrm{Uni}(m, \ \mathbb{F}_p)$ for some integer $m \leq n$ then $\mathbb{F}[V]^G$ is Cohen-Macaulay.*

PROOF : The inclusion $\mathbb{F}_p^m \hookrightarrow \mathbb{F}_p^n$ induces an inclusion $\mathrm{Uni}(m, \ \mathbb{F}_p) \subset \mathrm{GL}(n, \ \mathbb{F}_p)$ and

$$\mathbb{F}_p[z_1, \ldots, z_n]^{\mathrm{Uni}(m, \ \mathbb{F}_p)} = \mathbb{F}_p[\mathbf{h}_{m,0}, \ldots, \mathbf{h}_{m,m-1}, \ z_{m+1}, \ldots, z_n].$$

Thus $\mathbb{F}_p[V]^{\mathrm{Uni}(m, \ \mathbb{F}_p)}$ is Cohen-Macaulay and the result follows from 8.3.1. $\square$

The standard inclusion $\iota : \mathbb{F}_p^{n-1} \hookrightarrow \mathbb{F}_p^n$ which is given by $\iota(a_1, \ldots, a_{n-1}) = (a_1, \ldots, a_{n-1}, \ 0)$ induces a map

$$\iota^* : \mathbb{F}_p[z_1, \ldots, z_n]^{\mathrm{Uni}(n, \ \mathbb{F}_p)} \longrightarrow \mathbb{F}_p[z_1, \ldots, z_{n-1}]^{\mathrm{Uni}(n-1, \ \mathbb{F}_p)}$$

and one sees that

$$\iota^*(\mathbf{h}_{n,i}) = \begin{cases} 0 & i = 0 \\ \mathbf{h}_{n-1,i-1}^p & i = 1, \ldots, n-1. \end{cases}$$

This is in complete analogy with the behavior of the Dickson polynomials under restriction from $\mathbb{F}_p^n$ to $\mathbb{F}_p^{n-1}$.

8.4   Groups Generated by Pseudoreflections

If $\varrho : G \hookrightarrow \mathrm{GL}(n, \mathbb{F})$ is a representation of a finite group such that $\varrho(G)$ is generated by pseudoreflections, but the characteristic of $\mathbb{F}$ divides the order of G, then the ring of invariants may fail to be a polynomial algebra. For example the Coxeter group $W(\mathbf{F}_4)$ corresponding to the root system of type $\mathbf{F}_4$ does not have polynomial invariants at the prime 3 [249]. The group has order 1152. There are not too many positive results concerning the rings of invariants of arbitrary finite groups generated by pseudoreflections. One such result is:

THEOREM 8.4.1 (A. Dress): *Let $\varrho : G \hookrightarrow \mathrm{GL}(n, \mathbb{F})$ be a representation of a finite group. If $\varrho(G)$ is generated by pseudoreflections then $\mathbb{F}[V]^G$ is a unique factorization domain.*

PROOF : Let $f \in \mathbb{F}[V]^G$ and write

$$f = f_1^{k_1} \cdots f_m^{k_m} \in \mathbb{F}[V]$$

where $f_1, \ldots, f_m \in \mathbb{F}[V]$ are distinct irreducible polynomials. Since f is invariant the group G permutes the principal ideals $(f_1), \ldots, (f_m)$. We denote by $B_1, \ldots, B_k$ the orbits of this G-action and define polynomials $F_1, \ldots, F_k$ by

$$F_i = \prod_{f_j^{k_j} \in B_i} f_j \qquad i = 1, \ldots, k \,.$$

We claim that $F_1, \ldots, F_k \in \mathbb{F}[V]^G$. To see this fix an integer $j \in \{1, \ldots, m\}$. For an element $g \in G$ let $<g>$ denote the subgroup of G generated by g. Note that each of the ideals $(F_1), \ldots, (F_k)$ are stable under the action of G and hence

$$gF_i = \lambda(g)F_i \qquad i = 1, \ldots, k$$

where $\lambda : G \longrightarrow \mathbb{F}^\times$ is a homomorphism. If $\varrho(g)$ is a non-diagonalizable pseudoreflection than g has order p so $\lambda(g) = 1$ since $\mathbb{F}^\times$ contains no p^{th} root of unity other than 1. If $\varrho(g) = s$ is a diagonalizable pseudoreflection, then $F_i \in \mathbb{F}[V]_\lambda^{<g>}$. By the theorem of Springer and Stanley 7.6.4 $\mathbb{F}[V]_\lambda^{<g>} = \mathbb{F}[V]^{<g>} \cdot \ell_s^{e_s}$, since $|<g>|$ has order prime to p, where $\ell_s : V \longrightarrow \mathbb{F}$ is a linear functional with $\ker(\ell_s) = H_s$, the hyperplane of s, and $e_s \in \mathbb{N}$ is the order of of $\det(s) \in \mathbb{F}^\times$. We therefore have

$$F_i = \bar{F}_i \ell_s^{e_s} \qquad \bar{F}_i \in \mathbb{F}[V]^{<g>} \,.$$

Since $\deg(\ell_s) = 1$ the polynomial ℓ_s is irreducible. The ring $\mathbb{F}[V]$ is a unique factorization domain so it follows from the definition of $f_1, \ldots, f_m$ that up to a nonzero scalar factor $\ell_s = f_j$ for some $j \in \{1, \ldots, m\}$. In other words,

$\ell_s \in B_i$ for some $i \in \{1, \ldots, k\}$. By 7.1.2 $<s>$ consists of 1 together with all pseudoreflections in $\varrho(G)$ with reflecting hyperplane H_s, so if $t \in s_\varrho(G)$ is a pseudoreflection not in $<s>$ then $t\ell_s = \ell_s$, whereas $t\ell_s = \det(t^{-1})\ell_s$ by 7.1.6 if $t \in <s>$. Since $\varrho(G)$ is generated by pseudoreflections it follows that the principal ideal $(\ell_s) = (f_j)$ is invariant under the action of G, and hence $B_i = \{\ell_s\}$ and $F_i = \ell_s^{e_s} = f_j^{k_j}$.

Let us summarize what we have shown so far. For $f \in \mathbb{F}[V]^G$ we have

$$f = F_1 \cdots F_k \in \mathbb{F}[V],$$

where either $F_i \in \mathbb{F}[V]^G$, or $F_i = \ell_{s_i}^{e_{s_i}}$ for some pseudoreflection $s_i \in s_\varrho^\Delta(G)$, and moreover the reflecting hyperplanes of these pseudoreflections are pairwise distinct.

It is now an easy matter to see that in fact $F_1, \ldots, F_k \in \mathbb{F}[V]^G$. For suppose $F_i = \ell_{s_i}^{e_{s_i}} \notin \mathbb{F}[V]^G$ for some diagonalizable pseudoreflection in $\varrho(G)$. Then

$$f = s_i f = s_i(F_1 \cdots F_k) = s_i F_1 \cdot s_i F_2 \cdots s_i F_k$$

$$= \det(s_i)^{-e_{s_i}} F_1 \cdot F_2 \cdots F_k = \det(s_i)^{-e_{s_i}} f$$

since $sF_j = F_j$ for $j \in \{1, \ldots, \widehat{i}, \ldots, k\}$, either by 7.1.6 or because $F_j \in \mathbb{F}[V]^G$, and $sF_i = \det(s_i)^{-e_{s_i}} F_i$. Therefore $\det(s_i)^{-e_{s_i}} = 1$ and hence $sF_i = s\ell_{s_i}^{e_{s_i}} = \det(s_i)^{-e_{s_i}} F_i = F_i$, so $F_1, \ldots, F_k \in \mathbb{F}[V]^G$ as claimed. It is now a routine matter to complete the proof as in 1.5.7. $\square$

The study of the invariants of groups generated by pseudoreflections in the modular case is is the subject of several of Nakajima's papers [153] — [162], as well as [168], and [164], which amplifies work of Demazure [65], and Kac and Peterson [117].

Chapter 9
Polynomial Tensor Exterior Algebras

Let $\varrho : G \hookrightarrow \mathrm{GL}(n, \ \mathbb{F})$ be a representation of a finite group G in the n-dimensional vector space $V = \mathbb{F}^n$. The action of G on V extends not just to the polynomial algebra $\mathbb{F}[V]$ but also to the exterior algebra $E[V]$ (see section 6.2) and hence to their tensor product $\mathbb{F}[V] \otimes E[V]$. Taking the fixed points of the action of G on $\mathbb{F}[V] \otimes E[V]$ gives the ring of invariants $(\mathbb{F}[V] \otimes E[V])^G$ which is one of the subjects of this chapter.

A minor complication is caused by the fact that the exterior algebra $E[V]$ (and hence also $\mathbb{F}[V] \otimes E[V]$) is not commutative in the classical sense. The algebra $E[V]$ is however **commutative in the graded sense**, i.e.

$$\omega' \cdot \omega'' = (-1)^{i'i''} \omega'' \cdot \omega' \qquad \forall \, \omega' \in \Lambda^{i'}(V^*), \ \omega'' \in \Lambda^{i''}(V^*) \, .$$

This condition is often referred to as **skew commutativity** or **anti-commutativity**. The theorems in previous chapters that we have proven in the commutative case are also valid for algebras which are commutative in the graded sense. In fact the proofs (or minor modifications of them) that we have given are sufficient in this wider context.

One may think of $\mathbb{F}[V] \otimes E[V]$ as the algebra of polynomial differential forms on V by introducing the exterior derivative d into $\mathbb{F}[V] \otimes E[V]$. Alternatively one may think of $\mathbb{F}[V] \otimes E[V]$ as the Koszul complex, or, in the case of $\mathbb{F} = \mathbb{F}_p$, the Galois field of p elements for an odd prime p, one may regard $\mathbb{F}[V] \otimes E[V]$ as $H^*(V, \ \mathbb{F}_p)$, the cohomology of the elementary abelian p-group V with coefficients in $\mathbb{F}_p$. In section 9.4 we present an application of this viewpoint to the cohomology of the symmetric group. There are still further ways to think of $\mathbb{F}[V] \otimes E[V]$, see e.g. [6]. Each of these viewpoints leads to a natural way to grade $\mathbb{F}[V] \otimes E[V]$.

We will adopt a neutral approach to grading $\mathbb{F}[V] \otimes E[V]$ by bigrading it, thereby separating out the commuting and anti-commuting variables. (The grading most used by topologists is perhaps more appropriate, but probably only topologists would feel comfortable with it.) Both $\mathbb{F}[V]$ and $E[V]$ have natural gradings in which the elements of V^* have degree 1. For the tensor product we introduce the bigrading

$$(\mathbb{F}[V] \otimes E[V])_{i,j} = \mathbb{F}[V]_i \otimes E[V]_j \qquad i, j \in \mathbb{N}_0 \, .$$

The operation of G on $\mathbb{F}[V] \otimes E[V]$ preserves the bigrading and hence the ring of invariants $(\mathbb{F}[V] \otimes E[V])^G$ inherits a bigrading from $\mathbb{F}[V] \otimes E[V]$.

More generally, if A and B are positively graded algebras over the field $\mathbb{F}$ we may bigrade $A \otimes B$ by setting

$$(A \otimes B)_{i,j} = A_i \otimes B_j \qquad i, j \in \mathbb{N}_0 \, .$$

In this context it is natural to introduce as Poincaré series the double series

$$P(C, \, t, \, s) = \sum_{i,j=0}^{\infty} \dim_{\mathbb{F}}(C_{i,j}) \, t^i s^j$$

for a bigraded algebra C over $\mathbb{F}$. One of our first goals will be to obtain the analog of Molien's theorem for the algebra of invariants $(\mathbb{F}[V] \otimes E[V])^G$.

The **exterior derivative**

$$d : \mathbb{F}[V] \otimes E[V] \longrightarrow \mathbb{F}[V] \otimes E[V]$$

is defined by the requirements

(i) $\ d(z_i) = dz_i$ and $d(dz_i) = 0, \quad i = 1, \ldots, n \, ,$

(ii) $\ d(f \otimes 1) = \sum\limits_{i=1}^{n} \dfrac{\partial f}{\partial z_i} \otimes dz_i \, ,$

(iii) the usual Leibnitz rule for products

where $z_1, \ldots, z_n$ denotes a basis $x_1 \otimes 1, \ldots, x_n \otimes 1$ for $V^* \otimes 1$ and $dz_1, \ldots, dz_n$ the corresponding basis $1 \otimes x_1, \ldots, 1 \otimes x_n$ for $1 \otimes V^*$. This allows us to write elements of $\mathbb{F}[V] \otimes E[V]$ of bidegree (t, s) in the form

$$\sum_{i_1, \ldots, i_s} f_{i_1, \ldots, i_s} dz_{i_1} \cdots dz_{i_s} \, ,$$

where $f_{i_1, \ldots, i_s} \in \mathbb{F}[V]$ is homogeneous of degree t, and think of them as polynomial differential forms on V when convenient.

REMARK : Since $d^2 = 0$ we may regard $\mathbb{F}[V] \otimes E[V]$ equipped with the exterior derivative as a complex. If $\mathbb{F}$ has characteristic zero this complex is acyclic [247]. When suitably regraded (compare 6.2) it provides a free resolution of $\mathbb{F}$ as an $E[V]$-module. In characteristic p this is no longer the case. For example, if V is one-dimensional with basis vector z, then $z^{p-1}dz$

is a cycle which is not a boundary. (See for example [211] for a discussion of this and related complexes.)

The exterior derivative d commutes with the G-action and hence for $f \in \mathbb{F}[V]^G$ we receive $df \in (\mathbb{F}[V] \otimes E[V])^G$. This way of obtaining elements in $(\mathbb{F}[V] \otimes E[V])^G$ from the polynomial invariants of G is adequate to describe $(\mathbb{F}[V] \otimes E[V])^G$ when $\varrho(G)$ is generated by pseudoreflections and G has order prime to the characteristic of the ground field $\mathbb{F}$. This result is due to L. Solomon, as is most of the material in this chapter, and will be discussed in section 9.3.

9.1   Polynomial Invariants and Representation Theory

The invariant theory of $\mathbb{F}[V] \otimes E[V]$ was exploited by L. Solomon [226], [227] to study the representation of G afforded by $\mathbb{F}[V]_G$ when G is a Coxeter group. To see how $\mathbb{F}[V]_G$ and $(\mathbb{F}[V] \otimes E[V])^G$ are related, we recall some elementary facts about representations of finite groups as may be found in the first few pages of [202].

Denote by G a finite group and by $\mathbb{F}$ an algebraically closed field with characteristic prime to the order of G. If N is a finite dimensional G-representation, then N decomposes as a direct sum of irreducible G-representations. If M is an irreducible G-module then Schur's lemma says

$$N\Big|_M \cong \mathrm{Hom}_G(M,\, N)$$

is the largest subrepresentation of N which is isomorphic to a direct sum of copies of M. The notation $N\Big|_M$ was already introduced in special cases in section 1.2 as well as the terminology M-**isotypical summand** or M-**isotypical component** for $N\Big|_M$. One has

$$N \cong \bigoplus_{M \in \mathrm{Irr}(G)} N\Big|_M$$

where the sum ranges over the (isomorphism classes of) irreducible representations of G. This direct sum decomposition is called the **canonical decomposition** of the representation N.

If N is a **graded G-representation of finite type**, i.e. a graded module of finite type over the group algebra $\mathbb{F}(G)$, then, since the component N_i of degree i is a finite-dimensional G-representation, one obtains a **graded canonical decomposition** into **graded M-isotypical summands**. Both $\mathbb{F}[V]$ and $\mathbb{F}[V]_G$ are graded G-representations of finite type.

EXAMPLE 1 : Consider $\tau : \Sigma_3 \hookrightarrow GL(3, \mathbb{C})$, the tautological representation, obtained by letting Σ_3 permute the standard basis vectors of $\mathbb{C}^3 = V$. The representation τ is reducible and

$$V \cong \mathbb{C} \oplus \mathbf{L}$$

where $\mathbb{C}$ denotes a trivial one-dimensional representation given by the span of the vector $(1, 1, 1)$ and $\mathbf{L}$ is the orthogonal complement. In section 1.2 we saw that (n.b. $\mathbf{L}$ and V are self dual as Σ_3 representations, i.e. $\mathbf{L} \cong \mathbf{L}^*$ as $\mathbb{C}(\Sigma_3)$-modules and analogously for V and V^*)

$$
\begin{aligned}
(\mathbb{C}[V]_{\Sigma_3})_3 &\cong \det &&\cong \Lambda^3(V) \\
(\mathbb{C}[V]_{\Sigma_3})_2 &\cong \mathbf{L} &&\cong \Lambda^2(V) \\
(\mathbb{C}[V]_{\Sigma_3})_1 &\cong \mathbf{L} &&\cong \Lambda^1(V) \\
(\mathbb{C}[V]_{\Sigma_3})_0 &\cong \mathbb{C} &&\cong \Lambda^0(V)
\end{aligned}
$$

as Σ_3-representations.

The representations $\mathbb{C} = \Lambda^0(V)$, $\Lambda^1(V) = \mathbf{L} = \Lambda^2(V)$, $\Lambda^3(V) = \det$ are a complete list of the irreducible representations of Σ_3 and therefore the isotypical summands of $\mathbb{C}[V]_{\Sigma_3}$ are:

$$
\begin{aligned}
\mathbb{C}[V]_{\Sigma_3}\Big|_{\det} &\cong \Sigma^3(\det) \\
\mathbb{C}[V]_{\Sigma_3}\Big|_{\mathbf{L}} &\cong \Sigma^2(\mathbf{L}) \oplus \Sigma^1(\mathbf{L}) \\
\mathbb{C}[V]_{\Sigma_3}\Big|_{\mathbb{C}} &\cong \mathbb{C},
\end{aligned}
$$

where, for an ungraded module M, $\Sigma^d(M)$ denotes the graded module defined by

$$(\Sigma^d(M))_i = \begin{cases} M & i = d \\ 0 & \text{otherwise.} \end{cases}$$

$\Sigma^d(M)$ is the d-**fold shift** or d-**fold suspension** of the module M regarded as a graded module concentrated in degree zero. Thus the Poincaré series of the isotypical summands are

$$P(\mathbb{C}[V]_{\Sigma_3}\Big|_{\det}, t) = t^3$$

$$P(\mathbb{C}[V]_{\Sigma_3}\Big|_{\mathbf{L}}, t) = t + t^2$$

$$P(\mathbb{C}[V]_{\Sigma_3}\Big|_{\mathbb{C}}, t) = 1.$$

From 7.4.3 we have

$$\mathbb{C}[V] \cong \mathbb{C}[V]^{\Sigma_3} \otimes \mathbb{C}[V]_{\Sigma_3}$$

as $\mathbb{C}[V]^{\Sigma_3}$- and $\mathbb{C}(\Sigma_3)$-modules and therefore we obtain for the isotypical components of $\mathbb{C}[V]$:

$$\mathbb{C}[V]\Big|_{\det} \cong \mathbb{C}[V]^{\Sigma_3} \otimes \left(\mathbb{C}[V]_{\Sigma_3}\Big|_{\det}\right)$$

$$\mathbb{C}[V]\Big|_{\mathsf{L}} \cong \mathbb{C}[V]^{\Sigma_3} \otimes \left(\mathbb{C}[V]_{\Sigma_3}\Big|_{\mathsf{L}}\right)$$

$$\mathbb{C}[V]\Big|_{\mathbb{C}} \cong \mathbb{C}[V]^{\Sigma_3}$$

from which we obtain the following Poincaré series:

$$P(\mathbb{C}[V]\Big|_{\det}, t) = \frac{t^3}{(1-t)(1-t^2)(1-t^3)}$$

$$P(\mathbb{C}[V]\Big|_{\mathsf{L}}, t) = \frac{2(t+t^2)}{(1-t)(1-t^2)(1-t^3)}$$

$$P(\mathbb{C}[V]\Big|_{\mathbb{C}}, t) = \frac{1}{(1-t)(1-t^2)(1-t^3)}.$$

If N and W are finite dimensional vector spaces the natural map

$$W \otimes N^* \longrightarrow \mathrm{Hom}(N, W) \qquad (w \otimes f)(n) = f(n) \cdot w$$

is an isomorphism called the Hom $- \otimes$-*duality isomorphism*. It extends to graded vector spaces of finite type. For two finite dimensional G-representations M and V the Hom $- \otimes$- duality isomorphism for graded vector spaces of finite type is compatible with the G-action and gives

$$\mathrm{Hom}_G(M, \mathbb{F}[V]) \cong \mathrm{Hom}(M, \mathbb{F}[V])^G \cong (\mathbb{F}[V] \otimes M^*)^G.$$

This yields:

PROPOSITION 9.1.1 : *Let V and M be finite-dimensional representations of a finite group G with M irreducible over a field $\mathbb{F}$ of characteristic prime to the order of G. If $\mathbb{F}$ is a splitting field for G then the M-isotypical summand of $\mathbb{F}[V]$ is given by $\mathbb{F}[V]\Big|_M \cong (\mathbb{F}[V] \otimes M^*)^G$.* $\square$

From the trace formula 4.3.1 we obtain the following variation on the basic theorem of Molien [226].

THEOREM 9.1.2 (T. Molien - L. Solomon): *Let V, M be finite dimensional representations of a finite group G over a field $\mathbb{F}$ of characteristic zero. Then*

$$P((\mathbb{F}[V] \otimes M^*)^G, t) = \frac{1}{|G|} \sum_{g \in G} \frac{\chi_{M^*}(g)}{\det(1 - g^{-1}t)}$$

where $\chi_{M^*} : G \longrightarrow \mathbb{F}$ is the character of the dual representation M^* to M.

PROOF : Since $\mathbb{F}$ has characteristic zero we have the averaging operator

$$\pi^G = \frac{1}{|G|} \sum_{g \in G} g$$

which applied to any G-module N projects N onto the invariants N^G as a direct summand in N. Thus we have (compare section 4.3)

$$P((\mathbb{F}[V] \otimes M^*)^G, \ t) = \frac{1}{|G|} \sum_{g \in G} \sum_{i=0}^{\infty} \mathrm{tr}(g, \ \mathbb{F}[V]_i \otimes M^*)$$

$$= \frac{1}{|G|} \sum_{g \in G} \sum_{i=0}^{\infty} \mathrm{tr}(g, \ \mathbb{F}[V]_i) \cdot \chi_{M^*}(g)$$

and the proof may be completed as in the proof of Molien's theorem in section 4.3. $\square$

REMARK : As remarked in section 4.3, working with a Brauer lift of the characters shows that the preceding theorem is valid over any field $\mathbb{F}$ whose characteristic is prime to the order of G.

COROLLARY 9.1.3 : *Let V and M be finite dimensional representations of a finite group G over a field $\mathbb{F}$ of characteristic prime to the order of G. If $\mathbb{F}$ is a splitting field for G then the Poincaré series of the M-isotypical summand of $\mathbb{F}[V]$ is given by*

$$P(\mathbb{F}[V]\Big|_M, \ t) = \frac{1}{|G|} \sum_{g \in G} \frac{\chi_{M^*}(g^{-1})}{\det(1 - g^{-1}t)} \cdot$$

PROOF : This is immediate from the two preceding results. $\square$

9.2 A Variant of Molien's Theorem

Let $\varrho : G \hookrightarrow \mathrm{GL}(n, \ \mathbb{F})$ be a representation of a finite group G on $V = \mathbb{F}^n$. Then G acts on $\mathbb{F}[V]$ and $E[V]$ and hence on their tensor product $\mathbb{F}[V] \otimes E[V]$. The ring of invariants $(\mathbb{F}[V] \otimes E[V])^G$ will be referred to as the **ring of invariant polynomial differential forms on V**.

EXAMPLE 1 : Let $\lambda = \exp(2\pi i/k)$ and

$$D = \begin{bmatrix} \lambda & 0 \\ 0 & \lambda^{-1} \end{bmatrix} \in \mathrm{GL}(2, \ \mathbb{C}).$$

The matrix D defines a representation of the cyclic group $\mathbb{Z}/k$ of order k on $V = \mathbb{C}^2$ whose polynomial invariants were examined in section 4.3. We found

$$\mathbb{C}[x, y]^{\mathbb{Z}/k} = \overset{k-1}{\underset{i=0}{\oplus}} \mathbb{C}[x^k, y^k] \cdot (xy)^i \,.$$

If we let $\mathbb{Z}/k$ act on $\mathbb{C}[x, y] \otimes E[dx, dy]$ then the action of $\mathbb{Z}/k$ sends monomials to monomials and so in addition to the polynomial invariants we find as further invariants:

$$\overset{k-1}{\underset{i=1}{\oplus}} \mathbb{C}[x^k, y^k] \cdot x^{i-1} y^i dx$$
$$\overset{k-1}{\underset{i=1}{\oplus}} \mathbb{C}[x^k, y^k] \cdot x^i y^{i-1} dy$$
$$\mathbb{C}[x^k, y^k] \cdot x^{k-1} dx$$
$$\mathbb{C}[x^k, y^k] \cdot y^{k-1} dy$$
$$\mathbb{C}[x^k, y^k] \cdot dx \cdot dy$$

since a monomial

$$x^a y^b (dx)^\epsilon (dy)^\delta \qquad a,\, b \in \mathbb{N} \quad \epsilon,\, \delta = 0,\, 1$$

is invariant if and only if

$$a + \epsilon - b - \delta \equiv 0 \bmod k \,.$$

Therefore $(\mathbb{C}[x, y] \otimes E[dx, dy])^{\mathbb{Z}/k}$ is a free module over $\mathbb{C}[x^k, y^k]$ with basis

$$\left\{ (xy)^i,\ x^{j-1} y^j dx,\ x^j y^{j-1} dy,\ x^{k-1} dx,\ y^{k-1} dy,\ dx \cdot dy \ \Big|\ \begin{matrix} i = 0, \ldots, k-1 \\ j = 1, \ldots, k-1 \end{matrix} \right\} \,.$$

For the Poincaré series we thus obtain

$$P((\mathbb{C}[x, y] \otimes E[dx, dy])^{\mathbb{Z}/k}, t, s) =$$
$$\frac{(1 + t^2 + \cdots + t^{2k-2}) + 2(st + st^3 + \cdots + st^{2k-3}) + 2st^{k-1} + s^2}{(1 - t^k)^2} \,.$$

For $g \in G$ we introduce the generating function

$$\mathrm{tr}(g,\ \mathbb{F}[V] \otimes E[V])(t,\ s) = \sum_{i,j=0}^{\infty} \mathrm{tr}(g,\ \mathbb{F}[V]_i \otimes E[V]_j)\, t^i s^j \,.$$

The analog of 4.3.1 in this context is:

PROPOSITION 9.2.1 : *Let $\varrho : G \hookrightarrow \mathrm{GL}(n,\ \mathbb{F})$ be a representation of a finite group G over a field $\mathbb{F}$. Then*

$$\mathrm{tr}(g,\ \mathbb{F}[V] \otimes E[V])(t,\ s) = \frac{\det(1 + g^{-1}s)}{\det(1 - g^{-1}t)} \,.$$

PROOF : Without loss of generality we may assume $\mathbb{F}$ is algebraically closed so the matrix representing $g \in G$ may be assumed to be upper triangular with the eigenvalues of g along the diagonal. If $z_1, \ldots, z_n$ is a basis for V^* with respect to which the action of g is upper triangular with eigenvalues $\lambda_1, \ldots, \lambda_n$ on V^* (n.b. V^* not V) then the eigenvalues of g acting on $\Lambda^j(V^*)$ are the product of j of eigenvalues of g acting on V^* with distinct indices. Therefore $\operatorname{tr}(g, \Lambda^j(V^*))$, being the sum of the eigenvalues of g acting on $\Lambda^j(V^*)$, is the coefficient of s^j in the expansion of

$$(1 + \lambda_1 s)(1 + \lambda_2 s) \cdots (1 + \lambda_n s)$$

where $\lambda_1, \ldots, \lambda_n$ are the eigenvalues of g on V^*. Hence

$$\operatorname{tr}(g, \ E[V])(s) = \prod_{j=1}^{n} (1 + \lambda_j s) \, .$$

By the product formula for the Poincaré series of tensor products

$$\operatorname{tr}(g, \ \mathbb{F}[V] \otimes E[V])(t, \ s) = \operatorname{tr}(g, \ \mathbb{F}[V])(t) \cdot \operatorname{tr}(g, \ E[V])(s)$$

and the result follows from 4.3.1. $\Box$

As in section 4.3 the trace formula yields:

THEOREM 9.2.2 (L. Solomon): *Let* $\varrho : G \hookrightarrow \mathrm{GL}(n, \ \mathbb{F})$ *be a representation of a finite group* G *over a field* $\mathbb{F}$ *of characteristic zero. Then*

$$P((\mathbb{F}[V] \otimes E[V])^G, \ t, \ s) = \frac{1}{|G|} \sum_{g \in G} \frac{\det(1 + gs)}{\det(1 - gt)} \, . \quad \Box$$

EXAMPLE 2 : Consider the action of $\mathbb{Z}/k$ on $\mathbb{C}^2$ as in Example 1. Solomon's theorem 9.2.2 says

$$P((\mathbb{C}[x, \ y] \otimes E[dx, \ dy])^{\mathbb{Z}/k}, \ t, \ s) = \frac{1}{k} \sum_{i=0}^{k-1} \frac{\det(1 + D^i s)}{\det(1 - D^i t)}$$

$$= \frac{1}{k} \sum_{i=0}^{k-1} \frac{\det \begin{bmatrix} 1 + \lambda^i s & 0 \\ 0 & 1 + \lambda^{-i} s \end{bmatrix}}{\det \begin{bmatrix} 1 - \lambda^i t & 0 \\ 0 & 1 - \lambda^{-i} t \end{bmatrix}}$$

$$= \frac{1}{k} \left[\sum_{i=0}^{k-1} \frac{(1 + \lambda^i s)(1 + \lambda^{-i} s)}{(1 - \lambda^i t)(1 - \lambda^{-i} t)} \right] \, .$$

Comparing this with the formula obtained by direct computation in example 1 leads to the identity:

(*)
$$\frac{1}{k}\left[\sum_{i=0}^{k-1}\frac{(1+\lambda^i s)(1+\lambda^{-i}s)}{(1-\lambda^i t)(1-\lambda^{-i}t)}\right] =$$
$$\frac{(1+t^2+\cdots+t^{2k-2})+2(st+st^3+\cdots+st^{2k-3})+2st^{k-1}+s^2}{(1-t^k)^2}.$$

The formula (*) may be used as in section 4.3 to compute the Poincaré series for the dihedral group D_{2k} in the natural 2-dimensional representation.

EXAMPLE 3 : Consider the action of the dihedral group D_{2k} on $V = \mathbb{C}^2$ given by the matrices

$$D = \begin{bmatrix} \lambda & 0 \\ 0 & \lambda^{-1} \end{bmatrix} \qquad S = \begin{bmatrix} 1 & 0 \\ 0 & -1 \end{bmatrix}$$

where $\lambda = \exp(2\pi i/k)$. By the theorem of Solomon 9.2.2 we obtain for the Poincaré series of $(\mathbb{F}[x, y] \otimes E[dx, dy])^{D_{2k}}$ the expression

$$P((\mathbb{F}[x, y] \otimes E[dx, dy])^{D_{2k}}, t, s) = \frac{1}{2k}\sum_{g\in D_{2k}}\frac{(1+gs)}{(1-gt)}$$

$$= \frac{1}{2k}\left[\sum_{i=0}^{k-1}\frac{\det(1+D^i s)}{\det(1-D^i t)}+\sum_{i=0}^{k-1}\frac{\det(1+SD^i s)}{\det(1-SD^i t)}\right].$$

The first sum inside the brackets is evaluated by using the formula (*). To evaluate the second sum in the brackets notice that the eigenvalues of the reflection matrices SD^i are ± 1 for $i = 0, \ldots, k-1$, so

$$\frac{\det(1+SD^i s)}{\det(1-SD^i t)} = \frac{\det\begin{bmatrix} 1+s & 0 \\ 0 & 1-s \end{bmatrix}}{\det\begin{bmatrix} 1-t & 0 \\ 0 & 1+t \end{bmatrix}} = \frac{1-s^2}{1-t^2}.$$

Substituting into the formula obtained above from the theorem of Solomon gives for the Poincaré series $P((\mathbb{F}[V] \otimes E[V])^{D_{2k}}, t, s)$

$$\frac{1}{2k}\left\{k\left[\frac{(1+t^2+\cdots+t^{2k-2})+2(st+st^3+\cdots+st^{2k-3})+2st^{k-1}+s^2}{(1-t^k)^2}\right]+k\left[\frac{1-s^2}{1-t^2}\right]\right\}.$$

This expression may be further simplified to yield the formula

$$P((\mathbb{F}[x, y] \otimes E[dx, dy])^{D_{2k}}, t, s) = \frac{(1+st)(1+st^{k-1})}{(1-t^2)(1-t^k)}.$$

The numerator of this expression should be recognizable as the Poincaré series of an exterior algebra with two generators, one of bidegree $(1,1)$ and one of bidegree $(k-1,1)$. The rational function $\frac{1}{(1-t^2)(1-t^k)}$ is the Poincaré series of a polynomial algebra on two generators, one of bidegree $(2,0)$ and one of bidegree $(k,0)$. Recall from section 4.3 Example 2, that

$$\mathbb{C}[x,\ y]^{D_{2k}} = \mathbb{C}[f_1,\ f_2]$$

where $\deg(f_1) = 2$ and $\deg(f_2) = k$. Since

$$\mathbb{C}[x,\ y]^{D_{2k}} \subset (\mathbb{C}[x,\ y] \otimes E[dx,\ dy])^{D_{2k}}$$

and

$$df_1,\ df_2 \in (\mathbb{C}[x,\ y] \otimes E[dx,\ dy])^{D_{2k}}$$

we obtain a map

$$\varphi : \mathbb{C}[f_1,\ f_2] \otimes E[df_1,\ df_2] \longrightarrow (\mathbb{C}[x,\ y] \otimes E[dx,\ dy])^{D_{2k}}\ .$$

By definition of the exterior derivative and the Jacobian determinant we have

$$\varphi(df_1 \cdot df_2) = \det \begin{bmatrix} \frac{\partial f_1}{\partial x} & \frac{\partial f_1}{\partial y} \\ \frac{\partial f_2}{\partial x} & \frac{\partial f_2}{\partial y} \end{bmatrix} \cdot dx \cdot dy\ .$$

The Jacobian determinant of $f_1,\ f_2$ is nonzero because f_1 and f_2 are algebraically independent. Therefore φ is a monomorphism. In addition,

$$P(\mathbb{C}[f_1,\ f_2] \otimes E[df_1,\ df_2],\ t,\ s) = \frac{(1+st)(1+st^{k-1})}{(1-t^2)(1-t^k)}$$

$$= P((\mathbb{C}[x,\ y] \otimes E[dx,\ dy])^{D_{2k}},\ t,\ s)$$

so we conclude that φ is an isomorphism.

9.3   Groups Generated by Pseudoreflections

Suppose $\varrho : G \hookrightarrow \mathrm{GL}(n,\ \mathbb{F})$ is an irreducible pseudoreflection representation of a finite group G over a field $\mathbb{F}$ whose characteristic is prime to the order of G. By a result of Steinberg [39] (page 127, exercise 3 (d)) the representations $\Lambda^i(V^*)$, $i = 0, \ldots, n$, are all irreducible and therefore

$$\mathbb{F}[V]_G \Big|_{\Lambda^i(V)} \cong (\mathbb{F}[V]_G \otimes \Lambda^i(V^*))^G$$

by 9.1.1. From 7.4.3 we have

$$\mathbb{F}[V] \cong \mathbb{F}[V]^G \otimes \mathbb{F}[V]_G$$

as $\mathbb{F}(G)$- and $\mathbb{F}[V]^G$-modules. Therefore

$$\mathbb{F}[V] \otimes E[V] \cong \mathbb{F}[V]^G \otimes \mathbb{F}[V]_G \otimes E[V]$$

as $\mathbb{F}(G)$- and $\mathbb{F}[V]^G$-modules as well. Taking invariants on both sides yields

$$(\mathbb{F}[V] \otimes E[V])^G \cong (\mathbb{F}[V]^G \otimes \mathbb{F}[V]_G \otimes E[V])^G \cong \mathbb{F}[V]^G \otimes (\mathbb{F}[V]_G \otimes E[V])^G.$$

Applying the functor $\mathbb{F} \otimes_{\mathbb{F}[V]^G} -$ to this isomorphism yields [1]

$$(\mathbb{F}[V] \otimes E[V])^G // \mathbb{F}[V]^G \cong (\mathbb{F}[V]_G \otimes E[V])^G$$

and passing to the $\Lambda^i(V)$ isotypical component on both sides we obtain [226]:

THEOREM 9.3.1 (L. Solomon): *Let* $\varrho : G \hookrightarrow \mathrm{GL}(n, \mathbb{F})$ *be an irreducible representation of a finite group* G *over a field* $\mathbb{F}$ *whose characteristic is prime to the order of* G. *Then*

$$(\mathbb{F}[V]_G)\Big|_{\Lambda^i(V)} \cong (\mathbb{F}[V] \otimes \Lambda^i(V^*))^G // \mathbb{F}[V]^G$$

yields the $\Lambda^i(V)$-*isotypical summand of* $\mathbb{F}[V]_G$. □

This theorem of Solomon provides a clear connection between the invariants $(\mathbb{F}[V] \otimes E[V])^G$ and the canonical decomposition of the G-representation $\mathbb{F}[V]_G$ when G acts on V as a pseudoreflection group and $\mathbb{F}$ has characteristic prime to the order of G. With this as motivation our next goal is to compute $(\mathbb{F}[V] \otimes E[V])^G$ in the nonmodular pseudoreflection case.

Recall from the introduction to this chapter that $\mathbb{F}[V] \otimes E[V]$ is bigraded by requiring

$$\mathrm{bideg}(f \otimes \omega) = (\deg(f), \deg(\omega))$$

for $f \in \mathbb{F}[V]$ and $\omega \in E[V]$. The exterior derivative

$$d : \mathbb{F}[V] \otimes E[V] \longrightarrow \mathbb{F}[V] \otimes E[V]$$

commutes with the G-action and hence for $f \in \mathbb{F}[V]^G$ we receive the element $df \in (\mathbb{F}[V] \otimes E[V])^G$, its exterior derivative.

For $\varrho : G \hookrightarrow \mathrm{GL}(n, \mathbb{F})$ with $\varrho(G)$ generated by pseudoreflections and $|G| \in \mathbb{F}^\times$ the theorem of Shephard-Todd 7.4.1 says

$$\mathbb{F}[V]^G = \mathbb{F}[f_1, \dots, f_n]$$

and hence we obtain a natural map

$$\mathbb{F}[f_1, \dots, f_n] \otimes E[df_1, \dots, df_n] \longrightarrow (\mathbb{F}[V] \otimes E[V])^G.$$

This map was shown to be an isomorphism by L. Solomon [226] in characteristic zero.

---

[1] If A^* is a graded connected algebra over a graded connected algebra B^* then $A^*//B^*$ is an alternate notation for $\mathbb{F} \otimes_{B_*} A^*$ where $\mathbb{F}$ is the ground field.

THEOREM 9.3.2 (L. Solomon): *Let $\varrho : G \hookrightarrow \mathrm{GL}(n, \mathbb{F})$ be a representation of a finite group G. Assume that $\varrho(G)$ is generated by pseudoreflections and the order of G is prime to the characteristic of $\mathbb{F}$. Then*

$$(\mathbb{F}[V] \otimes E[V])^G \cong \mathbb{F}[f_1, \ldots, f_n] \otimes E[df_1, \ldots, df_n]$$

where (Shephard-Todd) $\mathbb{F}[V]^G \cong \mathbb{F}[f_1, \ldots, f_n]$ and d is the exterior derivative.

PROOF : We begin by collecting the tools we need in the proof. Let $\mathbb{F}[V]^G = \mathbb{F}[f_1, \ldots, f_n]$. Since $f_1, \ldots, f_n$ are a system of parameters in $\mathbb{F}[V]^G$ with degrees prime to p the Jacobian determinant is not zero by 6.5.2, so

$$J = \det \left[\frac{\partial f_i}{\partial z_j} \right] \neq 0 \in \mathbb{F}[V].$$

By the theorems of Steinberg 7.5.6 and Springer-Stanley 7.6.4 J divides any $\det^{-1}$-relative invariant of G.

We first show that the map

$$\mathbb{F}[f_1, \ldots, f_n] \otimes E[df_1, \ldots, df_n] \longrightarrow (\mathbb{F}[V] \otimes E[V])^G$$

is a monomorphism. To this end note that there are natural maps

$$\begin{array}{ccccc} \mathbb{F}[V] & \overset{i}{\longrightarrow} & \mathbb{F}[V] \otimes E[V] & \overset{\pi}{\longrightarrow} & \mathbb{F}[V] \\ f & \longmapsto & f \otimes 1, \quad f \otimes \omega & \longmapsto & \varepsilon(\omega)f \end{array}$$

where $\varepsilon : E[V] \longrightarrow \mathbb{F}$ is the augmentation. The maps i and π commute with the G-action and since $\pi \cdot i = \mathrm{id}$ it follows on taking invariants that

$$i : \mathbb{F}[V]^G \hookrightarrow (\mathbb{F}[V] \otimes E[V])^G$$

is an inclusion. Next note

$$df_1 \cdots df_n = J \cdot dz_1 \cdots dz_n$$

by definition of the exterior derivative and the Jacobian determinant. Since $J \neq 0$ it follows that $df_{i_1} \cdots df_{i_k}$ for $i_1 < \cdots i_k$ are linearly independent in $(\mathbb{F}[V] \otimes E[V])^G$. Therefore the natural map

$$\mathbb{F}[f_1, \ldots, f_n] \otimes E[df_1, \ldots, df_n] \longrightarrow (\mathbb{F}[V] \otimes E[V])^G$$

is an inclusion and it remains to show that $\mathbb{F}[f_1, \ldots, f_n] \otimes E[df_1, \ldots, df_n]$ are the only invariants.

Let $FF(\mathbb{F}[V]^G)$ denote the field of fractions of $\mathbb{F}[V]^G$. Taking account of 1.2.4 one checks that

$$FF((\mathbb{F}[V])^G) \otimes_{\mathbb{F}[V]^G} (\mathbb{F}[V] \otimes E[V])^G \cong (FF(\mathbb{F}[V]) \otimes E[V])^G \cong (\mathbb{F}(V) \otimes E[V])^G$$

and hence a dimension count shows any element $\omega \in (\mathbb{F}[V] \otimes E[V])^G$ has the form

$$\omega = \sum u_{(i_1,\ldots,i_m)} df_{i_1} \cdots df_{i_m}$$

where $u_{(i_1,\ldots,i_m)} \in \mathbb{F}(V)$. Applying the transfer $\pi^G = \frac{1}{|G|} \sum_{g \in G} g$ to this equation we may suppose $u_{(i_1,\ldots,i_m)} \in \mathbb{F}(V)^G$ since ω and the df_{i_j}, $j = 1, \ldots, m$, are invariant and $\pi^G(-)$ is always invariant. It therefore remains to show that $u_{(i_1,\ldots,i_m)} \in \mathbb{F}[V]^G$. Let $\{i_{m+1}, \ldots, i_n\}$ be the complementary set to $\{i_1, \ldots, i_m\}$ in $\{1, \ldots, n\}$. Then

$$(u_{(i_1,\ldots,i_m)} df_{i_1} \cdots df_{i_m}) df_{i_{m+1}} \cdots df_{i_n} = u_{(i_1,\ldots,i_m)} \cdot J \cdot dz_1 \cdots dz_n.$$

On the other hand

$$u_{(i_1,\ldots,i_m)} df_{i_1} \cdots df_{i_m} \in \mathbb{F}[V] \otimes E[V]$$

so

$$(u_{(i_1,\ldots,i_m)} df_{i_1} \cdots df_{i_m}) df_{i_{m+1}} \cdots df_{i_n} = h \cdot dz_1 \cdots dz_n$$

for some $h \in \mathbb{F}[V]^G$. Combining these equations gives

$$h \cdot dz_1 \cdots dz_n = u_{(i_1,\ldots,i_m)} \cdot J \cdot dz_1 \cdots dz_n.$$

Both sides are invariants and $dz_1 \cdots dz_n$ is a det-invariant. Hence h is a $\det^{-1}$-relative invariant and J divides h in $\mathbb{F}[V]$ which implies $u_{(i_1,\ldots,i_m)} \in \mathbb{F}[V]^G$ as required. $\square$

COROLLARY 9.3.3 (L. Solomon): *Let $\varrho : G \hookrightarrow \mathrm{GL}(n, \mathbb{F})$ be a representation of a finite group G. Assume that $\varrho(G)$ is generated by pseudoreflections and the order of G is prime to the characteristic of $\mathbb{F}$. Then*

$$P((\mathbb{F}[V] \otimes E[V])^G, t, s) = \frac{\displaystyle\prod_{j=1}^{n}(1 + st^{m_j})}{\displaystyle\prod_{i=1}^{n}(1 - t^{d_i})}$$

where $d_i = \deg(f_i)$, $i = 1, \ldots, n$, are the fundamental degrees of G and $m_i = d_i - 1$, $i = 1, \ldots, n$, are the exponents. $\square$

We have the two expressions 9.2.2 and 9.3.3 for the Poincaré series of the algebra $(\mathbb{F}[V] \otimes E[V])^G$ when G acts on V via pseudoreflections and the characteristic of $\mathbb{F}$ is prime to the order of G. Equating these two expressions and manipulating a bit leads to the following result.

THEOREM 9.3.4 (L. Solomon): *Let* $\varrho : G \hookrightarrow \mathrm{GL}(n,\ \mathbb{F})$ *be a representation of a finite group* G. *Assume that* $\varrho(G)$ *is generated by pseudoreflections and the order of* G *is prime to the characteristic of* $\mathbb{F}$. *Then*

$$\prod_{i=1}^{n}(m_i + X) = \sum_{g \in G} X^{\lambda(g)} \in \mathbb{Z}[X]$$

where $\lambda(g) = \dim_{\mathbb{F}}(V^g)$ *is the dimension of the fixed point set of* $g \in G$.

PROOF : By 9.2.2 we have

$$P((\mathbb{F}[V] \otimes E[V])^G,\ t,\ s) = \frac{1}{|G|} \sum_{g \in G} \frac{\det(1 + gs)}{\det(1 - gt)}$$

so by the preceding corollary we obtain

$$(*) \qquad \frac{1}{|G|} \sum_{g \in G} \frac{\det(1 + gs)}{\det(1 - gt)} = \frac{\displaystyle\prod_{j=1}^{n}(1 + st^{m_j})}{\displaystyle\prod_{i=1}^{n}(1 - t^{d_i})}.$$

Let $\lambda(g) = \ell$ so that g has a matrix representation

$$\begin{bmatrix} 1 & 0 & 0 & \cdots & \cdots & 0 \\ 0 & 1 & 0 & \cdots & \cdots & 0 \\ \vdots & \vdots & & \vdots & \vdots & \vdots \\ 0 & \vdots & 1 & 0 & \cdots & 0 \\ 0 & \vdots & 0 & \lambda_{\ell+1} & 0 & 0 \\ \vdots & \vdots & & \vdots & \vdots & \vdots \\ 0 & \vdots & \cdots & \cdots & 0 & \lambda_n \end{bmatrix}$$

over a splitting field for G with ℓ 1's along the diagonal. Thus

$$1 + gs = \begin{bmatrix} 1+s & 0 & 0 & \cdots & \cdots & 0 \\ 0 & 1+s & 0 & \cdots & \cdots & 0 \\ \vdots & \vdots & & \vdots & \vdots & \vdots \\ 0 & \vdots & 1+s & 0 & \cdots & 0 \\ 0 & \vdots & 0 & 1+s\lambda_{\ell+1} & 0 & 0 \\ \vdots & \vdots & & \vdots & \vdots & \vdots \\ 0 & \vdots & \cdots & \cdots & 0 & 1+s\lambda_n \end{bmatrix}$$

and

$$1 - gt = \begin{bmatrix} 1-t & 0 & 0 & \cdots & \cdots & 0 \\ 0 & 1-t & 0 & \cdots & \cdots & 0 \\ \vdots & \vdots & & \vdots & \vdots & \vdots \\ 0 & \vdots & 1-t & 0 & \cdots & 0 \\ 0 & \vdots & 0 & 1-t\lambda_{\ell+1} & 0 & 0 \\ \vdots & \vdots & & \vdots & \vdots & \vdots \\ 0 & \vdots & \cdots & \cdots & 0 & 1-t\lambda_n \end{bmatrix}.$$

Make the substitution $s = -1 + X(1 - t)$. The matrix for $1 + gs$ becomes

$$\begin{bmatrix} X(1-t) & 0 & 0 & \cdots & 0 \\ \vdots & & \vdots & \vdots & \vdots \\ 0 & X(1-t) & 0 & \cdots & 0 \\ 0 & \vdots & 1-\lambda_{\ell+1}+\lambda_{\ell+1}X(1-t) & 0 & 0 \\ \vdots & & \vdots & \vdots & \vdots \\ 0 & \vdots & \cdots & 0 & 1-\lambda_n+\lambda_n X(1-t) \end{bmatrix}.$$

Hence taking determinants gives

$$\det(1 + gs) = X^\ell (1 - t)^\ell \Delta''(t)$$
$$\det(1 - gt) = (1 - t)^\ell \Delta'(t)$$

where

$$\Delta''(t) = \det \begin{bmatrix} 1-\lambda_{\ell+1}+\lambda_{\ell+1}X(1-t) & 0 & \cdots & 0 \\ \vdots & \vdots & \vdots & \vdots \\ 0 & \cdots & \cdots & 1-\lambda_n+\lambda_n X(1-t) \end{bmatrix}$$

and

$$\Delta'(t) = \det \begin{bmatrix} 1-\lambda_{\ell+1}t & 0 & \cdots & 0 \\ \vdots & \vdots & \vdots & \vdots \\ 0 & \cdots & \cdots & 1-\lambda_n t \end{bmatrix}$$

so

$$\frac{\det(1 + gs)}{\det(1 - gt)} = X^\ell \frac{\Delta''(t)}{\Delta'(t)}.$$

Setting $t = 1$ and noting $\Delta''(1) = \Delta'(1)$ we get

$$\frac{\det(1 + gs)}{\det(1 - gt)} \bigg|_{\substack{s=-1+X(1-t) \\ t=1}} = X^\ell$$

and therefore

$$\frac{1}{|G|} \sum_{g \in G} \frac{\det(1 + gs)}{\det(1 - gt)} \Big|_{s = -1 + X(1-t) \atop t=1} = \frac{1}{|G|} \sum_{g \in G} X^{\lambda(g)} .$$

Performing the same substitution on the right hand side of (∗) leads successively to the formulae

$$\frac{\prod(1 + st^{m_i})}{\prod(1 - t^{d_i})} \Big|_{s = -1 + X(1-t) \atop t=1} = \frac{\prod(1 + (-1 + X(1-t))t^{m_i})}{\prod[(1 - t)(1 + t + \cdots + t^{m_i})]} \Big|_{t=1}$$

$$= \frac{\prod[(1 - t^{m_i}) + X(1-t)t^{m_i}]}{\prod[(1 - t)(1 + t + \cdots + t^{m_i})]} \Big|_{t=1}$$

$$= \frac{\prod[(1 - t)(1 + t + \cdots + t^{m_i - 1} + Xt^{m_i})]}{\prod[(1 - t)(1 + t + \cdots + t^{m_i})]} \Big|_{t=1}$$

$$= \frac{\prod(m_i + X)}{\prod(m_i + 1)}$$

so taking account of 4.4.3 we obtain

$$\sum_{g \in G} X^{\lambda(g)} = \prod_{i=1}^{n}(X + m_i)$$

as was to be shown. □

For a representation $\varrho : G \hookrightarrow \mathrm{GL}(n, \, \mathbb{F})$ of a finite group G with $\varrho(G)$ generated by pseudoreflections and $|G|$ prime to the characteristic of $\mathbb{F}$, Chevalley proved (7.5.2) that $\mathbb{F}[V]_G$ is the regular representation of G. Steinberg has shown [39] (page 127, exercise 3 (d)) that $\Lambda^i(\varrho)$ is irreducible. Therefore $\Lambda^i(V)$ occurs $\binom{n}{i} = \dim_{\mathbb{F}}(\Lambda^i(V))$ times as a direct summand in $\mathbb{F}[V]_G$. The following formula of Solomon locates these summands for us.

THEOREM 9.3.5 (L. Solomon): Let $\varrho : G \hookrightarrow \mathrm{GL}(n, \, \mathbb{F})$ be a representation of a finite group G. Assume that $\varrho(G)$ is generated by pseudoreflections and the order of G is prime to the characteristic of $\mathbb{F}$. Then

$$P(\mathbb{F}[V]_G \Big|_{\Lambda^i(V)}, \, t) = e_i(t^{m_1}, \ldots, t^{m_n})$$

where, as usual, $e_1, \ldots, e_n$ denote the elementary symmetric polynomials and $m_1, \ldots, m_n$ the exponents of $\varrho(G)$.

PROOF : By 9.3.2

$$(\mathbb{F}[V] \otimes E[V])^G \cong \mathbb{F}[f_1, \ldots, f_n] \otimes E[df_1, \ldots, df_n] \cong \mathbb{F}[V]^G \otimes E[df_1, \ldots, df_n]$$

so

$$(\mathbb{F}[V] \otimes E[V])^G // \mathbb{F}[V]^G \cong E[df_1, \ldots, df_n]$$

and taking the Poincaré series we find

$$P((\mathbb{F}[V] \otimes E[V])^G // \mathbb{F}[V]^G, \ t, \ s) = \prod_{i=1}^{n}(1 + st^{m_i})$$

$$= \sum_{i=0}^{n} e_i(t^{m_1}, \ldots, t^{m_n})s^i.$$

By 9.1.1 the coefficient of s^i in this expression is $P(\mathbb{F}[V]_G\big|_{\Lambda^i(V^*)}, \ t)$ and the result follows. $\square$

REMARK : The double duality property (see section 7.3 observation 7.3.2, [59] and [227]) for the exponents of a real reflection group, namely that $m_i + m_{n-i+1}$ is a constant independent of i implies that [227]

$$e_i(t^{m_1}, \ldots, t^{m_n}) = t^k e_{n-i}(t^{m_1}, \ldots, t^{m_n})$$

for some integer k, and hence apart from a dimension shift the $\Lambda^i(V)$ and the $\Lambda^{n-i}(V)$ isotypical summands of $\mathbb{F}[V]_G$ are isomorphic.

For example, suppose that

$$m_i + m_{n-i+1} = m \text{ for } 1 \leq i \leq \left[\frac{n}{2}\right].$$

Then for any $i \in \{1, \ldots, \left[\frac{n}{2}\right]\}$ we have

$$m_1 + \cdots m_n = k + m_1 + m_{n-i-1}$$

where

$$k = m_1 + \cdots + m_n - m.$$

Hence

$$t^k e_1(t^{m_1}, \ldots, t^{m_n}) = t^k(t^{m_1} + \cdots + t^{m_n}) = t^{k+m_1} + \cdots + t^{k+m_n}$$

$$= t^{m_1 + \cdots + \widehat{m_n}} + \cdots + t^{m_1 + \cdots + \widehat{m_i} + \cdots + m_n} + \cdots + t^{\widehat{m_1} + m_2 + \cdots + m_n}$$

$$= e_{n-1}(t^{m_1}, \ldots, t^{m_n}).$$

Moreover $\Lambda^i(V^*)$ and $\Lambda^{n-i}(V^*)$ are Poincaré dual in $E[V]$. The fundamental class of $E[V]$ is $dz_1 \cdots dz_n$ which is a det-relative invariant. The fundamental class of $\mathbb{F}[V]_G$ is a $\det^{-1}$-relative invariant. Since $\det = \det^{-1}$ in the real case it follows that the $\Lambda^i(V^*)$- and $\Lambda^{n-i}(V^*)$-isotypical summands of $\mathbb{F}[V]_G$ are Poincaré dual in $\mathbb{F}[V]_G$ also. Combining these two observations we see that if we write

$$P(\mathbb{F}[V]_G\big|_{\Lambda^i(V^*)}, \ t) = t^s(a_0 + a_1 t + \cdots + a_r t^r)$$

with $a_0 \neq 0$ then [227] $a_j = a_{r-j}$.

9.4   An Application to the Cohomology of the Symmetric Group

Let p be an odd prime, $n \in \mathbb{N}$, and Σ_m the symmetric group on m letters. In this section[2] we show how the results of section 9.3 may be applied to compute the cohomology groups $H^*(\Sigma_{np}; \mathbb{F}_p)$ provided that $n < p$. We assume that the reader is familiar with the cohomology of groups, as found for example in [38], [54] or [189]. The tools used in this section are Burnside's fusion theorem (as found for example in [96]) and transfer for the cohomology of groups.

In the following discussion $H^*(-; \mathbb{F})$ denotes the cohomology of the group − with coefficients in the field $\mathbb{F}$ acted upon trivially by the group. We begin by recalling some essential facts ([54] chapter XII section 8 and 9, especially (9.3) and (9.4)) concerning the transfer map in the cohomology of groups, and at the same time establish the notations we are going to employ.

Let $H < G$ be finite groups. The inclusion map $i : H \hookrightarrow G$ induces a map (the **restriction**)

$$\mathrm{Res}^G_H : H^*(G; \mathbb{F}) \longrightarrow H^*(H; \mathbb{F}),$$

which is also denoted by i^*, and a transfer map

$$\mathrm{Tr}^G_H : H^*(H; \mathbb{F}) \longrightarrow H^*(G; \mathbb{F}).$$

The composition

$$\mathrm{Tr}^G_H \cdot \mathrm{Res}^G_H : H^*(G; \mathbb{F}) \longrightarrow H^*(G; \mathbb{F})$$

is just multiplication by $|G : H| \in \mathbb{Z}$ ([54] loc.cit.). The composition

$$\mathrm{Res}^G_H \cdot \mathrm{Tr}^G_H : H^*(H; \mathbb{F}) \longrightarrow H^*(H; \mathbb{F})$$

is more complicated[3] to describe. To do so, we introduce the diagram

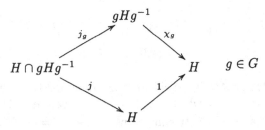

where j, j_g are the inclusions and $\chi_g(x) = gxg^{-1}$. The two ways around the diagram yield the two maps

$$\mathrm{Res}^{gHg^{-1}}_{H \cap gHg^{-1}} \cdot \chi^*_g, \ \mathrm{Res}^H_{H \cap gHg^{-1}} : H^*(H; \mathbb{F}) \longrightarrow H^*(H \cap gHg^{-1}; \mathbb{F}).$$

---

[2] The results of this section are the outgrowth of a discussion with David Benson.
[3] See e.g. [80]

An element $u \in H^*(H; \mathbb{F})$ is called **stable** if

$$\operatorname{Res}^{gHg^{-1}}_{H \cap gHg^{-1}} \cdot \chi_g^*(u) = \operatorname{Res}^H_{H \cap gHg^{-1}}(u)$$

for all $g \in G$. The set of stable elements in $H^*(H; \mathbb{F})$ is denoted by $H^*(H; \mathbb{F})^{\operatorname{st}_H^G}$. (N.b. The definition of stable is a relative one: relative to the inclusion $H \hookrightarrow G$.) If $H \lhd G$ is a normal subgroup then the stable elements $H^*(H; \mathbb{F})^{\operatorname{st}_H^G}$ coincide with the invariants $H^*(H; \mathbb{F})^G = H^*(H; \mathbb{F})^{G/Z_G(H)}$, where $Z_G(H)$ denotes the centralizer of H in G. We require ([54] loc.cit.):

THEOREM 9.4.1 (H. Cartan and S. Eilenberg): *Let $H < G$ be finite groups. Then*

$$\operatorname{Im}\{\operatorname{Res}^G_H : H^*(G; \mathbb{F}) \longrightarrow H^*(H; \mathbb{F})\} = H^*(H; \mathbb{F})^{\operatorname{st}_H^G}. \qquad \square$$

We next explain how the computation of $H^*(\Sigma_{np}; \mathbb{F}_p)$ for $n < p$ is related to invariant theory. Note that $(np)! = p^n \ell$, where ℓ is relatively prime to p. Hence $\operatorname{Syl}_p(\Sigma_{np})$ has order p^n. Let X be a set with np elements which we partition into n disjoint blocks $X_1, \ldots, X_n$ of p elements each. On each block X_i there is a cyclic permutation of order p. Since the blocks are disjoint these cycles commute with each other. Hence we obtain an inclusion $V = \bigoplus_n \mathbb{Z}/p \hookrightarrow \Sigma_{np}$. Since $|V| = p^n$ it follows that $V = \operatorname{Syl}_p(\Sigma_{np})$. Let $N = N_{\Sigma_{np}}(V)$ be the normalizer of V in Σ_{np}.

The symmetric group Σ_n acts on X by permuting the blocks $X_1, \ldots, X_n$. The subgroup

$$N = (\mathbb{Z}/p) \int \Sigma_n = V \rtimes \Sigma_n < \Sigma_{np},$$

where $\int$ denotes the wreath product, normalizes V. An element of Σ_{np} that normalizes V belongs to this subgroup so we thus obtain an exact sequence

$$1 \longrightarrow V \longrightarrow N \longrightarrow W \longrightarrow 1$$

defining the Weyl group $W = W_{\Sigma_{np}}(V)$ of V in Σ_{np}. One sees directly that $W_{\Sigma_{np}}(V) = G(p-1, 1, n)$ in the notations for groups of type 2a in the Shephard-Todd list (section 7.3 table 7.3.1) of complex pseudoreflection groups. The action of W on V is via the mod p reduction of the representation of $W = G(p-1, 1, n) \hookrightarrow \operatorname{GL}(n, \mathbb{Z}(\zeta_m))$ as a pseudoreflection group. Note that $p \nmid |W|$ so theorem 7.4.1 applies to ϱ reduced mod p. The invariants of these groups were discussed in section 7.4 example 1.

We may identify $H^*(V; \mathbb{F}_p)$ with $E[V] \otimes \mathbb{F}[V]$, where $V^* \subset E[V]$ is identified with $H^1(V; \mathbb{F}_p) \cong V^*$, $V^* \subset \mathbb{F}[V]$ is identified with $\beta(H^1(V; \mathbb{F}_p)) \subset$

$H^2(V; \; \mathbb{F}_p)$, and β is the Bockstein operator. The group W has order prime to p so the map

$$\mathrm{Tr}_V^N \cdot \mathrm{Res}_V^N : H^*(N; \; \mathbb{F}_p) \longrightarrow H^*(N; \; \mathbb{F}_p)$$

is multiplication by $|W| \neq 0 \in \mathbb{F}_p$. Therefore Res_V^N is a monomorphism. Since $V \lhd N$ is a normal subgroup $H^*(V; \; \mathbb{F}_p)^{\mathrm{st}_V^N}$ coincides with $H^*(V; \; \mathbb{F}_p)^W$.

PROPOSITION 9.4.2 : *With the preceding notations*

$$H^*(N; \; \mathbb{F}_p) \cong E[u_1, \dots, u_n] \otimes \mathbb{F}_p[f_1, \dots, f_n]$$

where

$$\mathrm{Res}_V^N(f_i) = e_i(t_1^{p-1}, \dots, t_n^{p-1}) \qquad i = 1, \dots, n$$

$$\beta(u_i) = f_i \qquad i = 1, \dots, n$$

and $t_1, \dots, t_n \in H^1(V; \; \mathbb{F}_p) = V^*$ *is a basis permuted by* Σ_n *in its tautological representation as a permutation group.*

PROOF : The preceding discussion shows

$$H^*(N; \; \mathbb{F}_p) \cong H^*(V; \; \mathbb{F}_p)^W .$$

Identifying $H^*(V; \; \mathbb{F}_p)$ with $E[V] \otimes \mathbb{F}_p[V]$ and applying 9.3.2 and the results of example 1 of section 7.3 yields the conclusion. $\square$

To draw conclusions about $H^*(\Sigma_{np}; \; \mathbb{F}_p)$ from information about $H^*(N; \; \mathbb{F}_p)$ we make use of Burnside's fusion theorem ([96] chapter 7 section 1 theorem 1.1). We first recall the concepts involved.

DEFINITION : *Given* $S \subset H < G$, H *a subgroup of* G, $S \subset H$ *a subset of* H. *We say that* H **controls** S**-fusion** *in* G *if, whenever* s', $s'' \in S$ *and there is a* $g \in G$ *with* $gs'g^{-1} = s''$, *then there is also an* $h \in H$ *with* $hs'h^{-1} = s''$.

In other words, if we let G and H act on S by conjugation, then the elements of S have the same orbits under both actions. This means that several distinct H-orbits of points in S cannot be *fused* to a single G-orbit.

THEOREM 9.4.3 (W. Burnside): *If* G *is a finite group such that* $\mathrm{Syl}_p(G)$ *is abelian, then* $N_G(\mathrm{Syl}_p(G))$ *controls* $\mathrm{Syl}_p(G)$*-fusion in* G. $\square$

THEOREM 9.4.4 : *Let* p *be an odd prime and* $n \in \mathbb{N}$ *with* $n < p$. *Then*

$$H^*(\Sigma_{np}; \; \mathbb{F}_p) \cong E[u_1, \dots, u_n] \otimes \mathbb{F}_p[f_1, \dots, f_n]$$

where

$$\mathrm{Res}_{\mathrm{Syl}_p(\Sigma_{np})}^{\Sigma_{np}}(f_i) = e_i(t_1^{p-1}, \dots, t_n^{p-1}) \qquad i = 1, \dots, n$$

$$\beta(u_i) = f_i \qquad i = 1, \dots, n$$

and $t_1, \ldots, t_n \in H^1(\mathrm{Syl}_p(\Sigma_{np}; \ \mathbb{F}_p)) = V^* = \mathrm{Hom}_{\mathbb{Z}}(\mathrm{Syl}_p(\Sigma_{np}), \ \mathbb{F}_p)$ *is a basis permuted by* Σ_n *in its tautological representation as a permutation group. If* W *is the Weyl group of* $V = \mathrm{Syl}_p(\Sigma_{np})$ *then the natural map*

$$\mathrm{Res}^{\Sigma_{np}}_{\mathrm{Syl}_p(\Sigma_{np})} : H^*(\Sigma_{np}; \ \mathbb{F}_p) \longrightarrow H^*(V; \ \mathbb{F}_p)^W$$

is an isomorphism.

PROOF : In view of the preceding results and discussions it suffices to show that
$$\mathrm{Res}^{\Sigma_{np}}_{\mathrm{Syl}_p(\Sigma_{np})} : H^*(\Sigma_{np}; \ \mathbb{F}_p) \longrightarrow H^*(V; \ \mathbb{F}_p)^W$$
is an isomorphism. To this end note that

$$\mathrm{Res}^{\Sigma_{np}}_{N_{\Sigma_{np}}(V)} : H^*(\Sigma_{np}; \ \mathbb{F}_p) \longrightarrow H^*(N_{\Sigma_{np}(V)}(V); \ \mathbb{F}_p)$$

is a monomorphism, because composing it with $\mathrm{Tr}^{\Sigma_{np}}_{N_{\Sigma_{np}}(V)}$ yields multiplication by $|\Sigma_{np} : N_{\Sigma_{np}}(V)|$, which is relatively prime to p. By the theorem of Cartan-Eilenberg (9.4.1),

$$\mathrm{Im}\{\mathrm{Res}^{\Sigma_{np}}_V : H^*(\Sigma_{np}; \ \mathbb{F}_p) \longrightarrow H^*(V; \ \mathbb{F}_p)\} = H^*(V; \ \mathbb{F}_p)^{\mathrm{st}^{\Sigma_{np}}_V}.$$

$V \hookrightarrow \Sigma_{np}$ is a p-Sylow subgroup and is abelian. Hence by Burnside's fusion theorem 9.4.3

$$H^*(V; \ \mathbb{F}_p)^{\mathrm{st}^{\Sigma_{np}}_V} = H^*(V; \ \mathbb{F}_p)^{\mathrm{st}^{N_{\Sigma_{np}}(V)}_V}$$

Since $V \triangleleft N_{\Sigma_{np}}(V)$ is normal

$$H^*(V; \ \mathbb{F}_p)^{\mathrm{st}^{N_{\Sigma_{np}}(V)}_V} = H^*(V; \ \mathbb{F}_p)^W$$

and the result follows from 9.4.2. $\square$

Chapter 10
Invariant Theory and Algebraic Topology

In chapter 8 we encountered the Dickson algebra: an algebra of *universal* invariants in the sense that $\mathbf{D}^*(n)$ is contained in the ring of invariants of any representation $\varrho : G \hookrightarrow \mathrm{GL}(n, \ \mathbb{F})$ when $\mathbb{F}$ is a Galois field. The Dickson algebra is a feature of invariant theory over finite fields not present over other fields. Another such feature is the Steenrod algebra. It is derived from another aspect of algebra over fields of nonzero characteristic, the Frobenius homomorphism. In section[1] 10.3 we develop the Steenrod algebra in a manner completely independent of any knowledge of algebraic topology, and indicate some of its elementary applications to invariant theory in section 10.4 and chapter 11. Some less elementary applications appear in sections 10.5 and 10.6.

The groups generated by psuedoreflections have played an important role in the interaction between algebraic topology and invariant theory. In section 10.2 we show how the classification of Shephard and Todd of the complex pseudoreflection groups can be extended to cover the classification of the pseudoreflection representations of these groups at the good primes.

We begin, however, by describing one way invariant theory enters algebraic topology. To do so we assume familiarity with the basic ideas of algebraic topology as found in the books of Massey [140], Spanier [229] and Milnor-Stasheff [146]. We also apply invariant theory to the determination of the torsion primes for Lie groups (see also [65], [116] and [165]) rederiving results of Borel and Demazure.

The reader interested only in invariant theory could skip sections 10.1, 10.5 and section 10.7. The material in sections 10.2 and 10.6 depends very much on [2], [75] and [223] and might also be skipped on a first reading. When reading section 10.5 a copy of [2] would be most useful.

---

[1] This section is independent of the other sections in this chapter. We make use of it in chapter 11. The reader interested only in invariant theory could read sections 10.3, for the definition and basic properties of the Steenrod algebra, 10.4, to see the connection to invariant theory, and then skip directly to chapter 11.

10.1   Spaces with Polynomial Cohomology

Fix a field $\mathbb{F}$. Suppose that $\tilde{X} \downarrow^{\pi} X$ is a finite covering of CW-complexes, and let W be the group of covering transformations, i.e.

$$W = \left\{ \varphi \quad \middle| \quad \begin{array}{ccc} \tilde{X} & \xrightarrow{\varphi} & \tilde{X} \\ \pi \searrow & © & \swarrow \pi \\ & X & \end{array} \right\}.$$

In other words, W is the group of all homeomorphisms $\varphi : \tilde{X} \longrightarrow \tilde{X}$ such that $\pi\varphi = \pi$. By functoriality of cohomology, we must have $\pi^*(H^*(X; \mathbb{F})) \subseteq H^*(\tilde{X}; \mathbb{F})^W$. If the covering is regular and the field $\mathbb{F}$ has characteristic prime to the order of W then we may apply a transfer argument (for a definition of the transfer see [213]) to conclude that $H^*(X; \mathbb{F}) \cong H^*(\tilde{X}; \mathbb{F})^W$. In particular, for a compact connected Lie group G with maximal torus $T \hookrightarrow G$, and normalizer $N_G(T)$, we have the regular covering

$$W_G \longrightarrow BT \longrightarrow BN_G(T)$$

where W_G is the Weyl group of G. So for a field of characteristic prime to the order of W_G one has $H^*(BN_G(T); \mathbb{F}) \cong H^*(BT; \mathbb{F})^{W_G}$.

Consider the left coset space G/T, and let $f_g : G/T \longrightarrow G/T$ be induced by left multiplication with the element $g \in G$. If $t \in T$ is a generator, then one easily sees that a fixed point of f_t is precisely a coset xT such that $t \in xTx^{-1}$. In other words $\mathrm{Fix}_{f_t}(G/T) = N_G(T)/T = W_G$. The map f_t is homotopic to the identity map (since G is arcwise connected), and hence applying the Lefschetz fixed point theorem as formulated by Dold [73] we obtain the result of Hopf [110] and Weil [258] that the Euler characteristic of G/T is equal to the order of the Weyl group W_G.

Consider next the diagram:

$$\begin{array}{ccccc} W_G & \longrightarrow & G/T & \longrightarrow & G/N_G(T) \\ \| & & \downarrow & & \downarrow \\ W_G & \longrightarrow & BT & \longrightarrow & BN_G(T) \\ \downarrow & & \downarrow & & \downarrow \\ * & \longrightarrow & BG & = & BG \end{array}$$

where the rows are principal coverings and the columns are fibrations. The Weyl group W_G acts on the space of left cosets G/T by right translation, i.e. $w(gT) = gTw = gwT$. If $w \in W_G$ then w has a fixed point on G/T if and only if $w = 1$. Since the Euler characteristic of G/T is $|W_G|$ we have

$$L(w, \mathbb{Z}) = \begin{cases} 0 & \text{if } w \neq 1 \\ |W_G| & \text{if } w = 1, \end{cases}$$

where $L(-, \mathbb{Z})$ denotes the Lefschetz number of —. The odd dimensional cohomology of G/T is zero [36], and hence the Lefschetz number $L(w, \mathbb{Z})$ is just the trace of the action of w on $H^*(G/T; \mathbb{Z})$. Therefore for any field $\mathbb{F}$ the character χ of the representation of W_G on $H^*(G/T; \mathbb{F})$ is (this result is originally due to Leray [134])

$$\chi(w) = \begin{cases} 0 & w \neq 1 \\ |W_G| & w = 1. \end{cases}$$

By character theory [202] the action of W_G on $H^*(G/T; \mathbb{F})$ is equivalent to the regular representation since we are assuming the characteristic of $\mathbb{F}$ is relatively prime to $|W_G|$. A transfer argument applied to the top row (which is a regular covering) shows that

$$H^*(G/N_G(T); \mathbb{F}) = H^*(G/T; \mathbb{F})^{W_G} = \begin{cases} 0 & \text{if } * > 0 \\ \mathbb{F} & \text{if } * = 0 \end{cases}$$

so the right hand column yields that $H^*(BN_G(T); \mathbb{F}) \cong H^*(BG; \mathbb{F})$. (Alternatively, one could apply the Becker-Gottlieb [25] transfer for fibrations with compact fibre to the right hand column after arguing that the Euler characteristic of $G/N_G(T)$ with $\mathbb{F}$-coefficients is zero by applying the usual transfer to the top covering.) Combining all this leads to the following important theorem of Borel [33].

THEOREM 10.1.1 (A. Borel): *Let G be a compact connected Lie group with maximal torus T and Weyl group W_G. If $\mathbb{F}$ is a field of characteristic relatively prime to $|W_G|$ then the projection map $BT \longrightarrow BG$ induces an isomorphism $H^*(BG; \mathbb{F}) \cong H^*(BT; \mathbb{F})^{W_G}$.* □

In the nonmodular case, the invariants of the action of the Weyl group W_G acting on $H^*(BT; \mathbb{F})$ are a polynomial algebra by 7.4.1 since $H^*(BT; \mathbb{F}) \cong \mathbb{F}[x_1, \ldots, x_n]$ where n is the rank of G, and W_G acts as a real reflection group. (N.b. Since we are now doing *topology* the variables $x_1, \ldots, x_n$ belong to $H^2(BT; \mathbb{F})$ and all have degree 2.)

The classifying spaces of compact connected Lie groups are important not only in algebraic topology, but also in other branches of mathematics. A long standing important problem in homotopy theory has been to characterize the classifying spaces of compact connected Lie groups among all CW-complexes. A first step in this direction is to understand which polynomial algebras over $\mathbb{F}$ can occur as the cohomology of a space. We should try to construct examples in a systematic way and then show that the construction yields all examples. There is no loss of generality (because of the universal coefficient theorem) in restricting attention to the fields $\mathbb{Q}$ and $\mathbb{F}_p$. Over $\mathbb{Q}$ every polynomial algebra on generators of even degree can occur as the rational cohomology of a CW -complex, and graded commutativity says these are all that can occur. For the Galois fields $\mathbb{F}_p$ the situation is considerably more complicated, and as of this writing the complete classification has not been obtained.

The theorem of Borel 10.1.1 shows how to express the rational cohomology of the classifying space of a compact connected Lie group in terms of the invariants of the Weyl group acting on the cohomology of the maximal torus. The idea of Clark-Ewing [56], building on work of Holzsager [108] and Sullivan [244], is to reverse this procedure to construct topological spaces with polynomial cohomology in characteristic p.

The starting point is the theorem of Shephard-Todd (see section 7.4). Given a complex pseudoreflection group $\varrho : G \hookrightarrow \mathrm{GL}(n,\ \mathbb{C})$, it follows from 7.3.1 that ϱ is already defined over the subring of $\mathbb{C}$ generated by $\mathbb{Z}$ and the roots of unity of characters of ϱ. Hence ϱ is defined over the ring of p-adic integers $\mathbb{Z}_p^\wedge$ for those values of p indicated in table 7.3.1 as GOOD PRIMES.

Let $k \in \mathbb{N}$ be a positive integer. There is a functor

$$K(-,\ k) : \mathbf{Ab} \longrightarrow \mathbf{TOP}$$

from the category $\mathbf{Ab}$ of abelian groups to the category $\mathbf{TOP}$ of topological spaces, that assigns to an abelian group A the Eilenberg-Mac Lane space $K(A,\ k)$. $K(A,\ k)$ is a CW -complex and is characterized up to homotopy among CW -complexes by the properties:

$$\pi_i(K(A,\ k)) = \begin{cases} A & \text{for } i = k \\ 0 & \text{otherwise.} \end{cases}$$

For example when $A = \bigoplus_n \mathbb{Z}$ and $k = 2$ the space $K(\bigoplus_n \mathbb{Z},\ 2)$ is homotopy equivalent to BT, where T is a torus of rank n. When $A = \bigoplus_n \mathbb{Z}_p^\wedge$ and $k = 2$ then $K(\bigoplus_n \mathbb{Z}_p^\wedge,\ 2)$ is in an appropriate sense a p-adic completion of the topological space BT. For this reason we could write $BT_p^\wedge$ for $K(\bigoplus_n \mathbb{Z}_p^\wedge,\ 2)$. We quote without proof the following important fact:

PROPOSITION 10.1.2 : Let $V = \mathrm{Hom}(\bigoplus_n \mathbb{Z}_p^\wedge, \mathbb{F}_p)$. There is a natural isomorphism $H^*(K(\bigoplus_n \mathbb{Z}_p^\wedge,\ 2);\ \mathbb{F}_p) \simeq \mathbb{F}_p[V]$. $\square$

If $G \longrightarrow \mathrm{GL}(n,\ \mathbb{Z}_p^\wedge)$ is a representation, then since $K(-,\ 2)$ is a functor, the group G acts on $K(\bigoplus_n \mathbb{Z}_p^\wedge,\ 2)$, and because of the naturality of the isomorphism

$$H^*(K(\bigoplus_n \mathbb{Z}_p^\wedge,\ 2);\ \mathbb{F}_p) \simeq \mathbb{F}_p[V]$$

the action of G on $H_*(K(\bigoplus_n \mathbb{Z}_p^\wedge,\ 2);\ \mathbb{F}_p)$ is via the composition of the representation ϱ with reduction mod p, $G \longrightarrow \mathrm{GL}(n,\ \mathbb{Z}_p^\wedge) \longrightarrow \mathrm{GL}(n,\ \mathbb{F}_p)$. There is no loss in generality in supposing that the action of G on $K(\bigoplus_n \mathbb{Z}_p^\wedge, \ 2)$ is free, because if it were not free we could always form $K(\bigoplus_n \mathbb{Z}_p^\wedge, \ 2) \times EG$, where EG is a free acyclic G-space and use the diagonal action, which is free. There is then the principal covering

$$G \longrightarrow K(\bigoplus_n \mathbb{Z}_p^\wedge, \ 2) \longrightarrow BX_\varrho$$

defining the space BX_ϱ (there are topological reasons for calling the space BX_ϱ rather than just say X_ϱ) uniquely up to homotopy type. A transfer argument for regular coverings leads to the following [54]:

THEOREM 10.1.3 (A. Clark and J. Ewing): *Let* $\varrho : G \hookrightarrow \mathrm{GL}(n, \ \mathbb{Z}_p^\wedge)$ *be a p-adic integral representation of a finite group* G *such that the mod p reduction* $\varrho : G \hookrightarrow \mathrm{GL}(n, \ \mathbb{Z}_p^\wedge) \longrightarrow \mathrm{GL}(n, \ \mathbb{F}_p)$ *is generated by pseudoreflections, and (the nonmodularity condition) the order of* G *is relatively prime to p. Then with the preceding notations there is an isomorphism*

$$H^*(BX_\varrho; \ \mathbb{F}_p) \simeq \mathbb{F}_p[V(G)]^G$$

and hence

$$H^*(BX_\varrho; \ \mathbb{F}_p) \simeq \mathbb{F}_p[f_1, \ldots, f_n]$$

where $\deg(f_i) = 2d_i$ *and* $\prod_{i=1}^n d_i = |G|$. $\square$

This theorem provides us with a large family of spaces with polynomial cohomology subject to the nonmodularity condition that the generators are all in degrees relatively prime to p. The prime 2 is special, as usual in topology, and there is a separate discussion for it that provides more examples: this is probably only of interest to topologists who are quite aware of how to proceed, so we do not belabor this point.

The theorem of Clark-Ewing has, however, one defect: it starts with a p-adic representation whose mod p reduction is generated by pseudoreflections. The results of Shephard-Todd [206] (see chapter 7) classify only rational, real and complex pseudoreflection subgroups of $\mathrm{GL}(n, \ \mathbb{C})$ up to automorphism. Thus, for example, given a group in the list and a corresponding prime, it is not apparent how many p-adic lifts of the mod p pseudoreflection representation there are, and hence how many different homotopy types BX_ϱ could arise for a given group. This is remedied by a theorem of Dwyer, Miller and Wilkerson [75] that we discuss in the next section.

10.2   Classification of Nonmodular Pseudoreflection Groups

Shephard and Todd [206] found all the irreducible finite pseudoreflection subgroups of $GL(n, \mathbb{C})$ up to automorphism. If $\varrho : G \hookrightarrow GL(n, \mathbb{C})$ is a pseudoreflection representation then by theorem 7.3.1 we may assume that $\varrho(G) \subset GL(n, \mathbb{F}_\chi)$, where $\mathbb{F}_\chi \subset \mathbb{C}$ is the character field of ϱ. If p is a good prime for ϱ then $\mathbb{F}_p$ is a quotient field of the ring of integers $\mathbb{Z}_\chi$ in $\mathbb{F}_\chi$, so by clearing denominators we may reduce ϱ modulo p, i.e we may define a representation
$$\varrho_p : G \hookrightarrow GL(n, \mathbb{Z}_\chi) \xrightarrow{r} GL(n, \mathbb{F}_p)$$
where r is the reduction map modulo a suitable prime ideal. The representation ϱ_p is a pseudoreflection representation. The following theorem of Dwyer, Miller and Wilkerson sets up a number of useful bijective correspondences when, in addition, $|G|$ is not divisible by p.

THEOREM 10.2.1   (W. G. Dwyer-H. Miller-C. W. Wilkerson): *Let p be an odd prime, d and n positive integers with $p \nmid d$. Then there are bijective correspondences between the following sets:*
  (i) *Conjugacy classes of subgroups $G < GL(n, \mathbb{F}_p)$ such that G is generated by pseudoreflections and $|G| = d$.*
  (ii) *Conjugacy classes of subgroups $G < GL(n, \mathbb{Z}_p^\wedge)$ such that G is generated by pseudoreflections and $|G| = d$.*
  (iii) *Conjugacy classes of subgroups $G < GL(n, \mathbb{C})$ such that G is generated by pseudoreflections, $|G| = d$ and the character field can be embedded in $\mathbb{Q}_p^\wedge$.*

PROOF : Let us first discuss why representations $\varrho : G \longrightarrow GL(n, \mathbb{F}_p)$ can be lifted to $GL(n, \mathbb{Z}_p^\wedge)$. This follows from the fact that the kernel of the reduction map $GL(n, \mathbb{Z}_p^\wedge) \longrightarrow GL(n, \mathbb{F}_p)$ is a pro-p-group. Specifically, let
$$\cdots \longrightarrow GL(n, \mathbb{Z}/p^k) \longrightarrow \cdots \longrightarrow GL(n, \mathbb{Z}/p^2) \longrightarrow GL(n, \mathbb{Z}/p)$$
be the inverse system induced by the reduction maps. The inverse limit of this tower is $GL(n, \mathbb{Z}_p^\wedge)$, the successive maps being epic with kernels finite p-groups. Thus, since we are assuming the order of G is prime to p we can successively lift ϱ to $GL(n, \mathbb{Z}/p^k)$, and hence to $GL(n, \mathbb{Z}_p^\wedge)$.

N.b. The restriction of the reduction map $GL(n, \mathbb{Z}_p^\wedge) \longrightarrow GL(n, \mathbb{F}_p)$ to a subgroup whose order is prime to p is monic (see lemma 10.7.1). In particular, if an element has order prime to p then its order is the same in both groups.

The uniqueness of such a lift is seen as follows. Suppose given two p-adic representations of G that become equivalent upon reducing mod p, say M' and M''. Let
$$\alpha : M' \otimes_{\mathbb{Z}_p^\wedge} \mathbb{F}_p \longrightarrow M'' \otimes_{\mathbb{Z}_p^\wedge} \mathbb{F}_p$$

be a G-isomorphism. We lift α to a p-adic isomorphism of $\mathbb{Z}_p^\wedge$-modules and average it over G, which is possible since $|G|$ is prime to p, to obtain a G-equivariant equivalence covering α.

Next recall that an element $g \in \mathrm{GL}(n, A)$, for any ring A, is a diagonalizable pseudoreflection if and only if g is of the form

$$g = B \begin{bmatrix} 1 & 0 & \cdots & \cdots & 0 \\ 0 & 1 & \cdots & \cdots & 0 \\ \vdots & \vdots & \vdots & \vdots & \vdots \\ 0 & 0 & \cdots & 1 & 0 \\ 0 & 0 & \cdots & 0 & \zeta \end{bmatrix} B^{-1} \qquad B \in \mathrm{GL}(n, A)$$

where ζ is a root of unity. If $g \in \mathrm{GL}(n, \mathbb{F}_p)$ is a pseudoreflection of order prime to p, then ζ is an mth root of unity for some m dividing $p - 1$. Lift B to $\tilde{B} \in \mathrm{GL}(n, \mathbb{Z}_p^\wedge)$ and ζ to $\tilde{\zeta} \in \mathbb{Z}_p^\wedge$: $\tilde{\zeta}$ also an mth root of unity. Then

$$\tilde{g} = \tilde{B} \begin{bmatrix} 1 & 0 & \cdots & \cdots & 0 \\ 0 & 1 & \cdots & \cdots & 0 \\ \vdots & \vdots & \vdots & \vdots & \vdots \\ 0 & 0 & \cdots & 1 & 0 \\ 0 & 0 & \cdots & 0 & \tilde{\zeta} \end{bmatrix} \tilde{B}^{-1} \in \mathrm{GL}(n, \mathbb{Z}_p^\wedge)$$

lifts g and is a pseudoreflection having the same order as g.

If $\hat{g}$ is any other lift of the pseudoreflection $g \in \mathrm{GL}(n, \mathbb{F}_p)$ of order prime to p, then by the uniqueness of lifting up to conjugation of elements of order prime to p, $\hat{g}$ is conjugate to a pseudoreflection, and hence is itself a pseudoreflection.

Thus a pseudoreflection representation $\varrho : G \hookrightarrow \mathrm{GL}(n, \mathbb{F}_p)$ of a finite group of order prime to p lifts to a pseudoreflection representation $\tilde{\varrho} : G \hookrightarrow \mathrm{GL}(n, \mathbb{Z}_p^\wedge)$ that is unique up to conjugation.

This is the equivalence (i) $\Longleftrightarrow$ (ii).

N.b. If $\varrho : G \hookrightarrow \mathrm{GL}(n, \mathbb{F}_p)$ is finite and generated by pseudoreflections but $p \mid |G|$ then the representation ϱ may fail to lift. For example if $n < p - 1$ then $\mathrm{GL}(n, \mathbb{Q}_p^\wedge)$ contains no element of order p. Thus lifting $\mathrm{GL}(2, \mathbb{F}_p)$ to a subgroup of $\mathrm{GL}(2, \mathbb{Q}_p^\wedge)$ necessitates that $2 \geq p - 1$. The only primes satisfying this condition are 2 and 3, and the tautological representations of both $\mathrm{GL}(2, \mathbb{F}_2)$ and $\mathrm{GL}(2, \mathbb{F}_3)$ lift to the respective p-adic general linear group.

In fact, things can be worse when p divides the order of G. For example it can happen that the pseudoreflection representation $\varrho : G \longrightarrow \mathrm{GL}(n, \mathbb{F}_p)$ lifts to the p-adics, *but not as a pseudoreflection representation*. Specifically, consider the group GL(2, $\mathbb{F}_3$) which is group number 12 in the Shephard-Todd list. The character field of GL(2, $\mathbb{F}_3$) is $\mathbb{Q}(\sqrt{-2})$, so the complex pseudoreflection representation ϱ is defined over $\mathbb{Q}(\sqrt{-2})$ which embeds in $\mathbb{Q}_3^\wedge$. (See the discussion in section 7.3.) The matrix

$$A = \begin{bmatrix} 1 & 1 \\ 0 & 1 \end{bmatrix} \in \text{GL}(2, \ \mathbb{F}_3)$$

has order 3 and generates the 3-Sylow subgroup of GL(2, $\mathbb{F}_3$) and SL(2, $\mathbb{F}_3$). A is a non-diagonal pseudoreflection. Since neither $\mathbb{Q}(\sqrt{-2})$ nor $\mathbb{Q}_3^\wedge$ contains a cube root of unity $\varrho(A) \in$ GL(2, $\mathbb{Q}(\sqrt{-2})$) cannot be a pseudoreflection. So $\varrho(A) \in$ GL(2, $\mathbb{Q}(\sqrt{-2})$) lifts A, and hence the subgroup generated by A, to characteristic zero but not as a pseudoreflection representation. Concretely, the matrix

$$\tilde{A} = \frac{-1}{2} \begin{bmatrix} 1 & 1 \\ -3 & 1 \end{bmatrix}$$

has order 3, lifts the transvection A from GL(2, $\mathbb{F}_3$) to[2] GL(2, $\mathbb{Z}[\frac{1}{2}]$), but not as a pseudoreflection.

The representation ϱ : GL(2, $\mathbb{F}_3$) $\hookrightarrow$ GL(2, $\mathbb{Q}(\sqrt{-2})$), described in section 7.3, lifting the tautological representation of GL(2, $\mathbb{F}_3$) to GL(2, $\mathbb{Q}(\sqrt{-2})$), also lifts the tautological representation of SL(2, $\mathbb{F}_3$) to SL(2, $\mathbb{Q}(\sqrt{-2})$). The group SL(2, $\mathbb{C}$) contains no pseudoreflections, so ϱ(SL(2, $\mathbb{F}_3$)) $\subset$ GL(2, $\mathbb{Q}(\sqrt{-2})$) contains no pseudoreflections at all, even though in its tautological representation SL(2, $\mathbb{F}_3$) $\subset$ GL(2, $\mathbb{F}_3$) the group SL(2, $\mathbb{F}_3$) is generated by pseudoreflections (in fact by transvections).

Another example of this kind is provided by the group GL(3, $\mathbb{F}_2$) which we discussed briefly in section 7.3. The tautological representation of this group on $\mathbb{F}_2^3$ lifts to GL(3, $\mathbb{Q}(\sqrt{-7})$) but not as a pseudoreflection group. The product with a copy of $\mathbb{Z}/2$ generated by $-I$, $\mathbb{Z}/2 \times$ GL(3, $\mathbb{F}_2$) $\subset$ GL(2, $\mathbb{Q}(\sqrt{-7})$), is generated by pseudoreflections. It is group number 24 in the Shephard-Todd list.

Clearly, for primes dividing the order of the group we do not yet know what is going on.

To deal with the equivalence of (ii) and (iii) first of all note that by 7.3.1 the Schur index of a pseudoreflection group is 1, so the character field $\mathbb{F}_\varrho$ of $\varrho : G \hookrightarrow$ GL(n, $\mathbb{C}$) is unambiguously defined either as the smallest subfield of $\mathbb{C}$ over which ϱ is defined, or as the smallest subfield of $\mathbb{C}$ containing all the characters of ϱ. Since this latter field is obtained by adjoining roots of unity

---

[2] We use the notation $\mathbb{Z}[\frac{1}{p}]$ for the subring of $\mathbb{Q}$ consisting of rational numbers whose denominator is a power of the prime p.

to $\mathbb{Q}$, it follows that $\mathbb{F}_\varrho$ is contained in a cyclotomic extension and hence that $\mathbb{F}_\varrho | \mathbb{Q}$ is an abelian extension, and in particular Galois. We need the following:

Rigidity Property: *Let $G < \mathrm{GL}(n, \mathbb{C})$ be a complex pseudoreflection group, and α a field automorphism of $\mathbb{C}$. Then G and $\alpha(G) < \mathrm{GL}(n, \mathbb{C})$ are conjugate as subgroups.*

N.b. The representations $\varrho : G \hookrightarrow \mathrm{GL}(n, \mathbb{C})$ and $\alpha\varrho : G \hookrightarrow \mathrm{GL}(n, \mathbb{C})$ may *fail* to be conjugate, even though the subgroups which are their images are conjugate. This can be seen for the dihedral group D_{10} of order 10, which has two distinct representations as a pseudoreflection group over $\mathbb{Q}(\sqrt{5})$ which are interchanged by the nontrivial automorphism of $\mathbb{Q}(\sqrt{5})$. This is reflected in the reductions of these representations mod 5: see example 1 in section 5.6. This phenomenon was already observed by Shephard and Todd [206] who do not distinguish between a representation and its complex conjugate.

The rigidity property is verified by classification (*this is not a proof: only a verification*) as follows. There is no loss of generality in supposing the representation $\varrho : G \hookrightarrow \mathrm{GL}(n, \mathbb{C})$ is irreducible. Then $\alpha\varrho(G) < \mathrm{GL}(n, \mathbb{C})$ is also irreducible of the same rank and order as G. The degrees of the polynomial generators of the rings of invariants are also identical. Inspecting the table 7.3.1 yields the conclusion.

We use this to set up the desired bijection. Suppose that $G < \mathrm{GL}(n, \mathbb{Z}_p^\wedge)$ is a subgroup generated by pseudoreflections with character field $\mathbb{F}_G$, $\mathbb{F}_G \mid \mathbb{Q}$ a finite Galois extension. Since the Schur index of pseudoreflection groups is 1, G is conjugate to a subgroup G' of $\mathrm{GL}(n, \mathbb{F}_G)$, which, since characters determine a representation over $\mathbb{F}_G$ up to conjugacy, is well defined up to conjugacy in $\mathrm{GL}(n, \mathbb{F}_G)$.

Next note that as $\mathbb{F}_G \mid \mathbb{Q}$ is Galois any two embeddings of $\mathbb{F}_G$ in $\mathbb{C}$ differ by a field automorphism of $\mathbb{C}$. Therefore the rigidity property shows that the conjugacy class of the subgroup $G < \mathrm{GL}(n, \mathbb{C})$ we have obtained is independent of the embedding of $\mathbb{F}_G$ in $\mathbb{C}$. This gives a function from the conjugacy classes in (ii) to those in (iii).

By reversing the procedure we obtain an inverse to this function. Specifically, suppose $G < \mathrm{GL}(n, \mathbb{C})$ has a character field $\mathbb{F}_G$ that embeds in $\mathbb{Q}_p^\wedge$. Choose an embedding $\mathbb{F}_G \hookrightarrow \mathbb{Q}_p^\wedge$. Since the representation $G < \mathrm{GL}(n, \mathbb{C})$ is defined over $\mathbb{F}_G$, this embedding induces a representation $G \hookrightarrow \mathrm{GL}(n, \mathbb{Q}_p^\wedge)$. The order of G is prime to p, so we can clear denominators and obtain a representation defined in $\mathrm{GL}(n, \mathbb{Z}_p^\wedge)$, which is unique up to conjugacy. $\square$

Thus corresponding to each pseudoreflection group G over $\mathbb{C}$ and a good prime (i.e. a prime that does not divide the order of G and for which the character field $\mathbb{F}_G$ embeds in $\mathbb{Q}_p^\wedge$), the Clark-Ewing construction yields a unique homotopy type BX_G such that:

$$H^*(BX_G; \ \mathbb{F}_p) \cong \mathbb{F}_p[V]^G \cong \mathbb{F}_p[f_1, \ldots, f_n]$$

where $V = \mathbb{F}_p{}^n$ is acted on by a mod p reduction of G as pseudoreflection group.

It still remains to determine for a given algebra $H^* := \mathbb{F}_p[f_1, \ldots, f_n]$ with $\deg(f_i) = 2d_i$ and $\prod_{i=1}^n d_i = |G| \not\equiv 0$ mod p whether it is a ring of invariants, and if so, if there is some *other* homotopy type apart from the one provided by the Clark-Ewing construction that realizes H^* as a mod p cohomology algebra. This latter question is a deep topological problem and the subject of [75]. The first question, namely when is $H^* := \mathbb{F}_p[f_1, \ldots, f_n]$ with $\deg(f_i) = 2d_i$ and $\prod_{i=1}^n d_i = |G| \not\equiv 0$ mod p a ring of invariants, is the subject of the next few sections.

10.3 The Steenrod Algebra

In discussing the converse (see section 10.5) of the theorem of Shephard and Todd there is no way to avoid using Steenrod operations and the Steenrod algebra. Quite apart from their importance in algebraic topology and homotopy theory, the Steenrod operations are a new feature of the invariant theory of finite groups in characteristic p, and would eventually have been discovered by representation theorists or invariant theorists. See for example [127].

To justify this remark, consider that the Frobenius map is a *new* feature of just about anything in finite characteristic as opposed to characteristic 0. The question is how to make use of it. Guided by algebraic topology we proceed as follows.

Let p be a prime and V an $\mathbb{F}_p$-vector space. Define[3]

$$P(\xi) : \mathbb{F}_p[V] \longrightarrow \mathbb{F}_p[V][[\xi]]$$

by the rules

- $P(\xi)(v) = v + v^p\xi$ for $v \in V$
- $P(\xi)(u \cdot w) = P(\xi)(u) \cdot P(\xi)(w)$ for $u, \ w \in \mathbb{F}_p[V]$
- $P(\xi)(1) = 1$.

To consider $P(\xi)$ as a ring homomorphism of degree 0 we give ξ the degree -1 when $p = 2$ and $-2(p - 1)$ when p is an odd prime, and likewise the

---

[3] If A is a ring then $A[[\xi]]$ denotes the ring of formal power series over A in the variable ξ.

elements of V the degree 1 when $p = 2$ and 2 when p is an odd prime. This agrees with the standard conventions of algebraic topology. By separating out homogeneous components we obtain $\mathbb{F}_p$-linear maps

$$\mathcal{P}^i : \mathbb{F}_p[V] \longrightarrow \mathbb{F}_p[V] \qquad \text{for } p \text{ odd}$$

$$\mathrm{Sq}^i : \mathbb{F}_p[V] \longrightarrow \mathbb{F}_p[V] \qquad \text{for } p = 2$$

by the requirement

$$P(\xi)(f) = \sum_{i=0}^{\infty} \mathcal{P}^i(f)\xi^i$$

for odd primes and

$$P(\xi)(f) = \sum_{i=0}^{\infty} \mathrm{Sq}^i(f)\xi^i$$

for $p = 2$.

In principle we should write $P_V(\xi)$ to indicate the dependence on V, but as $P(\xi)$ is natural with respect to linear maps $\varphi : V' \longrightarrow V''$, i.e.

$$\varphi^* P(\xi) = P(\xi)\varphi^*$$

where φ^* is the algebra homomorphism induced by φ, this is unnecessary.

The operations $\mathcal{P}^i$ are called the **Steenrod reduced power operations** and the Sq^i are referred to as **Steenrod squaring operations**. Collectively they are referred to as **Steenrod operations**. The Steenrod operations are $\mathbb{F}_p$-linear maps and in addition satisfy:

$$\mathcal{P}^i(u) = \begin{cases} u^p & 2i = \deg(u) \\ 0 & 2i > \deg(u) \end{cases}$$

$$\mathrm{Sq}^i(u) = \begin{cases} u^2 & i = \deg(u) \\ 0 & i > \deg(u) \end{cases}$$

which are called the **unstability conditions**. By the multiplicative property of $P(\xi)$, one has

$$\mathcal{P}^k(u \cdot v) = \sum_{i+j=k} \mathcal{P}^i(u) \cdot \mathcal{P}^j(v)$$

$$\mathrm{Sq}^k(u \cdot v) = \sum_{i+j=k} \mathrm{Sq}^i(u) \cdot \mathrm{Sq}^j(v)$$

which are called the **Cartan formulae**. For a linear polynomial $x \in \mathbb{F}_p[x_1, \ldots, x_n]$ these yield:

$$\mathcal{P}^i(x^j) = \binom{j}{i} x^{j+i(p-1)}$$

$$\mathrm{Sq}^i(x^j) = \binom{j}{i} x^{i+j} .$$

The Steenrod operations satisfy certain identities, such as for $p = 2$:

$$\mathrm{Sq}^1 \cdot \mathrm{Sq}^1 = 0$$
$$\mathrm{Sq}^1 \cdot \mathrm{Sq}^2 = \mathrm{Sq}^3 \, .$$

We verify the first of these by induction on the dimension of V. If V has dimension 1, with basis vector x, then by the preceding formulae

$$\mathrm{Sq}^1\mathrm{Sq}^1(x^s) = (s)(s+1)x^{s+2} = 0$$

since either s or $s+1$ is even. If $\dim_{\mathbb{F}_p}(V) = n$ then choose a basis $x_1, \ldots, x_n$ for V. If $f \in \mathbb{F}[V]$ then f is a sum of monomials of the form $x_1^s \cdot w$, where w is a monomial in $x_2, \ldots, x_n$, so it suffices to show $\mathrm{Sq}^1\mathrm{Sq}^1(x_1^s w) = 0$. By the inductive hypothesis we have

$$\mathrm{Sq}^1\mathrm{Sq}^1(x_1^s \cdot w) = \mathrm{Sq}^1(\mathrm{Sq}^1(x_1^s) \cdot w + x_1^s \cdot \mathrm{Sq}^1(w))$$
$$= \mathrm{Sq}^1(x_1^s)\mathrm{Sq}^1(w) + \mathrm{Sq}^1(x_1^s)\mathrm{Sq}^1(w) = 0$$

which proves the assertion. The second formula may be verified analogously.

We define[4] the **mod p Steenrod algebra**, denoted by $\mathcal{A}^*$ when $p = 2$ and $\mathcal{P}^*$ when p is odd, to be the graded free associative $\mathbb{F}_p$-algebra with 1 generated by the Sq^i respectively $\mathcal{P}^i$ modulo the ideal of relations that hold in every polynomial algebra $\mathbb{F}_p[x_1, \ldots, x_n]$ for $n = 1, 2, 3, \ldots$ With the grading conventions we are using in this section $\mathcal{P}^i$ has degree $2i(p-1)$ and Sq^i degree i.

The Steenrod algebra action on $\mathbb{F}_p[V]$ clearly commutes with the action of $\mathrm{GL}(V)$ on $\mathbb{F}_p[V]$. If $G \hookrightarrow \mathrm{GL}(n, \mathbb{F}_p)$ is a faithful representation of a finite group G then the ring of invariants $\mathbb{F}_p[x_1, \ldots, x_n]^G$ is an algebra over the Steenrod algebra satisfying the Cartan formula and the unstability condition. Therefore the Steenrod operations can be used to produce new invariants from old ones. (See section 10.4.) This is a new feature of invariant theory in characteristic p. An algebra together with a module structure over the Steenrod algebra which satisfies the unstability condition and the Cartan formula is called an **unstable algebra over the Steenrod algebra.**

From the discussion so far one can see that this imposes serious restrictions on which polynomial algebras can occur as a ring of invariants. For example, if $p = 2$ no the polynomial algebra $\mathbb{F}_2[w_1, \ldots, w_n]$, with $3 = \deg(w_1) \leq \cdots \leq \deg(w_n)$ can occur as a ring of invariants. For if it did, then it would admit an action of the Steenrod squares satisfying the identities above. But

$$w_1^2 = \mathrm{Sq}^3(w_1) = \mathrm{Sq}^1 \cdot \mathrm{Sq}^2(w_1) = \mathrm{Sq}^1(0) = 0$$

---

[4] Actually this is the algebra of **reduced powers** for p odd since we have not included the Bockstein.

since there is no element in $\mathbb{F}_2[w_1, \ldots, w_n]$ of degree 5 apart from 0. This would be a contradiction. (See the next section for a further discussion of the use of the Steenrod algebra in computing rings of invariants.)

The remarkable theorem of Adams-Wilkerson (see section 10.5) says that the reduced power operations provide a complete set of restrictions on a polynomial algebra for it to occur as the ring of invariants of a finite group whose order is prime to p.

The definition of the Steenrod algebra does not make explicit a set of defining relations, i.e. none are constructed or given. These are provided by the **Bullett-Macdonald identity** [42]. Introduce the variables $u = 1 + t + \cdots + t^{p-1}$ and $s = tu$. Note that in characteristic p we have $u = (1 - t)^{p-1}$. By successive application of $P(\xi)$ and substitution we obtain the two compositions

$$P(s)P(1), \; P(u)P(t^p) : \mathbb{F}[V] \longrightarrow \mathbb{F}[V][[s]] \, .$$

The Bullett-Macdonald identity is the assertion that these are equal. To verify this note that $P(1)$, $P(s)$, $P(u)$, and $P(t^p)$ are multiplicative, so we are only required to check the identity on $x \in \mathbb{F}_p[x]$. This follows from the fact that $P(\xi)(x) = x + \xi x^p$ and a short calculation.

The more traditional relations between the Steenrod operations are commutation rules for $\mathcal{P}^i \mathcal{P}^j$, respectively $\mathrm{Sq}^i \mathrm{Sq}^j$, and are called **Adem relations**. They were originally conjectured by Wu Wen-Tsün based on his study of the mod p cohomology of Grassmann manifolds [265]. For $p = 2$ they are:

$$\mathrm{Sq}^i \mathrm{Sq}^j = \sum_{k=0}^{[i/2]} \binom{j - k - 1}{i - 2k} \mathrm{Sq}^{i+j-k} \mathrm{Sq}^k$$

for all $i, j > 0$ such that $i < 2j$. For p an odd prime they are:

$$\mathcal{P}^i \mathcal{P}^j = \sum_{k=0}^{[i/p]} (-1)^{i+1} \binom{(p-1)(j-k) - 1}{i - pk} \mathcal{P}^{i+j-k} \mathcal{P}^k$$

for all $i, j > 0$ such that $i < pj$, where $[a/b]$ denotes the integral part of a/b.

To verify the Adem relations we amplify[5] on [42]. Using the notations for odd primes, direct calculation gives:

$$P(s)P(1) = \sum_{a, \, k} s^a P^a P^k$$

$$P(u)P(t^p) = \sum_{c, \, j} u^c t^{pj} P^{a+b-j} P^j$$

and the Bullett-Macdonald identity says these are equal. Recall

$$\frac{1}{2\pi i}\oint_\gamma z^m dz = \begin{cases} 1 & m=1 \\ 0 & \text{otherwise} \end{cases}$$

where γ is a small circle around $0 \in \mathbb{C}$. Therefore

$$\sum_k P^a P^k = \frac{1}{2\pi i}\oint_\gamma \frac{P(s)P(1)}{s^{a+1}}ds$$

$$= \frac{1}{2\pi i}\oint_\gamma \frac{P(u)P(t^p)}{s^{a+1}}ds$$

$$= \frac{1}{2\pi i}\sum_{c,\,j}\oint_\gamma \frac{u^{c-j}t^{pj}}{s^{a+1}}ds\, P^c P^j$$

The formula $s = t(1-t)^{p-1}$ gives $ds = (1-t)^{p-2}(1-pt)dt$, so putting $c = a+b$ and substituting gives

$$\frac{u^{a+b-j}t^{pj}}{s^{a+1}}ds = \frac{(1-t)^{(p-1)(a+b-j)}t^{pj}(1-t)^{p-2}(1-pt)}{[t(1-t)^{p-1}]^{a+1}}dt$$

$$= (1-t)^{b-j-1)(p-1)+(p-2)}t^{pj-a-1}(1-pt)dt$$

$$= (1-t)^{(b-j)(p-1)-1}t^{pj-a-1}(1-pt)dt$$

$$= \left[\sum_k (-1)^k \binom{(b-j)(p-1)-1}{k}t^k\right]t^{k+pj-a-1}(1-pt)dt$$

$$= \sum_k (-1)^k \binom{(b-j)(p-1)-1}{k}\left[t^{k+pj-a-1} - pt^{k+pj-a}\right]dt.$$

Therefore

$$P^a P^b = \sum_j \left[\frac{1}{2\pi i}\oint_\gamma \frac{u^{a+b-j}t^{pj}}{s^{a+1}}ds\right]P^{a+b-j}P^j$$

$$= \sum_j \frac{1}{2\pi i}\sum_k (-1)^k \binom{(b-j)(p-1)-1}{k}\left[t^{k+pj-a-1} - pt^{k+pj-a}\right]dt\, P^{a+b-j}P^j.$$

Only the terms where

$$k+pj-a-1 = -1 \quad (k = a-pj)$$

$$k+pj-a = -1 \quad (k = a-pj-1)$$

contribute anything to the sum, so

$$P^a P^b = \sum_j \left[(-1)^{a-pj}\binom{(b-j)(p-1)-1}{a-pj} + (-1)^{a-pj-1}p\binom{(b-j)(p-1)-1}{a-pj-1}\right]P^{a+b-j}P^j$$

and since

$$\binom{(b-j)(p-1)-1}{a-pj} - p\binom{(b-j)(p-1)-1}{a-pj-1} \equiv \binom{(b-j)(p-1)-1}{a-pj} \bmod p$$

we conclude

$$P^a P^b = \sum_j (-1)^{j+1} \binom{(b-j)(p-1)-1}{a-pj} P^{a+b-j} P^j .$$

The proof for $p = 2$ is identical.

Thus there is a surjective map from the free associative algebra with 1 generated by the Steenrod operations modulo the Adem relations, which we denote by $\mathcal{B}^*$, onto the Steenrod algebra. In fact, this map, $\mathcal{B}^* \longrightarrow \mathcal{P}^*$ when p is odd and $\mathcal{B}^* \longrightarrow \mathcal{A}^*$ when $p = 2$, is an isomorphism and hence the Adem relations are a complete set of defining relations for the Steenrod algebra. To prove this we first introduce some notations and definitions.

Given a sequence $I = (i_1, i_2, \ldots, i_k)$ of non-negative integers we write

$$\mathcal{P}^I = \mathcal{P}^{i_1} \mathcal{P}^{i_2} \cdots \mathcal{P}^{i_k}$$

for p odd and for $p = 2$

$$\mathrm{Sq}^I = \mathrm{Sq}^{i_1} \mathrm{Sq}^{i_2} \cdots \mathrm{Sq}^{i_k} .$$

These iterations of Steenrod operations are called **basic monomials**. We call k the **length** of I and write $\ell(I)$ for the length of I. A basic monomial is called **admissible** if $i_s \geq p i_{s+1}$ for $s \geq 1$. To show that the Adem relations are a complete set of defining relations for the Steenrod algebra it suffices to show:

THEOREM 10.3.1 : *The admissible monomials span $\mathcal{B}^*$ as an $\mathbb{F}_p$-vector space. The images of the admissible monomials in the Steenrod algebra are linearly independent.*

PROOF : We begin by showing that the admissible monomials span $\mathcal{B}^*$.

For a sequence $I = (i_1, i_2, \ldots, i_k)$, the **moment** of I, denoted by $m(I)$, is defined by $m(I) = \sum_{s=1}^{k} s \cdot i_s$. We first show that an inadmissible monomial is a sum of monomials of smaller moment. Granted this it follows by induction over the moment that the admissible monomials span $\mathcal{B}^*$. To keep from doubling the proofs we give the details in the notations for p odd.

Suppose that $\mathcal{P}^I$ is an inadmissible monomial. Then there is a smallest s such that $i_s < p i_{s+1}$, i.e.

$$\mathcal{P}^I = \mathcal{Q}' \mathcal{P}^{i_s} \mathcal{P}^{i_{s+1}} \mathcal{Q}''$$

where Q', Q'' are basic monomials, and Q' is admissible. It is therefore possible to apply an Adem relation to $\mathcal{P}^I$ to obtain

$$\mathcal{P}^I = \sum_j a_j Q' \mathcal{P}^{i_s + i_{s+1} - j} \mathcal{P}^j Q''$$

for certain coefficients $a_j \in \mathbb{F}_p$. The terms on the right hand side all have smaller moment than $\mathcal{P}^I$ and so by induction we may express $\mathcal{P}^I$ as a sum of admissible monomials. (N.b. The admissible monomials are *reduced* in the sense that no Adem relation can be applied to them.)

We next show that the admissible monomials are linearly independent as elements of the Steenrod algebra (as we have defined it above). This we do by adapting an argument of Serre [197] and Cartan [53].

Let $e_n = x_1 x_2 \cdots x_n \in \mathbb{F}_p[x_1, \ldots, x_n]$ be the nth elementary symmetric function. Set $\xi = 1$ in $P(\xi)$ and denote the resulting operator by P. Then,

$$P(e_n) = P(\prod_{i=1}^n x_i) = \prod_{i=1}^n P(x_i)$$

$$= \prod_{i=1}^n (x_i + x_i^p) = \prod_{i=1}^n x_i \cdot \prod_{i=1}^n (1 + x_i^{p-1})$$

$$= e_n(x_1, \ldots, x_n) \cdot \left(\sum_{i=1}^n e_i(x_1^{p-1}, \ldots, x_n^{p-1}) \right).$$

so separating out homogeneous components and writing St^i for Sq^i when $p = 2$ and $\mathcal{P}^i$ when p is an odd prime, we obtain a formula of Wu Wen-Tsün:

$$\mathrm{St}^i(e_n) = e_n \cdot e_i(x_1^{p-1}, \ldots, x_n^{p-1}).$$

We claim that the monomials

$$\left\{ \mathrm{Sq}^I(e_n) \mid \mathrm{Sq}^I \text{ admissible and } \deg(\mathrm{Sq}^I) \leq n \right\}$$

when $p = 2$, and for odd primes p

$$\left\{ \mathcal{P}^I \mid \mathcal{P}^I \text{ admissible and } \deg(\mathcal{P}^I) \leq 2n \right\}$$

are linearly independent in $\mathbb{F}_p[x_1, \ldots, x_n]$. To see this note that in case $\ell(I) \leq n$, each entry in I is at most n (so the following formula makes sense), and

$$\mathrm{St}^I(e_n) = e_n \cdot \prod_{j=1}^s e_{i_j}(x_1^{p-1}, \ldots, x_n^{p-1}) + \cdots$$

where $I = (i_1, \ldots, i_s)$, $\mathrm{St}^I = \mathrm{St}^{i_1} \cdots \mathrm{St}^{i_s}$ and the remaining terms are lower in the lexicographic ordering. So $e_n \cdot \prod_{j=1}^{s} e_{i_j}(x_1^{p-1}, \ldots, x_n^{p-1})$ is the largest monomial in $\mathrm{St}^I(e_n)$ in the lexicographic order. Thus

$$\left\{ \mathrm{Sq}^I(e_n) \mid \mathrm{Sq}^I \text{ admissible and } \deg(\mathrm{Sq}^I) \leq n \right\}$$

when $p = 2$, and for odd primes p,

$$\left\{ \mathcal{P}^I(e_n) \mid \mathcal{P}^I \text{ admissible and } \deg(\mathcal{P}^I) \leq 2n \right\},$$

have distinct largest monomials, so are linearly independent.

By letting $n \longrightarrow \infty$ we obtain the assertion, completing the proof. $\quad\square$

Thus the Steenrod algebra may be regarded (this is one traditional definition) as the graded free associative algebra with 1 generated by the Sq^i respectively $\mathcal{P}^i$ modulo the ideal generated by the Adem relations.

10.4   Steenrod Operations and Invariants

One aspect of the discussion in the preceding section may be formulated as follows:

PROPOSITION 10.4.1 : *Let $\varrho : G \hookrightarrow \mathrm{GL}(n, \mathbb{F}_p)$ be a representation of a finite group G over the Galois field $\mathbb{F}_p$ of p elements. Then $\mathbb{F}_p[V]^G$ is an unstable algebra over the mod p Steenrod algebra.* $\quad\square$

The additional structure as unstable algebra over the Steenrod algebra is an important extra feature of $\mathbb{F}_p[V]^G$ in characteristic p (see for example [130], [220]). The Steenrod operations can be used to produce new invariants from old ones: for if $f \in \mathbb{F}_p[V]^G$, then $\mathcal{P}^i(f) \in \mathbb{F}_p[V]^G$.

EXAMPLE 1 : Consider for example $\varrho : G \hookrightarrow \mathrm{GL}(n, \mathbb{Z})$ a real crystallographic group, and let

$$\varrho_p : G \hookrightarrow \mathrm{GL}(n, \mathbb{Z}) \xrightarrow{r_p} \mathrm{GL}(n, \mathbb{Z}/p)$$

where r_p denotes reduction modulo p. If p is an odd prime then by 10.7.1 ϱ_p is again faithful. The representation ϱ, and hence also ϱ_p, leaves invariant a quadratic form $q \in \mathbb{F}_p[V]^G$. From this quadratic form we may produce other invariants. Concretely, when $G = W(\mathbf{F}_4)$ is the Coxeter group associated to the root system of type $\mathbf{F}_4$, then with respect to a suitable basis $\{z_1, z_2, z_3, z_4\}$ for V^*, $W(\mathbf{F}_4)$ leaves invariant the quadratic form (see section 7.2 example 4)

$$q = z_1^2 + z_2^2 + z_3^2 + z_4^2 \in \mathbb{F}_p[z_1, z_2, z_3, z_4].$$

When $p = 3$ it follows that[6]

$$z_1{}^4 + z_2{}^4 + z_3{}^4 + z_4{}^4 = -\mathcal{P}^1{}_3(q) \in \mathbb{F}_3[V]^{W(\mathbf{F_4})}$$

is also invariant under the action of $W(\mathbf{F_4})$. Hence $\mathbb{F}_3[z_1, \ z_2, \ z_3, \ z_4]^{W(\mathbf{F_4})}$ contains at least two linearly independent invariants of degree 8 (remember, since we are doing topology and 3 is an odd prime the variables z_1, z_2, z_3, z_4 have degree 2), namely q^2 and $\mathcal{P}^1_3(q)$. Since the Poincaré series of $\mathbb{Q}[V]^{W(\mathbf{F_4})}$ begins with $1 + t^4 + t^8 + \cdots$ this shows that $p = 3$ is a bad prime for $W(\mathbf{F_4})$. This explains the ad hoc computation in section 7.4 example 4 showing that 3 is a bad prime for $W(\mathbf{F_4})$. (A similar analysis may be applied to $W(\mathbf{E_8})$ at the prime $p = 5$.)

Here is another more extensive example of how Steenrod operations may be used in invariant theory.

    EXAMPLE 2 : Let $p \equiv 3 \bmod 4$ be a prime and $D_{2(p+1)} \hookrightarrow GL(2, \ \mathbb{F}_p)$ the representation of the dihedral group of order $2(p+1)$ described in section 5.6 example 6. Recall that this representation is given by a pair of matrices

$$T = \begin{bmatrix} a & b \\ -b & a \end{bmatrix}, \quad S = \begin{bmatrix} 0 & 1 \\ 1 & 0 \end{bmatrix} \in GL(2, \ \mathbb{F}_p)$$

where $a, \ b \in \mathbb{F}_p$, $T^{p+1} = 1 = S^2$ and $a^2 + b^2 = 1$.

Consider the quadratic polynomial function

$$f : \mathbb{F}_p \oplus \mathbb{F}_p \longrightarrow \bar{\mathbb{F}}_p$$

defined by

$$f(\alpha, \beta) = [\alpha, \ \ \beta] \begin{bmatrix} a & b \\ -b & a \end{bmatrix} \begin{bmatrix} \alpha \\ \beta \end{bmatrix} = [a\alpha - b\beta, \ \ b\alpha + a\beta] \begin{bmatrix} \alpha \\ \beta \end{bmatrix}$$

$$= a\alpha^2 - b\alpha\beta + b\alpha\beta + a\beta^2 = a(\alpha^2 + \beta^2).$$

If $a \neq 0 \in \mathbb{F}_p$ then $f \in \mathbb{F}_p[V]$ is a nonzero quadratic polynomial. The action of T on f is given by

$$(T \circ f)(\alpha, \ \beta) = f(T^{-1} \begin{bmatrix} \alpha \\ \beta \end{bmatrix}) = \left(T^{-1} \begin{bmatrix} \alpha \\ \beta \end{bmatrix} \right)^{\mathrm{tr}} T \left(T^{-1} \begin{bmatrix} \alpha \\ \beta \end{bmatrix} \right)$$

$$= [\alpha, \ \ \beta] (T^{-1})^{\mathrm{tr}} T T^{-1} \begin{bmatrix} \alpha \\ \beta \end{bmatrix} = [\alpha, \ \ \beta] T \begin{bmatrix} \alpha \\ \beta \end{bmatrix} = f(\alpha, \ \beta)$$

since $T^{-1} = T^{\mathrm{tr}}$. Therefore f is invariant under the action of T. Since f is unaltered by the interchange of α and β, f is also invariant under the action of S, so $f \in \mathbb{F}_p[x, \ y]^{D_{2(p+1)}}$.

---

[6] We follow a standard practice in algebraic topology to indicate that the base field is the Galois field $\mathbb{F}_p$ by writting $\mathcal{P}^i{}_p$ for the i-th Steenrod reduced power over $\mathbb{F}_p$.

If $a = 0 \in \mathbb{F}_p$ then the matrix

$$T^{\mathrm{tr}} = \begin{bmatrix} 0 & -b \\ b & 0 \end{bmatrix} \in \mathrm{GL}(2, \ \mathbb{F}_p)$$

has order $p+1$ and $b^2 = \det(T^{\mathrm{tr}}) = 1$. So we must have $b = \pm 1$. The matrices

$$\begin{bmatrix} 0 & -1 \\ 1 & 0 \end{bmatrix}, \quad \begin{bmatrix} 0 & 1 \\ -1 & 0 \end{bmatrix} \in \mathrm{GL}(2, \ \mathbb{F}_p)$$

are inverse to each other and have order $4 = p + 1$ so $p = 3$. Indeed the matrices

$$\begin{bmatrix} 0 & -1 \\ 1 & 0 \end{bmatrix}, \quad \begin{bmatrix} 0 & 1 \\ 1 & 0 \end{bmatrix} \in \mathrm{GL}(2, \ \mathbb{F}_3)$$

generate a dihedral group D_8 of order 8 which is the mod 3 reduction of the symmetry group of the square studied in section 3.2 example 3. The quadratic polynomial function

$$f : \mathbb{F}_3 \oplus \mathbb{F}_3 \longrightarrow \bar{\mathbb{F}}_3 \qquad f(\alpha, \ \beta) = \alpha^2 + \beta^2$$

is nonzero and invariant under D_8 when $p = 3$.

Thus, to summarize, for any prime $p \equiv 3 \bmod 4$ the quadratic polynomial $f = x^2 + y^2 \in \mathbb{F}_p[x, \ y]$ is $D_{2(p+1)}$-invariant for a suitable choice of basis $\{x, \ y\}$ for V^*. Therefore

$$\mathcal{P}^1(f) = \mathcal{P}^1(x^2 + y^2) = 2x\mathcal{P}^1(x) + 2y\mathcal{P}^1(y)$$
$$= 2xx^p + 2yy^p = 2(x^{p+1} + y^{p+1})$$

belongs to $\mathbb{F}_p[x, \ y]^{D_{2(p+1)}}$ also. The polynomials $x^2 + y^2$, $x^{p+1} + y^{p+1} \in \mathbb{F}_p[x, \ y]$ are algebraically independent. To verify this we apply 5.6.1. We have

$$\det \begin{bmatrix} \frac{\partial}{\partial x}(x^2 + y^2) & \frac{\partial}{\partial y}(x^2 + y^2) \\ \frac{\partial}{\partial x}(x^{p+1} + y^{p+1}) & \frac{\partial}{\partial y}(x^{p+1} + y^{p+1}) \end{bmatrix} = \det \begin{bmatrix} 2x & 2y \\ x^p & y^p \end{bmatrix}$$

$$= 2xy^p - 2yx^p \neq 0.$$

The degree of $x^2 + y^2$ is 2 and that of $x^{p+1} + y^{p+1}$ is $p + 1$. The product of the degrees is $2(p + 1) = |D_{2(p+1)}|$ and hence by 7.4.2 we conclude that

$$\mathbb{F}_p[x, \ y]^{D_{2(p+1)}} = \mathbb{F}_p[x^2 + y^2, \ x^{p+1} + y^{p+1}] = \mathbb{F}_p[f, \ \mathcal{P}^1(f)].$$

This was first observed by Wilkerson [260] in connection with a problem of Steenrod. The computation using Steenrod operations is more direct than the discussion using field extensions in section 5.6 example 6.

This example contains the germ of a general result. To formulate this we require some preliminaries.

The Steenrod algebra contains a family of derivations generalizing $\mathcal{P}^1$ defined inductively by the formulae:

$$\mathcal{P}^{\Delta_i} = \begin{cases} \mathcal{P}^1 & \text{if } i = 1 \\ [\mathcal{P}^{\Delta_{i-1}}, \mathcal{P}^{p^{i-1}}] & \text{for } i > 1 \end{cases}$$

where $[-, -]$ denotes the commutator of the two arguments.

LEMMA 10.4.2 : Let p be a prime, $V = \mathbb{F}_p^n$ and $x \in V^*$. Then $\mathcal{P}^{\Delta_i}(x) = x^{p^i}$ for $i = 1, 2, \dots$.

PROOF : For $i = 1$ this is the defining property of the operation $\mathcal{P}^{\Delta_1} = \mathcal{P}^1$, so we may proceed inductively and suppose that $i > 1$. We then have

$$\mathcal{P}^{\Delta_i}(x) = \mathcal{P}^{p^{i-1}}\mathcal{P}^{\Delta_{i-1}}(x) - \mathcal{P}^{\Delta_{i-1}}\mathcal{P}^{p^{i-1}}(x)$$

$$= \mathcal{P}^{p^{i-1}}(x^{p^{i-1}}) - 0 = \binom{p^{i-1}}{p^{i-1}} x^{p^{i-1}+p^{i-1}(p-1)} = x^{p^i}$$

since $\mathcal{P}^{p^{i-1}}(x) = 0$, because $p^{i-1} > 1 = \deg(x)$ as $i - 1 > 0$. $\square$

Lemma 10.4.2 could be used to define the derivations, which on linear forms are just the iterations of the Frobenius map $x \mapsto x^p$. We need only require

(i) $\mathcal{P}^{\Delta_0}(x) = \frac{\deg(x)}{2} \cdot x$,

(ii) $\mathcal{P}^{\Delta_i}(x) = x^{p^i}$ for $\deg(x) = 2$, and $i > 0$,

(iii) $\mathcal{P}^{\Delta_i}$ is a derivation.

This is equivalent to choosing a basis $z_1, \dots, z_n$ for V^* and requiring that

$$\mathcal{P}^{\Delta_i} = \sum_{j=1}^n z_j^{p^i} \frac{\partial}{\partial z_j}.$$

The proof of 10.4.2 then shows

$$\mathcal{P}^{\Delta_i} = \begin{cases} \mathcal{P}^1 & \text{if } i = 1 \\ [\mathcal{P}^{\Delta_{i-1}}, \mathcal{P}^{p^{i-1}}] & \text{for } i > 1. \end{cases}$$

The Steenrod operations are tightly bound to the Dickson polynomials through the derivations $\mathcal{P}^{\Delta_0}, \mathcal{P}^{\Delta_1}, \dots$, as the following result shows.

PROPOSITION 10.4.3 : If $V = \mathbb{F}_p^n$, then

$$-\mathcal{P}^{\Delta_n} = \mathbf{d}_{n,0}\mathcal{P}^{\Delta_0} + \mathbf{d}_{n,1}\mathcal{P}^{\Delta_1} + \cdots + \mathbf{d}_{n,n-1}\mathcal{P}^{\Delta_{n-1}}$$

where $\mathbf{d}_{n,0}, \dots, \mathbf{d}_{n,n-1} \in \mathbb{F}_p[V]$ are the Dickson polynomials (see section 8.1).

PROOF : Define the derivation Δ by

$$\Delta = \mathbf{d}_{n,0}\mathcal{P}^{\Delta_0} + \mathbf{d}_{n,1}\mathcal{P}^{\Delta_1} + \cdots + \mathbf{d}_{n,n-1}\mathcal{P}^{\Delta_{n-1}} + \mathcal{P}^{\Delta_n}.$$

It is enough to show that $\Delta(x) = 0$ for all linear polynomials $x \in \mathbb{F}_p[V]$. By lemma 10.4.2

$$\begin{aligned}
\Delta(x) &= \mathbf{d}_{n,0}\mathcal{P}^{\Delta_0}(x) + \mathbf{d}_{n,1}\mathcal{P}^{\Delta_1}(x) + \cdots + \mathbf{d}_{n,n-1}\mathcal{P}^{\Delta_{n-1}}(x) + \mathcal{P}^{\Delta_n}(x) \\
&= \mathbf{d}_{n,0}x + \mathbf{d}_{n,1}x^p + \cdots + \mathbf{d}_{n,n-1}x^{p^{n-1}} + x^{p^n} \\
&= \prod_{v^* \in V^*} (x + v)
\end{aligned}$$

by the definition of the Dickson polynomials in section 8.1. Since $-x \in V^*$ the right hand side is zero and result follows. □

With these preliminaries out of the way we introduce the definition of the orthogonal group of a quadratic form.

DEFINITION : *Let $\mathbb{F}$ be a field with algebraic closure $\bar{\mathbb{F}}$, $n \in \mathbb{N}$ an integer and $Q : \mathbb{F}^n \longrightarrow \bar{\mathbb{F}}$ a quadratic form. The **orthogonal group of** Q, denoted by $O(n, \mathbb{F}, Q)$, is defined by $O(n, \mathbb{F}, Q) = \{g \in GL(n, \mathbb{F}) \mid TQ = Q\}$.*

PROPOSITION 10.4.4 : *Let p be an odd prime, $n \in \mathbb{N}$ and $Q : \mathbb{F}_p{}^n \longrightarrow \bar{\mathbb{F}}_p$ a non-degenerate quadratic form on the vector space $V = \mathbb{F}_p{}^n$ over the Galois field $\mathbb{F}_p$ with p elements. If $G < O(n, \mathbb{F}_p, Q)$ then*

$$Q, \mathcal{P}^{\Delta_1}(Q), \ldots, \mathcal{P}^{\Delta_{n-1}}(Q) \in \mathbb{F}_p[V]^G$$

are algebraically independent in $\mathbb{F}_p[V]^G$.

PROOF : With no loss of generality we may choose a basis $z_1, \ldots, z_n$ for V^* such that [71]

$$Q = \sum_{i=1}^{n} \lambda_i z_i^2$$

where $\lambda_1, \ldots, \lambda_n \in \mathbb{F}_p$ and the product $\lambda_1 \cdots \lambda_n \neq 0 \in \mathbb{F}_p$. We then have

$$\mathcal{P}^{\Delta_j}(Q) = 2 \sum_{i=1}^{n} \lambda_i z_i^{p^j+1} \quad j = 1, \ldots, n-1.$$

Therefore

$$\frac{\partial \mathcal{P}^{\Delta_j}(Q)}{\partial z_i} = 2\lambda_i z_i^{p^j} \quad i = 1, \ldots, n, \ j = 0, \ldots, n-1.$$

Hence

$$
\det\left[\frac{\partial \mathcal{P}^{\Delta_j}(Q)}{\partial z_i}\right] = \det \begin{bmatrix} 2\lambda_1 z_1 & \cdots & 2\lambda_n z_n \\ 2\lambda_1 z_1^p & \cdots & 2\lambda_n z_n^p \\ \vdots & \vdots & \vdots \\ 2\lambda_1 z_1^{p^{n-1}} & \cdots & 2\lambda_n z_n^{p^{n-1}} \end{bmatrix}
$$

$$
= (2\lambda_1 \cdots \lambda_n)^n \det \begin{bmatrix} z_1 & \cdots & z_n \\ z_1^p & \cdots & z_n^p \\ \vdots & \vdots & \vdots \\ z_1^{p^{n-1}} & \cdots & z_n^{p^{n-1}} \end{bmatrix}
$$

$$
= (2\lambda_1 \cdots \lambda_n)^n [z_1 z_2^p \cdots z_n^{p^{n-1}} + \cdots] \neq 0
$$

because the remaining terms are lower in the lexicographic ordering. Therefore the invariants Q, $\mathcal{P}^{\Delta_1}(Q), \ldots, \mathcal{P}^{\Delta_{n-1}}(Q)$ are algebraically independent 5.6.1.
□

REMARK : When $n = 2$ it is not hard to show that Q, $\mathcal{P}^1(Q) \in \mathbb{F}_p[x, y]^G$ is a system of parameters when $G < \mathbb{O}(n, \mathbb{F}_p, Q)$. Since rings of invariants of 2-dimensional representations are Cohen-Macaulay $\mathbb{F}_p[x, y]^G$ is free and finitely generated over $\mathbb{F}_p[Q, \mathcal{P}^1(Q)]$. For $G = \mathbb{O}(2, \mathbb{F}_p, Q)$ one has (example 2) $\mathbb{F}_p[x, y]^{\mathbb{O}(2, \mathbb{F}_p, Q)} = \mathbb{F}_p[Q, \mathcal{P}^1(Q)]$. Indeed. this remains true when $\mathbb{F}_p$ is replaced by an arbitrary Galois field (see section 11.1).

10.5 The Theorem of Adams-Wilkerson

The theorem of Adams-Wilkerson discussed in this section is a converse to the theorem of Shephard-Todd-Chevalley 7.4.1 in characteristic p and represents an application of algebraic topology to invariant theory making considerable use of the Steenrod algebra and the Dickson algebra. In characteristic p the Shephard-Todd-Chevalley theorem says that the ring of invariants of a finite pseudoreflection group of order prime to p is a polynomial subalgebra on generators of degree also prime to p. The theorem of Adams-Wilkerson addresses the question of when a polynomial subalgebra

$$
\mathbb{F}_p[f_1, \ldots, f_n] \subseteq \mathbb{F}_p[x_1, \ldots, x_n]
$$

with generators of degrees relatively prime to p is the ring of invariants of a finite pseudoreflection group of order prime to p. The basic reference for this section is [2]. The exposition here is based on [223] and unpublished notes from a seminar jointly directed with R. M. Switzer in Göttingen in 1981/1982.

From the discussion of the previous sections it is apparent that if $\mathbb{F}_p[f_1, \ldots, f_n]$ is a ring of invariants, then it must admit an unstable action of the Steenrod algebra. The Adams-Wilkerson theorem says that in the *nonmodular* case this is enough.

There are separate discussions required for the prime 2 and all other primes. For convenience of notations and to avoid doubling the exposition we confine our exposition to the case p odd. The main result of Adams-Wilkerson [2] for odd primes may be stated as follows:

THEOREM 10.5.1 (J. F. Adams and C. W. Wilkerson): *Let p be an odd prime and*

$$H^* \cong \mathbb{F}_p[u_1, \ldots, u_n] \qquad \deg u_i = 2d_i, \ 1 \leq i \leq n$$

be an unstable algebra over the mod p Steenrod algebra satisfying

$$p \nmid d_1 d_2 \cdots d_n \,.$$

Then there exists a finite group $W < \mathrm{GL}(n, \mathbb{F}_p)$ of order $d_1 \cdots d_n$ generated by pseudoreflections such that $H^* = \mathbb{F}_p[x_1, \ldots, x_n]^W$.

Recall that an **unstable algebra over the Steenrod algebra** is a graded commutative algebra over the Steenrod algebra that satisfies the Cartan formula and the unstability condition. If just the Cartan formula holds then we speak simply of an **algebra over the Steenrod algebra**. In this section we will be dealing exclusively with integral domains and we denote by $\mathrm{UnId}/\mathcal{P}^*$ the category of unstable graded integral domains over the Steenrod algebra $\mathcal{P}^*$. We follow the exposition of [223] in which the Dickson algebra plays a central role.

There are four key results that we cite from the literature. The first is a result of Wilkerson [262], building on a result of Serre [199], which says that polynomial algebras on 2-dimensional generators are closed under integral extensions in $\mathrm{UnId}/\mathcal{P}^*$.

THEOREM 10.5.2 (**Integral Closure Theorem**): *A polynomial algebra $A^* = \mathbb{F}_p[x_1, \ldots, x_n]$ on generators x_i of degree 2 is integrally closed in $\mathrm{UnId}/\mathcal{P}^*$, i.e. if $A^* \leq B^*$ an integral extension in $\mathrm{UnId}/\mathcal{P}^*$, then $A^* = B^*$.*

An object $A^* \in \mathrm{UnId}/\mathcal{P}^*$ has a graded field of fractions, (see for example section 5.5) which we denote by $FF(A^*)$. Introduce as before the **total Steenrod operation**

$$\mathcal{P}(\xi) : A^* \longrightarrow A^*[[\xi]] \qquad \deg(\xi) = -2(p-1)$$

by the formula[7]

$$\mathcal{P}(\xi)(a) = \sum_{k=0}^{\infty} \mathcal{P}^k(a)\xi^k \,.$$

---

[7] We set $\mathcal{P}^0 = \mathrm{Id}$.

One verifies as a consequence of the Cartan formula that $\mathcal{P}(\xi)$ is a graded algebra homomorphism. For an element $a/b \in FF(A^*)$ we may set

$$\mathcal{P}(\xi)(a/b) = \frac{\mathcal{P}(\xi)(a)}{\mathcal{P}(\xi)(b)},$$

which makes sense since $\mathcal{P}(\xi)(b)$ is a formal power series beginning with $b \neq 0$. Since $FF(A^*)$ has elements of negative degree the unstability condition no longer holds, and $FF(A^*)$ is simply an algebra over the Steenrod algebra. In [261] Wilkerson shows:

LEMMA 10.5.3 (**Separable Extension Lemma**): *Let $\mathbb{K}^*$ be a graded field over $\mathcal{P}^*$ and $\mathbb{L}^* \geq \mathbb{K}^*$ a separable field extension. Then there is a unique extension to $\mathbb{L}^*$ of the $\mathcal{P}^*$ action on $\mathbb{K}^*$.*

A major role in the proof of Adams-Wilkerson is played by derivations. Recall from § 10.4 that the Steenrod algebra contains an infinite family of derivations defined inductively by the formulae:

$$\mathcal{P}^{\Delta_i} = \begin{cases} \mathcal{P}^1 & \text{if } i = 1 \\ [\mathcal{P}^{\Delta_{i-1}}, \mathcal{P}^{p^{i-1}}] & \text{for } i > 1. \end{cases}$$

The operation $\mathcal{P}^{\Delta_i}$ is of degree $2(p^i - 1)$. The following result is a summary of Lemmas 5.3-5.5 of [2] which are the result of an orgy of calculations. (Section 2 of [2] also contains a number of elementary facts about the Steenrod algebra, some of which we will also need. The reader can take this section of [2] as prototypical for the manipulations often encountered when working with the Steenrod algebra.)

THEOREM 10.5.4 (**Δ-Theorem**): *Suppose $H^* \in \mathrm{UnId}/\mathcal{P}^*$ has finite Krull dimension. Then there exists an integer n with the following properties: any n distinct derivations $\mathcal{P}^{\Delta_{i_1}}, \ldots, \mathcal{P}^{\Delta_{i_n}}$ are linearly independent over H^*, but any $n + 1$ derivations $\mathcal{P}^{\Delta_{i_0}}, \ldots, \mathcal{P}^{\Delta_{i_n}}$ are linearly dependent over H^*. Thus there exist elements $h_0, \ldots, h_n \in H^*$, all nonzero, such that*

$$h_0 \mathcal{P}^{\Delta_0} + \cdots + h_n \mathcal{P}^{\Delta_n} \equiv 0$$

is identically zero on H^*. $\square$

Let n be the integer of the Δ-theorem and $h_0, \ldots, h_n \in H^*$ nonzero elements such that the derivation $\Delta = h_0 \mathcal{P}^{\Delta_0} + \cdots + h_n \mathcal{P}^{\Delta_n}$ is identically zero on H^*. Note that $\deg(h_i) = 2(p^n - p^i)$ by homogeneity. (In fact we will see later (compare 10.4.3) that h_i may be chosen as the Dickson polynomial $\mathbf{d}_{n,i}$, but first we must find a Dickson algebra inside H^*.) Introduce an indeterminate X of degree 2 and consider the polynomial

$$\Delta(X) = h_0 X + h_1 X^p + \cdots + h_n X^{p^n} \in H^*[X].$$

Let E^* be the splitting field of $\Delta(X)$ over the field of fractions $FF(H^*)$. Since the formal derivative $\Delta'(X) = h_0 \neq 0$ does not vanish, $\Delta(X)$ is separable, so that $E^*|FF(H^*)$ is a separable field extension. By the Separable Extension Lemma there is a unique extension of the Steenrod algebra structure of $FF(H^*)$ to E^*.

Let $V = \{v \in E^* \mid \Delta(v) = 0\}$. Since $\Delta(X)$ is separable V contains p^n elements, and since $\Delta(X)$ is a p polynomial, it is a linear transformation, so its kernel V is an $\mathbb{F}_p$-vector subspace of E^* of dimension n over $\mathbb{F}_p$. Finally note

$$\Delta(X) = h_n \cdot \prod_{v \in V} (X + v).$$

We collect some facts about V.

PROPOSITION 10.5.5 : With the preceding notations we have:
(a) If $t_1, \ldots, t_n$ is a basis for V, then $t_1, \ldots, t_n$ are algebraically independent.
(b) The elements of V are unstable, i.e. satisfy the unstability condition, and hence $P^* = \mathbb{F}_p[t_1, \ldots, t_n]$ is an unstable integral domain over the Steenrod algebra.
(c) The action of $\mathcal{P}^*$ on P^* commutes with the action of $\mathrm{GL}(V)$, so the Dickson algebra $\mathbf{D}^*(n) = \mathbb{F}_p[V]^{\mathrm{GL}(V)}$ is a $\mathcal{P}^*$-subalgebra of P^*.
(d) Every element of P^* is integral over H^*.
PROOF : We take these assertions in order.
(a) Suppose that

$$\det \left[\mathcal{P}^{\Delta_i}(t_j) \right] = 0 \in E^*.$$

Then there are coefficients $a_1, \ldots, a_n \in E^*$, not all zero, such that

$$\partial' = a_1 \mathcal{P}^{\Delta_1} + \cdots + a_n \mathcal{P}^{\Delta_n}$$

is identically zero on P^*. Hence the elements of V are all roots of the polynomial

$$f(X) = a_1 X^p + \cdots + a_n X^{p^n} \in E^*.$$

Therefore $\Delta(X)$ divides $f(X)$ and

$$f(X) = a_n \prod_{v \in V} (X - v) = (\frac{a_n}{h_n}) \Delta(X).$$

Comparing coefficients of X on both sides we obtain $0 = (\frac{a_n}{h_n}) \cdot h_0$, so $a_n = 0$. But then $f(X)$ is identically zero, i.e. $a_1 = \cdots = a_n = 0$, which is a contradiction. Therefore

$$\det \left[\mathcal{P}^{\Delta_i}(t_j) \right] \neq 0 \in E^*$$

and hence $t_1, \ldots, t_n$ are algebraically independent by the 5.6.1.

(b) This follows from Lemmas 5.6 - 5.8 of [2].

(c) is obvious.

(d) The integrality of $x \in P^*$ over H^* is lifted from another argument of [2]: one considers the H^*-module Q^* of all derivations $\partial : H^* \longrightarrow H^*$ of the form

$$\sum_{k=0}^{n} a_k \mathcal{P}^{\Delta_k} \qquad a_0, \ldots, a_n \in H^*.$$

Let $y_1, \ldots, y_s$ be a finite set of generators of H^* as an algebra and define

$$\Phi : Q^* \longrightarrow \bigoplus_{i=1}^{s} H^*$$

by $\Phi(\partial) = (\partial y_1, \ldots, \partial y_s)$. Then Φ is injective, and since H^* is Noetherian Q^* must be a finitely generated H^*-module. Thus there is an integer m such that $\mathcal{P}^{\Delta_m}$ is a H^*-linear combination of $\mathcal{P}^{\Delta_0}, \ldots, \mathcal{P}^{\Delta_{m-1}}$, say

$$(*) \qquad d_0 \mathcal{P}^{\Delta_0} + \cdots + d_{m-1} \mathcal{P}^{\Delta_{m-1}} + \mathcal{P}^{\Delta_m} \equiv 0$$

on H^* for suitable $d_0, \ldots, d_{m-1} \in H^*$. By the uniqueness of extensions of derivations over separable extensions ([266] volume I chapter 2 section 17) $(*)$ must hold on E^* as well and thus on V. Hence each $v \in V$ satisfies

$$d_0 v + d_1 v^p + \cdots + d_{m-1} v^{p^{m-1}} + v^{p^m} = 0.$$

It follows that any $x \in P^* = \mathbb{F}_p[t_1, \ldots, t_n]$ is integral over H^*. $\quad\square$

PROOF OF THE THEOREM OF ADAMS-WILKERSON (10.5.1) : We continue to employ the previous notations. Let A^* be the algebra obtained by adjoining $t_1, \ldots, t_n$ to H^*. Then A^* is an unstable integral domain over the Steenrod algebra and we have inclusions

$$H^* \subseteq A^* \supseteq P^* \supseteq \mathbf{D}^*(n).$$

Suppose we knew that $\mathbf{D}^*(n) \leq A^*$ were an integral extension. Then $P^* \leq A^*$ would also be an integral extension, so from the Integral Closure Theorem it would follow that $A^* = P^*$, hence that $H^* \leq P^*$ and we could complete the proof as follows.

Consider the Galois group W of E^* over $FF(H^*)$. Clearly $W \leq GL(n, \mathbb{F}_p) = GL(V)$ because the elements of W define linear transformations on V (since $\Delta(X)$ is additive) and a Galois automorphism is determined by its effect on the roots of $\Delta(X)$, namely V.

We next show that $H^* = P^{*W} = \mathbb{F}_p[t_1, \ldots, t_n]^W$. Evidently $H^* \leq P^{*W}$. On the other hand, because $E^* > FF(H^*)$ is a Galois extension, every $x \in P^{*W}$

lies in $FF(H^*)$ and, by proposition 10.5.5 (d), is integral over H^*. Since H^* is a polynomial algebra it is integrally closed in its field of fractions, so $x \in H^*$ and $P^{*W} \leq H^*$. Combining these inclusions yields $H^* = \mathbb{F}_p[t_1, \ldots, t_n]^W$.

It finally remains to show that $A^* \geq \mathbf{D}^*(n)$ is an integral extension.

From our discussion of Dickson's theorem 8.1.5 we have:

$$\prod_{v \in V}(X - v) = X^{p^n} - \mathbf{d}_{n, \, n-1}X^{p^{n-1}} + \cdots + (-1)^n \mathbf{d}_{n, \, 0}X$$

where $\mathbf{d}_{n, \, n-1}, \ldots, \mathbf{d}_{n, \, 0}$ are the Dickson polynomials, and

$$\mathbb{F}_p[t_1, \ldots, t_n]^{\mathrm{GL}(n, \mathbb{F}_p)} = \mathbb{F}_p[\mathbf{d}_{n, \, n-1}, \ldots, \mathbf{d}_{n, \, 0}] = \mathbf{D}^*(n).$$

We require a computational lemma from [2].

LEMMA 10.5.6 : *Let L^* be an algebra over the Steenrod algebra and $x \in L^*$ an unstable element (i.e. the unstable conditions are valid for x). For $d, \, r > 0$ there exists an operation $a \in \mathcal{P}^*$ such that*

$$\left(\mathcal{P}^{\Delta s}(x)\right)^{p^r}(x) = \mathcal{P}^{\Delta_{r+s}}(ax)$$

for each $s \geq 0$ □

This is Lemma 2.3 in [2]. It is proved by several pages of computations. N.b. In [2] the operations $\mathcal{P}^{\Delta_i}$ are denoted by Q^i.

From this we may conclude that $\mathbf{d}_{n, \, n-1}, \ldots, \mathbf{d}_{n, \, 0} \in A^*$ is a system of parameters. To do so we show $A^*/(\mathbf{d}_{n, \, n-1}, \ldots, \mathbf{d}_{n, \, 0})$ is finite dimensional over $\mathbb{F}_p$ as follows. If $z \in A^{2d}$ with $d \not\equiv 0 \bmod p$, then by the preceding lemma there is an $a \in \mathcal{P}^*$ with

$$dz^{p^n} = d^{p^n}z^{p^n} = (\mathcal{P}^{\Delta_0}(z))^{p^n} = \mathcal{P}^{\Delta_n}(az).$$

So setting $b = d^{-1}a \in \mathcal{P}^*$ we have

$$z^{p^n} = \mathcal{P}^{\Delta_n}(bz).$$

The derivation $\mathbf{d}_{n, \, 0}\mathcal{P}^{\Delta_0} + \cdots + \mathbf{d}_{n, \, n-1}\mathcal{P}^{\Delta_{n-1}} + \mathcal{P}^{\Delta_n}$ is just

$$\frac{1}{h_n}(h_0\mathcal{P}^{\Delta_0} + \cdots + h_n\mathcal{P}^{\Delta_n}) = \frac{1}{h_n} \cdot \Delta$$

and hence is zero on E^*. Thus we have

$$z^{p^n} = \mathcal{P}^{\Delta_n}(bz) = -\mathbf{d}_{n, \, 0}\mathcal{P}^{\Delta_0}(bz) - \cdots - \mathbf{d}_{n, \, n-1}\mathcal{P}^{\Delta_{n-1}}(bz)$$

which belongs to $(\mathbf{d}_{n, \, 0}, \ldots, \mathbf{d}_{n, \, n-1})$. But A^* is generated as an algebra by the algebra generators $x_1, \ldots x_n$ of H^* and $t_1, \ldots, t_n$, all of which

lie in gradings $\not\equiv 0$ modulo p. If $z_1, \ldots, z_q$ denote these algebra generators of A^*, then we see that $A^*/(\mathbf{d}_{n,\,n-1}, \ldots, \mathbf{d}_{n,\,0})$ is spanned over $\mathbb{F}_p$
by the monomials $\{z_1^{i_1} \cdots z_q^{i_q} \mid i_k < p^n,\ 1 \leq k \leq q\}$. This set being
finite implies $A^*/(\mathbf{d}_{n,\,n-1}, \ldots, \mathbf{d}_{n,\,0})$ is finite dimensional over $\mathbb{F}_p$, and
hence $\mathbf{d}_{n,\,n-1}, \ldots, \mathbf{d}_{n,\,0} \in A^*$ is a system of parameters by the Noether
Normalization Theorem (5.3.3). Therefore by 5.1.1 $A^* \geq \mathbf{D}^*(n)$ is integral,
which as previously noted completes the proof of the theorem of Adams and
Wilkerson. $\square$

Note that *a posteriori* one sees that the integer n of the Δ-Theorem is the
same as the number of polynomial generators of H^*.

If $H^* \in \mathrm{UnId}/\mathcal{P}^*$ then by the Δ-theorem the coefficients of the Δ-polynomial
generate a Dickson algebra $\mathbf{D}^*(n) \subset H^*$. Therefore we have a diagram

$$\mathbf{D}^*(n) \hookrightarrow H^*$$
$$\Big\uparrow \Big\downarrow$$
$$\mathbb{F}_p[V].$$

By a theorem of Lannes [132] the algebras $\mathbb{F}_p[V] \in \mathrm{UnId}/\mathcal{P}^*$ are injective
objects. Hence we receive a homomorphism of algebras over the Steenrod
algebra $H^* \longrightarrow \mathbb{F}_p[V]$. This leads to a more conceptual proof of the theorem
of Adams and Wilkerson [132] at the expense of considerably more input:
namely, the proof that $\mathbb{F}_p[V]$ is an injective object in $\mathrm{UnId}/\mathcal{P}^*$ [132].

10.6   Another Proof of Dickson's Theorem

The methods developed to prove the theorem of Adams and Wilkerson yield
as a bonus another proof [223] of Dickson's theorem 8.1.5. In addition we
obtain a number of interesting and useful formulae for the action of the
Steenrod algebra on the Dickson polynomials. These formulae have been used
by Landweber and Stong [130] in their study of the codimension of rings of
invariants and in [224] to determine which Dickson algebras can be realized as
the mod p cohomology of a topological space. As in previous sections of this
chapter, to simplify the exposition, we will suppose that p is an odd prime.
The corresponding results for $p = 2$ are routine modifications of those for p
odd.

We denote by V the n-dimensional vector space over the field of p elements $\mathbb{F}_p$.
The full general linear group $\mathrm{GL}(n,\ \mathbb{F}_p)$ acts on V and $\mathbb{F}[V]^{\mathrm{GL}(n,\ \mathbb{F}_p)} = \mathbf{D}^*(n)$
is the Dickson algebra. Introduce the polynomial

$$\varphi(X) = \prod_{v^* \in V^*} (X + v^*) \in \mathbb{F}_p[V][X]$$

where X is an indeterminate of grading 2. The polynomial $\varphi(X)$ is equal to $X \cdot \varphi_{\{V^* \smallsetminus \{0\}\}}(X)$ where $\varphi_{\{V^* \smallsetminus \{0\}\}}(X)$ is the orbit polynomial of the orbit $V^* \smallsetminus \{0\}$ of the $\mathrm{GL}(n, \mathbb{F}_p)$ action. The polynomial $\varphi(X)$ is a p-polynomial and hence as in section 8.1 we may write $\varphi(X)$ in the form

$$\varphi(X) = X^{p^n} + y_1 X^{p^{n-1}} + \cdots + y_n X$$

where $y_i = c_{p^n - p^{n-i}}(V^* \smallsetminus \{0\}) \in \mathbf{D}^*(n)$ is of degree $2(p^n - p^{n-i})$. (The polynomials y_i are of course just the Dickson polynomials (see section 8.1) $\mathbf{d}_{n,n-i}$ in another notation.)

LEMMA 10.6.1 : *The derivations*

$$\mathcal{P}^{\Delta_1}, \ldots, \mathcal{P}^{\Delta_n} : \mathbb{F}_p[V] \longrightarrow \mathbb{F}_p[V]$$

are linearly independent over $\mathbb{F}_p[V]$.

PROOF :   Choose a basis $x_1, \ldots, x_n$ for V^* and consider (apply 10.4.2)

$$\det \left[\mathcal{P}^{\Delta_i}(x_j) \right] = \det \begin{bmatrix} x_1^p & \cdots & x_n^p \\ \vdots & \vdots & \vdots \\ x_1^{p^n} & \cdots & x_n^{p^n} \end{bmatrix}.$$

Note that upon expanding the coefficient of $x_1^p x_2^{p^2} \cdots x_n^{p^n}$ is 1 so the determinant is nonzero and therefore $\mathcal{P}^{\Delta_1}, \ldots, \mathcal{P}^{\Delta_n}$ are linearly independent over $\mathbb{F}_p[V]$. $\square$

LEMMA 10.6.2 : *The derivation*

$$\mathcal{P}^{\Delta_n} + y_1 \mathcal{P}^{\Delta_{n-1}} + \cdots + y_n \mathcal{P}^{\Delta_0}$$

is identically zero on $\mathbb{F}_p[V]$.

PROOF :   The roots of the polynomial $\varphi(X) \in \mathbb{F}_p[V][X]$ are the elements of V^*. Hence for any $x \in V^*$ we have by 10.4.2

$$\begin{aligned} 0 = \varphi(x) &= x^{p^n} + y_1 x^{p^{n-1}} + \cdots + y_n x \\ &= \mathcal{P}^{\Delta_n}(x) + y_1 \mathcal{P}^{\Delta_{n-1}}(x) + \cdots + y_n \mathcal{P}^{\Delta_0}(x) \\ &= (\mathcal{P}^{\Delta_n} + y_1 \mathcal{P}^{\Delta_{n-1}} + \cdots + y_n \mathcal{P}^{\Delta_0})(x). \end{aligned}$$

Since V^* generates $\mathbb{F}_p[V]$ as an algebra the result follows. $\square$

LEMMA 10.6.3 : *With the preceding notations, the elements* $y_1, \ldots, y_n$ *are a system of parameters.*

PROOF :   For $x \in V^*$ we have

$$-x^{p^n} = -\mathcal{P}^{\Delta_n} = y_1 \mathcal{P}^{\Delta_{n-1}}(x) + \cdots + y_n \mathcal{P}^{\Delta_0}(x)$$

$$= y_1 x^{p^{n-1}} + y_2 x^{p^{n-2}} + \cdots + y_n \in (y_1, \ldots, y_n).$$

Therefore $\sqrt{(y_1, \ldots, y_n)}$ is the augmentation ideal of $\mathbb{F}_p[V]$ and hence $\mathbb{F}_p[V]/(y_1, \ldots, y_n)$ is totally finite and the result follows.   □

LEMMA 10.6.4 :   *The derivations $\mathcal{P}^{\Delta_i}$ satisfy the commutation relations*

$$[\mathcal{P}^{\Delta_i}, \mathcal{P}^{\Delta_j}] = \begin{cases} 0 & \text{for } i, \ j \neq 0 \\ \mathcal{P}^{\Delta_i} & \text{for } i \neq 0, \ j = 0 \\ -\mathcal{P}^{\Delta_j} & \text{for } i = 0, \ j \neq 0 \\ 0 & \text{for } i = j = 0. \end{cases}$$

PROOF :   By the definition of the Steenrod algebra in section 10.3 it suffices to check this formula on forms $x \in V^* \subset \mathbb{F}_p[V]$. An application of lemma 10.4.2 yields the result.   □

LEMMA 10.6.5 :   *Suppose given $z_0, \ldots, z_n \in \mathbb{F}_p[V]$ such that*

(∗) $z_n \mathcal{P}^{\Delta_0} = z_{n-1} \mathcal{P}^{\Delta_1} + \cdots + z_0 \mathcal{P}^{\Delta_n}$

as derivations on $\mathbb{F}_p[V]$. If z_n is a pth power and $r \in \mathbb{N}$ then

$$z_n \mathcal{P}^{\Delta_r} = \mathcal{P}^{\Delta_r}(z_{n-1}) \mathcal{P}^{\Delta_1} + \cdots + \mathcal{P}^{\Delta_r}(z_0) \mathcal{P}^{\Delta_n}$$

as derivations on $\mathbb{F}_p[V]$.

PROOF :   Since z_n is a pth power $\mathcal{P}^{\Delta_r}(z_n) = 0$. Let $f \in \mathbb{F}_p[V]$ be arbitrary. Precompose with $\mathcal{P}^{\Delta_r}$ on both sides of (∗) and evaluate the resulting differentials on f to obtain

$$z_n \mathcal{P}^{\Delta_r} \mathcal{P}^{\Delta_0}(f) = \mathcal{P}^{\Delta_r}(z_n \mathcal{P}^{\Delta_0}(f))$$

$$= \mathcal{P}^{\Delta_r}(z_{n-1} \mathcal{P}^{\Delta_1} + \cdots + z_0 \mathcal{P}^{\Delta_n})(f) = \mathcal{P}^{\Delta_r}(z_{n-1} \mathcal{P}^{\Delta_1}(f) + \cdots + z_0 \mathcal{P}^{\Delta_n}(f))$$

$$= \mathcal{P}^{\Delta_r}(z_{n-1}) \mathcal{P}^{\Delta_1}(f) + \cdots + \mathcal{P}^{\Delta_r}(z_0) \mathcal{P}^{\Delta_n}(f)$$

$$+ z_{n-1} \mathcal{P}^{\Delta_r} \mathcal{P}^{\Delta_1}(f) + \cdots + z_0 \mathcal{P}^{\Delta_r} \mathcal{P}^{\Delta_n}(f)$$

$$= (\mathcal{P}^{\Delta_r}(z_{n-1}) \mathcal{P}^{\Delta_1} + \cdots + \mathcal{P}^{\Delta_r}(z_0) \mathcal{P}^{\Delta_n})(f)$$

$$+ z_{n-1} \mathcal{P}^{\Delta_1} \mathcal{P}^{\Delta_r}(f) + \cdots + z_0 \mathcal{P}^{\Delta_n} \mathcal{P}^{\Delta_{-r}}(f)$$

(by lemma 10.6.4)

$$= (\mathcal{P}^{\Delta_r}(z_{n-1}) \mathcal{P}^{\Delta_1} + \cdots + \mathcal{P}^{\Delta_r}(z_0) \mathcal{P}^{\Delta_n})(f) + (z_{n-1} \mathcal{P}^{\Delta_1} + \cdots + z_0 \mathcal{P}^{\Delta_n})(\mathcal{P}^{\Delta_r}(f))$$

(by (∗) applied to $\mathcal{P}^{\Delta_r}(f)$)

$$= (\mathcal{P}^{\Delta_r}(z_{n-1})\mathcal{P}^{\Delta_1} + \cdots + \mathcal{P}^{\Delta_r}(z_0)\mathcal{P}^{\Delta_n}(f)) + z_n \mathcal{P}^{\Delta_0}\mathcal{P}^{\Delta_r}(f).$$

Rearranging gives

$$(\mathcal{P}^{\Delta_r}(z_{n-1})\mathcal{P}^{\Delta_1} + \cdots + \mathcal{P}^{\Delta_r}(z_0)\mathcal{P}^{\Delta_n})(f) = z_n \mathcal{P}^{\Delta_r}\mathcal{P}^{\Delta_0}(f) - z_n \mathcal{P}^{\Delta_0}\mathcal{P}^{\Delta_r}(f)$$
$$= z_n[\mathcal{P}^{\Delta_r}, \mathcal{P}^{\Delta_0}](f) = z_n \mathcal{P}^{\Delta_r}(f)$$

by lemma 10.6.2. □

LEMMA 10.6.6 : *Suppose given $z_0, \ldots, z_n \in \mathbb{F}_p[V]$ such that*

(∗) $$z_n \mathcal{P}^{\Delta_0} = z_{n-1}\mathcal{P}^{\Delta_1} + \cdots + z_0 \mathcal{P}^{\Delta_n}$$

as derivations on $\mathbb{F}_p[V]$. If z_n is a pth power then for $j = 1, \ldots, n$

$$\mathcal{P}^{\Delta_i}(z_j) = \begin{cases} z_n & \text{for } i+j = n \\ 0 & \text{otherwise.} \end{cases}$$

PROOF : From lemma 10.6.5 we have

$$z_n \mathcal{P}^{\Delta_i} = \mathcal{P}^{\Delta_i}(z_{n-1})\mathcal{P}^{\Delta_1} + \cdots + \mathcal{P}^{\Delta_i}(z_0)\mathcal{P}^{\Delta_n}.$$

Rearranging gives

$$0 = \mathcal{P}^{\Delta_i}(z_{n-1})\mathcal{P}^{\Delta_1} + \cdots + (\mathcal{P}^{\Delta_i}(z_{n-i}) - z_n)\mathcal{P}^{\Delta_i} + \cdots + \mathcal{P}^{\Delta_i}(z_0)\mathcal{P}^{\Delta_n}.$$

By lemma 10.6.1 the derivations $\mathcal{P}^{\Delta_1}, \ldots, \mathcal{P}^{\Delta_n}$ are linearly independent over $\mathbb{F}_p[V]$. Hence the coefficients of the preceding equation all vanish and the result follows. □

PROPOSITION 10.6.7 : *Let $n \in \mathbb{N}$ and*

$$\varphi(X) = \prod_{v^* \in V^*} (X + v^*) = X^{p^n} + y_1 X^{p^{n-1}} + \cdots + y_n X \in \mathbf{D}^*(n)[X]$$

then

$$\mathcal{P}^{\Delta_i}(y_k) = \begin{cases} -y_n & \text{for } i+k = n \\ y_k y_n & \text{for } i = n \\ 0 & \text{otherwise.} \end{cases}$$

PROOF : The element $y_n = \displaystyle\prod_{v^* \in V^* \smallsetminus \{0\}} v^*$ is a $(p-1)^{st}$ power, say

$y_n = u^{p-1}$. From 10.6.2 we have

$$0 = y_n \mathcal{P}^{\Delta_0} + y_{n-1} \mathcal{P}^{\Delta_1} + \cdots + y_1 \mathcal{P}^{\Delta_{n-1}} + \mathcal{P}^{\Delta_n}.$$

Multiply by u and rearrange to give

$$-z_n \mathcal{P}^{\Delta_0} = z_{n-1} \mathcal{P}^{\Delta_1} + \cdots + z_1 \mathcal{P}^{\Delta_{n-1}} + z_0 \mathcal{P}^{\Delta_n}$$

where

$$z_i = \begin{cases} u & \text{for } i = 0 \\ u y_i & \text{for } i = 1, \ldots, n. \end{cases}$$

The element $-z_n$ is a pth power, so we may apply lemma 10.6.6 to obtain

$$\mathcal{P}^{\Delta_i}(z_j) = z_n, \qquad i + j = n.$$

Since $\mathcal{P}^{\Delta_i}$ is a derivation we obtain

$$\mathcal{P}^{\Delta_i}(y_k) = \mathcal{P}^{\Delta_i}\Big(\frac{-z_k}{z_0}\Big) = \frac{-z_0 \mathcal{P}^{\Delta_i}(z_k) + z_k \mathcal{P}^{\Delta_i}(z_0)}{z_0^2}$$

$$= \begin{cases} -\frac{z_0 z_n}{z_0^2} = -y_n & \text{for } i = n - k \\ \frac{z_k z_n}{z_0^2} = y_k y_n & \text{for } i = n \\ 0 & \text{otherwise} \end{cases}$$

as claimed. $\square$

THEOREM 10.6.8 (L. E. Dickson): *Let $n \in \mathbb{N}$ and*

$$\varphi(X) = \prod_{v^* \in V^*} (X + v^*) = X^{p^n} + y_1 X^{p^{n-1}} + \cdots + y_n X \in \mathbf{D}^*(n)[X]$$

then

$$\mathbf{D}^*(n) = \mathbb{F}_p[y_1, \ldots, y_n].$$

PROOF : We apply the preceding proposition to evaluate the determinant:

$$\det \begin{bmatrix} y_n & y_{n-1} & \cdots & y_1 \\ \mathcal{P}^{\Delta_1} y_n & \mathcal{P}^{\Delta_1} y_{n-1} & \cdots & \mathcal{P}^{\Delta_1} y_1 \\ \vdots & \vdots & \vdots & \vdots \\ \mathcal{P}^{\Delta_{n-1}} y_n & \mathcal{P}^{\Delta_{n-1}} y_{n-1} & \cdots & \mathcal{P}^{\Delta_{n-1}} y_1 \end{bmatrix} = \det \begin{bmatrix} y_n & y_{n-1} & \cdots & y_1 \\ 0 & -y_n & \cdots & 0 \\ \vdots & \vdots & \vdots & \vdots \\ 0 & \cdots & 0 & -y_n \end{bmatrix}$$

$$= (-1)^{n-1} y_n^{\,n} \neq 0.$$

Therefore by 5.6.1 $y_1, \ldots, y_n \in \mathbf{D}^*(n)$ are algebraically independent and by 10.6.3 they are a system of parameters. Since $\frac{1}{2} \deg(y_i) = p^n - p^{n-i}$ for $i = 1, \ldots, n$ we also have

$$\prod \frac{\deg(y_i)}{2} = \prod_{i=1}^{n} (p^n - p^{n-i}) = |\mathrm{GL}(n, \ \mathbb{F}_p)|$$

and the result follows form 7.4.2. $\square$

In the notations for the Dickson polynomials $\mathbf{d}_{n,0}, \ \mathbf{d}_{n,1}, \ldots, \mathbf{d}_{n,n-1} \in \mathbb{F}_p[V]$ of section 8.1 the formulae of 10.6.7 take the form

$$\mathcal{P}^{\Delta_i}(\mathbf{d}_{n,n-k}) = \begin{cases} -\mathbf{d}_{n, \ n} & \text{for } i + k = n \\ \mathbf{d}_{n, \ n-k}\mathbf{d}_n & \text{for } i = n \\ 0 & \text{otherwise.} \end{cases}$$

10.7　Torsion Primes for Weyl Groups

Suppose that $\varrho : W \hookrightarrow \mathrm{GL}(n, \ \mathbb{Q})$ is a finite crystallographic group. If we choose a fundamental system of roots (see section 7.2) as basis for $\mathbb{Q}^n$, then the action of W preserves the integral lattice they span. Thus the representation ϱ is equivalent to an integral representation, and so has a mod$p$ reduction for every prime p.

LEMMA 10.7.1 : *Let $G \hookrightarrow \mathrm{GL}(n, \ \mathbb{Z})$ be a faithful representation of a finite group G and p an odd prime. Then the composition*

$$G \hookrightarrow \mathrm{GL}(n, \ \mathbb{Z}) \xrightarrow{r_p} \mathrm{GL}(n, \ \mathbb{Z}/p)$$

is a monomorphism, where r_p is induced by reduction mod p.

PROOF : Suppose that $I \neq T \in \mathrm{GL}(n, \ \mathbb{Z})$ has finite order k and $T \equiv I \bmod p$. Write $T = I + p^e A$, $A \not\equiv 0 \bmod p$. Then

$$0 = T^k - I = (I + p^e A)^k - I = kp^e A + \binom{k}{2} p^{2e} A^2 + \cdots$$

which implies (after factoring out p^e)

$$0 = kA + \binom{k}{2} p^e A^2 + \cdots \equiv kA \bmod p$$

and hence $k \equiv 0 \bmod p$. Write $k = pk'$. Replacing T by $T^{k'}$ we may suppose that T has order p, i.e. that $k = p$. Then

$$I = T^p = (I + p^e A)^p = I + \sum_{j=1}^{p} \binom{p}{j} p^{je} A^j$$

which yields

$$(*) \qquad\qquad p^{e+1} A = -\sum_{j=2}^{k} \binom{p}{j} p^{je} A^j \, .$$

For $j = 2, 3, \ldots, p$ we have $ej \geq e + 1$ with equality if and only if $e = 1$ and $j = 2$. For p odd however, $\binom{p}{2} \equiv 0 \bmod p$. So the right hand side of $(*)$ is divisible by p^{e+2}. Hence the left hand side must also by divisible by p^{e+2} so $A \equiv 0 \bmod p$ which is a contradiction and the result follows. $\quad\square$

Since W is crystallographic it is isomorphic to the Weyl group of a compact simply connected (semi-simple) Lie group H (see [241]) (and also to the Weyl group of any quotient of H by a finite central subgroup). Choose such a compact simply connected Lie group H and let $T \hookrightarrow H$ be a maximal torus. The Weyl group W_H of H acts on $H^2(BT; \mathbb{Z})$ and under the identification $W \leftrightarrow W_H$ the rational representation ϱ of W corresponds to the action of W_H on $H^2(BT; \mathbb{Q})$. Thus we receive an integral representation $\varrho_\mathbb{Z} : W \hookrightarrow \mathrm{GL}(n, \mathbb{Z})$ whose rationalization is ϱ. This integral representation is equivalent to the representation obtained from the root system of W without recourse to H (see [240]).

By the preceding lemma the representation

$$W \hookrightarrow \mathrm{GL}(n, \mathbb{Z}) \xrightarrow{r_p} \mathrm{GL}(n, \mathbb{F}_p)$$

is faithful for each odd prime p, where $\mathbb{F}_p$ is the Galois field of p elements. This representation corresponds to the action of W_H on $H^2(BT; \mathbb{F}_p)$. Denote by $V = \mathbb{F}_p^n$ the vector space on which W acts via the above representation. If $p \nmid |W|$ then the theorem of Shephard-Todd-Chevalley 7.4.1 assures that $\mathbb{F}_p[V]^W$ is a polynomial algebra. If $p \mid |W|$ then it can happen that $\mathbb{F}_p[V]^W$ fails to be a polynomial algebra (see for example section 10.3 example 1 and section 7.4 example 4). In [65] Demazure provides a condition, expressed purely algebraically, assuring for a prime p dividing $|W|$ that $\mathbb{F}_p[V]^W$ is a polynomial algebra. His proof is also completely algebraic, but for good measure he interprets his condition topologically in terms of the cohomology of the classifying space BH. In this form the result was already known to A. Borel and is implicit in [33] § 26 and § 27 (see also [35] page 416). We choose to formulate the result as follows:

THEOREM 10.7.2 (A. Borel, M. Demazure): *Let H be a compact connected Lie group with maximal torus $T \hookrightarrow H$ and Weyl group W. Let p be an odd prime and $\varrho : W \hookrightarrow \mathrm{GL}(n, \mathbb{Z})$ be the integral representation of W afforded by the action of W on $H^2(BT; \mathbb{Z})$. If $H_*(H; \mathbb{Z})$ has no p-torsion then:*

(i) $\mathbb{Z}_{(p)}[V]^W$ is a polynomial algebra for the representation $W \hookrightarrow \mathrm{GL}(n, \mathbb{Z}) \hookrightarrow \mathrm{GL}(n, \mathbb{Z}_{(p)})$,

(ii) $\mathbb{F}_p[V]^W$ is a polynomial algebra for the representation $W \hookrightarrow \mathrm{GL}(n, \mathbb{Z}) \xrightarrow{r_p} \mathrm{GL}(n, \mathbb{F}_p)$.

Here $\mathbb{Z}_{(p)}$ denotes the integers localized at the prime ideal $(p) \subset \mathbb{Z}$. Thus $\mathbb{Z}_{(p)}$ is the subring of $\mathbb{Q}$ consisting of all fractions a/b with $(b, p) = 1$. In the statement of part (i) V denotes $H^2(BT; \mathbb{Z}_{(p)})$ and in part (ii) $H^2(BT; \mathbb{F}_p)$.

We collect a number of results needed for the proof.

LEMMA 10.7.3 : *Let M, N be finitely generated free $\mathbb{Z}_{(p)}$-modules and $\varphi : M \longrightarrow N$ a $\mathbb{Z}_{(p)}$-module homomorphism. Then φ is an isomorphism if and only if $\varphi \otimes_{\mathbb{Z}_{(p)}} \mathbb{Q}$ and $\varphi \otimes_{\mathbb{Z}_{(p)}} \mathbb{F}_p$ are isomorphisms.*

PROOF : One direction is clear: if φ is an isomorphism then so are $\varphi \otimes_{\mathbb{Z}_{(p)}} \mathbb{Q}$ and $\varphi \otimes_{\mathbb{Z}_{(p)}} \mathbb{F}_p$. Conversely suppose that $\varphi \otimes_{\mathbb{Z}_{(p)}} \mathbb{Q}$ and $\varphi \otimes_{\mathbb{Z}_{(p)}} \mathbb{F}_p$ are isomorphisms. As $\varphi \otimes_{\mathbb{Z}_{(p)}} \mathbb{Q}$ is an isomorphism and M and N are free it follows φ is a monomorphism. Introduce the exact sequence

$$0 \longrightarrow M \xrightarrow{\varphi} N \longrightarrow L \longrightarrow 0$$

where L is the cokernel of φ. Since $\varphi \otimes_{\mathbb{Z}_{(p)}} \mathbb{Q}$ is an isomorphism $L \otimes_{\mathbb{Z}_{(p)}} \mathbb{Q} = 0$ so L is a torsion module. Therefore

$$L \otimes_{\mathbb{Z}_{(p)}} \mathbb{F}_p \cong L/p \cdot L.$$

However $\varphi \otimes_{\mathbb{Z}(p)} \mathbb{F}_p$ is also an isomorphism so $L \otimes_{\mathbb{Z}_{(p)}} \mathbb{F}_p = 0$ and hence $L = p \cdot L$. Since L is finitely generated this implies $L = 0$. □

LEMMA 10.7.4 : *Let p be a prime and A^* a graded commutative algebra of finite type over $\mathbb{Z}_{(p)}$ such that*

(i) *A^j is a free $\mathbb{Z}_{(p)}$-module for $j = 0, 1, \ldots$,*

(ii) *$A^* \otimes_{\mathbb{Z}_{(p)}} \mathbb{Q}$ is an exterior algebra on odd dimensional generators,*

(iii) *$A^* \otimes_{\mathbb{Z}_{(p)}} \mathbb{F}_p$ is an exterior algebra on odd dimensional generators.*

Then A^* is an exterior algebra on odd dimensional generators over $\mathbb{Z}_{(p)}$.

PROOF : The Poincaré series of $A^* \otimes_{\mathbb{Z}_{(p)}} \mathbb{Q}$ has the form

$$\prod_{i=1}^{n} (1 + t^{2m_i - 1}).$$

Let us call n the **rank** of the algebra A^*. If the rank of A^* is 1 then

$$A^j = \begin{cases} \mathbb{Z}_{(p)} & \text{for } j = 0, \, 2m - 1 \\ 0 & \text{otherwise,} \end{cases}$$

so choosing a generator x for A^{2m-1} we see $A^* \cong E(x)$. Thus we may assume inductively that the lemma is true for all graded commutative algebras A^* of finite type and rank at most $n-1$ satisfying (i) – (iii). Let A^* be a graded commutative algebra of finite type and rank n satisfying (i) – (iii). Denote by $2m-1$ the smallest integer such that

$$A^j = 0 \text{ for } 0 < j < 2m-1$$
$$A^{2m-1} \neq 0.$$

Choose a basis $x_1, \dots, x_s$ for the free finitely generated $\mathbb{Z}_{(p)}$-module A^{2m-1}. The natural map $A^* \hookrightarrow A^* \otimes_{\mathbb{Z}_{(p)}} \mathbb{Q}$ is a monomorphism so $x_1{}^2 = 0$. Next we claim $x_1 y = 0$ if and only if y is a $\mathbb{Z}_{(p)}$-multiple of x_1. To see this, suppose y is not a multiple of x_1 and $x_1 y = 0$. Since A^* is torsion free and finite type we may suppose that y is indivisible. The equation $x_1 y = 0 \in A^* \otimes_{\mathbb{Z}_{(p)}} \mathbb{F}_p$ and (iii) implies that $y = 0 \in A^* \otimes_{\mathbb{Z}_{(p)}} \mathbb{F}_p$ because y is not a multiple of x_1. But then y is divisible by p, a contradiction, unless $y = 0$ as claimed.

Let $B^* \subset A^*$ be the subalgebra of A^* generated by $x_2, \dots, x_s$ and the remaining algebra generators of A^* of degree greater than $2m-1$. Then B^* satisfies (i) – (iii) and has rank $n-1$. Hence by the inductive hypothesis

$$B^* \cong E(x_2, \dots, x_s, x_{s+1}, \dots, x_n).$$

The preceding discussion shows that the map

$$\varphi : E(x_1) \otimes_{\mathbb{Z}_{(p)}} E(x_2, \dots, x_n) \longrightarrow A^*$$

induced by multiplication is a monomorphism and $\varphi \otimes_{\mathbb{Z}_{(p)}} \mathbb{Q}$ and $\varphi \otimes_{\mathbb{Z}_{(p)}} \mathbb{F}_p$ are isomorphisms. Hence by the preceding lemma, φ is an isomorphism, establishing the result by induction. $\square$

THEOREM 10.7.5 (A. Borel [33]): *Let H be a compact connected Lie group and p an odd prime. If $H_*(H; \mathbb{Z})$ has no p-torsion then $H^*(BH; \mathbb{F}_p)$ and $H^*(BH; \mathbb{Z}_{(p)})$ are polynomial algebras.*

PROOF : By a theorem of H. Hopf [109] (see also the Appendix of [145]) $H^*(H; \mathbb{Q})$ is a primitively generated exterior algebra on odd dimensional generators. Since $H^*(H; \mathbb{Z})$ has no p torsion $H^*(H; \mathbb{Q})$ and $H^*(H; \mathbb{F}_p)$ have the same Poincaré series. By the Borel structure theorem [145] for Hopf algebras over $\mathbb{F}_p$ the Hopf algebra $H^*(H; \mathbb{F}_p)$ is a tensor product of Hopf algebras on a single generator. Since the Poincaré series is that of an exterior algebra we conclude that $H^*(H; \mathbb{F}_p)$ is a primitively generated exterior algebra on odd dimensional generators. By the lemma $H^*(H; \mathbb{Z}_{(p)})$ is also an exterior algebra on odd dimensional generators.

Let $\mathbb{F}$ denote either the Galois field $\mathbb{F}_p$ or the local ring $\mathbb{Z}_{(p)}$. Consider the Milnor-Moore spectral sequence, $\{E_r, d_r\}$, [52] exposeé 7 and [143], with coefficients[8] in $\mathbb{F}$ of the fibration

$$H \hookrightarrow EH \longrightarrow BH .$$

We have

$$E_r \Longrightarrow H^*(BH; \ \mathbb{F})$$

$$E_2 = \mathrm{Ext}_{H_*(H; \ \mathbb{F})}(\mathbb{F}, \ \mathbb{F}) .$$

Since $H_*(H; \ \mathbb{F})$ is an exterior algebra on odd dimensional generators $\mathrm{Ext}_{H_*(H; \ \mathbb{F})}(\mathbb{F}, \ \mathbb{F})$ is a polynomial algebra on generators of even total degrees. The differentials d_r lower total degree by one so the checkerboard effect of E_2 implies that $E_2 = E_\infty$. Since E_∞ is a free graded commutative algebra there is no extension problem. $\square$

THEOREM 10.7.6 (R. Bott [36]): *Let H be a compact connected Lie group with maximal torus $T \hookrightarrow H$. Then $H_{2i+1}(H/T; \ \mathbb{Z}) = 0$ for $i = 1, 2, \ldots$, and hence $H_*(H/T; \ \mathbb{Z})$ is torsion free.* $\square$

This result is proven by Morse theory and it would take us very far afield indeed to prove it here.

THEOREM 10.7.7 (H. Hopf, A. Weil): *Let H be a compact connected Lie group with maximal torus T and Weyl group W. Then the Euler characteristic of H/T is $|W|$.* $\square$

This is a consequence of the Lefschetz fixed point formula and we sketched a proof in section § clew. See [1] theorem 4.21 for a detailed proof.

COROLLARY 10.7.8 : *Let H be a compact connected Lie group with maximal torus $T \hookrightarrow H$ and Weyl group W. Let $p \in \mathbb{N}$ be a prime. If $H_*(H; \ \mathbb{Z})$ has no p-torsion then the mod p Serre spectral sequence of the fibration*

$$H/T \hookrightarrow BT \longrightarrow BH$$

with coefficients in $\mathbb{F}_p$ or $\mathbb{Z}_{(p)}$ collapses and $H^*(BT; \ \mathbb{F})$ is a free $H^*(BH; \ \mathbb{F})$-module of rank $|W|$ for $\mathbb{F} = \mathbb{F}_p$ or $\mathbb{F} = \mathbb{Z}_{(p)}$.

---

[8] The construction of Milnor shows that the Milnor-Moore spectral sequence may be constructed for any coefficient ring R, and has the expected E^2 term provided $H^*(H; \ R)$ is a free graded R-module of finite type.

PROOF : Let $\mathbb{F}$ denote either the Galois field $\mathbb{F}_p$ or the local ring $\mathbb{Z}_{(p)}$. By theorems 10.7.5, 10.7.6 the E_2 term of the Serre spectral sequence

$$E_2 = H^*(BH;\ \mathbb{F}) \otimes H^*(H/T;\ \mathbb{F})$$

is concentrated in even total degrees. By the checkerboard effect we have $E_2 = E_\infty$. By theorem 10.7.7 E_2 is a free $H^*(BH;\ \mathbb{F})$-module of rank $|W|$ and hence the same must be true for $H^*(BT;\ \mathbb{F})$. $\square$

PROOF OF THE THEOREM OF BOREL-DEMAZURE 10.7.2 : Consider the fibration

$$H/T \hookrightarrow BT \xrightarrow{\pi} BH\ .$$

Let $\mathbb{F}$ denote either the Galois field $\mathbb{F}_p$ or the local ring $\mathbb{Z}_{(p)}$. The Weyl group W of H acts on BT and the map π is equivariant when BH is given the trivial W action. Hence by 10.7.8

$$\pi^* : H^*(BH;\ \mathbb{F}) \longrightarrow H^*(BT;\ \mathbb{F})$$

is a monomorphism and $\mathrm{Im}(\pi^*) \subset H^*(BT;\ \mathbb{F})^W$. By theorem 10.7.5 it will be enough to show that $\mathrm{Im}(\pi^*) = H^*(BT;\ \mathbb{F})^W$, i.e. that π^* maps $H^*(BH;\ \mathbb{F})$ onto the invariants.

Let $FF(-)$ denote as usual the field of fractions functor. Identify $H^*(BH;\ \mathbb{F})$ with its image under π^*. We claim $FF(H^*(BH;\ \mathbb{F})) = FF(H^*(BT;\ \mathbb{F})^W)$. To verify this note

$$FF(H^*(BT;\ \mathbb{F})^W) \le FF(H^*(BT;\ \mathbb{F})) = \mathbb{F}(t_1,\ldots,t_n)$$

is a Galois extension with Galois group W and $t_1,\ldots,t_n$ is an $\mathbb{F}$ basis for $H^2(BT;\ \mathbb{F})$. Hence $\mathbb{F}(t_1,\ldots,t_n)$ has dimension $|W|$ as vector space over $FF(H^*(BT;\ \mathbb{F})^W)$. By corollary 10.7.8 $\mathbb{F}(t_1,\ldots,t_n)$ also has dimension $|W|$ as a vector space over $FF(H^*(BH;\ \mathbb{F}))$. Since $FF(H^*(BH;\ \mathbb{F})) \subseteq FF(H^*(BT;\ \mathbb{F})^W)$ the claim is established.

Next, note by theorem 5.1.1 that the extension $H^*(BH;\ \mathbb{F}) \hookrightarrow H^*(BT;\ \mathbb{F})$ is integral. Hence every element $f \in H^*(BT;\ \mathbb{F})^W$ satisfies:
 — f is integral over $H^*(BH;\ \mathbb{F})$
 — $f \in FF(H^*(BH;\ \mathbb{F}))$.
$H^*(BH;\ \mathbb{F})$ is a polynomial algebra and therefore integrally closed in its field of fractions. Hence $f \in H^*(BH;\ \mathbb{F})$ and π^* is an epimorphism as required. $\square$

For further connections between invariant theory and torsion primes for Lie groups we refer to [65], [117] and [165].

Chapter 11
The Steenrod Algebra and Modular Invariant Theory

In this chapter we continue the study of modular invariants begun in chapter 8, using as tools the Steenrod algebra and the Dickson algebra. As we noted in chapter 10 the ring of invariants $\mathbb{F}_p[V]^G$ of a representation $\varrho : G \hookrightarrow GL(n, \mathbb{F}_p)$ over the Galois field $\mathbb{F}_p$ is an unstable algebra over the Steenrod algebra. We first indicate how to extend the definition of the Steenrod operations to an arbitrary Galois field and then exploit this extra structure.

We study ideals in rings of invariants that are stable under the action of the Steenrod algebra and apply the results of this study to the transfer map Tr : $\mathbb{F}_q[V] \longrightarrow \mathbb{F}_q[V]^G$. The image of the transfer is an ideal $\mathrm{Im}(\mathrm{Tr}) \subset \mathbb{F}_q[V]^G$ stable under the action of the Steenrod algebra. The case $G = GL(n, \mathbb{F}_q)$ is a universal example, in the sense that classes in $\mathrm{Im}(\mathrm{Tr}^{GL(n, \mathbb{F}_q)})$ lie in $\mathrm{Im}(\mathrm{Tr}^G)$ for any $\varrho : G \hookrightarrow GL(n, \mathbb{F}_q)$. We will show that the radical of the ideal $\mathrm{Im}(\mathrm{Tr}^{GL(n, \mathbb{F}_q)})$ is the principal ideal generated by the top Dickson class $\mathbf{d}_{n,0}$, a result first proved by M. Feshbach for the prime 2 by explicit computations.

Unless explicitly stated to the contrary all algebras considered in this chapter will be graded and connected over the ground field. Since sign conventions play no role in this chapter we prefer to use the grading conventions of algebraists, i.e. the elements of $V^* \subset \mathbb{F}[V]$ are to have degree 1. Algebras over the Steenrod algebra are also to be commutative.

The category of $\mathbb{F}_q$-vector spaces is denoteed by $\mathrm{Vect}_{\mathbb{F}_q}$ and $\mathrm{GCA}_{\mathbb{F}_q}$ denotes the category of graded connected commutative $\mathbb{F}_q$-algebras.

11.1   The Steenrod Algebra over Galois Fields

Let p be a prime, $q = p^s$, $s \in \mathbb{N}$, and V an $\mathbb{F}_q$-vector space. We indicate how the constructions of section 10.3 may be modified to apply when an arbitrary

Galois field is the ground field. Define[1]

$$P(\xi) : \mathbb{F}_q[V] \longrightarrow \mathbb{F}_q[V][[\xi]]$$

by the rules

— $P(\xi)(v) = v + v^q \xi$ for $v \in V^*$
— $P(\xi)(u \cdot w) = P(\xi)(u) \cdot P(\xi)(w)$ for $u, w \in \mathbb{F}_q[V]$
— $P(\xi)(1) = 1$.

To consider $P(\xi)$ as a ring homomorphism of degree 0 we give ξ the degree $(1 - q)$ and the elements of V the degree 1. (Since sign conventions for commutation rules play no role in what follows we employ the grading conventions preferred by algebraists and not those of topologists.) By separating out homogeneous components we obtain $\mathbb{F}_q$-linear maps[2]

$$\mathcal{P}^i : \mathbb{F}_q[V] \longrightarrow \mathbb{F}_q[V]$$

by the requirement

$$P(\xi)(f) = \sum_{i=0}^{\infty} \mathcal{P}^i(f)\xi^i.$$

The operations $\mathcal{P}^i$ are called the **Steenrod reduced power operations** over $\mathbb{F}_q$, and when $q = 2$, also denoted by Sq^i and referred to as **Steenrod squaring operations**. Collectively they are referred to as **Steenrod operations**. They define endomorphisms of the functor

$$\mathbb{F}_q[-] : \mathrm{Vect}_{\mathbb{F}_q} \longrightarrow \mathrm{GCA}_{\mathbb{F}_q}$$

that assigns to a vector space V over $\mathbb{F}_q$ the graded algebra of homogeneous polynomial functions $\mathbb{F}_q[V]$ on V. The **Steenrod algebra** is the $\mathbb{F}_q$-subalgebra of the algebra of all endomorphisms of the functor $\mathbb{F}_q[-]$ generated by 1 and $\{\mathcal{P}^i \mid i = 1, 2, \ldots, \}$. (See [127] for an elaboration of this definition.)

The Steenrod operations satisfy:

$$\mathcal{P}^i(u) = \begin{cases} u^q & i = \deg(u) \\ 0 & i > \deg(u), \end{cases}$$

which are called the **unstability conditions**. By the multiplicative property of $P(\xi)$, one has

$$\mathcal{P}^k(u \cdot v) = \sum_{i+j=k} \mathcal{P}^i(u) \cdot \mathcal{P}^j(v),$$

---

[1] If A is a ring then $A[[\xi]]$ denotes the ring of formal power series over A in the variable ξ. We repeat some of the material from section 10.3 for the convenience of those who skipped chapter 10.
[2] If there is need to indicate the ground field we write $\mathcal{P}_q^i$ for the operations over $\mathbb{F}_q$.

which are called the **Cartan formulae**. For a linear polynomial $x \in \mathbb{F}_q[x_1, \ldots, x_n]$ these reduce to:

$$\mathcal{P}^i(x^j) = \binom{j}{i} x^{j+i(q-1)}.$$

As in the case of the prime field (see section 10.3) introduce the variables $u = 1 + t + \cdots + t^{q-1}$ and $s = tu$. Note that as $a \mapsto a^q$ is an automorphism of $\mathbb{F}_q$ we have $u = (1-t)^{q-1}$. By successive application of $P(\xi)$ and substitution we obtain the two compositions

$$P(s)P(1), \; P(u)P(t^p) : \mathbb{F}[V] \longrightarrow \mathbb{F}[V][[t]].$$

Since $P(1)$, $P(s)$, $P(u)$, and $P(t^q)$ are multiplicative, we obtain just as in the case of the prime field

$$P(s)P(1) = P(u)P(t^q),$$

This identity leads to commutation rules for $\mathcal{P}^i \mathcal{P}^j$ called **Adem relations**:

$$\mathcal{P}^i \mathcal{P}^j = \sum_{k=0}^{[i/q]} (-1)^{i+1} \binom{(q-1)(j-k)-1}{i-qk} \mathcal{P}^{i+j-k} \mathcal{P}^k$$

for all $i, j > 0$ such that $i < qj$, where $[a/b]$ denotes the integral part of a/b. The remarkable thing about these relations is that the coefficients always lie in the prime subfield $\mathbb{F}_p < \mathbb{F}_q$. These may be verified as in section 10.3.

Thus there is a surjective map from the free associative algebra with 1 generated by the Steenrod operations modulo the Adem relations, which we denote by $\mathcal{B}^*$, onto the Steenrod algebra. In fact, this map, $\mathcal{B}^* \longrightarrow \mathcal{P}^*$ when $q \neq 2$, and $\mathcal{B}^* \longrightarrow \mathcal{A}^*$ when $q = 2$, is an isomorphism, and hence the Adem relations are a complete set of defining relations for the Steenrod algebra. To prove this we modify some of the notations and definitions of section 10.3 to apply to an arbitrary Galois field as ground field.

Given a sequence $I = (i_1, i_2, \ldots, i_k)$ of non-negative integers we write

$$\mathcal{P}^I = \mathcal{P}^{i_1} \mathcal{P}^{i_2} \ldots \mathcal{P}^{i_k}.$$

These iterations of Steenrod operations are called **basic monomials**. We call k the **length** of I and denote it by $\ell(I)$. A basic monomial is called **admissible** if $i_s \geq q i_{s+1}$ for $s \geq 1$. To show that the Adem relations are a complete set of defining relations for the Steenrod algebra it suffices to show:

— *The admissible monomials span $\mathcal{B}^*$ as an $\mathbb{F}_q$-vector space.*

— *The images of the admissible monomials in the Steenrod algebra are linearly independent.*

This is done as in section 10.3 theorem 10.3.1 and therefore the Steenrod algebra over a Galois field may be regarded as the graded free associative algebra with 1 generated by the $\mathcal{P}^i$ modulo the ideal generated by the Adem relations.

The Steenrod algebra action on $\mathbb{F}_q[V]$ clearly commutes with the action of $GL(V)$ on $\mathbb{F}_q[V]$. If $G \hookrightarrow GL(n, \mathbb{F}_q)$ is a faithful representation of a finite group G then the ring of invariants $\mathbb{F}_q[x_1, \ldots x_n]^G$ is an algebra over the Steenrod algebra satisfying the Cartan formulae and the unstability condition. An algebra A^* over $\mathbb{F}_q$ together with a module structure over the Steenrod algebra which satisfies the unstability condition and the Cartan formulae is called an **unstable algebra over the Steenrod algebra**. The unstability condition is a strong nontriviality requirement for the action of $\mathcal{P}^*$ on A^*.

11.2 $\mathcal{P}^*$-Invariant Ideals

Fix a prime p, let $q = p^s$, $s \in \mathbb{N}$, and let $\varrho : G \hookrightarrow GL(n, \mathbb{F}_q)$ be a representation of a finite group. The ring of invariants $\mathbb{F}_q[V]^G$ is an unstable algebra over the Steenrod algebra $\mathcal{P}^*$. In this section we study ideals $I \subset \mathbb{F}_q[V]^G$ which are invariant [3] with respect to the action of $\mathcal{P}^*$. More generally we employ the following terminology.

DEFINITION : *Let H^* be an algebra over the Steenrod algebra $\mathcal{P}^*$. An ideal $I \subset H^*$ is called $\mathcal{P}^*$-**invariant** or -**stable** if $\mathcal{P}^* \cdot I \subset I$.*

Let H^* be an unstable algebra over the Steenrod algebra and $I \subset H^*$ an ideal, not necessarily $\mathcal{P}^*$-invariant. Define

$$J(I) = \{x \in I \mid \mathcal{P}^i(x) \in I \ \forall i \geq 0\}.$$

Note that $J(I) = I$ if indeed I is $\mathcal{P}^*$-invariant. If I is not $\mathcal{P}^*$-invariant then $J(I)$ is closer to being $\mathcal{P}^*$-invariant than I was, but may fail to be $\mathcal{P}^*$-invariant since $\mathcal{P}^i(x) \in I$ need not imply $\mathcal{P}^j\mathcal{P}^i(x) \in I$. For an example see [167].

LEMMA 11.2.1 (S. P. Lam [129]): *Let H^* be an unstable algebra over the Steenrod algebra and $I \subset H^*$ an ideal. Then*
  (i) *$J(I) \subset H^*$ is an ideal, and*
  (ii) *if I is prime so is $J(I)$.*
PROOF : Clearly $J(I)$ is closed under addition. If $x \in H^*$ and $y \in J(I)$ then $y \in I$ so $xy \in I$. By the Cartan formula

$$\mathcal{P}^k(xy) = \sum_{i+j=k} \mathcal{P}^i(x) \cdot \mathcal{P}^j(y).$$

---

[3] Much of the material in this section is based on published and unpublished material of P.S. Landweber, whose influence will be clearly discerned.

Since $y \in J(I)$ it follows that $\mathcal{P}^j(y) \in I$ for $j \geq 0$. Hence the right hand side of the above equation belongs to I, which says that $\mathcal{P}^k(xy) \in I$ for $k \geq 0$, so $xy \in J(I)$. Hence $J(I)$ is an ideal establishing (i).

To prove (ii) we suppose that I is an ideal in H^* and $z \in H^*$, $z \notin I$. Since H^* is unstable $\mathcal{P}^k(z) = 0$ for $k >> 0$. Therefore there exists an integer $h(z) \in \mathbb{N}$ such that $\mathcal{P}^{h(z)}(z) \notin I$ but $\mathcal{P}^k(z) \in I$ for $k > h(z)$.

Next, suppose that I is a prime ideal in H^*, x, $y \in H^*$, x, $y \notin J(I)$, but $xy \in J(I)$. Then by the Cartan formula

$$\mathcal{P}^{h(x)+h(y)}(xy) = \sum\nolimits_{i+j=h(x)+h(y)} \mathcal{P}^i(x)\mathcal{P}^j(y)$$
$$= \sum_{\substack{r+s=h(x)+h(y)\\r<h(x)\ s>h(y)}} \mathcal{P}^r(x)\cdot\mathcal{P}^s(y) \;+\; \mathcal{P}^{h(x)}(x)\cdot\mathcal{P}^{h(y)}(y) \;+\; \sum_{\substack{r+s=h(x)+h(y)\\r>h(x)\ s<h(y)}} \mathcal{P}^r(x)\cdot\mathcal{P}^s(y).$$

The terms

$$\sum_{\substack{r+s=h(x)+h(y)\\r<h(x)\ s>h(y)}} \mathcal{P}^r(x)\cdot\mathcal{P}^s(y), \qquad \sum_{\substack{r+s=h(x)+h(y)\\r>h(x)\ s<h(y)}} \mathcal{P}^r(x)\cdot\mathcal{P}^s(y)$$

belong to I, since in either of the sums one of the factors in the product $\mathcal{P}^r(x)\cdot\mathcal{P}^s(y)$ belongs to I by the definition of $h(x)$ and $h(y)$. Therefore rearranging the terms we see that $\mathcal{P}^{h(x)}(x)\cdot\mathcal{P}^{h(y)}(y) \in I$. But, by definition of $h(x)$, $h(y)$, we have $\mathcal{P}^{h(x)}(x)$, $\mathcal{P}^{h(y)}(y) \notin I$ which contradicts the fact that I is a prime ideal. $\square$

For an ideal I in an unstable algebra H^* over the Steenrod algebra we may define a descending chain of ideals

$$I = J_0(I) \supseteq J_1(I) \supseteq \cdots$$

by setting $J_i(I) = J(J_{i-1}(I))$ for $i > 0$.

LEMMA 11.2.2 : *Let H^* be an unstable algebra over the Steenrod algebra and $I \subset H^*$ an ideal. Set $J_\infty(I) = \bigcap J_i(I)$. Then*

(i) $J_\infty(I)$ is an ideal stable under the action of the Steenrod algebra, and

(ii) if I is a prime ideal so is $J_\infty(I)$.

PROOF : Suppose n, $k \in \mathbb{N}$ and $x \in J_{n+k}(I)$. Let $\mathcal{P}^{i_1} \cdots \mathcal{P}^{i_k} \in \mathcal{P}^*$ be an admissible monomial of length k. We claim $\mathcal{P}^{i_1} \cdots \mathcal{P}^{i_k}(x) \in J_n(I)$. For $k = 1$ this is immediate from the fact that $J_{n+1}(I) = J(J_n(I))$ and the definition of $J(-)$. For $k > 1$ the result follows inductively by noting

$$\mathcal{P}^{i_1} \cdots \mathcal{P}^{i_k}(x) = \mathcal{P}^{i_1}(\mathcal{P}^{i_2} \cdots \mathcal{P}^{i_k}(x)),$$

and $\mathcal{P}^{i_2} \cdots \mathcal{P}^{i_k}(x) \in J_{n+1}(I)$. Hence if $x \in J_\infty(I)$ then $\mathcal{P}^{i_1} \cdots \mathcal{P}^{i_k}(x) \in J_\infty(I)$ for any admissible monomial, so $J_\infty(I)$ is a $\mathcal{P}^*$-invariant ideal. This proves (i) and (ii) follows from the fact that the intersection of a descending chain of prime ideals is again prime. $\square$

REMARK : If H^* is an unstable algebra over the Steenrod algebra notice that $I \subset H^*$ is a $\mathcal{P}^*$-invariant ideal if and only if $J_\infty(I) = \cdots = J(I) = I$.

PROPOSITION 11.2.3 : If $(0) \neq I \subset H^*$ is a $\mathcal{P}^*$-invariant ideal in an unstable Noetherian algebra H^* over the Steenrod algebra and $\mathfrak{p} \subset H^*$ is an isolated prime ideal of I, then $\mathfrak{p}$ is a $\mathcal{P}^*$-invariant ideal.

PROOF : We have $\mathfrak{p} \supset I \neq (0)$. Apply $J_\infty(-)$ to this inclusion to get

$$\mathfrak{p} \supset J_\infty(\mathfrak{p}) \supset J_\infty(I) = I \neq (0),$$

where $J_\infty(I) = I$ because I is $\mathcal{P}^*$-invariant. By lemma 11.2.1 $J_\infty(\mathfrak{p})$ is a nonzero prime ideal, contained in $\mathfrak{p}$ and containing I. Since $\mathfrak{p}$ is an isolated prime ideal of I it is a minimal prime over I and hence $\mathfrak{p} = J_\infty(\mathfrak{p})$ so $\mathfrak{p}$ is $\mathcal{P}^*$-invariant. $\square$

In [169] it is shown that all the associated prime ideals of a $\mathcal{P}^*$-invariant ideal are $\mathcal{P}^*$-invariant. Here by Macaulay's unmixedness theorem 6.7.5 we only obtain:

COROLLARY 11.2.4 : Let H^* be an unstable algebra over the Steenrod algebra. If H^* is Cohen-Macaulay and $I \subseteq H^*$ is a $\mathcal{P}^*$-invariant ideal generated by $\mathrm{ht}(I)$ elements then all the associated prime ideals of H^* are also $\mathcal{P}^*$-invariant. $\square$

PROPOSITION 11.2.5 : If H^* is an unstable Noetherian algebra over the Steenrod algebra and I is a $\mathcal{P}^*$-invariant ideal, then so is $\sqrt{I}$.

PROOF : By the Lasker-Noether theorem

$$\sqrt{I} = \bigcap_{\mathfrak{p} \in \mathrm{Isol}(I)} \mathfrak{p}$$

where the intersection runs over the set $\mathrm{Isol}(I)$ of isolated prime ideals of I. By 11.2.3 the isolated prime ideals of I are $\mathcal{P}^*$-invariant. Since the intersection of $\mathcal{P}^*$-invariant ideals is $\mathcal{P}^*$-invariant the result follows. $\square$

11.3 Going Up and Down with $\mathcal{P}^*$-Invariant Ideals

In this section we prove the classical going up/down theorems 5.4.2/5.4.5 for $\mathcal{P}^*$-invariant prime ideals. These have been optimally generalized by M. D. Neusel in [167].

THEOREM 11.3.1 : *Suppose that $A' \supseteq A''$ is a finite integral extension of unstable algebras over the Steenrod algebra $\mathcal{P}^*$.*

(i) **(Lying $\mathcal{P}^*$-invariantly over)** *If $\mathfrak{p}'' \subset A''$ is a $\mathcal{P}^*$-invariant prime ideal then there is a $\mathcal{P}^*$-invariant prime ideal $\mathfrak{p}' \subset A'$ with $\mathfrak{p}' \cap A'' = \mathfrak{p}''$. There are no strict inclusions between such prime ideals $\mathfrak{p}'$. In this situation we say that $\mathfrak{p}'$ lies $\mathcal{P}^*$-invariantly over $\mathfrak{p}''$.*

(ii) **(Going $\mathcal{P}^*$-invariantly up)** *If $\mathfrak{p}''_1 \supset \mathfrak{p}''_0$ are $\mathcal{P}^*$-invariant prime ideals in A'' and $\mathfrak{p}'_0$ is a $\mathcal{P}^*$-invariant prime ideal in A' lying $\mathcal{P}^*$-invariantly over $\mathfrak{p}''_0$ then there is an $\mathcal{P}^*$-invariant prime ideal $\mathfrak{p}'_1$ in A' lying $\mathcal{P}^*$-invariantly over $\mathfrak{p}''_1$ with $\mathfrak{p}'_1 \supset \mathfrak{p}'_0$.*

PROOF : Let $\mathfrak{p}'' \subset A''$ be a $\mathcal{P}^*$-invariant prime ideal. By the usual going up theorem 5.4.2 there exists a prime ideal $\mathfrak{p} \subset A'$ lying over $\mathfrak{p}''$. Let $\mathfrak{p}' = J_\infty(\mathfrak{p})$. Note that $\mathfrak{p}' \supset \mathfrak{p}''$ since $\mathfrak{p}''$ is $\mathcal{P}^*$-invariant. By 11.2.2 $\mathfrak{p}'$ is a $\mathcal{P}^*$-invariant prime ideal in A'. Since

$$\mathfrak{p}'' \subset A'' \cap \mathfrak{p}' \subset A'' \cap \mathfrak{p} = \mathfrak{p}''$$

it follows that $\mathfrak{p}'$ lies $\mathcal{P}^*$-invariantly over $\mathfrak{p}''$. This proves (i).

To prove (ii) just apply $J_\infty(-)$ to the result of the usual going up theorem. Since $J_\infty(-)$ preserves inclusions the result (ii) follows from 11.2.2. $\square$

THEOREM 11.3.2 : *Suppose that $A' \supseteq A''$ is a finite extension of commutative integral domains over the Steenrod algebra $\mathcal{P}^*$, A'' is integrally closed, and the corresponding extension $\mathbb{L}' \supseteq \mathbb{L}''$ of fields of fractions is normal (i.e. an irreducible polynomial over $\mathbb{L}''$ with a root in $\mathbb{L}'$ splits completely in $\mathbb{L}'$, but is not necessarily separable).*

(i) **($\mathcal{P}^*$-Invariant transitivity)** *The Galois group $G = \mathrm{Gal}(\mathbb{L}' \mid \mathbb{L}'')$ acts transitively on the $\mathcal{P}^*$-invariant prime ideals $\mathfrak{p}'$ of A' lying $\mathcal{P}^*$-invariantly over a given prime ideal $\mathfrak{p}''$ of A''.*

(ii) **(Going $\mathcal{P}^*$-invariantly down)** *If $\mathfrak{p}''_0 \subset \mathfrak{p}''_1$ are $\mathcal{P}^*$-invariant prime ideals in A'' and $\mathfrak{p}'_1$ is a $\mathcal{P}^*$-invariant prime ideal of A' lying over $\mathfrak{p}''_1$ then there is a $\mathcal{P}^*$-invariant prime ideal $\mathfrak{p}'_0$ in A' lying over $\mathfrak{p}''_0$, with $\mathfrak{p}'_0 \subset \mathfrak{p}'_1$.*

PROOF : By 5.4.5 the Galois group acts transitively on all the prime ideals lying over a given prime ideal. Since the Galois action sends $\mathcal{P}^*$-invariant ideals to $\mathcal{P}^*$-invariant ideals (i) follows. To establish (ii) just apply the functor $J_\infty(-)$ to the result of the usual going down theorem 5.4.5. $\square$

11.4 $\mathcal{P}^*$-Invariant Ideals in Rings of Invariants

In this section we specialize the preceding discussions to the case of a ring of invariants[4] $\mathbb{F}_q[V]^G$ of a representation $\varrho : G \hookrightarrow \mathrm{GL}(n, \, \mathbb{F}_q)$ of a finite group. The first basic result is due to J. -P. Serre. The proof we present is due to C.W. Wilkerson.

PROPOSITION 11.4.1 (J. -P. Serre): *If* $\mathfrak{p} \subset \mathbb{F}_q[V]$ *is a* $\mathcal{P}^*$-*invariant prime ideal, then* $\mathfrak{p}$ *is generated by* $\mathfrak{p} \cap V^*$, *i.e.* $\mathfrak{p}$ *is generated by the linear forms that it contains.*

PROOF : Consider the quotient map

$$\pi : \mathbb{F}_q[V] \longrightarrow \mathbb{F}_q[V]/\mathfrak{p},$$

and let $z_1, \ldots, z_m \in \pi(V^*)$ be a basis. Then $z_1, \ldots, z_m$ generate $\mathbb{F}_q[V]/\mathfrak{p}$ as an algebra. Since $\deg(z_1) = \cdots = \deg(z_m) = 1$ we have[5]

$$\mathcal{P}^{\Delta_i}(z_j) = z_j^{q^i} \qquad j = 1, \ldots, m \quad i = 1, \ldots$$

and hence

$$\det(\mathcal{P}^{\Delta_i}(z_j)) = \det(z_j^{q^i}).$$

The lift to $\mathbb{F}[V]$ of this determinant was evaluated by L. E. Dickson in [68] (see also [2] lemma 5.9 and [39] chap. V § 5 exercise 6) who showed that in $\mathbb{F}[V]$ it is the product of one nonzero linear form from each line in $\mathrm{Span}_{\mathbb{F}} \{z_1, \ldots, z_m\}$, and hence not in $\mathfrak{p}$. Thus $z_1, \ldots, z_m$ are algebraically independent by 5.6.1. Hence $\mathbb{F}_q[V]/\mathfrak{p} \cong \mathbb{F}[z_1, \ldots, z_m]$ from which the result follows. $\square$

PROPOSITION 11.4.2 : *Let* $\varrho : G \hookrightarrow \mathrm{GL}(n, \, \mathbb{F}_q)$ *be a representation of a finite group* G. *If* $\mathfrak{p} \subset \mathbb{F}_q[V]^G$ *is a nonzero* $\mathcal{P}^*$-*invariant prime ideal, then* $\mathfrak{p}$ *contains the top Dickson class* $\mathbf{d}_{n,0} = \prod\limits_{0 \neq v \in V^*} v.$

PROOF : By theorem 11.3.1 we may choose a $\mathcal{P}^*$-invariant prime ideal $\tilde{\mathfrak{p}} \subset \mathbb{F}_q[V]$ lying over $\mathfrak{p}$. By 11.4.1 $\tilde{\mathfrak{p}}$ contains a nonzero element $u \in V^*$. Therefore

$$\mathbf{d}_{n,0} = u \cdot \prod_{\substack{v \in V^* \\ 0 \neq v \neq u}} v$$

also belongs to $\tilde{\mathfrak{p}}$. Since $\mathbf{d}_{n,0}$ is stable under $\mathrm{GL}(n, \, \mathbb{F}_q)$ it also belongs to $\mathbb{F}_q[V]^G$. Hence $\mathbf{d}_{n,0} \in \tilde{\mathfrak{p}} \cap \mathbb{F}_q[V]^G = \mathfrak{p}$ as required. $\square$

---

[4] I.e we study ideals in $\mathbb{F}_q[V]^G$ which are stable under the action of the Steenrod algebra. This should explain the title of this section.

[5] Remember we have reverted to the grading conventions of algebra, so the elements of V^* have degree 1 and the Steenrod operations $\mathcal{P}^i$ have degree $i(q-1)$.

COROLLARY 11.4.3 : *Let* $\varrho : G \hookrightarrow \mathrm{GL}(n, \ \mathbb{F}_q)$ *be a representation of a finite group* G. *If* $I \subset \mathbb{F}_q[V]^G$ *is a* $\mathcal{P}^*$*-invariant radical ideal, i.e.* $\sqrt{I} = I$, *then* $\mathbf{d}_{n,0} \in I$.

PROOF : By the Lasker-Noether theorem

$$I = \sqrt{I} = \bigcap_{\mathfrak{p} \in \mathrm{Isol}(I)} \mathfrak{p}$$

since I is equal to its own radical. By 11.2.3 the isolated primes of I are $\mathcal{P}^*$-invariant since I is $\mathcal{P}^*$-invariant. The result then follows from 11.4.2. □

COROLLARY 11.4.4 : *Let* $\varrho : G \hookrightarrow \mathrm{GL}(n, \ \mathbb{F}_q)$ *be a representation of a finite group. If* $I \subset \mathbb{F}_q[V]^G$ *is a nonzero* $\mathcal{P}^*$*-invariant ideal, then* I *contains a power of the top Dickson class* $\mathbf{d}_{n,0}$.

PROOF : By 11.2.5 $\sqrt{I}$ is a $\mathcal{P}^*$-invariant ideal. Since $\sqrt{\sqrt{I}} = \sqrt{I}$ it is also a radical ideal. Therefore $\mathbf{d}_{n,0} \in \sqrt{I}$ by 11.4.3, and hence there is a $k \in \mathbb{N}$ such that $\mathbf{d}_{n,0}^k \in I$ by the definition of $\sqrt{I}$. □

COROLLARY 11.4.5 : *Let* $\varrho : G \hookrightarrow \mathrm{GL}(n, \ \mathbb{F}_q)$ *be a representation of a finite group. If* $\mathbb{F}_q[V]^G$ *is Cohen-Macaulay and* $I \subset \mathbb{F}_q[V]^G$ *an ideal* $\mathcal{P}^*$*-invariant under the action of the Steenrod algebra and generated by* $\mathrm{ht}(I)$ *elements, then the associated prime ideals of* I *are also* $\mathcal{P}^*$*-invariant.* □

THEOREM 11.4.6 (P. S. Landweber): *The only* $\mathcal{P}^*$*-invariant prime ideals in the Dickson algebra* $\mathbf{D}^*(n)$ *are*

$$(\mathbf{d}_{n,0}, \ \mathbf{d}_{n,1}, \ldots, \mathbf{d}_{n,i}) \qquad i = 0, \ldots, n-1.$$

PROOF : For $n = 1$ this is clear, so we may proceed by induction. Let $\dim_{\mathbb{F}_q}(V) = n$ and choose a codimension 1 subspace $W \subset V$. Let

$$i^* : \mathbb{F}_q[V] \longrightarrow \mathbb{F}_q[W]$$

be the map induced by the inclusion. Recall (see the proof of 8.1.6)

$$(*) \qquad i^*(\mathbf{d}_{n,i}) = \begin{cases} \mathbf{d}_{n-1, \ i-1}^q & \text{for } 1 \le i \le n-1 \\ 0 & \text{for } i = 0. \end{cases}$$

Since the Frobenius map

$$f \mapsto f^q \qquad f \in \mathbb{F}_q[V]$$

is an algebra homomorphism it follows

$$\mathbf{d}_{n-1, \ i-1}^q(t_1, \ldots, t_{n-1}) = \mathbf{d}_{n-1, \ i-1}(t_1^q, \ldots, t_{n-1}^q).$$

In other words (this is the fractal property of the Dickson algebra)

$$\mathrm{Im}(\{\mathbf{D}^*(n) \xrightarrow{i^*} \mathbb{F}_q[W]\}) \subset \mathbb{F}_q[t_1^q, \ldots, t_{n-1}^q]^{\mathrm{GL}(n-1, \ \mathbb{F}_q)},$$

while clearly $\ker(i^*) = (\mathbf{d}_{n,0})$. The Frobenius

$$\zeta : \mathbb{F}_q[t_1, \ldots, t_{n-1}] \longrightarrow \mathbb{F}_q[t_1^q, \ldots, t_{n-1}^q]$$

is $\mathrm{GL}(n-1, \ \mathbb{F}_q)$ equivariant, and provides an isomorphism (fractal property)

$$\zeta_* : \mathbf{D}^*(n) \longrightarrow \mathbb{F}_q[t_1^q, \ldots, t_{n-1}^q]^{\mathrm{GL}(n-1, \ \mathbb{F}_q)}$$

multiplying degrees by q. Moreover, from the Cartan formulae,

$$\zeta(\mathcal{P}^i(f)) = \mathcal{P}^{iq}(\zeta(f))$$

and

$$\mathcal{P}^j(\zeta(f)) = 0 \qquad j \not\equiv 0 \bmod q$$

for any $f \in \mathbb{F}_q[t_1, \ldots, t_{n-1}]$. Therefore, the fractal property and the induction hypothesis imply that the only $\mathcal{P}^*$-invariant prime ideals in $\mathbb{F}_q[t_1^q, \ldots, t_{n-1}^q]^{\mathrm{GL}(n-1, \ \mathbb{F}_q)}$ are

$$(\mathbf{d}_{n-1,0}^q, \ldots, \mathbf{d}_{n-1,j}^q) \qquad j = 0, \ldots, n-2.$$

The surjective map

$$i^* : \mathbf{D}^*(n) \longrightarrow \mathbb{F}_q[t_1^q, \ldots, t_{n-1}^q]^{\mathrm{GL}(n-1, \ \mathbb{F}_q)}$$

commutes with the action of the Steenrod algebra, so sets up a bijective correspondence between $\mathcal{P}^*$-invariant prime ideals in $\mathbb{F}_q[t_1^q, \ldots, t_{n-1}^q]^{\mathrm{GL}(n-1, \ \mathbb{F}_q)}$ and $\mathcal{P}^*$-invariant prime ideals in $\mathbf{D}^*(n)$ containing the principal ideal $(\mathbf{d}_{n,0}) = \ker(i^*)$. The result is then immediate from $(*)$. $\square$

COROLLARY 11.4.7 : If $I \subset \mathbf{D}^*(n)$ is a $\mathcal{P}^*$-invariant ideal, then $\sqrt{I} = (\mathbf{d}_{n,0}, \ldots, \mathbf{d}_{n,i})$ for some $i \in \{0, \ldots, n-1\}$ and hence $\sqrt{I}$ is a prime ideal. It is the unique isolated prime ideal of I.

PROOF : By the Lasker-Noether theorem

$$\sqrt{I} = \bigcap_{\mathfrak{p} \in \mathrm{Isol}(I)} \mathfrak{p}.$$

If $\mathfrak{p}$ is an isolated prime ideal of I then by 11.2.3 $\mathfrak{p}$ is $\mathcal{P}^*$-invariant, and hence by 11.4.6 $\mathfrak{p} = (\mathbf{d}_{n,0}, \ldots, \mathbf{d}_{n,i})$ for some $i \in \{0, \ldots, n-1\}$. The isolated primes of I are minimal primes in $\mathbf{D}^*(n)/I$ so there are no proper inclusions between them. The ideals $(\mathbf{d}_{n,0}, \ldots, \mathbf{d}_{n,i})$ for $i \in \{1, \ldots, n-1\}$ are ordered by inclusion and hence I can have only one isolated prime ideal. $\square$

Let $\varrho : G \hookrightarrow \mathrm{GL}(n, \mathbb{F}_q)$ be a representation of a finite group G and set $V = \mathbb{F}_q^n$ as usual. For any subspace $W \subseteq V$ we have the composition

$$q_W : \mathbb{F}_q[V]^G \hookrightarrow \mathbb{F}_q[V] \longrightarrow \mathbb{F}_q[W].$$

Since $\mathrm{Im}(q_W) \subset \mathbb{F}_q[W]$ is an integral domain it follows that $\mathfrak{p}_W = \ker(q_W) \subset \mathbb{F}_q[V]^G$ is a prime ideal. Since q_W commutes with the action of the Steenrod algebra $\mathcal{P}^*$ the ideal $\mathfrak{p}_W \subset \mathbb{F}_q[V]^G$ is $\mathcal{P}^*$-invariant.

Conversely, if $\mathfrak{p} \subset \mathbb{F}_q[V]^G$ is a $\mathcal{P}^*$-invariant prime ideal, then by the $\mathcal{P}^*$-invariant lying over theorem 11.3.1 there is a $\mathcal{P}^*$-invariant prime ideal $\tilde{\mathfrak{p}} \subset \mathbb{F}_q[V]$ lying $\mathcal{P}^*$-invariantly over $\mathfrak{p}$. By the proposition of Serre 11.4.1 $\tilde{\mathfrak{p}} = (U^*)$ for some subspace $U^* \subset V^*$. Let $W \subset V$ be the subspace annihilated by U^*. Then unraveling the definitions we see $\mathfrak{p} = \mathfrak{p}_W$, and $\mathrm{ht}(\mathfrak{p}) = \mathrm{codim}_{\mathbb{F}_q}(W)$.

THEOREM 11.4.8 : *Let $\varrho : G \hookrightarrow \mathrm{GL}(n, \mathbb{F}_q)$ be a representation of a finite group. Then there is a bijective correspondence between $\mathcal{P}^*$-invariant prime ideals $\mathfrak{p} \subset \mathbb{F}_q[V]^G$ and the orbits of G on the total Graßmann* $\mathrm{Graß}(V)$ *of V. The correspondence is order reversing.*

N.b. The ordering on $\mathcal{P}^*$-invariant prime ideals is by inclusion. The ordering on the orbit space $\mathrm{Graß}(V)/G$ is defined by $B' \leq B''$ if and only if for every $W' \in B'$ there exists $W'' \in B''$ with $W' \subseteq W''$. In other words, $B' \leq B''$ if and only if every subspace of B' is a subspace of some subspace in B''. (Since G acts transitively on the orbits it is enough to require this condition for one $W' \in B'$.)

PROOF : Let $\mathfrak{p} \subset \mathbb{F}_q[V]^G$ be a $\mathcal{P}^*$-invariant prime ideal. By the preceding discussion the $\mathcal{P}^*$-invariant prime ideals $\mathcal{P}^*$-invariantly lying over $\mathfrak{p}$ are all of the form $\mathfrak{p}_W = \ker(q_W)$ for some $W \leq V$. The equivariant transitivity theorem 11.3.2 says that G permutes transitively these ideals, which via Serre's proposition 11.4.1 translates to: G permutes transitively the W with $\mathfrak{p}_W$ lying equivariantly over $\mathfrak{p}$, and the result follows. $\square$

COROLLARY 11.4.9 : *Let $\varrho : G \hookrightarrow \mathrm{GL}(n, \mathbb{F}_q)$ be a representation of a finite group. If $\varrho(G)$ contains $\mathrm{SL}(n, \mathbb{F}_q)$, then the only $\mathcal{P}^*$-invariant prime ideals in $\mathbb{F}[V]^G$ are*

$$(L_n^e, \mathbf{d}_{n,1}, \ldots, \mathbf{d}_{n,i})$$

for some $i \in \{1, \ldots, n-1\}$, where $L_n^{q-1} = \mathbf{d}_{n,0}$ is the top $\mathrm{SL}(n, \mathbb{F}_q)$-invariant and $e = |G : \mathrm{SL}(n, \mathbb{F}_q)|$.

PROOF : The group $\mathrm{SL}(n, \mathbb{F}_q)$ acts transitively on the set of all k-dimensional subspaces of V (see [71]) and hence the result follows as in 11.4.6, 11.4.8 using 8.1.8. $\square$

COROLLARY 11.4.10 : Let $\varrho : G \hookrightarrow \mathrm{GL}(n, \ \mathbb{F}_q)$ be a representation of a finite group. If $\varrho(G)$ contains $\mathrm{SL}(n, \ \mathbb{F}_q)$, and I is a $\mathcal{P}^*$-invariant ideal in $\mathbb{F}[V]^G$ then

$$\sqrt{I} = (L^e, \ \mathbf{d}_{n,1}, \ldots, \mathbf{d}_{n,i})$$

for some $i \in \{1, \ldots, n-1\}$, where L is the top $\mathrm{SL}(n, \ \mathbb{F}_q)$-invariant, $L_n^{q-1} = \mathbf{d}_{n,0}$, and $e = |G : \mathrm{SL}(n, \ \mathbb{F}_q)|$.

PROOF : This follows from 8.1.8 and the preceding corollary. $\square$

11.5 Applications to the Transfer

Let $\varrho : G \hookrightarrow \mathrm{GL}(n, \ \mathbb{F}_q)$ be a representation of a finite group G. In this section we will examine the image of the transfer (see section 2.4),

$$\mathrm{Im}(\mathrm{Tr}^G : \mathbb{F}_q[V] \longrightarrow \mathbb{F}_q[V]^G),$$

using the results on $\mathcal{P}^*$-invariant ideals developed in the preceding sections. Of course we have seen in section 2.4 that Tr^G is surjective if $p \nmid |G|$, however this fails to be the case when $p \mid |G|$. Before looking at an example we make a general observation.

LEMMA 11.5.1 : Let $\varrho : G \hookrightarrow \mathrm{GL}(n, \ \mathbb{F})$ be a representation of a finite group over a field of characteristic p, and suppose that p divides $|G|$. Then $\mathrm{Tr}^G = \sum_{g \in G} (g - 1)$.

PROOF : For $f \in \mathbb{F}[V]$ we have

$$\mathrm{Tr}^G(f) = \sum_{g \in G} gf = (\sum_{g \in G} gf) - |G| \cdot f = \sum_{g \in G} (g - 1)(f)$$

as required. $\square$

EXAMPLE 1 : Let $\mathbb{F}$ be a field of nonzero characteristic p and suppose $T \in \mathrm{GL}(n, \ \mathbb{F})$ is a transvection. Let $\mathbb{Z}/p \hookrightarrow \mathrm{GL}(n, \ \mathbb{F})$ be the representation afforded by T. Consider the action of T on V^* (n.b. V^* not V) and write

$$T(z) = z + \ell(z) \cdot x \qquad \forall \, z \in V^*$$

where $\ell : V^* \longrightarrow \mathbb{F}$ is a linear functional with $\ker(\ell) = H_T = \ker(T - I)$, the hyperplane of T, and $0 \neq x \in \mathrm{Im}(T - I) \subset H_T$ a transvector for T. Let $y \in V^*$ with $\ell(y) = 1$. Note that

$$(*) \qquad\qquad (T^i - I)(z) = i\ell(z) \cdot x \qquad \forall \, z \in V^*.$$

If $z_1, \ldots, z_{n-1}$ is a basis for H_T, then $z_1, \ldots, z_{n-1}, y$ is a basis for V^*. If $z_1{}^{e_1} \cdots z_{n-1}{}^{e_{n-1}} y^{e_n}$ is a monomial in this basis, then by lemma 11.5.1

$$\mathrm{Tr}^{\mathbb{Z}/p}(z_1{}^{e_1} \cdots z_{n-1}{}^{e_{n-1}} y^{e_n}) = \sum_{i=0}^{p-1} (T^i - I)(z_1{}^{e_1} \cdots z_{n-1}{}^{e_{n-1}} y^{e_n})$$

$$= \sum_{i=0}^{n-1} ((T^i - I)(z_1))^{e_1} \cdots ((T^i - I)(z_{n-1}))^{e_{n-1}} ((T^i - I)(y))^{e_n} \,,$$

(n.b. in the preceding sum the terms with $i = 0$ are zero since $T^0 = I$) which in view of $(*)$ will vanish if $(e_1, \ldots, e_{n-1}) \neq (0, \ldots, 0)$, in which case

$$\mathrm{Tr}^{\mathbb{Z}/p}(y^{e_n}) = \left(\sum_{i=1}^{p-1} i^{e_n}\right) x^{e_n} \,.$$

Since the monomials $\{z_1{}^{e_1} \cdots z_{n-1}{}^{e_{n-1}} y^{e_n} \mid e_i \in \mathbb{N}, \ i = 1, \ldots, n\}$ are an additive basis for $\mathbb{F}[V]$ it follows that $\mathrm{Im}(\mathrm{Tr}^{\mathbb{Z}/p}) = (x^s)$, the principal ideal generated by x^s, where s is the smallest integer such that $x^s \in \mathrm{Im}(\mathrm{Tr}^{\mathbb{Z}/p})$. This integer s is just $p - 1$. To see this, note that s is the smallest integer e such that

$$\sum_{i=1}^{p-1} i^e \not\equiv 0 \bmod p \,.$$

To see $s \leq p - 1$, note

$$\sum_{i=1}^{p-1} i^{p-1} = \sum_{i=1}^{p-1} 1 = (p-1) = -1 \bmod p \,.$$

On the other hand, if $e < p - 1$, let $\zeta \in \mathbb{F}_q^{\times}$ be a generator. Then

$$\sum_{i=1}^{p-1} i^e = 1^e + 2^e + \cdots + (p-1)^e = \zeta^e + (\zeta^2)^e + \cdots + (\zeta^{p-1})^e$$

$$= \zeta^e + \zeta^{2e} + \cdots + \zeta^{(p-1)e} = \zeta + \zeta^2 + \cdots + \zeta^{p-1}$$

$$= 1 + 2 + \cdots + (p-1) \equiv 0 \bmod p \,,$$

because for $e < p - 1$ the integers $e, \ 2e, \ldots, \ (p-1)e$ are a complete set of residues without repetition mod $p - 1$.

Thus we have shown that $\mathrm{Im}(\mathrm{Tr}^{\mathbb{Z}/p}) = (x^{p-1})$, the principal ideal in $\mathbb{F}[V]^{\mathbb{Z}/p}$ generated by x^{p-1}. Note that $x \in \mathbb{F}[V]^{\mathbb{Z}/p}$, since $x \in H_T = V^{\mathbb{Z}/p}$. If $p \neq 2$, then $x \notin \mathrm{Im}(\mathrm{Tr}^{\mathbb{Z}/p})$. Moreover, we know from section 8.2 that $\mathbb{F}[V]^{\mathbb{Z}/p} = \mathbb{F}[x_1, \ldots, x_{n-1}, y^p]$, where $x = x_1, \ldots, x_{n-1}$ is a basis for H_T. So $\mathrm{Im}(\mathrm{Tr}^{\mathbb{Z}/p})$ is very sparse in $\mathbb{F}[V]^{\mathbb{Z}/p}$.

EXAMPLE 2 : Let $E(r)$ denote the elementary abelian p-group $\underset{r}{\oplus}\mathbb{Z}/p$ with p^r elements. Let $\mathbb{F}$ be a field of characteristic $p \neq 0$ and for $n \in \mathbb{N}$ consider the matrices

$$
S_i = \begin{bmatrix} 1 & 0 & \cdots & & 0 \\ 0 & 1 & 0\cdots & & 0 \\ \vdots & \ddots & \cdots & & \\ 0 & \cdots & 1\cdots 0 & 1 & \\ \vdots & \cdots & \cdots & & \vdots \\ 0 & \cdots & & 0 & 1 \end{bmatrix} \in \mathrm{GL}(n,\ \mathbb{F}) \quad i = 1,\ldots, n-1,
$$

where the off diagonal 1 is in the ith row and the last column. In other words, if $E_{i,j}$ denotes the $n \times n$ matrix with a 1 in the ith row and j^{th} column, and zeros elsewhere for i, $j \in \{1,\ldots,n\}$, then $S_i = I + E_{i,n}$ for $i = 1,\ldots, n-1$. These matrices are transvections with a common hyperplane $H = \mathbb{F}^{n-1}$, the codimension 1 subspace spanned by the first $n-1$ standard basis vectors. The matrices $S_1,\ldots, S_{n-1}$ commute with each other, have order p, and afford a faithful representation of $E(n-1)$. For $r = 1,\ldots, n-1$ denote by $\sigma_r : E(r) \hookrightarrow \mathrm{GL}(n,\ \mathbb{F})$ the representation of $E(r)$ afforded by $S_1,\ldots, S_r$. In section 8.2 there is a detailed account of the invariant theory of $E(r)$ and we recall 8.2.6:

$$
\mathbb{F}[z_1,\ldots, z_n]^{E(r)} = \mathbb{F}[f_1,\ldots, f_r,\ z_{r+1},\ldots, z_n]
$$

where

$$
f_i = \prod_{\lambda \in \mathbb{F}_p} (z_i + \lambda z_n) = c_p([z_i])
$$

is the p-th Chern class of the orbit of z_i for $i = 1,\ldots, r$. (In the preceding product $\mathbb{F}_p < \mathbb{F}$ is the prime subfield.)

Let $T \in \mathrm{GL}(n,\ \mathbb{F})$ be a transvection with transvector $x \in V$. Denote by $<T> \subset \mathrm{GL}(n,\ \mathbb{F})$ the subgroup generated by T. From example 1

$$
\mathrm{Im}(\mathrm{Tr}^{<T>}) = (x^{p-1}).
$$

Since

$$
\mathrm{Tr}^{E(r)} = \mathrm{Tr}^{E(r)}_{E(r-1)} \mathrm{Tr}^{E(r-1)}
$$

we obtain

$$
\mathrm{Im}(\mathrm{Tr}^{E(r)}) = \mathrm{Tr}^{E(r)}_{E(r-1)}(\mathrm{Im}(\mathrm{Tr}^{E(r-1)})).
$$

Inductively we may assume $\mathrm{Im}(\mathrm{Tr}^{E(r-1)}) = (x^{p-1})$. Then

$$\mathrm{Tr}^{E(r)}_{E(r-1)}(x^{p-1}) = \sum_{i=0}^{p-1} S_r^i(x^{p-1}) = \sum_{i=0}^{p-1}(S_r^i x)^{p-1}$$

$$= \sum_{i=0}^{p-1}(ix)^{p-1} = (1^{p-1} + 2^{p-1} + \cdots + (p-1)^{p-1})x^{p-1}$$

$$= -(p-1)x^{p-1} = x^{p-1}.$$

We obtain $\mathrm{Im}(\mathrm{Tr}^{E(r)}) = (x^{p-1})$, again a very thin ideal in $\mathbb{F}[z_1, \ldots, z_n]^{E(r)}$.

The worst conceivable case, i.e. $\mathrm{Im}(\mathrm{Tr}^G) = (0)$, however cannot occur as the following lemma shows.

LEMMA 11.5.2 : *Let $\varrho : G \hookrightarrow \mathrm{GL}(n, \mathbb{F})$ be a representation of a finite group. Then $\mathrm{Im}(\mathrm{Tr}^G) \neq (0)$.*

PROOF : If we pass to fields of fractions it is enough to show that

$$\mathrm{Tr}^G : \mathbb{F}(V) \longrightarrow \mathbb{F}(V)^G$$

is nonzero. The field extension $\mathbb{F}(V)^G \subseteq \mathbb{F}(V)$ being Galois is separable ([58] volume 2 chapter 3 proposition 6.1). A standard result of field theory (see for example [58] volume 2 chapter 3 propositions 9.4) then implies $\mathrm{Tr}^G \neq 0$ as required. □

The following theorem and corollary have appeared several times in the literature, see for example [130], [147], [181], each time with a different proof. Our proof is no exception, it too is different.

THEOREM 11.5.3 : *Let $\varrho : G \hookrightarrow \mathrm{GL}(n, \mathbb{F}_q)$ be a representation of a finite group. Then some power of the top Dickson class $\mathbf{d}_{n,0}$ lies in $\mathrm{Im}(\mathrm{Tr}^G)$, i.e. $\mathbf{d}_{n,0} \in \sqrt{\mathrm{Im}(\mathrm{Tr}^G)}$.*

PROOF : It follows from the definition of the transfer and lemma 11.5.2 that $\mathrm{Im}(\mathrm{Tr}^G) \subset \mathbb{F}[V]^G$ is a nonzero $\mathcal{P}^*$-invariant ideal so the conclusion follows from 11.4.4. □

COROLLARY 11.5.4 : *Let $\varrho : G \hookrightarrow \mathrm{GL}(n, \mathbb{F}_q)$ be a representation of a finite group. Then there is a class $h_G \in \mathbb{F}_q[V][\mathbf{d}_{n,0}^{-1}]$ such that $\mathrm{Tr}^G(h_G) = 1 \in \mathbb{F}_q[V]^G[\mathbf{d}_{n,0}^{-1}]$.*

PROOF : By 11.5.3 there is a class $f \in \mathbb{F}_q[V]$ with $\mathrm{Tr}^G(f) = \mathbf{d}_{n,0}^k$ for some $k \in \mathbb{N}$. Set $h_G = \dfrac{f}{\mathbf{d}_{n,0}^k}$. □

COROLLARY 11.5.5 (S. A. Mitchell, P. S. Landweber and R. E. Stong): Let G be a finite group and $\varrho : G \hookrightarrow \mathrm{GL}(n, \ \mathbb{F}_q)$ a representation. Then $\mathbb{F}_q[V]^G[\mathbf{d}_{n,0}^{-1}]$ is a projective $\mathbf{D}^*(n)[\mathbf{d}_{n,0}^{-1}]$-module.

PROOF : We have the inclusions

$$\mathbf{D}^*(n) \hookrightarrow \mathbb{F}_q[V]^G \hookrightarrow \mathbb{F}_q[V] \, .$$

The module $\mathbb{F}_q[V]$ is a free $\mathbf{D}^*(n)$-module by Dickson's theorem 8.1.5 and Macaulay's theorem 6.7.7. If we localize by inverting $\mathbf{d}_{n,0}$, then by 11.5.4 the transfer

$$\mathrm{Tr} : \mathbb{F}_q[V]^G[\mathbf{d}_{n,0}^{-1}] \underset{\mathrm{Tr}^G}{\overset{\hookrightarrow}{\leftarrow}} \mathbb{F}_q[V][\mathbf{d}_{n,0}^{-1}]$$

provides a splitting to the inclusion as $\mathbb{F}_q[V]^G[\mathbf{d}_{n,0}^{-1}]$-modules, and hence a fortiori as $\mathbf{D}^*(n)[\mathbf{d}_{n,0}^{-1}]$-modules. Therefore $\mathbb{F}_q[V]^G[\mathbf{d}_{n,0}^{-1}]$ is a direct summand of a free $\mathbf{D}^*(n)[\mathbf{d}_{n,0}^{-1}]$-module and hence projective. $\quad\square$

LEMMA 11.5.6 : Let $\varrho : G \hookrightarrow \mathrm{GL}(n, \ \mathbb{F}_q)$ be a representation of a finite group and $f \in \mathrm{Im}(\mathrm{Tr}^G)$. If $H \leq G$ is a subgroup then $f \in \mathrm{Im}(\mathrm{Tr}^H)$, i.e. $\mathrm{Im}(\mathrm{Tr}^G) \subset \mathrm{Im}(\mathrm{Tr}^H)$.

PROOF : Let $g_1, \ldots, g_{|G:H|}$ be a transversal of H in G, i.e. a complete set of left coset representatives of H in G. Let $f = \mathrm{Tr}^G(F)$. Then

$$f = \mathrm{Tr}^G(F) = \sum_{g \in G} gF = \sum_{\substack{i=1 \\ h \in H}}^{|G:H|} hg_i F$$

$$= \sum_{h \in H} h \sum_{i=1}^{|G:H|} g_i F = \mathrm{Tr}^H \Big(\sum_{i=1}^{|G:H|} g_i F \Big)$$

which shows that $f \in \mathrm{Im}(\mathrm{Tr}^H)$. $\quad\square$

If $\mathbb{F}$ is a finite field and $\varrho : G \hookrightarrow \mathrm{GL}(n, \ \mathbb{F})$ is a representation of a finite group G then it follows that $\mathrm{Im}(\mathrm{Tr}^{\mathrm{GL}(n, \ \mathbb{F})}) \subset \mathrm{Im}(\mathrm{Tr}^G)$. Therefore $\mathrm{Im}(\mathrm{Tr}^{\mathrm{GL}(n, \ \mathbb{F})})$ consists of polynomials that are universally in the image of the transfer for any representation $\varrho : G \hookrightarrow \mathrm{GL}(n, \ \mathbb{F})$.

PROPOSITION 11.5.7 (M. Feshbach): Let $\varrho : G \hookrightarrow \mathrm{GL}(n, \ \mathbb{F}_q)$ be a representation of a finite group and $f \in \mathrm{Im}(\mathrm{Tr}^G) \subset \mathbb{F}_q[V]^G$. If $z_1, \ldots, z_n$ is a basis for V^* and $g \in G$ an element of order p, then f belongs to the ideal $I_g = (gz_1 - z_1, \ldots, gz_n - z_n) \bigcap \mathbb{F}[V]^G \subset \mathbb{F}[V]^G$.

PROOF : Let $H \leq G$ be the subgroup generated by g. By lemma 11.5.6 $f \in \mathrm{Im}(\mathrm{Tr}^H)$. Let $f = \mathrm{Tr}^H(F)$. The polynomial F is a linear combination of monomials $z_1^{e_1} \cdots z_n^{e_n}$, where $e_1 + \cdots + e_n = \deg(f)$, so it suffices to show that $\mathrm{Tr}^H(z_1^{e_1} \cdots z_n^{e_n})$ lies in the indicated ideal. To this end note[6] that for $h \in \mathbb{F}_q[V]$ we have

$$\mathrm{Tr}^H(h) = h + gh + \cdots + g^{p-1}h = (1 + g + \cdots + g^{p-1})(h) = (g-1)^{p-1}(h) \, .$$

If $h = z_i \bar{h}$ for some $i \in \{1, \dots, n\}$ then

$$(g-1)(h) = (g-1)(z_i \bar{h}) = [(g-1)(z_i)][(g-1)(\bar{h})] \in (gz_i - z_i)$$

and therefore $\mathrm{Tr}^H(h) \in (gz_i - z_i) \cap \mathbb{F}[V]^G$ and the result follows. □

The following result was originally proven by M. Feshbach (unpublished) [83] for the field $\mathbb{F}_2$. A published proof sketch of the case $\mathbb{F} = \mathbb{F}_q$ appears in [84]. There is a nice proof in [128] using ideas based on Gröbner bases, and H. Derkson [66] has a proof using geometric invariant theory, a variant of which appears in [220].

THEOREM 11.5.8 : *Let p be a prime and V an n-dimensional vector space over $\mathbb{F}_q$. Then the image of the transfer map $\mathrm{Tr}^{\mathrm{GL}(n, \, \mathbb{F}_q)}$: $\mathbb{F}[V] \longrightarrow \mathbf{D}^*(n)$ is an ideal, and its radical is $\sqrt{\mathrm{Im}(\mathrm{Tr}^{\mathrm{GL}(n, \, \mathbb{F}_q)})} = (\mathbf{d}_{n,0})$. In other words, if $f \in \mathbf{D}^*(n)$ then some power of f lies in the image of the transfer $\mathrm{Tr}^{\mathrm{GL}(n, \, \mathbb{F}_q)}$ if and only if $\mathbf{d}_{n,0}$ divides f.*

PROOF : By 11.5.3 $\mathbf{d}_{n,0} \in \sqrt{\mathrm{Im}(\mathrm{Tr}^{\mathrm{GL}(n, \, \mathbb{F}_q)})}$. Without loss of generality we may suppose that $n \geq 2$. Let $0 \neq z \in V^*$ and extend z to a basis $z = z_1, z_2, \dots, z_n$ for V^*. The element $g \in \mathrm{GL}(n, \mathbb{F}_q)$ which acts on V^* by

$$g(z_i) = \begin{cases} z_i & \text{for } i \neq 2 \\ z_1 + z_2 & \text{for } i = 2, \end{cases}$$

has order p, and

$$(g-1)(z_i) = \begin{cases} 0 & \text{for } i \neq 2 \\ z_1 & \text{for } i = 2. \end{cases}$$

Therefore by 11.5.7 $(z) \cap \mathbf{D}^*(n)$ contains some power f^k of f, $k \neq 0$, so z divides f. The group $\mathrm{GL}(n, \mathbb{F}_q)$ acts transitively on $V^* \setminus \{0\}$, so every nonzero element $z \in V^*$ divides f. Since $\mathbf{d}_{n,0}$ is the product of the nonzero elements of V^* it follows $\mathbf{d}_{n,0}$ divides a suitably high power of f. But, $\mathbf{d}_{n,0} \in \mathbf{D}^*(n)$ is a prime element and therefore must divide f. □

---

[6] Since we are in characteristic p, one has $(x-1)^q = x^q - 1 = (x-1)(x^{q-1} + \cdots + x + 1)$ so $(x-1)^{q-1} = (x^{q-1} + \cdots + x + 1)$.

In [181] Priddy and Wilkerson write down (the last formula in the paper) the formula

$$\mathrm{Tr}^{\mathrm{GL}(n,\ \mathbb{F}_q)}\Big(\det(x_i^{p^j})\prod_{i=0}^{n-1} x_{i+1}^{p^n-p^i-1}\Big) = \mathbf{d}_{n,0}^n$$

This shows that $\mathbf{d}_{n,0}^n \in \mathrm{Im}(\mathrm{Tr}^G)$ for any $\varrho : G \hookrightarrow \mathrm{GL}(n,\ \mathbb{F}_q)$. The essential steps in our proof of 11.5.8 are 11.5.3, which holds for any finite group, are:

(i) G acts transitively on V^*,

(ii) for any basis $z_1, \ldots, z_n$ of V^* the element $\in \mathrm{GL}(n,\ \mathbb{F}_q)$ defined by

$$g(z_i) = \begin{cases} z_i & \text{for } i \neq 2 \\ z_1 + z_2 & \text{for } i = 2, \end{cases}$$

belongs to $\varrho(G)$, and

(iii) $\mathbf{d}_{n,0} \in \mathbf{D}^*(n)$ is a prime element.

The matrix g in (ii) is triangular with 1's along the diagonal, so $g \in \mathrm{SL}(n,\ \mathbb{F}_q)$. The element $\mathbf{d}_{n,0} \in \mathbb{F}_q[V]^{\mathrm{SL}(n,\ \mathbb{F}_q)}$ is no longer prime, being the $(p-1)^{st}$-power of the class $L_n \in \mathbb{F}_q[V]^{\mathrm{SL}(n,\ \mathbb{F}_q)}$ (see 8.1.8). Therefore the argument used to prove 11.5.8 shows:

THEOREM 11.5.9 : Let $\varrho : G \hookrightarrow \mathrm{GL}(n,\ \mathbb{F}_q)$ be a representation of a finite group such that $\varrho(G) \supseteq \mathrm{SL}(n,\ \mathbb{F}_q)$. Then

$$(\mathbf{d}_{n,0}) \subseteq \sqrt{\mathrm{Im}(\mathrm{Tr}^G)} \subseteq (L_n).$$

In particular $\sqrt{\mathrm{Im}(\mathrm{Tr}^G)}$ is a power of the prime ideal (L_n). $\quad\square$

References

[1] J. F. Adams, *Lectures on Lie Groups,* W. A. Benjamin, New York, Amsterdam, 1969.

[2] J. F. Adams and C. W. Wilkerson, *Finite H-spaces and Algebras over the Steenrod Algebra,* Annals of Math. 111 (1980), 95–143.

[3] J. F. Adams and C. W. Wilkerson, *Finite H-spaces and Algebras over the Steenrod Algebra: a correction,* Annals of Math. 113 (1981), 621–622.

[4] A. Adem, J. Maginnis and R. J. Milgram, *Symmetric Invariants and Cohomology of Groups,* Math. Ann. 287 (1990), 391–411.

[5] A. Adem and R.J. Milgram, *Cohomology of Finite Groups,* Springer-Verlag, Heidelberg, Berlin, New York, 1995.

[6] J. Aguadé and L. Smith, *Modular Cohomology Algebras,* Amer. J. of Math. 107 (1985), 507–530.

[7] E. E. Allen, *A Conjecture of Procesi and a New Basis for the Decomposition of the Graded Left Regular Representation of S_n,* Adv. in Math. 100 (1993), 262–292.

[8] G. Almkvist, *Invariants of Z/pZ in Characteristic p,* In: Invariant Theory (Proceedings of the 1982 Montecatini Conference), Lecture Notes in Math.996, Springer-Verlag, Heidelberg, Berlin, 1983, 109–117.

[9] G. Almkvist, *Invariants, Mostly Old,* Pac. J. of Math. 86 (1980), 1–13.

[10] G. Almkvist, *Some Formulas in Invariant Theory,* J. of Algebra 77 (1982), 338–359.

[11] G. Almkvist and R. Fossum. *Decompositions of Exterior and Symmetric Powers of Indecomposable Z/pZ-Modules in Characteristic p and Relations to Invariants,* Séminaire d'Algèbre Paul Dubreil, Proceedings, Paris 1976/7, Lecture Notes in Math.641, Springer-Verlag, Heidelberg, Berlin, 1978.

[12] E. Artin, *Geometric Algebra,* Interscience Tracts in Pure Math. 3, Interscience Publishers, New York, 1957.

[13] E. Artin and J. Tate, *A Note on Finite Ring Extensions,* J. of the Japanese Mathematical Society 3 (1951), 74–77.

[14] M. F. Atiyah and I. G. Macdonald, *Introduction to Commutative Algebra,* Addison-Wesley, Reading, Mass., 1969.

[15] M. Auslander, *On the Purity of the Branch Locus,* Amer. J. of Math. 84 (1962), 116–125.

[16] M. Auslander and D. A. Buchsbaum, *Unique Factorization in Regular Local Rings*, Proc. Nat. Acad. Sci. USA 45 (1959), 733–734.

[17] M. Auslander and D. A. Buchsbaum, *On Ramification Theory in Noetherian Rings*, Amer. J. of Math. 81 (1959), 749–765.

[18] M. Auslander and I. Reiten, *Grothendieck Groups of Algebras and Orders*, J. of Pure and Appl. Algebra 39 (1986), 1–51.

[19] L. L. Avramov, *Pseudoreflection Group Actions on Local Rings*, Nagoya Math. J. 88 (1982), 161–180.

[20] H. F. Baker, *Note Introductory to the Study of Klein's Group of Order 168*, Math. Proc. Camb. Phil. Soc. 31 (1935), 468–481.

[21] S. Balcerzyk and T. Józefiak, *Commutative Noetherian and Krull Rings*, Polish Scientific Publishers, Warsaw, 1989.

[22] S. Balcerzyk and T. Józefiak, *Commutative Rings: dimension, multiplicity and homological methods*, Polish Scientific Publishers, Warsaw, 1989.

[23] G. Barbançon and M. Raïs, *Sur le théorème de Hilbert différentiable pour les groupes linéaires finis*, Ann. Scient. Éc. Norm. Sup. 4e série t. 16 (1983), 355–373.

[24] P. F. Baum and L. Smith, *The Real Cohomology of Differentiable Fibre Bundles*, Comm. Math. Helv. 42 (1967), 171–179.

[25] J. C. Becker and D. H. Gottlieb, *Transfer Maps for Fibrations and Duality*, Comp. Math. 33 (1976), 107–133.

[26] D. Benson, *Polynomial Invariants of Finite Groups*, Cambridge Univ. Press, London, 1993.

[27] E. R. Berlekamp, *An Analog of the Discriminant over Fields of Characteristic Two*, J. of Algebra 38 (1976), 315–317.

[28] M. Bernard, *Schur Indices and Splitting Fields of Unitary Reflection Groups*, J. of Algebra 38 (1976), 318–342.

[29] M. -J. Bertin, *Anneau des invariants du groupe alterné en caractéristique 2*, Bull. Sci. Math. de France 94 (1970), 65–72.

[30] M. -J. Bertin, *Anneaux d'invariants d'anneaux de polynômes en caractéristique p*, C. R. Acad. Sci. Paris t. 277 (Série A) (1973), 691–694.

[31] W. Beyon and G. Lusztig, *Some Numerical Results on Characters of Exceptional Lie Groups*, Math. Proc. Camb. Phil. Soc. 84 (1978), 417–426.

[32] M. Bôcher, *Higher Algebra*, MacMillian and Co., New York, Chicago, Boston, 1907.

[33] A. Borel, *Sur la cohomologie des espaces fibrés principaux et des éspaces homogènes des groupes de Lie compacts*, Annals of Math. (2) 57 (1953), 115–207.

[34] A. Borel, *Sur l'homologie et la cohomologie des groupes de Lie compacts connexes*, Amer. J. of Math. 76 (1954), 273–342.

[35] A. Borel, *Topology of Lie Groups and Characteristic Classes*, Bull. of the Amer. Math. Soc. 61 (1955), 397–432.

[36] R. Bott, *An Application of Morse Theory to the Topology of Lie Groups*, Bull. Soc. Math. de France 84 (1956), 251–281.

[37] D. Bourguiba and S. Zarati, *Depth and Steenrod Operations*, Inventiones Math. (to appear).

[38] K. S. Brown, *Cohomology of Groups*, Graduate Texts in Math. 87, Springer-Verlag, New York, 1982.

[39] N. Bourbaki, *Groupes et algèbres de Lie, Ch. 4, 5 et 6*, Masson, Paris, 1981.

[40] C. Broto, *Àlgebres d'invariants i àlgebres sobre l'àlgebra de Steenrod*, Bull. Soc. Cat. Cièn. VIII (1) (1986), 117–145.

[41] C. Broto, L. Smith and R. E. Stong, *Thom Modules and Pseuporeflection Groups*, J. of Pure and Appl. Algebra 60 (1989), 1–20.

[42] S. R. Bullett and I. G. Macdonald, *On the Adem Relations*, Topology 21 (1982), 329–332.

[43] H. E. A. Cambell and I. P. Hughes, *2-Dimensional Invariants of $GL_2(F_p)$ and some of its Subgroups over the Field F_p*, Preprint, Queens University, 1993.

[44] H. E. A. Campbell and R. D. Pollack, *Vector Invariants of $U_2(p)$*, Adv. in Math., (to appear).

[45] H. E. A. Campbell, I. P. Hughes and R. D. Pollack, *Vector Invariants of Symmetric Groups*, Can. Math. Bull. 33 (1990), 391–397.

[46] H. E. A. Campbell, I. P. Hughes and R. D. Pollack, *Rings of Invariants and p-Sylow Subgroups*, Can. Math. Bull. 34 (1991), 42–47.

[47] D. Carlisle and P. Kropholler, *Modular Invariants of Finite Symplectic Groups*, Preprint, Mancheseter, 1992.

[48] D. Carlisle and P. Kropholler, *Rational Invariants of Certain Orthogonal and Unitary Groups*, Proc. London. Math. Soc. 24 (1992), 57–60.

[49] D. Carlisle and P. Kropholler, *Invariants of some Finite Classical Groups over GF(2)*, Preprint, Manchester, 1993.

[50] H. Cartan, *Algèbres d'Eilenberg-Mac Lane et Homotopie*, *Séminaire Henri Cartan*, 1954/55, W. A. Benjamin, New York, 1967.

[51] H. Cartan, *Quotient d'un éspace analytique par un groupe d'automorphismes*, in: Algebraic Geometry and Topology, A Symposium in Honor of S. Lefschetz, Ed.: R.H. Fox, D.C. Spencer and A.W. Tucker, Princeton Uni. Press, Princeton 1957.

[52] H. Cartan, *Périodicité des Groupes d'Homotopie Stables des Groupes Classique, d'áprès Bott*, *Séminaire Henri Cartan*, 1959/60, W. A. Benjamin, New York, 1967.

[53] H. Cartan, *Sur l'iteration des opérations de Steenrod*, Comm. Math. Helv. 29 (1955), 40–58.

[54] H. Cartan and S. Eilenberg, *Homological Algebra*, Princeton University Press, Princeton, 1956.

[55] C. Chevalley, *Invariants of Finite Groups Generated by Reflections*, Amer. J. of Math. 67 (1955), 778–782.

[56] A. Clark and J. Ewing, *The realization of Polynomial Algebras as Cohomology Rings*, Pacific J. of Math. 50 (1974), 425–434.

[57] A. M. Cohen, *Finite Complex Reflection Groups*, Ann. Scient. Éc. Norm. Sup.4e série t. 9 (1976), 379–436.

[58] P. M. Cohen, *Algebra, Volumes I, II, III*, John Wiley & Sons, Chichester, New York, 1989.

[59] A. J. Coleman, *The Betti Numbers of the Simple Lie Groups*, Can. J. of Math. 10 (1958), 344–356.

[60] J. H. Conway and N. J. A. Sloane (et al.), *Sphere Packings, Lattices and Groups*, Springer-Verlag, New York, 1988.

[61] D. Cox, J. Little and D. O'Shea, *Ideals, Varieties, and Algorithms*, Springer-Verlag, New York, 1992.

[62] H. S. M. Coxeter, *Discrete Groups Generated by Reflections*, Annals of Math. 35 (1934), 588–621.

[63] H. S. M. Coxeter and W. O. J. Moser, *Generators and Relations for Discrete Groups*, Springer-Verlag, Heidelberg, Berlin, 1957.

[64] C. W. Curtis and I. Reiner, *Representation Theory of Finite Groups and Associative Algebras*, Interscience Publishers, New York, 1962.

[65] M. Demazure, *Invariants symétriques entiers des groupes de Weyl et torsion*, Inventiones Math. 21 (1973), 287–301.

[66] H. Derkson, *private communication*, Dagstuhl Workshop on Computational Invariant Theory, 1996.

[67] W. Dicks and E. Formanek, *Poincaré Series and a Problem of S. Montgomery*, Linear and Multilinear Algebra 12 (1982), 21–30.

[68] L. E. Dickson, *A Fundamental System of Invariants of the General Modular Linear Group with a Solution of the Form Problem*, Trans. of the Amer. Math. Soc. 12 (1911), 75–98.

[69] L. E. Dickson, *Binary Modular Groups and their Invariants*, Amer. J. of Math. 33 (1911), 175–192.

[70] L. E. Dickson, *On Finite Algebras*, Nachr. v. d. Ges. d. Wiss. zu Göttingen (1905), 358–393.

[71] L. E. Dickson, *Linear Groups*, Dover Publications Inc., New York 1958.

[72] J. Dieudonné and J. B. Carrell, *Invariant Theory, Old and New*, Academic Press, New York, 1971.

[73] A. Dold, *Fixed Point Index and Fixed Point Theorem for Euclidean Neighborhood Retracts*, Topology 4 (1965), 1–8.

[74] A. Dress, *On Finite Groups Generated by Pseudoreflections*, J. of Algebra 11 (1969), 1–5.

[75] W. G. Dwyer, H. Miller, and C. W. Wilkerson, *Homotopy Uniqueness of BG*, private communication, Chicago, 1986.

[76] W. G. Dwyer, H. Miller, and C. W. Wilkerson, *Homotopy Uniqueness of Classifying Spaces*, Topology 31 (1992), 29–45.

[77] D. Eisenbud, *Commutative Algebra*, Springer-Verlag, Heidelberg, Berlin, 1995.

[78] G. Ellingsrud and T. Skjelbred, *Profondeur d'anneaux d'invariants en caracteristique p*, Comp. Math. 41 (1980), 233–244.

[79] D. Engelmann, *Optimal, Pseudooptimal and Perfect Homogeneous Systems of Parameters for Rings of Invariants*, Preprint, Humboldt University, Berlin, 1996.

[80] L. Evens, *The Cohomology of Groups*, Claerndon Press, Oxford, 1991.

[81] M. M. Feldstein, *Invariants of the Linear Group Modulo p^k*, Trans. of the Amer. Math. Soc. 25 (1923), 223–238.

[82] M. Feshbach, *The Image of $H^*(BG; \mathbb{Z})$ in $H^*(BT; \mathbb{Z})$ for G a compact Lie Group with Maximal Torus T*, Topology 20 (1981), 93–95.

[83] M. Feshbach, *The Image of the Trace in the Ring of Invariants*, Preprint, University of Minnesota, 1981.

[84] M. Feshbach, *p-Subgroups of Compact Lie Groups and Torsion of Infinite Height in $H^*(BG;\ \mathbb{F}_p)$*, Mich. Math. J. 29 (1982), 299–306.

[85] L. Flatto. *Invariants of Finite Reflection Groups*, L'Enseign. de Math. 24 (1978), 235–292.

[86] P. Fleischmann, *On the Ring of Vector Invariants for the Symmetric Group*, Preprint, Institute for Experimental Mathematics, Essen, 1996.

[87] P. Fleischmann and W. Lempken, *On Generators of Modular Invariant Rings of Finite Groups*, Preprint, Institute for Experimental Mathematics, Essen, 1996.

[88] R. M. Fossum and P. A. Griffith, *Complete Local Factorial Rings which are not Cohen-Macaulay in Characteristic p*, Ann. Scient. Éc. Norm. Sup. 4e série, t. 8 (1975), 189–200.

[89] A. K. Garsia and D. Stanton, *Group Actions on Stanley-Reisner Rings and Invariants of Permutation Groups*, Adv. in Math. 51 (1984), 107–201.

[90] W. Gaschütz, *Fixkörper von p−Automorphismengruppen reintranszendenter Körpererweiterungen von p−Charakteristik*, Math. Zeit. 71 (1959), 466–468.

[91] A. V. Geramita and C. Small, *Introduction to Homological Methods in Commutative Rings*, Queens Papers in Pure and Applied Math. № 43, Queens University, Kingston Ontario Canada, 1976.

[92] O. E. Glenn, *Modular Invariant Processes*, Bull. of the Amer. Math. Soc. 21 (1914-15), 167 – 173.

[93] M. Göbel, *Computing Bases for Permutation Invariant Polynomials*, J. of Symbolic Computation 19 (1995), 285–291.

[94] N. L. Gordeev, *Finite Linear Groups whose Algebra of Invariants are Complete Intersections*, Math. USSR Izvestiya 28 (1987), 335–379.

[95] N. L. Gordeev, *Coranks of Elements of Linear Groups and Complexity of Algebraic Invariants*, Leningrad Math. J. 2 (1991), 245 – 267.

[96] D. Gorenstein, *Finite Groups*, Harpur and Row, New York, 1968.

[97] L. C. Grove and C. T. Benson, *Finite Reflection Groups, second edition*, Graduate Texts in Math. 99, Springer-Verlag, Heidelberg, Berlin, 1985.

[98] J. Herzog and E. Kunz, Eds., *Der kanonische Modul eines Cohen-Macaulay-Rings*, Lecture Notes in Math.238, Springer-Verlag, Heidelberg, Berlin, 1971.

[99] J. Herzog, E. Marcos and R. Waldi, *On the Grothendieck Group of a Quotient Singularity Defined by a Finite Abelian Group,* J. of Algebra 149 (1992), 122–138.

[100] J. Herzog and H. Sanders, *The Grothendieck Group of Invariant Rings and of Simple Hypersurface Singularities,* In: Singularities, Representation of Algebras, and Vector Bundles, Proceedings, Lambrecht 1985, Lecture Notes in Math.1273, Springer-Verlag, Heidelberg, Berlin, 1987, 134–149.

[101] D. Hilbert, *Über die Theorie der Algebraischen Formen,* Math. Ann. 36 (1890), 473–534.

[102] D. Hilbert, *Über die vollen Invariantensystem,* Math. Ann. 42 (1893), 313–373.

[103] D. Hilbert, *Hilbert's Invariant Theory Papers,* In: Lie Groups: History, Frontiers and Applications, Volume VIII, translated by M. Ackerman, commented by R. Hermann, Math. Sci. Press, Brookline, Mass., 1978.

[104] D. Hilbert, *Theory of Algebraic Invariants,* translated by Reinhard C. Laudenbacher, Cambridge University Press, Cambridge, 1994.

[105] H. Hiller, *Geometry of Coxeter Groups,* Pitman Books Ltd., London, 1982.

[106] H. Hiller and L. Smith, *On the Realization and Classification of Cyclic Extensions of Polynomial Algebras over the Steenrod Algebra,* Proc. of the Amer. Math. Soc. 100 (1987), 731–738.

[107] M. Hochster and J. A. Eagon, *Cohen-Macaulay Rings, Invariant Theory, and the Generic Perfection of Determinantal Loci,* Amer. J. of Math. 93 (1971), 1020–1058.

[108] R. Holzsager, *H-spaces of Category ≤ 2,* Topology 9 (1970), 211-216.

[109] H. Hopf, *Über der Topologie der Gruppen-Mannigfaltigkeiten und ihre Verallgemeinerungen,* Annals of Math. (2) 42 (1941), 22–52.

[110] H. Hopf, *Maximale Toroide und singuläre Elemente in geschlossen Lieschen Gruppe,* Comm. Math. Helv. 15 (1943), 59–70.

[111] R. Howe, *The Classical Groups and Invariants of Binary Forms,* Proc. Symp. Pure Math. 48 (1988), Amer. Math. Soc., Providence, 1988.

[112] Shou-Jen Hu and Ming-chang Kang, *Efficient Generation of Rings of Invariants,* J. of Algebra 180 (1996), 341–363.

[113] W. C. Huffman and N. J. A. Sloane, *Most Primitive Groups have Messy Invariants,* Adv. in Math. 32 (1979), 118–127.

[114] J. E. Humphreys, *Introduction to Lie algebras and Representation Theory*, Graduate Texts in Math. 9, Springer-Verlag, Berlin, New York, 1972.

[115] F. Ischebeck, *Eine Dualität zwischen den Funktoren Ext und Tor*, J. of Algebra 11 (1969), 510–531.

[116] V. G. Kac, *Torsion in Cohomology of Compact Lie Groups and Chow Rings of Reductive Algebraic Groups*, Inventiones Math. 80 (1985), 69–79.

[117] V. G. Kac and D. H. Peterson, *Generalized Invariants of Groups Generated by Reflections*, In: Progress in Math. 60: Geometry of Today, Roma 1984, Birkhauser Verlag, Boston, 1985, 231–249.

[118] R. Kane, *Poincaré Duality and the Ring of Coinvariants*, Can. J. of Math. (to appear).

[119] M. -C. Kang, *Picard Groups of Some Rings of Invariants*, J. of Algebra 58 (1979), 455–461.

[120] G. Kemper, *Calculating Invariant Rings of Finite Groups over Arbitrary Fields*, J. of Symbolic Computation (to appear).

[121] G. Kemper and G. Malle, *The Finite Irreducible Linear Groups with Polynomial Ring of Invariants*, Transformation Groups (to appear).

[122] M. Kervaire and T. Vust, *Fractions rationnelles invariantes par un groupe fini: quelques exemples*, In: Algebraische Transformationsgruppen und Invariantentheorie, Ed. H. Kraft, P. Slodowy and T. A. Springer. DMV Seminar 13, Birkhäuser Verlag, Basel, 1989.

[123] N. Killius, *Some Modular Invariant Theory of Finite Groups with particular Emphasis on the Cyclic Group*, Diplomarbeit, Göttingen 1996.

[124] F. Klein, *Vorlesung über das Ikosaeder und die Auflösung der Gleichung vom fünften Grad*, Leipzig, Teubner, 1884.

[125] F. Klein, *Lectures on Math.: The Evanston Colloquium*, MacMillan and Co. New York, 1894.

[126] H. Kraft. *Geometrische Methoden in der Invariantentheorie*, Aspects of Math., Vieweg Verlag, Braunschweig, 1984.

[127] N. J. Kuhn, *Generic Representations of the Finite General Linear Groups and the Steenrod Algebra*, Amer. J. of Math. 116 (1994), 327–360.

[128] K. Kuhnigk, *Das Transferhomomorphismus für Ringen von Invarianten*, Diplomarbeit, Göttingen 1997.

[129] S. P. Lam, *Unstable Algebras over the Steenrod Algebra and Cohomology of Classifying Spaces*, Proc. Conf on Algebraic Topology, Aarhus 1982, Springer-Verlag, Heidelberg, Berlin, 1983.

[130] P. S. Landweber and R. E. Stong, *The Depth of Rings of Invariants over Finite Fields*, Proc. New York Number Theory Seminar, 1984, Lecture Notes in Math.1240 Springer-Verlag, New York 1987.

[131] P. S. Landweber and R.E. Stong, *Invariants of Finite Groups*, unpublished correspondence, 1981–1993.

[132] J. Lannes, *Sur les espaces fonctionnels dont la source est le classifiant d'un p-groupe abélien élementaire*, Publ. Math. de l'I.H.E.S. 75 (1992), 135–244.

[133] J. Lannes, *Adams-Wilkerson Theory Revisited*, private communication, Baltimore, 1987.

[134] J. Leray, *Sur l'homologie des groupes de Lie, des éspaces homogènes, et des espaces fibré principeaux*, Coll. de Topologie algèbrique, Bruxelles 1950, 101–115.

[135] I. G. Macdonald, *Symmetric Functions and Hall Polynomials*, Clarendon Press, Oxford, 1995.

[136] S. Mac Lane, *Homology*, Springer-Verlag, Heidelberg, Berlin, 1974.

[137] F. J. MacWilliams, A. M. Odlyzko, N. J. A. Sloane and H. N. Ward, *Self-dual Codes over $GF(4)$*, J. of Computational Theory A25 (1978), 288–318.

[138] Z. Mahmud and L. Smith, *Invariant Theory and Steenrod Operations: Maps Between Rings of Invariants*, Math. Proc. Camb. Phil. Soc. 120 (1996), 103–116.

[139] C. L. Mallows and N. J. A. Sloane, *On the Invariants of a Linear Group of order 336*, Math. Proc. Camb. Phil. Soc. 74 (1973), 435–440.

[140] W. S. Massey, *A Basic Course in Algebraic Topology*, Graduate Texts in Math. 127, Springer-Verlag, New York, 1991.

[141] H. Matsumura, *Commutative Ring Theory*, Cambridge University Press, Cambridge, 1986.

[142] R. J. Milgram and S. B. Priddy, *Invariant Theory and $H^*(\mathrm{GL}_n(\mathbf{F}_p); \mathbf{F}_p)$*, J. of Pure and Appl. Algebra 44 (1987), 291–302.

[143] J. W. Milnor, *Construction of Universal Bundles II*, Annals of Math. (2) 63 (1956), 430–436.

[144] J. W. Milnor, *The Steenrod algebra and its Dual*, Annals of Math. (2) 67 (1958), 150–171.

[145] J. W. Milnor and J. C. Moore, *The Structure of Hopf Algebras*, Annals of Math. (2) 81 (1965) , 211–264.

[146] J. W. Milnor and J. D. Stasheff, *Characteristic Classes*, Annals of Math. Studies 76, Princeton University Press, Princeton, 1974.

[147] S. A. Mitchell, *Finite Complexes with A(n) Free Cohomology*, Topology 24 (1985), 227–248.

[148] T. Molien, *Über die Invarianten der linearen Substitutionsgruppen*, Sitzungsber. König. Preuss. Akad. Wiss. (1897), 1152–1156.

[149] H. Morikawa, *On the Invariants of Finite Nilpotent Groups*, Osaka Math. J. 10 (1958), 53–56.

[150] H. Mui, *Modular Invariant Theory and Cohomology Algebras of Symmetric Groups*, J. of Fac. Sci. Univ. Tokyo Sect. 1A Math. 22 (1975), 319–369.

[151] M. Nagata, *On the 14-th problem of Hilbert*, Amer. J. of Math. 81 (1959), 766–772.

[152] M. Nagata, *Local Rings*, John Wiley & Sons, New York, 1962.

[153] H. Nakajima, *Invariants of reflection groups in Positive Characteristics*, Proc. Japan Acad. Ser. A 55 (1979), 219–221.

[154] H. Nakajima, *Invariants of Finite Groups Generated by Pseudoreflections in Positive Characteristic*, Tsukuba J. of Math. 3 (1979), 109–122.

[155] H. Nakajima, *Invariants of Finite Abelian Groups Generated by Transvections*, Tokyo J. Math. 3 (1980), 201–214.

[156] H. Nakajima, *On some Invariant Subrings of Polynomial Rings in positive Characteristics*, Proc. 13th Symp. on Ring Theory, Okayama (1980), 91–107.

[157] H. Nakajima, *Modular Representations of p-Groups with Regular Rings of Invariants*, Proc. Japan Acad. Ser. A 56 (1980), 469–473.

[158] H. Nakajima, *Modular Representations of Abelian Groups with Regular Rings of Invariants*, Nagoya Math. J. 86 (1982), 229–248.

[159] H. Nakajima, *Relative Invariants of Finite Groups*, J. of Algebra 79 (1982), 218–234.

[160] H. Nakajima, *Rings of Invariants of Finite Groups which are Hypersurfaces*, J. of Algebra 80 (1983), 279–294.

[161] H. Nakajima, *Regular Rings of Invariants of Unipotent Groups*, J. of Algebra 85 (1983), 253–286.

[162] H. Nakajima, *Rings of Invariants of Finite Groups which are Hypersurfaces II*, Adv. in Math. 65 (1987), 39–64.

[163] A. Neeman, *The connection between a conjecture of Carlisle and Kropholler, now a theorem of Benson and Crawley-Bovey, and Grothendieck's Riemann-Roch and duality theorems*, Comment. Math. Helv. 70 (1995), 339 – 349.

[164] F. Neumann, M. D. Neusel, and L. Smith, *Rings of Generalized and Stable Invariants for Pseudoreflections and Pseudoreflection Groups*, J. of Algebra 182 (1996), 85–122.

[165] F. Neumann, M. D. Neusel, and L. Smith, *Rings of Generalized and Stable Invariants and Classifying Spaces of Compact Lie Groups*, Preprint, Göttingen, 1994.

[166] M. D. Neusel, *Invariants of some abelian p-Groups in Characteristic p*, Proc. of the Amer. Math. Soc. (to appear).

[167] M. D. Neusel, *Integral Extensions of Unstable Algebras over the Steenrod Algebra*, Royal Institute of Technology, Stockholm, preprint, 1996.

[168] M. D. Neusel and L. Smith, *Configurations of Hyperplanes, Groups Generated by Pseudoreflections and their Invariants*, J. of Pure and Appl. Algebra (to appear).

[169] M. D. Neusel and L. Smith, *The Lasker-Noether Theorem for $\mathcal{P}^*$-invariant ideals*, Forum Math. (to appear).

[170] E. Noether, *Der Endlichkeitssatz der Invarianten endlicher Gruppen*, Math. Ann. 77 (1916), 89–92.

[171] E. Noether, *Idealtheorie in Ringenbereichen*, Math. Ann. 83 (1921), 24–66.

[172] E. Noether, *Der Endlichkeitssatz der Invarianten endlicher linearer Gruppen der Characteristik p*, Nachr. v. d. Ges. d. Wiss. zu Göttingen (1926), 28–35.

[173] D. G. Northcott. *An Introduction to Homological Algebra*, Cambridge University Press, Cambridge, 1960.

[174] J. E. Olson, *A Combinatorial Problem on Finite Abelian Groups I, II*, J. of Number Theory 1 (1969), 8–10, 195–199.

[175] O. Ore, *On a Special Class of Polynomials*, Trans. of the Amer. Math. Soc. 35 (1933), 559–584.

[176] P. Orlik and L. Solomon, *The Hessian Map in the Invariant Theory of Reflection Groups*, Nagoya Math. J. 109 (1988), 1–21.

[177] P. Orlik and L. Solomon, *Discriminants in the Invariant Theory of Reflection Groups*, Nagoya Math. J. 109 (1988), 23–45.

[178] P. Orlik and H. Terao, *Arrangemants of Hyperplanes*, Springer-Verlag, Heidelberg, Berlin, 1992.

[179] S. Papadima, *Rigidity Properties of Compact Lie Groups Modulo Maximal Tori*, Math. Ann. 275 (1986), 637–652.

[180] K. Pommerening, *Invariants of Unipotent Groups, a Survey*, In: Invariant Theory, Ed. Koh, Lecture Notes in Math.1278, Springer-Verlag, Heidelberg, Berlin, 1987, 8–17.

[181] S. B. Priddy and C. W. Wilkerson, *Hilbert's Theorem 90 and the Segal Conjecture for Elementary Abelian p-Groups*, Amer. J. of Math. 107 (1985), 117–785.

[182] V. Reiner, *On Göbel's Bound for Invariants of Permutation Groups*, Arch. der Math. 65 (1995), 475 – 480.

[183] V. Reiner and L. Smith, *Systems of Parameters for Rings of Invariants* Preprint, Göttingen, 1996.

[184] P. Revoy, *Anneau des invariants du groupe alterné*, Bull. Sci. Math. de France 106 (1982), 427–431.

[185] D. R. Richman, *On Vector Invariants over Finite Fields*, Adv. in Math. 81 (1990), 30–65.

[186] D. R. Richman, *Invariants of Finite Groups over Fields of Characteristic p*, Adv. in Math. 124 (1996), 25–48.

[187] D. R. Richman, *Explicit Generators of the Invariants of Finite Groups*, Adv. in Math. 124 (1996), 49–76.

[188] J. Riordan, *An Introduction to Combinatorial Analysis*, John Wiley & Sons, New York, 1958.

[189] J. J. Rotman, *An Introduction to Homological Algebra*, Academic Press, New York, 1979.

[190] D. Rusin, *The Depth of Rings of Invariants and Cohomology Rings*, Preprint, University of Oklahoma, 1986.

[191] D. J. Saltman, *Noether's Problem over an Algebraically Closed Field*, Inventiones Math. 77 (1984), 71–84.

[192] P. Samuel, *Lectures on Unique Factorization Domains*, Tata Institute Lecture Notes, Bombay, 1964.

[193] B. J. Schmid, *Generating Invariants of Finite Groups*, C. R. Acad. Sci. Paris t. 308, Série I (1989), 1–6.

[194] B. J. Schmid, *Finite Groups and Invariant Theory*, Séminar d'Algèbre P. Dubriel et M.-P. Malliavin 1989-1990, Lecture Notes in Math.1478, New York, 1991.

[195] I. Schur, *Vorlesung über Invariantentheorie*, Bearbeitet und herausgegeben von Helmut Grunsky, Springer-Verlag, Heidelberg, Berlin, 1968.

[196] L. Schwartz, *Lectures on Lannes Technology*, University of Chicago Press, Chicago, 1994.

[197] J. -P. Serre, *Cohomologie modulo 2 des complexes d'Eilenberg-Mac Lane*, Comm. Math. Helv. 27 (1953), 198–232.

[198] J. -P. Serre, *Faisceaux algébriques cohérents*, Annals of Math. 61 (1955), 197–278.

[199] J. -P. Serre, *Sur la dimension cohomologique des groupes profinis*, Topology 3 (1964), 413–420.

[200] J. -P. Serre, *Algèbre locale—multiplicités*, Lecture Notes in Math.11, Springer-Verlag, Heidelberg, Berlin, 1965.

[201] J. -P. Serre, *Groupes finis d'automorphismes d'anneaux locaux réguliers*, Colloq. d'Alg. Éc. Norm. Sup. de Jeunes Filles, Paris, 1967, 8-01—8-11.

[202] J. -P. Serre, *Représentations linéaires des groupes finis*, Hermann, Paris, 1967.

[203] K. Shoda, *Über die Invarianten endlicher Gruppen linearer Substitutionen im Körper der Charakteristik p*, Jap. J. of Math. 17 (1940), 109–115.

[204] G. C. Shephard, *Unitary Groups Generated by Reflections*, Can. J. of Math. 5 (1953), 364–383.

[205] G. C. Shephard, *Abstract Definitions for Reflection Groups*, Can. J. of Math. 9 (1957), 373–376.

[206] G. C. Shephard and J. A. Todd, *Finite Unitary Reflection Groups*, Can. J. of Math. 6 (1954), 274–304.

[207] W. M. Singer, *The Transfer in Homological Algebra*, Math. Zeit. 202 (1989), 493 – 523.

[208] B. Singh, *Invariants of Finite Groups Acting on Local Unique Factorization Domains*, J. of the Indian Math. Soc. 34 (1970), 31–38.

[209] N. J. A. Sloane, *Error Correcting Codes and Invariant Theory*, Amer. Math. Monthly 84 (1977), 82–107.

[210] L. Smith, *Linear Algebra (corrected second edition)*, Springer-Verlag, New York, 1992.

[211] L. Smith, *On the Characteristic Zero Cohomology of Free Loop Spaces*, Amer. J. of Math. 103 (1981), 887–910.

[212] L. Smith, *A Note on the Realization of Complete Intersection Algebras by the Cohomology of a Space*, Quart. J. of Math. Oxford (2) 33 (1982), 379–384.

[213] L. Smith, *Transfer and Ramified Coverings*, Math. Proc. Camb. Phil. Soc. 93 (1983), 485–493.

[214] L. Smith, *On the Invariant Theory of Finite Pseudoreflection Groups*, Arch. Math. 44 (1985), 225–228.

[215] L. Smith, *Finite Loop Spaces with Maximal Tori have Finite Weyl Groups*, Proc. of the Amer. Math. Soc. 119 (1993), 299–302.

[216] L. Smith, *Some Rings of Invariants that are Cohen-Macaulay*, Canad. Math. Bull. 39 (1996), 238 – 240.

[217] L. Smith *The e-Invariant and Finite Coverings II*, Trans. of the Amer. Math. Soc. 347 (1995), 5009–5021.

[218] L. Smith *P*-Invariant Ideals in Rings of Invariants*, Forum Math. 8 (1996), 319–342.

[219] L. Smith *Noether's Bound in the Invariant Theory of Finite Groups*, Arch. der Math. 66 (1996), 89–92.

[220] L. Smith, *Polynomial Invariants of Finite Groups: A Survey of Recent Developments*, Preprint, Göttingen, 1996.

[221] L. Smith and R. E. Stong, *On the Invariant Theory of Finite Groups: Orbit Polynomials and Splitting Principles*, J. of Algebra 110 (1987), 134–157.

[222] L. Smith and R. E. Stong, *Invariants of Finite Groups*, unpublished correspondence, 1979–1993.

[223] L. Smith and R. M. Switzer, *Polynomial Algebras over the Steenrod Algebra*, Proc. Edinburgh Math. Soc. 27 (1984), 11–19.

[224] L. Smith and R. M. Switzer, *Realizability and Nonrealizability of Dickson Algebras as Cohomology Rings*, Proc. of the Amer. Math. Soc. 89 (1983), 303–313.

[225] W. Smoke, *Dimension and Multiplicity for Graded Algebras*, J. of Algebra 21 (1972), 149–173.

[226] L. Solomon, *Invariants of Finite Reflection Groups*, Nagoya J. of Math. 22 (1963), 57–64.

[227] L. Solomon, *Invariants of Euclidean Reflection Groups*, Trans. of the Amer. Math. Soc. 113 (1964) 274–286.

[228] L. Solomon, *Partition Identities and Invariants of Finite Groups*, J. of Combinatorial Theory (A) 23 (1977), 148–175.

[229] E. H. Spanier, *Algebraic Topology*, McGraw Hill, New York 1966.

[230] T. Sperlich. *Automorphisms of $P[V]_G$*, Proc. of the Amer. Math. Soc. 120 (1994) 5-11.

[231] T. A. Springer, *Invariant theory*, Lecture Notes in Math.585, Springer-Verlag, Heidelberg, Berlin, 1977.

[232] T. A. Springer, *On the Invariant Theory of SU_2*, Indag. Math. 42 (1980), 339–345.

[233] R. P. Stanley, *Relative invariants of Finite Groups generated by Pseudoreflections*, J. of Algebra 49 (1977), 134–148.

[234] R. P. Stanley, *Hilbert Functions of Graded Algebras*, Adv. in Math. 28 (1978), 57–83.

[235] R. P. Stanley, *Invariants of Finite Groups and their Applications to Combinatorics*, Bull. of the Amer. Math. Soc. (3) 1 (1979), 475–511.

[236] N. E. Steenrod and D. B. A. Epstein, *Cohomology Operations*, Annals of Math. Studies 50, Princeton University Press, Princeton, 1962.

[237] R. Steinberg, *Invariants of Finite Reflection Groups*, Can. J. of Math. 12 (1960), 616–618.

[238] R. Steinberg, *Differential Equations Invariant under Finite Reflection Groups*, Trans. of the Amer. Math. Soc. 112 (1964), 392–400.

[239] R. Steinberg, *On Dickson's Theorem on Invariants*, J. of Fac. Sci. Univ. Tokyo Sect. 1A Math. 34 (1987), 699–707.

[240] E. Stiefel, *"Uber eine Beziehung zwischen geschlossenen Lie'schen Gruppen und diskontinuierlichen Bewegungsgruppen euklidischer Räume und ihre Anwendung auf die Aufzählung der einfachen Lie'schen Gruppen*, Comm. Math. Helv. 14 (1941-42), 350–380.

[241] E. Stiefel, *Kristallographische Bestimmung der Charaktere der geschlossenen Lie'schen Gruppen*, Comm. Math. Helv. 17 (1944-45), 160–200.

[242] R. E. Stong, *unpublished correspondence, 1982-1993.*

[243] B. Sturmfels, *Algorithms in Invariant Theory*, Springer-Verlag, Heidelberg, Berlin, Vienna, 1993.

[244] D. Sullivan, *Genetics of Homotopy Theory and the Adams Conjecture*, Annals of Math. 100 (1974), 1–79.

[245] R. G. Swan, *Invariant Rational Functions and a Problem of Steenrod*, Inventiones Math. 7 (1969), 148–158.

[246] T. Tamagawa, *Dickson's Theorem on Invariants of Finite General Linear Groups*, private communication, Yale University, 1990.

[247] J. Tate, *Homology of Noetherian Rings and Local Rings*, Ill. J. of Math. 1 (1957), 14–27.

[248] H. Terao, *The Jacobian and Descriminants of Finite Reflection Groups*, Tôhoku Math. Journ. 41 (1989), 237–247.

[249] H. Toda, *Cohomology mod 3 of the Classifying Space BF_4 of the Exceptional Group F_4*, J. of Math. Kyoto Univ. 13 (1973), 97–115.

[250] J. G. Thompson, *Invariants of Finite Groups*, J. of Algebra 69 (1981), 143–145.

[251] J. S. Turner, *A Fundamental System of Invariants of a Modular Group of Transformations*, Trans. of the Amer. Math. Soc. 24 (1922), 129–134.

[252] B. L. van der Waerden. *Modern Algebra*, Ungar, New York, 1949.

[253] N. R. Wallach, *Invariant Differential Operators on a Reductive Lie Algebra and Weyl Group Representations*, J. of the Amer. Math. Soc.6 (1993), 779–816.

[254] K. Watanabe, *Certain Invariant Subrings are Gorenstein I*, Osaka J. of Math. 11 (1974), 1–8.

[255] K. Watanabe, *Certain Invariant Subrings are Gorenstein II*, Osaka J. of Math. 11 (1974), 379–388.

[256] K. Watanabe, *Invariant Subrings which are Complete Intersections I (Invariant Subrings of Finite Abelian Groups)*, Nagoya Math. J. 77 (1980), 89–98.

[257] H. Weber, *Lehrbuch der Algebra*, 2te Auflage, Vieweg Verlag, Braunschweig, 1899.

[258] A. Weil, *Demonstration topologique d'un théorème fondamental de Cartan*, C. R. Acad. Sci. Paris t. 200 (1935), 518–520.

[259] H. Weyl, *The Classical Groups*, second edition, Princeton Univ. Press, Princeton, 1946.

[260] C. W. Wilkerson, *Some Polynomial Algebras over the Steenrod Algebra*, Bull. of the Amer. Math. Soc. 79 (1973), 1274–1276.

[261] C. W. Wilkerson, *Classifying Spaces, Steenrod Operations and Algebraic Closure*, Topology 16 (1977), 227–237.

[262] C. W. Wilkerson, *Integral Closure of unstable Steenrod Algebra Actions*, J. of Pure and Appl. Algebra 13 (1978), 49–55.

[263] C. W. Wilkerson, *A Primer on the Dickson Invariants*, Proc. of the Northwestern Homotopy Theory Conference, Contemp. Math. 19, Amer. Math. Soc.1983, 421–434.

[264] R. M. W. Wood, *An Introduction to the Steenrod Algebra through Differential Operators*, Manchester Uni. Preprint, 1995.

[265] Wu Wen-Tsün, *Sur les puissance de Steenrod*, Colloque de Topologie de Strasbourg, 1951.

[266] O. Zariski and P. Samuel, *Commutative Algebra, Volumes I,II*, Graduate Texts in Math. 28, 29, Springer-Verlag, Berlin, New York, 1975.

Notations

$\hookrightarrow$ denotes an injective map

$\delta_{i,j} = \begin{cases} 1 & i = j \\ 0 & \text{otherwise} \end{cases}$

$\emptyset$ = the empty set

$—^G$ = G-invariant elements in the G-object — (set, module, algebra, etc.)

$<x \mid y>$ = scalar product of vectors x and y

$f(t)\big|_{t=a}$ = residue of the function $f(t)$ at $t = a$

$\binom{a}{b}$ = binomial coefficient $(= \frac{a!}{b!(a-b)!})$

$|X|$ = number of elements in X

$|G : H|$ = index of the subgroup $H < G$

$G \int H$ = wreath product of the groups G and H, i.e., $G^n \rtimes H$, where H acts via a permutation representation on G^n

$\mathbb{K} \mid \mathbb{L} = \mathbb{K}$ is a field extension of $\mathbb{L}$

$|\mathbb{K} : \mathbb{L}|$ = degree of the field extension $\mathbb{K} \mid \mathbb{L}$

A_n = alternating subgroup of Σ_n

$\text{Ass}(I)$ = set of associated primes of the ideal I

$A//B = \mathbb{F} \otimes_B A$ for A and B graded connected algebras over the field $\mathbb{F}$, and A a-B algebra

$\bar{A}$ = augmentation ideal of the graded connected algebra A

$\mathbb{C}$ = complex numbers

$c_i(B)$ = i-th Chern class of the G-invariant set B

$\text{codim}(A)$ = homological codimension of the algebra A

D_{2n} = dihedral group of order $2n$

$\mathbf{D}^*(n)$ = Dickson algebra

$\mathbf{d}_{n,i}$ = i-th Dickson polynomial in

$\mathbf{D}^*(n)$

$\deg(f)$ = degree of the polynomial f

$\deg(A)$ = degree of the algebra A

$\dim_{\mathbb{F}}(V)$ = dimension of the vector space V over the field $\mathbb{F}$

$\dim(A)$ = Krull dimension of the algebra A

$E[V]$ = alternating functions or exterior algebra on the vector space V

e_i = i-th elementary symmetric polynomial

$\mathbb{F}^\times$ = nonzero elements of the field $\mathbb{F}$

$\mathbb{F}_q$ = Galois field with q elements

$FF(—)$ = field of fractions functor

$\mathbb{F}[X]$ = polynomial algebra generated by the set X

$\mathbb{F}[V]$ = polynomial functions on the vector space V

$\mathbb{F}[V]^G$ = G-invariant polynomial functions on the vector space V

$\mathbb{F}[V]^G_\chi$ = χ-relative invariant polynomial functions on the vector space V

$\mathbb{F}[V]^G_n$ = G-invariant homogeneous polynomial functions of degree n on the vector space V

$\mathbb{F}[V]_G$ = algebra of coinvariants of the group G acting on V

$\text{Fun}(—, \mathbb{F})$ = vector space of functions from — to $\mathbb{F}$

$\mathcal{F}(\mathbf{D})$ = fundamental system of roots of the Dynkin diagram $\mathbf{D}$

$\text{GL}(n, A)$ = invertible $n \times n$ matrices over A

$\text{GL}(V)$ = self isomorphisms of the vector space V

$\text{Gal}(\mathbb{K} \mid \mathbb{L})$ = Galois group of the field extension $\mathbb{K} \mid \mathbb{L}$

hom–dim(A) = homological dimension of the algebra A

$\mathcal{H}(G) = \mathcal{H}_\varrho(G)$ = set of reflecting hyperplanes in $\varrho(G)$

$\mathfrak{Im}(z)$ = imaginary part of the complex number z

$\Lambda^i(V)$ = i-th exterior of the vector space V

$\mathrm{Mat}_n(A)$ = $n \times n$ matrices over A
M_d = degree d component of the graded module M

$\mathbb{N}$ = natural numbers (i.e. strictly positive integers)
$\mathbb{N}_0$ = natural numbers together with 0
$N_G(H)$ = normalizer in the group G of the subgroup H

p_i = i-th power sum polynomial
$P(M, t)$ = Poincaré series of the graded module M

$\mathbb{Q}$ = rational numbers

$\mathbb{R}$ = real numbers
$\mathfrak{Re}(z)$ = real part of the complex number z
$\mathcal{R}(G)$ = set of unit root vectors of real reflection group G
RX = free R-module on the set X
$\mathcal{R}(\mathbf{D})$ = root system of the Dynkin diagram $\mathbf{D}$

Σ_n = symmetric group on n-elements
$\mathbb{SP}^n(X)$ = n-th symmetric product of the set X

$\mathrm{Span}_F\{-\}$ = $\mathbb{F}$-linear span of —
$S^d(V^*)$ = d-th symmectric power of the vector space V^*
sgn = signum (sign of a permutation)
$s(G) = s_\varrho(G)$ = set of pseudo-reflections in $\varrho(G)$
$s_\Delta(G)$ = set of diagonalizable pseudo-reflections in $\varrho(G)$
$s_{\not\Delta(G)}$ = set of nondiagonalizable pseudoreflections in $\varrho(G)$

$\mathrm{Tot}(-)$ = totalization of the graded object —
Tr^G = transfer homorphism associated to G
Tr^G_H = relative transfer homorphism associated to $H \leq G$

V^* = dual of the vector spave V
V^G = G-fixed vectors in the vector space V
V_G = G-covariants of the vector space V

$W(H)$ = Weyl group $(N_G(H)/Z_G(H))$ of $H \leq G$
$W(\mathcal{R})$ = Coxeter group of the root system $\mathcal{R}$
$W(\mathbf{D})$ = Coxeter group associated to the Dynkin diagram $\mathbf{D}$

X^G = fixed point set of G on X
X^g = fixed point set of $g \in G$ on X
X/G = orbit space of G acting on X

$\mathbb{Z}$ = integers
$\mathbb{Z}/n$ = cyclic group of order n
ZG = center of the group G
$Z_H(G)$ = centerlizer in G of the subgroup H

Index

Printed in the United States
by Baker & Taylor Publisher Services